T0332691

Real-Time Embedded Systems

Real-Time Embedded Systems
Design Principles and Engineering Practices

Xiaocong Fan

AMSTERDAM • BOSTON • HEIDELBERG • LONDON
NEW YORK • OXFORD • PARIS • SAN DIEGO
SAN FRANCISCO • SINGAPORE • SYDNEY • TOKYO
Newnes is an imprint of Elsevier

ELSEVIER

Newnes

Newnes is an imprint of Elsevier
The Boulevard, Langford Lane, Kidlington, Oxford OX5 1GB, UK
225 Wyman Street, Waltham, MA 02451, USA

British Library Cataloguing in Publication Data
A catalogue record for this book is available from the British Library

Library of Congress Cataloging-in-Publication Data
A catalog record for this book is available from the Library of Congress

ISBN: 978-0-12-801507-0

For information on all Newnes publications
visit our website at http://store.elsevier.com/

Typeset by SPi Global, India

Printed and bound in USA
15   16   17   18   10   9   8   7   6   5   4   3   2   1

Contents

Preface

An embedded system is an electronic system that is designed to perform a dedicated function within a larger system. Real-time systems are those that can provide guaranteed worst-case response times to critical events, as well as acceptable average-case response times to noncritical events. When a real-time system is designed as an embedded component, it is called a real-time embedded system. Real-time embedded systems are widespread in consumer, industrial, medical, and military applications.

As more and more of our daily life depends on embedded technologies, the demand for engineers with the skill set for the development of real-time embedded software has soared in recent years. As a consequence, preparing students for the design and implementation of embedded software is becoming increasingly important. This textbook is written especially for advanced undergraduates or master-level students who are pursuing a major in software engineering, computer engineering, or a related discipline. The textbook may also benefit practicing engineers with a concentration in embedded software development.

This book takes a synergetic approach to introducing ideas and topics from real-time systems, embedded systems, and software development principles. Readers will not only gain a thorough understanding of concepts related to microprocessors, interrupts, and the cross-platform development process, and appreciate the importance of real-time modeling and scheduling, they will also be trained in good software engineering practices such as model documentation, model analysis, design patterns, and system standard conformance.

This textbook features three aspects that are essential for the development of real-time embedded software.

First, developing software for real-time embedded systems involves many activities, including specification of requirements, timing analysis, architecture design, multitasking design, and cross-platform testing and debugging. This book covers the whole process of embedded software development, with some topics fully explained and others only briefly mentioned (e.g., debugging and testing). In particular, this book presents various embedded software architectures in a systematic way, with a focus on a real-time operating system, which is the most advanced architecture adopted in large real-time embedded systems. Moreover, we have chosen to place significant emphasis on reusable design solutions. As shown in Table 0.1, this

Table 0.1 Summary of design patterns

Category	Pattern Name	Where in the Book
ISR	ISR-Pattern-min	Section 4.5.1
	ISR-Pattern-server	Section 4.5.2
	Interrupt chaining	Figure 4.7 in Section 4.5.3
	Interrupt cascading	Figure 4.9 in Section 4.5.4
	Interrupt disabling	Figure 4.11 in Section 4.5.5
	Double buffering	Figure 4.12 in Section 4.5.5
	Honor first request	Figure 12.17 in Section 12.3.2
Subclassing	Abstraction-occurrence	Figure 6.25 in Section 6.3.4
	General hierarchy	Figure 6.27 in Section 6.3.4
Software architecture	Round-robin DAS	Figure 12.10 in Section 12.2.2
	Round robin with interrupts	Figure 12.16 in Section 12.3.2
	FIFO queuing	Figure 12.20 in Section 12.4.1
	Priority queuing	Figure 12.21 in Section 12.4.2
	Serial port design pattern	Figure 14.5 in Section 14.2.2.1
Static task scheduler	Clock based	Section 15.2
	Frame based	Section 15.3
	Timing wheel	Section 22.3
Semaphore/mutex	Rendezvous synchronization pattern	Figure 18.8 in Section 18.3.1
	Multi-instance resource protection	Figure 18.19 in Section 18.4.1
Condition variable	Barrier synchronization pattern	Figure 18.24 in Section 18.5.1
	Producer-consumer pattern	Figure 18.25 in Section 18.5.2
	Read-write lock pattern	Figure 18.30 in Section 18.5.3
Message queue	Unidirectional queuing pattern	Figure 19.5 in Section 19.3.1
	Acked-unidirectional queuing pattern	Figure 19.6 in Section 19.3.2
	Bidirectional queuing pattern	Figure 19.7 in Section 19.3.3
	Client-server queuing pattern	Figure 19.10 in Section 19.3.4
Pipe	Unidirectional piping pattern	Figure 20.3 in Section 20.3
	Bidirectional piping pattern	Figure 20.3 in Section 20.3
Deadlock avoidance	Hierarchical messaging pattern	Figure 21.8 in Section 21.7.3

DAS, detect-acknowledge-service; FIFO, first in first out; ISR, interrupt service routine.

book introduces many design patterns, which represent the best practices that can be reused in a wide range of real-time embedded systems.

Second, Unified Modeling Language (UML) is a graphical language for specifying, visualizing, constructing, and documenting software systems. UML is useful in a variety of engineering problems, from single-process, embedded systems and stand-alone user applications to concurrent, distributed systems. This text features UML 2.4, the latest UML standard as of this writing. Throughout the book, UML diagrams are used for both system designs and concept illustrations. In particular, the UML real-time profile is carefully presented so that students can learn how to document their designs of real-time systems in a professional way.

Third, POSIX (for "portable operating system interface") is an open operating system interface standard that has been developed to promote interoperability and portability of applications across variants of Unix operating systems. Software systems built upon one real-time operating system can be easily ported to other POSIX-compliant operating systems. This text features POSIX.1-2008 (2013 edition). The operating system services and concepts covered in this book are fully compatible with the POSIX.1-2008 standard. The example codes provided in this book have been tested in QNX—a real-time operating system widely adopted in industry. Since QNX is POSIX compliant, the programs may also be compiled, without changing the source code, for execution on another POSIX-compliant operating system.

Briefly, this textbook consists of four parts:

- Part I is dedicated to a basic introduction to real-time embedded systems and the iterative development process. Although our emphasis is on the software aspects, complete isolation from the underlying hardware is neither feasible nor desirable. For such a reason, this part also contains two chapters on microprocessors and interrupts—fundamental topics for software engineers who wish to build embedded systems.
- Part II is dedicated to modeling techniques for real-time systems. In particular, we introduce the modeling tools covered by UML—a standard widely adopted in both academia and the software industry. Moreover, we introduce real-time UML—a profile for specifying real-time-related constraints in system models. UML diagrams are consistently used throughout the book to illustrate key concepts and design patterns.
- Part III is dedicated to the design of software architectures for real-time embedded systems. We start with generic architectures, which lead us to the most complicated architecture—a real-time operating system. The focus is then switched to multitasking and real-time scheduling—two critical issues to be addressed by any designers of real-time embedded systems.
- Part IV is dedicated to system implementation. We especially focus on those mechanisms available on any POSIX-compliant operating systems; this means that the design/implementation patterns given in this book are applicable to other POSIX-compliant operating systems as well.

The four parts together have 23 chapters. A one-semester course can use selected chapters/ sections to suit the interests of the instructor and students. For instance, some microprocessor types in Chapter 3 can be skipped in order to fit the materials in one or two lecture time. If UML basic modeling concepts have been covered in a prerequisite course on software engineering principles, chapters 6, 7, and 8 can be used as self-reading assignments or simply used as a reference. Depending on the students' familiarity with basic concepts of operating systems, some topics covered in Part IV (say, message queue, pipe, and signals) can be treated differently.

To aid instructors and students in using this text, we provide a supplements package on Elsevier's companion website: http://booksite.elsevier.com/9780128015070. This package includes PowerPoint slides and source code.

In this text, I have not been able to cover every major topic concerned with real-time embedded systems. I have exercised my best judgement in deciding which topics are suitable for software engineers, which emphasize, and which to omit. Seriously interested readers may refer to other textbooks for different perspectives.

Comments from colleagues are encouraged and welcome. Please feel free to send suggestions to Xiaocong Fan, Behrend College, Pennsylvania State University, Erie, PA 16563, USA (e-mail: xfan@psu.edu). I look forward to hearing from you about your experiences with the text.

Erie, PA, August 2014   *Xiaocong Fan*

Acknowledgments

First of all, I am grateful for the opportunity to learn from many excellent texts available on real-time systems or embedded systems, the authors including Jane W.S. Liu, David E. Simon, Rob Williams, Qing Li, and Bruce P. Douglass, to mention only a few. They may recognize their influence in some parts of the book, directly or indirectly.

I am indebted to the reviewers for numerous comments and enlightening suggestions. I would like to thank my students, who have been influential in shaping my thoughts on how best to organize and teach a course on real-time embedded systems.

The final manuscript was ably edited by SPi. A most special thanks go to Tim Pitts, Charlie Kent and Nicky Carter at Elsevier. This book would not be possible without their graceful management and great patience.

*To my wonderful wife, Yan,
and our precious sons—Mutian and Aaron*

Acronyms

AIC	Advanced interrupt controller
ANSI	American National Standards Institute
BSP	Board support package
CAN	Controller area network
CISC	Complex instruction set computing
COFF	Common Object File Format
COM	Serial communication port
CPU	Central processing unit
CSPR	Current program status register
DAS	Detect-acknowledge-service
DMA	Direct memory access
DSP	Digital signal processor
DTCM	Data tightly coupled memory
EDF	Earliest deadline first
EEPROM	Electrically erasable programmable read-only memory
ELF	Executable and Linking Format
EOI	End of interrupt
EPROM	Erasable programmable read-only memory
FIFO	First in first out
GPR	General-purpose register
GRM	General Resource Modeling
GRM	Graphical user interface
HLP	Highest locker protocol
HNERT	Highest-priority nonempty ready-thread list
I2C	Inter-Integrated Circuit
IC	Integrated circuit
ICE	In-circuit emulator
ICR	Interrupt command/status register
IEEE	Institute of Electrical and Electronics Engineers

IMR	Interrupt mask register
INV	Interrupt vector number
IP	Internet Protocol
IQR	Interrupt request
IRR	Interrupt request register
ISR	Interrupt service routine
LNA	Low-noise amplifier
LSb	Least significant bit
LSB	Least significant byte
MMU	Memory management unit
MOF	Meta Object Facility
MSB	Most significant byte
NVM	Nonvolatile memory
OCL	Object Constraint Language
OMG	Object Management Group
OOM	Object-oriented modeling
OOP	Object-oriented programming
OS	Operating system
PC	Personal computer
PCP	Priority ceiling protocol
PCR	Program counter register
PDF	Probability distribution function
PIC	Programmable interrupt controller
PIT	Programmable interval timer
POSIX	Portable Operating System Interface
PRF	Pulse repetition frequency
QoS	Quality of service
RAM	Random-access memory
RF	Radio frequency
RISC	Reduced instruction set computing
RMA	Rate-monotonic assignment
ROM	Read-only memory
RTOS	Real-time operating system
RT-UML	Real-time Unified Modeling Language
SFR	Special-function register
SPI	Serial peripheral interface
SRAM	Static read-access memory
TCM	Tightly coupled memory
UART	Universal asynchronous receiver-transmitter
UML	Unified Modeling Language
UTC	Coordinated Universal Time

PART I

Introduction

CHAPTER 1

Introduction to Embedded and Real-Time Systems

Contents

Part of the joy of looking at art is getting in sync in some ways with the decision-making process that the artist used and the record that's embedded in the work.

Chuck Close

1.1 Embedded Systems

In computing disciplines (computer engineering, software engineering, computer science, information systems and technology), the term "embedded system" is used to refer to an electronic system that is designed to perform a dedicated function and is often embedded within a larger system.

Embedded systems differ from general-purpose computing devices mainly in two aspects:

- First, an embedded system is designed simply for a specific function, whereas a general-purpose computing device, such as smartphone, laptop, or desktop computer, is not; they can be used as Web servers or data warehouses, or can be used for writing articles, reading news, playing games, or running scientific experiments, to mention only a few applications.
- Second, an embedded system is traditionally built together with the software intended to run on it. Such a parallel model of developing hardware and software together is known as hardware-software co-design. Recently, there has been a trend where an embedded system

Real-Time Embedded Systems. http://dx.doi.org/10.1016/B978-0-12-801507-0.00001-8

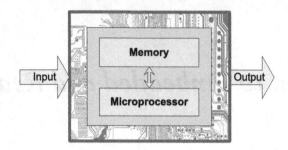

Figure 1.1
A generic embedded system architecture.

is built with a *well-defined interface* open to third-party embedded software providers. In contrast, a general-purpose computing device is often built independently from the software applications that may run on it.

An embedded system is a combination of computer hardware and software, and sometimes mechanical components as well. Figure 1.1 gives a bird's-eye view of a generic embedded system architecture, where the microprocessor and the memory blocks are the heart and the brain, respectively. Embedded software is commonly stored in nonvolatile memory devices such as read-only memory (ROM), erasable programmable ROM (EPROM), and flash memory. The microprocessor also needs another type of memory—random-access memory (RAM)—for its run-time computation.

When an embedded system is powered on, its microprocessor reads software instructions stored in memory, executes the instructions to process input information from peripheral components (through sensors, signals, buttons, etc.), and produces output to meet the needs of the external *embedding* system.

Given that the hardware components are chosen, most of the design effort is in the software, including application, device drivers, and sometimes an operating system. In many cases, it is possible to build a customized integrated circuit (IC) that is functionally equivalent to an embedded system. An IC-based solution is a hardwired solution that does not contain software and a microprocessor. However, the embedded system solution is more flexible and less expensive, especially when the product needs to be frequently upgraded to accommodate new changes. In response to a new change, for the hardwired solution, a new circuit needs to be designed, constructed, and delivered. In contrast, for the embedded system solution, software patches can be rapidly developed, and the upgrading process can be done over the Internet and may typically take just a few seconds.

Embedded systems are widespread in consumer, industrial, medical, and military applications. Just look around your home or workplace, and you may realize that almost every aspect of our everyday life has been wonderfully touched by embedded systems: dishwasher, garage door opener, TV remote control, microwave oven, programmable thermostat, Xbox controller, and USB memory card reader. This list goes on and on.

Someone said that there are more computers in our homes and offices than there are people who live and work there. If this is true, then there are even more embedded systems that have been and will continue changing every part of our lives. One proof of this statement is that about 98% of microprocessors go into embedded systems, whereas less than 2% of microprocessors are used in computers.

1.2 Real-Time Systems

There are systems that need to respond to a service request within a certain amount of time: they are called real-time systems [22, 45]. To a real-time system, each incoming service request imposes a task (job) that is typically associated with a real-time computing constraint, or simply called its timing constraint.

The timing constraint of a task is normally specified in terms of its deadline, which is the time instant by which its execution (or service) is required to be completed. Depending on how serious missing a task deadline is, a timing constraint can be either a hard or a soft constraint:

- A timing constraint is *hard* if the consequence of a missed deadline is fatal. A late response (completion of the requested task) is useless, and sometimes totally unacceptable.
- A timing constraint is *soft* if the consequence of a missed deadline is undesirable but tolerable. A late response is still useful as long as it is within some acceptable range (say, it occurs occasionally with some acceptably low probability).

Actual systems may have both hard and soft timing constraints. A system in which all tasks have soft timing constraints is a soft real-time system. A system is a hard real-time system if its key tasks have hard timing constraints.

1.2.1 Soft Real-Time Systems

A soft real-time system offers *best-effort* services; its service of a request is *almost always* completed within a known finite time. It may occasionally miss a deadline, which is usually considered tolerable. It is worth noting that although missing a deadline will not cause catastrophic effects, the usefulness of a result may degrade after its deadline, thereby degrading the system's quality of service.

Soft timing constraints are typically expressed in probabilistic or statistical terms, such as average performance and standard deviation. Table 1.1 gives some example soft real-time systems.

Table 1.1 Example soft real-time systems

Example System	Example Timing Constraint	Consequence of Missed Deadlines
Digital camera	Shutter speed, shown in seconds or fractions of a second, is a measurement of the time the shutter is open. When the shutter speed is set to 0.5 s, the shutter open time should be (0.5 ± 0.125) s 99.9% of the time	Unsatisfied users may switch to other models
Global positioning system	Upon identifying a waypoint, it can remind the driver at a latency of 1.5 s	The driver misses the waypoint
Robot-soccer player	Once it has caught the ball, the robot needs to kick the ball within 2 s, with the probability of breaking this deadline being less than 10%	Its team may lose the game
Wireless router	The average number of late/lost frames is less than 2/min	The user has bad Web surfing experience

1.2.2 Hard Real-Time Systems

In a hard real-time system, missing some deadlines is completely unacceptable, because this could result in catastrophic effects such as safety hazards or serious financial consequences.

A hard real-time system offers *guaranteed* services. Hence, the correctness of a hard real-time system is twofold: functional correctness and timing correctness. Here, the timing correctness of a system means that its service of a request is guaranteed to be completed within a strict deadline. Actually, in most cases timing correctness is even more important than functional correctness, because a partially functional system may be used as is and still has its values, whereas a fully functional system is useless if the offered services have no guaranteed service completion time.

Since breaking a hard timing constraint is unaffordable, it is normally a requirement that the designers/developers of a hard real-time system should *validate rigorously* that the system can meet its hard timing constraints. In the literature, the proof techniques include design-time schedulability analysis, exhaustive simulation, combinatorial performance testing, and symbolic reasoning tools based on temporal logics (e.g., model checking). While many of these techniques are beyond the scope of this book, we will cover basic approaches to schedulability analysis when it comes to real-time scheduling.

Hard timing constraints are typically expressed in deterministic terms. Table 1.2 gives some example hard real-time systems.

What if a system at run time anticipates that a deadline might be missed? A hard real-time system will try its very best to avoid such a bad thing happening. For example, an antimissile system may have implemented two ways of locking onto an incoming missile: one takes a longer but can calculate precise firing coordinates, and the other takes less time but can

Table 1.2 Example hard real-time systems

Example System	Example Timing Constraint	Consequence of Missed Deadlines
Antilock braking system	The antilock braking system should apply/ release braking pressure 15 times per second A wheel that locks up should stop spinning in less than 1 s	Loss of human lives
Antimissile system	It never needs more that 30 s to intercept a missile after it reenters the atmosphere (in the terminal phase of its trajectory)	Loss of human lives, huge financial loss
Cardiac pacemaker	The pacemaker waits for a ventricular beat after the detection of an atrial beat. The lower bound of the waiting time is 0.1 s, and the upper bound of the waiting time is 0.2 s	Loss of human life
FTSE 100 Index	It is calculated in real time and published every 15 s	Financial catastrophe

compute only an approximate firing range, which demands more weapons being activated to cover the firing range. Since the first approach consumes fewer resources, it is the default approach employed by the system to handle incoming missiles. However, the system would switch to the second approach if it predicts that waiting for the precise calculation would take too much time for the incoming missiles to be safely destroyed.

In contrast, for a soft real-time system, it typically does not take any corrective actions until the bad thing really happens. For example, a DVD player has to synchronize the video stream and the audio stream. A missed deadline happens when, owing to data loss or decoding latency, the timing difference of the two streams exceeds a certain tolerable threshold. After a missed deadline has been detected, a DVD player may selectively discard the decoding of some video/audio frames to resynchronize the two streams.

1.2.3 Spectrum of Real-Time Systems

A real-time system is called a *real-time embedded system* if it is designed to be embedded within some larger system.

Figure 1.2 shows the scope of embedded systems and real-time systems, as well as the spectrum of real-time embedded systems. Note that there is no precise boundary between soft and hard real-time systems. Systems that lie in between the two are often called *firm real-time systems*. A system, such as the radar system studied in the next section, can be a soft or a hard real-time system, depending on where and how it is to be used. For example, a radar system is definitely a hard real-time system when it is part of a military surveillance system; it can be deemed a soft real-time system when it is used for weather forecasting or for monitoring meteorological precipitation.

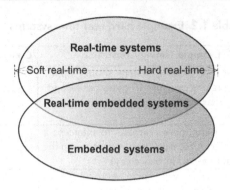

Figure 1.2
System classification.

1.3 Case Study: Radar System

Radar (*ra*dio *d*etection *a*nd *r*anging) is an object-detection system that uses radio waves or microwaves to determine the distance range, altitude, direction, or speed of objects. Radar has been widely used in many areas, including airport traffic control systems, aviation control systems, military surveillance systems, antimissile systems, and meteorological precipitation monitoring.

Radar is a very complex electronic and electromagnetic system [51, 58, 70]. Below we briefly describe the functionality of each subsystem. Readers may choose to skip the details for now, but will find them useful when we use radar as an example to explain advanced modeling concepts later in the book.

As shown in Figure 1.3, a radar system is typically composed of several subsystems.

The *radio frequency (RF) subsystem* generally consists of an antenna, an antenna feed, a duplexer, and some preselector filters. The antenna is the interface with the medium of radio wave propagation, and it is the first stage during signal reception and the last stage during signal transmission. The antenna feed collects energy as it is received from the antenna, or transmits energy as it is transmitted to the antenna. The duplexer switches the radar system between transmit mode and receive mode. The switch operation must be extremely rapid—say, within 5 ms (a real-time constraint)—in order to detect or track fast-moving targets. The switch operation frequency can be adjusted by an operator through the controller, which is part of the data processing subsystem. Preselector filters are used to attenuate out-of-band signals and images during the signal reception phase, and during transmission they are used to attenuate harmonics and images.

The *transmitter subsystem* consists of a digital waveform generator, an upconverting mixer, and a power amplifier. The digital waveform generator reads a desired waveform design and uses a D/A converter to produce analog signals in the baseband frequency range (intermediate

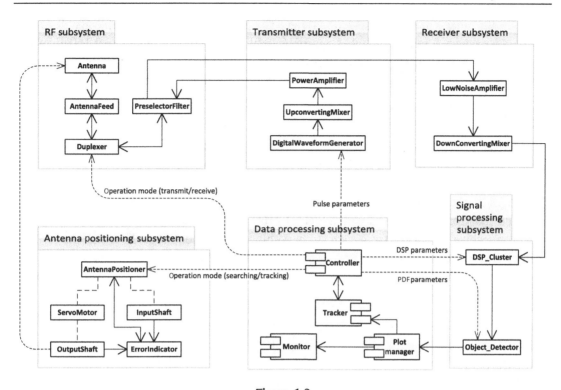

Figure 1.3

A radar system: its components, information flow (solid lines), and control flow (dashed lines).

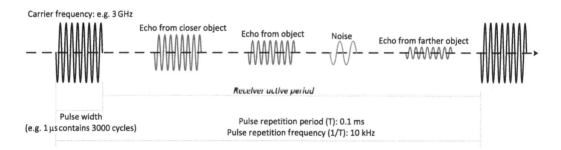

Figure 1.4

The radar signal in the time domain.

frequency). The generated signals must conform to the timing constraints specified in the design. The upconverting mixer is used to transform the baseband frequency signals into RF signals. The power amplifier is used to amplify the RF signals for transmission.

Figure 1.4 illustrates the radar signal in the time domain. The pulse width (or pulse duration) of the transmitted signal has to be long enough so that the radar can emit sufficient energy to ensure that the reflected pulse is detectable by the receiver. In such a sense, the pulse width constrains the radar's maximum detection range. The pulse width also determines the dead zone at close range. A radar echo takes approximately 1 µs to return from a target 150 m away. This means if the radar pulse width is 1 µs, the radar will not be able to detect targets closer than 150 m because the receiver is blanked while the transmitter is active.

The pulse repetition frequency (PRF) is the rate of pulse transmission. Most radar systems emit pulses multiple times in each direction so that multiple echo signals can be received. While a radar system might not be able to detect a target on the basis of simply one echo signal, the integration of multiple reflected pulses (of similar amplitudes) can reinforce the return signals and make detection easier. PRF can range from a few kilohertz to tens or hundreds of kilohertz. For example, a radar with a 1° horizontal beamwidth that sweeps the entire 360° horizon every second with a PRF of 2160 Hz can emit six pulses over each 1° arc; it thus can receive six echoes from each target within the detection range. Note that the (unambiguous) detection range is reduced as the PRF is increased because the receiver active period is shortened. A lower PRF gives a longer detection range, but often suffers from poorer target painting and velocity ambiguity.

Modern radar systems often use a technique called *PRF staggering*. With staggered PRF, the transmitter can use different PRFs to produce packets: a packet of pulses is transmitted with a fixed interval between pulses, and then another packet is transmitted with a slightly different interval. By PRF staggering, a radar can force the "jamming" signals from other radar systems to jump around erratically, inhibiting integration and thus suppressing their impacts on target detection. The PRF parameters can be set up by human operators and automatically adjusted by the controller of the data processing subsystem.

The *receiver subsystem* consists of a low-noise amplifier (LNA) and a downconverting mixer. After signal transmission, a radar system typically positions its antenna in that direction for a few milliseconds up to a few hundreds of milliseconds to collect the echo signals (*such a dwell time is another example of a real-time constraint*). Echo signals often contain noises The LNA is used to separate desired signals from undesired signals, such as thermal noise, clutter (echoes returned from objects that are not targets of interest, such as precipitation and birds), and interference (signals originating from active sources other than the radar). The LNA can also boost the power of the desired signals without introducing undesired distortions. The downconverting mixer is used to transform the RF signals into baseband frequency signals. Finally, the receiver subsystem uses A/D converters to sample an analog signal and save the digital values in shared memory accessible to the signal processing subsystem.

The *antenna position subsystem* is used to automatically rotate and position the radar antenna. The error indicator detects the error angle between the output shaft and the input shaft, and applies a feedback signal to the antenna positioner, which causes the servomotor to exert a torque on the output shaft in a direction to reduce the error. Through the controller, an operator can change the operation mode of the antenna positioner (say, from searching to tracking a target), or can alter the scan strategy (say, a conical scan, a unidirectional sector scan, or a circular scan).

Echoes from targets must be detected and processed before the transmitter emits the next pulse. After transmission of a pulse, if there is a reflective object (say, an unmanned aerial vehicle) at a distance of x meters from the antenna, the echo signal reflected by the object returns to the antenna in approximately $2x/c$ s, where $c = 3 \times 10^8$ m/s is the speed of light in a vacuum. To measure the target distance, the radar's detection range is divided into many *range intervals* or range bins, the length of which is equal to the desired range resolution (say, 200 m). The antenna's pulse repetition period can then be divided into many time frames, where the length of a time frame equals the time it takes the radio signal to propagate exactly one range interval. The digital values collected within each time frame form one sample of the situation in the corresponding range bin. Since the radar can transmit pulses multiple times in each direction, multiple samples for each range bin can be collected, and these are the inputs to the *signal processing subsystem*.

In general, the number of range bins can be in the hundreds or even thousands, and PRFs can range from a few to tens or hundreds of kilohertz. Such a high-demanding real-time constraint typically requires many digital signal processors (DSP) to form a computing cluster. The DSP cluster needs to produce a discrete Fourier transform of the samples for each range bin, and calculate the frequency spectrum of the echoes.

The *object detector* can determine whether there are objects in the direction in which the antenna is pointing and their positions and velocities, if applicable. If there is a moving object, the frequency of the reflected signal must be different from the frequency of the transmitted signal, which is called Doppler shift. The amount of Doppler shift is proportional to the velocity of the object.

A statistical hypothesis testing approach can be used to determine whether or not a measurement represents the influence of a target or merely interference [58]. The decision rule is simple: if the calculated "likelihood ratio" exceeds the detection threshold, declare a target to be present; otherwise, declare that a target is not present. Notice that the object detector relies on several unknown parameters that need to be replaced by their maximum likelihood estimates. The probability distribution functions that form the likelihood ratio may depend on one or more parameters. The position of the threshold can be dynamically adjusted to maximize the detection probability while keeping the false-alarm probability under a certain acceptable level. These parameters can be adjusted automatically by the controller. Again, when the number of range bins is large and the PRF is high, the object detector may need to make many thousands to millions of detection decisions per second.

The *data processing subsystem* is used to track targets of interest, interact with human operators, generate commands to control the radar, and adjust parameters of the other subsystems.

The controller module is the kernel of the radar system; it is not only responsible for synchronizing the behavior of the whole system to meet various timing constraints, but is also critical in tuning the performance of each component by adjusting certain parameters. For instance, it can alter the signal processing parameters such as the detection threshold and transform types, as well as the parameters for the statistical models used by the object detector. In addition, it can regulate the behavior of the digital waveform generator by changing the pulse parameters (such as the pulse width, pulse frequency, PRF, etc.).

The role of the plot manager is to manage real-time updates (also called plots) received from the object detector (say, once every 5 s) of the signal processing subsystem. The monitor module can be connected to a decision-support system so that human operators can make sense of the current situation and make real-time decisions.

The role of the tracker module is to process plots and to determine which plot belongs to which target, while rejecting any false alarms. Tracking is a computationally intensive activity, which typically has several steps:

- *Track prediction.* The tracker keeps a record (called a track) for each active target of interest. Each track has a unique ID, a state (including position, acceleration, speed, and heading), and possibly a target motion model (which is constructed to fit the motion history of the target). In this step, for each track the tracker needs to employ the associated motion model to predict its new state. It is particularly difficult when the target movement is very unpredictable or when the density of targets is high or the number of false returns is large. In such a situation, the tracker can employ a multiple-hypothesis approach where a track can be branched into many possible directions, with the most unlikely potential branches removed over time to reduce computational cost.
- *Track gating.* The gating process tentatively assigns a plot to an established track if the plot is within a threshold distance away from the predicted position of the track.
- *Track association.* In practice, it is very likely that a plot is assigned to more than one established track or more than one plot is assigned to one track. Such an ambiguous assignment could be exploited by a multiple-hypothesis tracker as mentioned above. However, when this feature is not desired, the ambiguity has to be resolved. Algorithms such as the nearest-neighbor approach can be effective in many cases to form one-to-one relationships between plots and tracks. The tracker will then invoke some smoothing algorithm (such as the Kalman filter) to combine a track with the associated plot to produce an improved estimate of the target state as well as a revision to the target motion model.

- *Track initiation.* After track association, a number of plots may remain unassociated with existing tracks. The tracker will create a new track for each of the unassociated plots. A new track is typically given the status of "tentative" until plots from subsequent radar updates have been successfully associated with the new track. Before being reinforced and confirmed by subsequent radar updates, tentative tracks are transparent to the operator so that false alarms can be washed away from the screen display.
- *Track maintenance.* Some existing tracks may be missing updates for a while (say, for the past five consecutive radar update opportunities). There is a chance that the target may no longer be there. Depending on the nature of the applications, such a track can be terminated—say, if the target was not seen for the past five out of the 10 most recent update opportunities.

Being a real-time system, a radar system typically operates with many timing constraints at different levels. For example, the object detector may produce batches of radar updates every 5 s, and it is thus necessary for the tracker to process all those radar updates within 5 s. The cluster of DSPs may need to complete its processing at the millisecond level per round, while the receiver may need to get its job done at the microsecond level per echo signal.

Problems

1.1 What is an embedded system? Identify some embedded systems used in your everyday life.
1.2 Are the laboratory computers embedded systems? Why or why not?
1.3 Is Google Glass an example of embedded systems? Why or why not?
1.4 What is a real-time system?
 (a) Identify some hard real-time systems, and for each, identify a few hard timing constraints.
 (b) Identify some soft real-time systems, and for each, identify a few soft timing constraints.
1.5 What would a hard real-time system do if it anticipates that a deadline might be missed?
1.6 Explain how a radar system could be either a hard real-time system or a soft real-time system. How do you classify a system as a hard real-time system instead of a soft real-time system?

Cross-Platform Development

Contents

Great computer hardware is only a doorstop without great software.

Charles King

Real-Time Embedded Systems. http://dx.doi.org/10.1016/B978-0-12-801507-0.00002-X

2.1 Cross-Platform Development Process

Compared with nonembedded software, the development of embedded software renders many unique challenges to software engineers:

- For embedded systems, especially for real-time embedded systems, timing correctness is equally important as functional correctness. Sometimes, a timing constraint is so important that it becomes an integral part of the functional requirements. For example, although a GPS navigation system can always identify waypoints correctly (functional correctness), it is useless if it is not able to report the waypoints before it is too late for the user to take actions.

- Most embedded systems are "dumb" devices that cannot run a debugger; this makes it hard to detect and clear program defects. For example, the processing system embedded inside a refrigerator can handle inputs from its touchpad and door sensor, provide output to a digital display, and control the cooling and icemaker machinery. It, however, may not have "luxurious" resources reserved for debugging.

- Most embedded systems are required to offer high reliability. For example, if a system has a reliability requirement of four nines (i.e., 99.99% availability), it is not tolerable if the downtime is greater than 9 s per day. High reliability is not a trivial objective especially for an embedded system that may be operating in a hostile or an unexpected environment. Unpredictable event patterns from the environment may significantly change an embedded system's sequence of execution.

- Efficient utilization of the memory space is another challenge. More memory means more monetary cost. Making software is a creative activity; making software that can fit into the available memory space demands even more creativity.

- Power management is critical to prolong the operating time of an embedded system. Being able to switch to a low-power state when inactive is a must-have feature for many embedded systems.

Owing to the uniqueness of embedded software, we need to distinguish two terms: host platform and target platform. The term "host platform" refers to the computing environment (i.e., the processor architecture and, if applicable, the operating system) upon which software is developed and its executable artifact is built. In contrast, the term "target platform" refers to the computing environment upon which software (actually its executable artifact) is intended to run.

For most software systems running on general-purpose computers, the host platform is the same as the target platform, and it is not necessary to distinguish the two. However, for embedded software, its target platform is typically different from its host platform [71].

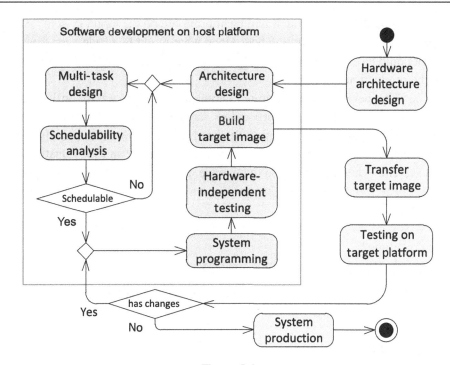

Figure 2.1
Real-time embedded systems development process.

As a road map, Figure 2.1 shows the cross-development process for real-time embedded systems. At a high level, some activities need to be conducted on the host platform, while others need to be performed on the target platform. We next explain each of those activities.

2.2 Hardware Architecture

As far as embedded systems are concerned, a hardware architecture is an abstract representation of an electronic or an electromechanical device that contains microprocessors, memory chips, and some peripherals (e.g., clocks, controllers, I/O devices, sensors, and actuators) as needed to fulfill its expected functionality.

Some common design factors related to hardware architecture design are as follows:

- *Processing power.* How fast a microprocessor will be required? A closely related factor is the processor's power consumption, typically expressed in terms of millions of instructions per second per milliwatt. A more powerful processor generally means higher power consumption, and a higher price too.
- *Memory requirements.* What types of random-access memory (RAM) and nonvolatile memory (NVM) will be needed? For each type, how much memory will be appropriate on

the evaluation board, and how much will be appropriate on the target system? A design with more memory on the evaluation board than what might be needed can often save a project that could have failed otherwise. However, more than necessary will certainly increase the production cost.

- *Peripherals.* What peripherals are required? Some advanced processor designs, also called microcontrollers or system-on-chip processors, have many built-in peripherals. Such a processor, if chosen, may be able to meet the needs for peripherals already. In general, it is advised to design a debugging interface (such as a serial port) on the evaluation board or the target system. When choices need to be made, performance always has higher priority than cost. A proven statement says that never use a $1.00 chip and expect the performance of a $10.00 chip.
- *Reliability.* Should the system be fail-proof? May it fail occasionally? What is the tolerable downtime?
- *Future upgrade.* How will field upgrades be performed?

We will learn some hardware fundamentals closely related to microprocessors and interrupts in Chapters 3 and 4, respectively.

2.3 Software Development

As shown in Figure 2.1, software development activities are performed on the host platform.

2.3.1 Software Design

Software design includes software architecture design, task design, and schedulability analysis:

- *Software architecture design.* How are different functional modules interrelated and synchronized? This topic is covered in Chapters 12 and 13.
- *Task design.* How many tasks should be considered? How should one assign priorities to those tasks? How should one document the task design and the timing constraints associated with the tasks? This is covered in Chapters 6–11 and 14.
- *Schedulability analysis.* As far as the identified timing constraints are concerned, is the set of tasks schedulable? This topic is covered in Chapters 15–17.

2.3.2 System Programming Language C/C++

A wide range of programming languages are in use today. However, for implementing real-time, resource-constrained embedded systems, the language chosen is required to have language-level direct access to low-level hardware. Moreover, different processor types accept different formats of binary code (machine instructions). The source code written in the chosen programming language is not executable until it is transformed into binary code by a computer utility called a compiler. This raises another requirement: a compiler has to be available for the chosen programming language such that the source code can be transformed into efficient binary code that is understandable to the processor selected in hardware design.

In practice, C and C++ are the de facto programming languages for embedded systems.[1] It is estimated that among all the embedded systems, about half of them are implemented in C, and about one third are implemented in C++. Their popularity is partially because C/C++ compilers are available for almost every processor type on the market.

The syntax of C/C++ programming is well beyond the scope of this book. Assuming the reader's familiarity with C, we will learn in Chapters 18–23 a few implementation patterns that utilize some operating system kernel objects such as timers, semaphores, and message queues.

In the rest of this section, we briefly cover a few concepts (semantics) that are relevant to binary code generation.

```
1  //-----start of a.h--------
   extern int a;
3  static int sa=5;
   //-----end of a.h----------
5
   //-----start of b.h--------
7  int b(int p);
   static int sb=10;
9  //-----end of b.h----------

11 //-----start of c.h--------
   extern int c;
13 static int sc=3;
   //-----end of c.h----------
15
   //-----start of one.c------
17 #include "a.h"
   #include "b.h"
19 int c = 2;
   int main(void)
21 {   int temp = 10;
     sb = sb + temp;
23   printf("Invoke from %d: %d\n", a, sb);
     b(temp+sb);
25   sb = sb + temp;
     printf("Invoke from %d: %d\n", a, sb);
27   b(temp+sb);
     return 0;
29 }
```

---

[1] A small portion of low-level code may be written in assembly languages.

```
     //-----end of one.c-----------
31
     //-----start of two.c--------
33 #include "c.h"
   #include "b.h"
35 int a = 1;
   int b(int p2)
37 {   static int temp = 2;
     sc++;
39   sb = sb + 2;
     temp = p2+temp+sc+sb;
41   if (temp < 50) goto done;
     temp = 100;
43 done:
     printf("Output from %d: %d\n", c, temp);
45   return temp;
   }
47 //-----end of two.c----------
```

Listing 2.1
An example C code.

A C/C++ program (application) can consist of numerous source files, each of which usually contains some *#include* directives that refer to header files. A compiler processes one source file at a time; it merges those headers with the source file to produce a transitory source file, which is called a *translation unit* or a compilation unit.

In Listing 2.1, there are two source files and three header files. The source file one.c together with a.h and b.h forms one translation unit, while the source file two.c together with b.h and c.h forms another translation unit.

2.3.2.1 Declarations and definitions

Each source or header file can define or declare many names, such as names for variables, names for functions, statement labels, tags for structures/unions/enumerations, identifiers for constants, and names for namespaces and classes in C++. The following discussion focuses on variable names and function names only.

We distinguish name declarations from name definitions. A name declaration simply states that the name belongs to the current translation unit. More than a name declaration, a name definition also imposes a requirement for storage.

- A variable declaration is a statement specifying a type for a variable name. A variable declaration is a variable definition if it contains no "extern" keyword, or it contains both an "extern" keyword and an initializer. For example, the variable a is defined in two.c and declared in a.h; similarly the variable c is defined in one.c and declared in c.h. Note that the variable `temp` is defined, independently, in both functions `main()` and `b()`. The statement "`extern int k = 0`" is a definition of variable k because it has an initializer.

- A function declaration is a statement containing a function prototype (function name, return type, the types of parameters and their order). A function declaration is a function definition if the function prototype is also followed by a brace-enclosed body, which generates storage in the code space. For example, the function `b()` in b.h is a function declaration, while the function `b()` in two.c is a function definition.

A name declared in C/C++ may have attributes. For example, a variable name has a type, a scope, a storage duration, and a linkage; a function name has all those attributes except storage duration.

The type of a variable determines its size and memory address alignment, the values it can take, and the operations that can be performed on it. A function's type specifies the function's parameter list and return type.

We next examine scope, storage duration, and linkage in detail.

2.3.2.2 Scope regions

A name's scope is that portion of a translation unit in which the name is visible. A name in an inner scope can hide a name from an outer scope. C and C++ each support five different kinds of scope regions [60]:

- *File/namespace scope.* In C, a name has file scope if it is declared in the outermost level of a translation unit. In C++, a name has namespace scope if it has file scope (global scope) or is declared in a namespace definition. In Listing 2.1, names with file scope include a, b, c, sa, sb, sc, and `main`.
- *Function scope.* A statement label has function scope. A label can be defined only in the body of a function definition and is in scope everywhere in that body. In Listing 2.1, the label `done` has a function scope.
- *Function prototype scope.* A name has function prototype scope if it is declared in the function parameter list of a function declaration without a body. In Listing 2.1, the parameter p of function `b()` in b.h has function prototype scope.
- *Block scope.* A name has block scope (called local scope in C++) if it is declared within a function definition or a block nested therein. In Listing 2.1, for the function `b()` defined in two.c, the parameter p2 and the variable `temp` have block scope.
- *Class scope.* A name in C++ has class scope if it is declared within the body of a class (including structure and union) definition. In C, there is no corresponding notion of class scope: each structure or union has a separate namespace for its members.

2.3.2.3 Storage duration

The storage duration of a variable determines the lifetime of the storage for that variable. Each variable in C/C++ has one of the following three storage durations:

- *Static.* For a variable with a static storage duration, its storage size and address are determined at compile time (before the program starts running); the lifetime of its storage is the entire program execution time. A variable declared at file/namespace scope has a static storage duration.
- *Automatic.* A local variable declared at block scope normally has an automatic storage duration. Local variables are stored in a run-time stack. Allocating storage for local variables usually takes just one machine instruction. Each time a function is called, a stack frame (a block of memory in the stack) is allocated for the function's local variables, and the stack frame is deallocated when the function returns. Thus, for a variable with an automatic storage duration, the lifetime of its storage begins upon entry into the block immediately enclosing the object's declaration and ends upon exit from the block.
- *Dynamic.* A local variable declared at block scope can have a dynamic storage duration if its storage is allocated by calling an allocation function, such as `malloc()` in C or the operator `new` in C++. Dynamic memory allocation allows a user to manage memory very economically. The drawback is that it is much slower than automatic allocation because it typically involves tens or hundreds of instructions. For a variable with a dynamic storage duration, the lifetime of its storage lasts until the memory is deallocated explicitly—say, by a `free` function in C or the operator `delete` in C++.

2.3.2.4 Linkage

Linkage determines whether name declarations in different scopes can refer to the same name definition. The linkage attribute applies to both variable names and function names.

C and C++ support three levels of linkage:

- *No linkage.* A name defined in block scope or structure/class scope normally has no linkage. In such a case, it can be referenced by name only within its scope. Outside its scope, declarations of the same name refer to different entities. Function parameters and local variables normally have no linkage.
- *Internal linkage.* A name defined in file/namespace scope can have internal linkage. In such a case, it can be referenced by name only within the same translation unit. Outside the translation unit, declarations of the same name refer to different entities.
- *External linkage.* A name defined in file/namespace scope or class scope can have external linkage. In such a case, it can be referenced by name within the same translation unit, and

can be referenced from other translation units. Declarations of the same name always refer to the same entity. Global variables and functions normally have external linkage by default.

2.3.2.5 Storage-class specifiers

A variable declaration in C/C++ can be preceded by a storage-class specifier, which is used to change the way of creating the memory storage for the variable [61]. Both storage duration and linkage can be affected by the use of a storage-class specifier.

A storage-class specifier can be one of the following keywords:

- *Auto.* This specifier can be used only in declarations of local variables with block scope. All local variables in C/C++ are of the "auto" type by default, so the keyword "auto" is very rarely used explicitly. The "auto" specifier indicates a variable with an automatic storage duration. If it is uninitialized, such a variable has a garbage value.
- *Register.* This specifier can be used only in declarations of variables with block scope. The "register" specifier basically requests the compiler to store the variable in a register; this allows faster access than an "auto" variable, which is stored in the main memory. You are not allowed to take the address of a register variable. "Register" is the only storage-class specifier that can be used for function parameters.
- *Extern.* This specifier can be used in declarations of both variables and functions.
    - A variable with the "extern" specifier has external linkage, which means that it can be referenced from other translation units. This also avoids unnecessary passing of variables as arguments during function calls.
    - A variable with the "extern" specifier has a static storage duration, which means that its memory is allocated at compile time and it exists and retains its value as long as the program runs. If it is uninitialized, such a variable is set to 0.
    - A local variable (block scope) normally has an automatic storage duration and no linkage. With the "extern" specifier, a local variable will have a static storage duration and external linkage.
    - With or without the "extern" specifier, a global variable (file/namespace scope) by default has a static storage duration and external linkage.
    - With or without the "extern" specifier, a function by default has external linkage, which means that it can be called from other translation units.
- *Static.* This specifier can be used in declarations of both variables and functions.
    - A variable with the "static" specifier has a static storage duration, which means that its memory is allocated at compile time and it exists and retains its value as long as the program runs. If it is uninitialized, such a variable is set to 0.

- A local variable (block scope) normally has an automatic storage duration. With the "static" specifier, a local variable will have a static storage duration. As a consequence, its value persists between different function calls.[2] This will not affect its linkage (no linkage) and scope attributes.
- With the "static" specifier, a global variable (file/namespace scope) has its linkage attribute changed to "internal linkage." For example, in Listing 2.1, the global variable sb defined in b.h has a "static" specifier. The header b.h is included by both one.c and two.c. Although each of the two translation units uses the same name sb, it actually refers to a distinct entity in each translation unit.
- A function normally has external linkage. With the "static" specifier, a function will have internal linkage, which means that it may be called only within the translation unit in which it is defined.

By default, a local variable without a specifier is treated the same as one with "auto," and a global variable without a specifier is treated the same as one with "extern" [62].

Table 2.1 summarizes how a variable's linkage and storage duration may be affected by a storage class specifier. Note that the only storage class specifiers allowed in a file/namespace scope declaration are "static" and "extern."

Table 2.1 A variable's linkage and storage duration depend on the storage class specifier

Storage Class Specifier	Block Scope		File/Namespace Scope		Structure Member/Class Scope	
	Linkage	Storage Duration	Linkage	Storage Duration	Linkage	Storage Duration
None	No linkage	Automatic	Normally external linkage	Static	No linkage	Same as enclosing object
Auto	No linkage	Automatic	NA	NA	NA	NA
Register	No linkage	Automatic	NA	NA	NA	NA
Extern	Normally external linkage	Static	Normally external linkage	Static	NA	NA
Static	No linkage	Static	Internal linkage	Static	External linkage in C++ only	Static in C++ only

NA, not applicable.

---

[2] When the modifier "static" specifies a variable or a function declaration, it makes the variable (function) *local* to the unit in which it is declared; in other words, the variable (function) cannot be referenced outside that unit. For instance, a static global variable (function) is limited to the file in which it is declared, and a static local variable inside a function is limited to that function. The lifetime of a static variable, global or local, is until the program terminates. This explains why a static variable can retain its value between function calls.

2.3.3 Test Hardware-Independent Modules

In practice, most modules of an embedded software are hardware independent. Hence, the skills for testing "general-purpose" software systems are equally applicable to the testing of those hardware-independent modules. For instance, test stubs, as used in the top-down testing approach, can be used to simulate the functionality of the software components that directly interact with hardware devices. As a simulator, a test stub of a module implements the interface of the actual module but simply provides *canned responses* to calls made during testing.

The topic of testing is beyond the scope of this book, and the interested reader is referred to textbooks on testing such as [30, 31, 54, 55].

2.4 Build Target Images

In this section, we will look at the toolchain for building target images, and introduce a standard object file format—ELF. As a case study, we will also examine the image building process for embedded systems that rely on the QNX operating system.

2.4.1 Cross-Development Toolchain

Figure 2.2 shows the code transformation process involving a chain of cross-development tools: cross compiler/assembler, linker, and dynamic linker.

2.4.1.1 Cross compiler/assembler

A compiler is called a *native compiler* if its output is intended to directly run on the host platform (or the same type of environment) where the compiler runs. A cross compiler is a compiler capable of generating executable code for a *target platform* that is different from the host platform. This statement applies to a cross assembler as well, except that it processes source files written in assembly languages.

A cross compiler/assembler is useful in a number of situations:

- The target platform (e.g., an embedded system) has extremely limited resources. It is not powerful enough to run a native compiler to generate executable code by itself.
- A source project is intended to run on different target platforms—say, different processor types, different versions of an operating system, or even different operating systems. By use of a cross compiler, a single host platform (development environment) can be set up to compile for each of these targets.

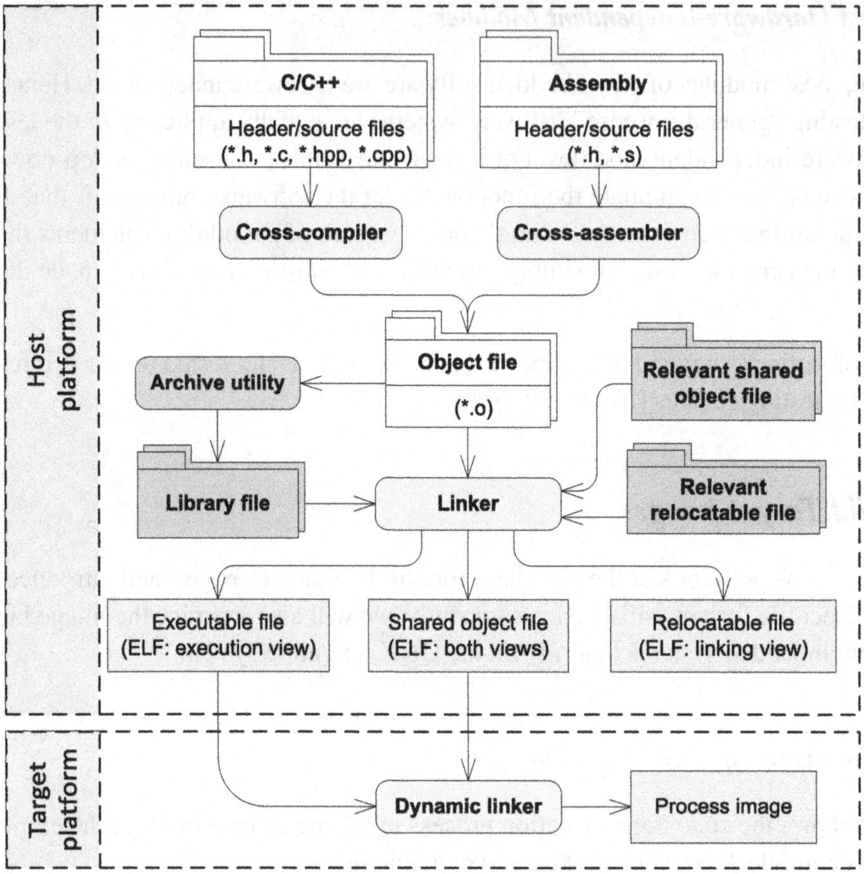

Figure 2.2
Cross-development toolchain.

For each compilation unit, the compiler/assembler generates an object file.[3] For instance, the C compiler can produce two object files—one.o and two.o—for the example code given in Listing 2.1.

Each object file contains a symbol table. A symbol table is an array-like data structure consisting of entries about the global symbols (i.e., names of global variables and nonstatic functions) defined in the compilation unit, as well as the external symbols (with "external linkage") referenced in the compilation unit. When the compiler encounters a symbol declaration, it stores that symbol and its attributes in the symbol table of the object file.[4]

---

[3] The term "object" used in this context is different from the one used in object-oriented modeling/programming.

[4] In object files, a symbol refers to a memory location, the content of which is either data for a variable or code for a function.

For the example code given in Listing 2.1, we have the following:

- In one.c, there are four global symbol definitions: global variables c, sa, and sb, and a function main(). It references two external symbols a and b().
- In two.c, there are four global symbol definitions: global variables a, sb, and sc, and a function b(). It references the external symbol c. The symbol table also has an entry for temp, which is a local variable with a static storage duration.

An object file's symbol table holds information that is needed by a compiler/linker to locate and relocate a program's symbolic definitions and references.

2.4.1.2 Linker

A complete program (application) is typically composed of multiple object files, each of which may cross-reference the definitions for data or functions defined in the other object files. The process of combining multiple object files into a single object file is called static linking, which is performed by a computer program known as a linker or link editor. The compiler automatically invokes the linker as the last step of compiling.

A linker has two major jobs:

1. *Symbol resolution.* While a linker is processing the application-specific object files, it needs to analyze each object file and determine where symbols with external linkage are defined. External symbol references may also involve application-independent object files—those that define commonly used functions (say, printf) for a wide range of applications (some known as archived library files, some known as .so, or sharable object files). It is worth noting that two variables of the same name can be defined in different scopes. In Listing 2.1, a static variable sb is defined in both one.c and two.c. This will not confuse the linker because the compiler has treated them as distinct symbols. Essentially the compiler uses "namespace" to distinguish variables: a local variable name is tagged by the function to which it belongs and a global variable name is tagged by the file name.
2. *Symbol relocation.* The final object file produced by a linker contains all the symbol definitions originating from the input object files, as well as those from static library files. As the linker merges the input object files and inserts code from the library files, the symbol offsets are changed. By symbol relocation, the linker, via a relocation table, modifies the binary code of the final object file so that each symbol reference reflects the actual address assigned to that symbol.

A linker can generate three types of object files: relocatable file, executable file, and shared object file:

- *Relocatable file.* A relocatable file holds code and data suitable for linking with other object files to create an executable file or a shared object file.

- *Executable file.* An executable file holds code and data suitable for execution. In an operating system, the basic unit of execution is called a process or *process image*, which is dynamically created from an executable file (say, as a result of an exec or spawn call). For systems supporting virtual memory, each process is put into its own address space.
- *Shared object file.* A shared object file holds code and data suitable for further linking. It can be combined with other relocatable and shared object files to create another object file. Some shared object files are *dynamically linkable*, and are intended to be loaded at run time and can be simultaneously shared by multiple process images.

2.4.1.3 Dynamic linker

In order to create a process image for an object file, the object file has to be loaded from its storage location into RAM. This job is performed by a program interpreter, which

- itself is an executable file or a shared object file;
- is self-interpreted (it does not need another program interpreter);
- is able to receive control from the system and provide a running environment for the object being loaded.

A *program loader* is a program interpreter that simply loads a program into memory and transfers the execution control to it. This approach works for stand-alone executable files. An executable file is stand-alone when it contains one copy for each library routine used in the program. This makes the executable object easier to distribute to diverse target environments, at the cost of larger memory space.

Another type of program, instead of having an embedded copy for each library routine, contains only reference information of sharable library routines. Obviously, such a program requires the presence of library files on the target system. It also demands a *dynamic linker*—a program interpreter that is more powerful than a simple program loader. A dynamic linker, once it has gained control from the system, will first load the whole program into memory to form an initial process image, then resolve symbolic references dynamically by loading and binding external shared libraries to form a complete process image, and finally transfer control to the process. This procedure is also called dynamic linking.

An advantage of dynamic linking is that multiple applications can share a single copy of a library. This also implies that a deployed system can automatically benefit from bug fixes and upgrades to libraries.

2.4.2 Executable and Linking Format

When tools from different vendors are used in the development process, some standard format ought to be adopted in the linking and dynamic linking process. ELF [2, 14] is a widely adopted object file format.[5] It supports cross-compilation, initializer/finalizer (e.g., the constructor and destructor in C++), dynamic linking, and other advanced system features.

An ELF object file has an ELF header, a section header table and/or a program header table, and a series of sections or segments. As shown in Figure 2.3, the ELF header of an object file resides at the beginning; it serves as a "road map" for the rest of the object file. An ELF header has the following fields:

- The first four bytes mark the file as an ELF object file. The "class" byte can take a value of 1 or 2, indicating the support for 32-bit or 64-bit processor architectures, respectively. The "data" byte can take a value of 1 or 2, specifying ELFDATA2LSB or ELFDATA2MSB as the data-encoding format. The encoding ELFDATA2LSB specifies 2's complement values,[6] with the least significant byte occupying the lowest address. The encoding ELFDATA2MSB specifies 2's complement values, with the most significant byte occupying the lowest address. The "pad" bytes are unused and are reserved for future changes.
- The "type" field indicates the object file type, which can take a value of 1, 2, or 3, representing relocatable file, executable file, or shared object file, respectively.
- The "machine" field specifies the required architecture for this object file. For example, the value 3 is reserved for Intel 80386 processors, and 40 is reserved for ARM processors.
- The "version" field identifies the version of the object file format, which is fixed to 1 for the current standard.
- The "entry" field gives the memory address of an entry point to which the system first transfers control. A system may have many entry points—say, for reset, interrupts, and software interrupts. However, an executable file can have one and only one entry point.
- The "phoff" field holds the byte offset of the program header table in the object file. It is zero if the file has no program header table.
- The "shoff" field holds the byte offset of the section header table in the object file. It is zero if the file has no section header table.
- The "flags" field holds processor-specific flags associated with the file. It is zero for the 32-bit Intel architecture because it defines no flags.
- The "ehsize" field holds the ELF header's size in bytes.
- The "phentsize" field holds the size in bytes of one entry in the file's program header table; all entries are of the same size.
- The "phnum" field holds the number of entries in the program header table.

---

[5] Another standard format is named common object file format (COFF).

[6] In a 2's-complement binary number representation, the weight of each bit is a power of 2, except for the most significant bit, whose weight is the negative of the corresponding power of 2. In general, the value v of an N-bit integer $a_{N-1}a_{N-2}\cdots a_0$ is given by $v = -a_{N-1}2^{N-1} + \sum_{i=0}^{N-2} a_i 2^i$.

- The "shentsize" field holds the size in bytes of one entry in the file's section header table; all entries (section headers) are of the same size.
- The "shnum" field holds the number of entries in the section header table.
- The "shstrndx" field holds an index in the section header table, giving the entry associated with the *section-header string table*.

The ELF standard provides two parallel views of an object file's contents. On the one hand, an ELF object file can be viewed as *a series of named sections*. This "linking" view is taken by compilers/assemblers and linkers. Sections are intended for further processing by a linker. On the other hand, an ELF object file can be viewed as *a series of named segments*. Segments are intended to be mapped into memory to create a process image. This "execution" view is taken by dynamic linkers or program loaders.

2.4.2.1 Linking view

Figure 2.3 shows the linking view of ELF, where it is optional to have a program header table. For example, there is no program header table in a relocatable file, whereas a shared object file may or may not have a program header table.

Table 2.2 lists some default sections with predefined names. Each section contains a single type of information, such as program code (instructions), data, symbol, or relocation information. Typically, an object file may have

- a .text section for executable instructions;
- a .data section for initialized data;
- a .bss (for "block started by symbols") section for uninitialized data;
- a .symtab section (symbol table) for static linking;

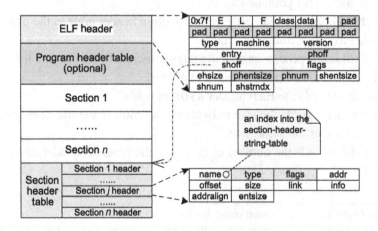

Figure 2.3
The linking view of ELF.

Table 2.2 Some predefined section headers, their types, and their attributes

Type	Name	Meanings	Flags
NOBITS	.bss	Uninitialized data (global, static)	ALLOC + WRITE
PROGBITS (program contents)	.comment	Extra information such as version control	
	.data	Initialized data (global, static)	ALLOC + WRITE
	.data1	More initialized data	ALLOC + WRITE
	.debug	Symbolic debugging information	
	.fini	Process termination code	ALLOC + EXECINSTR
	.got	Global offset table (each entry contains a symbol with an absolute address to be determined in dynamic linking)	
	.init	Process initialization code	ALLOC + EXECINSTR
	.interp	Path name of a program interpreter—say, a dynamic linker (to load the program and to link shared libraries if necessary)	ALLOC + EXECINSTR
	.line	Line number information for symbolic debugging	
	.plt	Procedure linkage table	
	.rodata	Read-only data (constant)	ALLOC
	.rodata1	More read-only data	ALLOC
	.text	Executable instructions	ALLOC + EXECINSTR
DYNAMIC	.dynamic	A table of entries for dynamic linking	ALLOC
DYNSYM	.dynsym	Symbol table for dynamic linking	[ALLOC]
HASH	.hash	Symbol hash table for dynamic linking	[ALLOC]
REL	.rel[name]	A table of relocation entries without explicit addends ([name] can be .text, .data, etc.)	[ALLOC]
RELA	.rela[name]	A table of relocation entries with explicit addends	[ALLOC]
STRTAB	.shstrtab	Section-header string table (section names)	
	.strtab	Symbol string table (for names associated with symbol table entries)	[ALLOC]
SYMTAB	.symtab	Symbol table for static linking	[ALLOC]

- a .strtab section holding a symbol string table;
- a .shstrtab section holding a section-header string table; and
- a few .rel[name] sections each containing the relocation information for a specific section with the name [name].

Each section in an object file occupies one contiguous (possibly empty) sequence of bytes, and has exactly one section header describing it. A section header has the following fields:

- The "name" field specifies the name of the section. Its value is an index in the .shstrtab section (section-header string table), giving the location of a null-terminated string.
- The "type" field categorizes the section's contents. The first column of Table 2.2 lists some of the supported types.
- The "flags" field supports one-bit flags: 0x1 for WRITE (the section contains writable data), 0x2 for ALLOC (the section occupies memory during process execution), and 0x4 for EXECINSTR (the section contains executable machine instructions).
- The "addr" field, if not zero, gives the *load address*[7]; that is, the starting address in the NVM at which the section's first byte resides.
- The "offset" field gives the byte offset from the beginning of the file to the first byte in the section.
- The "size" field gives the section's size in bytes.
- The "link" field, for a section related to dynamic linking or relocation, holds a section header index of a string table or symbol table.
- The "info" field, for a .rel or .rela section, holds the section header index of the section to which the relocation applies.
- The "addralign" field holds a value of 0 or positive integral powers of 2, specifying a constraint for section address alignment. If the value of addralign is greater than 1, the value of addr must be congruent to 0, modulo the value of addralign.
- The "entsize" field, if not zero, gives the size in bytes of an entry of a table section.

2.4.2.2 Execution view

Figure 2.4 shows the execution view of ELF, where it is optional to have a section header table. In this view, an object file is a set of segments described by a program header table, which is meaningful only for executable and shared object files. As the system creates a process image from an object file, a *file segment* (such as .text, .data) is logically copied/mapped into virtual memory to form a *memory segment*.

---

[7] The concept of load address is different from that of run address. A load address typically refers to a location in a NVM chip, whereas a run address always refers to a location in RAM. In some embedded systems, the run address can be the same as the load address. In such a case, the embedded application is directly downloaded into the target system memory for immediate execution without any code or data transfer from one memory type or location to another.

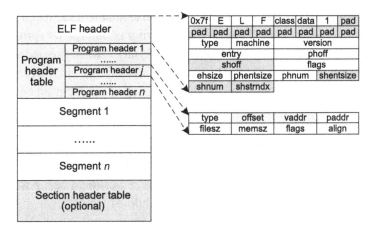

Figure 2.4
The execution view of ELF.

A file segment usually consists of several sections. Each file segment is described by an entry of the program header table, which has the following fields:

- offset. This gives the offset from the beginning of the file at which the first byte of this file segment resides.
- filesz. This gives the size in bytes of this file segment; it may be zero.
- vaddr. This gives the virtual memory address at which the first byte of the memory segment resides. This is also called the segment's *run address*, and it should be within the RAM range.
- memsz. This gives the size in bytes of the memory segment; it may be zero.
- paddr. This is reserved for the segment's physical address. When used, this is the starting address in the *NVM* at which the first byte of this file segment resides. This is also called the segment's *load address*. For an object capable of in situ execution, its segments can have the same load addresses and run addresses.
- type. This field specifies the segment type, which can be one of the following:
  - NULL. This segment is unused and should be ignored.
  - LOAD. This is a loadable segment. A corresponding memory segment is created in the process image, and the bytes from the file (specified by filesz) are mapped to the beginning of the memory segment. If the specified memory size is larger than the file size (memsz > filesz), the "extra" bytes in the memory segment hold the value 0.
  - DYNAMIC. This is a dynamic segment, specifying information for dynamic linking. This segment starts with a *.dynamic section*, followed by a few other sections such as a hash table section, symbol table section, a string table section, and a relocation table section.

- INTERP. This segment is meaningful only for executable files, specifying the location and size of a null-terminated path name to a program interpreter (program loader or dynamic linker). If it is present, it must precede any loadable segment.
- NOTE. This segment specifies the location and size of auxiliary information.
- PHDR. This segment specifies the location and size of the program header table itself. It may occur only if the program header table is part of the process image. If it is present, it must precede any loadable segment.

- flags. This gives flags (readable, writable, executable) relevant to the segment.
- align. If align > 1, this gives the value to which the file segment and the corresponding memory segment should be aligned. The value of align should be a positive, integral power of 2, and vaddr should equal offset, modulo align.

2.4.3 Memory Mapping

By default, especially when an executable file is intended to run on an operating system, each loadable segment of the object file is mapped by the linker to a virtual address location (specified by the vaddr field). For each program to be executed, the operating system typically allocates a protected memory space in RAM, where a process image is formed by loading the program from the file system (via a file descriptor).

An operating system is simply a "big" application. An operating system itself needs to be loaded from somewhere when the system starts. This job is done by a "starter" program, also known as an initial operating system loader. Where is such an operating system loader found when the system is powered on? Moreover, there are many embedded systems running without an operating system. Where should the instructions be loaded when such an embedded system starts? This brings us to memory mapping.

Via a script file, a linker can be directed to map each segment of an executable file to a specific memory location available on the target system. First of all, an operating system loader or an embedded application without an operating system needs to be mapped directly to specific locations on the memory chips of the target system. When the system starts, a booter is able to locate the loader or application image stored on the NVM and, if necessary, copy some segments from the NVM into appropriate RAM locations (see Chapter 5 for details). The loader or application gains control afterward and its execution follows.

As an example, Listing 2.2 gives a linker script written for an operating system loader. The TARGET command specifies the format of the input object files. The input format used in the example is 32-bit ELF that is customized to support little-endian ARM processors. The OUTPUT_FORMAT command specifies the format of the output object file. In the example, the output format is the same as the input format.

The ENTRY command specifies a symbol name "_start," which indicates the first executable instruction in an output file (its entry point).

The MEMORY command describes memory regions on the target board, their sizes, and their locations (start addresses of the regions in physical memory). The example has three named memory regions defined: stack, ram, and rom. In particular, the rom block starts at address 0x00300000 and has 16K (0x4000) bytes.

```
 1  TARGET(elf32-littlearm)
    OUTPUT_FORMAT(elf32-littlearm)
 3  ENTRY(_start)

 5  MEMORY
    {
 7    stack : ORIGIN = 0x0030E000,  LENGTH = 0x2000
      ram   : ORIGIN = 0x00304000,  LENGTH = 0x6000
 9    rom   : ORIGIN = 0x00300000,  LENGTH = 0x4000
    }
11
    SECTIONS
13  {
      .text :
15    {
        *(.text)
17      *(.rodata*)
        *(.glue_7)
19      *(.glue_7t)
      } > rom
21    _etext = .;

23    .data :
      {
25      *(.data)
        *(.sdata)
27    } > ram

29    .bss :
      {
31          *(.bss)
            *(.sbss)
33    } > ram
    }
```

Listing 2.2
Example linker script for an operating system loader.

The SECTIONS command describes the placement of named output sections, and which input sections go into which output sections. In the example, three output sections are defined: .text, .data, and .bss.

The content of an output section consists of one or more statements of the form filename (section, section, ...). An asterisk can be used to represent all input object files. In the example, the .text section contains four statements:

- *(.text): copy the .text section from each input object file into the .text section of the output object file.
- *(.rodata*): copy all the sections with the prefix .rodata from each input object file into the .text section of the output object file.
- *(.glue_7): copy the .glue_7 section from each input object file into the .text section of the output object file. The glue_7 segment contains glue functions generated by the compiler for the 32-bit ARM mode.
- *(.glue_7t): copy the .glue_7t section from each input object file into the .text section of the output object file. The glue_7t segment contains glue functions for the ARM thumb mode.

The linker does not shuffle sections to fit into the available memory regions. The statement ">region" at the end of a section command assigns the section to a specific memory region (it is reflected in the segment's `vaddr` field). In the example, the .text section goes to rom, while the .data and .bss sections go to ram.

The statement "_etext = .;" defines a global symbol "_etext" just after the last byte of the .text section.

Note that the .sdata and .data sections both contain initialized data, but the size of a .sdata session is less than a threshold (say, 64 KB), *smaller* than a .data section (with a size above the threshold). Likewise, a .sbss section is similar to a .bss section, except that it is smaller than a .bss section.

Figure 2.5 illustrates a possible memory mapping, given that the linker follows the script in Listing 2.2 to process the two input object files shown on the left in Figure 2.5.

2.4.4 Case Study: Building a QNX Image

In this section, the term "QNX image" refers to a holistic executable file object containing the user applications and the underlying QNX operating system.

Figure 2.6 shows a typical procedure for building a QNX image for a specific target board.

To develop software for a particular target board, it is critical to start with a board support package (BSP), which is a software component specifically tailored for supporting the hardware design of the target board. A commercial evaluation board typically comes with a sample BSP, which can save developers a great deal of time and effort.

A QNX BSP for a board, if available, comes with an initial program loader for loading the QNX operating system, as well as a startup module that contains all the device drivers for the board.

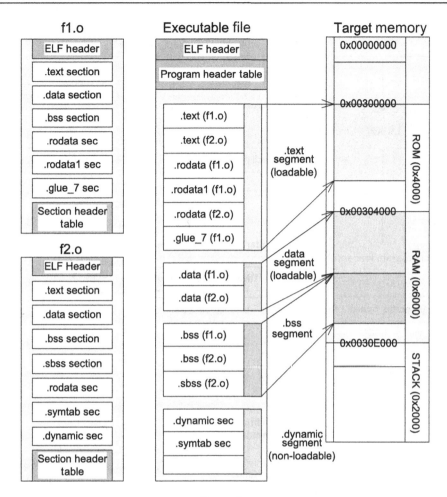

Figure 2.5
Mapping segments into target memory.

A system builder project is needed to glue all the pieces together [6]. A system builder project can be processed by a utility called mkifs (make image filesystem) to build a bootable QNX image. The behavior of mkifs is governed by a build-script file (.bld), which can be configured such that the QNX image produced contains

- the startup binary generated from the BSP;
- the QNX kernel and relevant shared library objects;
- the application-specific binaries generated from user C/Photon projects; and
- some relevant libraries and utilities.

The final QNX image object for an embedded system has two parts: an initial program loader (generated from the BSP) and the bootable QNX image.

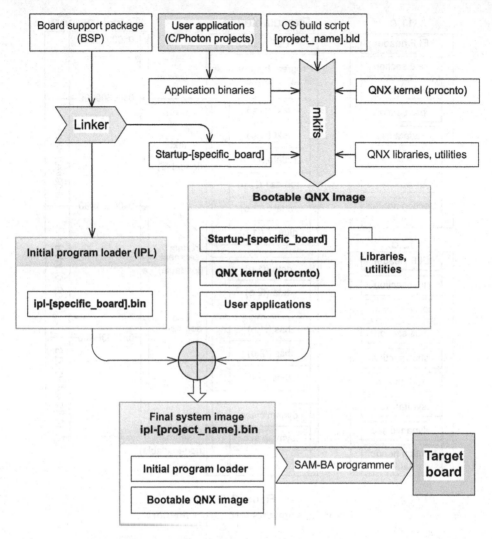

Figure 2.6
QNX operating system image building process.

2.5 *Transfer Executable File Object to Target*

Normally, a utility known as "programmer" is used to transfer (say, through high-voltage programming) an executable file object to the NVM (ROM or flash) available on the target system. For instance, in Figure 2.6, a programmer called SAM-BA is used to transfer the final QNX image to the target board. Once the executable file object is accessible to the processor of the target board, it is ready to run on the target platform.

The detailed boot process is covered in Chapter 5.

2.6 Integrated Testing on Target

The software modules that interact with hardware devices are yet to be tested. Many uncertain factors can make it difficult and challenging to test/debug on the target system. For instance, owing to jitters or some other reasons, it might be extremely hard to reproduce a time-critical input situation. The once useful techniques such as inserting printf() statements[8] become futile or even inadvertently interfere with the run-time behavior of the system [66].

Testing at this step, if necessary, can benefit from devices such as oscilloscopes and in-circuit emulators. The testing activities mainly involve the following aspects:

- *Device drivers.* They are tightly coupled with the corresponding devices.
- *Portability issue.* It can be time-consuming if the software has been written to support more than one target platform.
- *Shared-resource issues.* It is critical and should be thoroughly tested when there are data objects shared between interrupt service routines and user task code.
- *Racing conditions.* When multiple tasks use mutually exclusive resources, they may race each other to produce nondeterministic results. It is very hard to reproduce and debug a race condition because its occurrence is highly dependent on the relative timing between interfering tasks.
- *Timing correctness.* This can be tested only on the target system itself.

In order to resolve the problems detected in testing, it might be necessary to change the system design and implementation, and the development process would come back again to the host platform (see Figure 2.1).

2.7 System Production

When the system is ready for production, cost becomes the major concern. While it is advised to use more speed and memory in the development phase, it is now the time to consider reducing the production cost by downgrading the microprocessor and/or memory devices. Of course, the whole system has to be thoroughly tested after each downgrade.

In conclusion, this chapter serves as a road map for the whole book, where all the major activities involved in the process of cross-platform development are explained. Our focus was on the toolchain for the generation of target images and ELF—a widely adopted object file format. While knowing the detail of a standard object file format such as ELF is the business of those vendors who have an interest in producing interoperable development tools, knowing the big picture allows embedded software engineers to understand how an executable image is

---

[8] A statement such as printf() itself takes time to complete, which may completely change the system timing.

mapped into the target embedded system for static storage, as well as for run-time loading and execution.

A final note is that the development of many embedded systems demands teamwork among engineers with different skill sets. For instance, a complex embedded system might involve hardware engineers, software engineers, and systems engineers. Hardware engineers deal with hardware components, software engineers deal with software components, and systems engineers ensure that the software and hardware can cooperate smoothly to offer the desired services to the external environment.

Problems

2.1 What makes it challenging to develop an embedded system?

2.2 How many scope regions are defined in C/C++?

2.3 What is a static storage duration? What is an automatic storage duration?

2.4 How may the storage-class specifiers "extern" and "static" affect a variable's linkage and storage duration?

2.5 What is a cross compiler? Why are cross compilers needed in developing embedded systems?

2.6 What is a program interpreter? What is a program loader? What is a dynamic linker?

2.7 Why is it important to define a standard object file format such as ELF?

2.8 According to the ELF standard, what information is contained in a .data section, a .bss section, a .text section, and a .rodata section, respectively?

2.9 In embedded systems, why should we distinguish run address from load address?

2.10 In a linker script file, how are related sections grouped together into one segment?

2.11 Some circuit boards feature a serial debug port, which can output debugging information when the system is up and running. Why should a terminal set up a specific baud rate in order to correctly receive the output from the debug port?

Microprocessor Primer

Contents

There are only 10 types of people in the world: Those who understand binary and those who don't.

Anonymous

3.1 Introduction to Microprocessors

A microprocessor is a general-purpose central processing unit (CPU) manufactured on a single *integrated circuit*. To be useful, a microprocessor has to work with other components, such as a memory system for storing instructions and data, and external circuits to communicate with the peripheral environment.

A microprocessor is merely the processing core of an application system. Sometimes, a microprocessor and some commonly used circuits and components (e.g., memory, parallel I/O, serial I/O, clock circuit, etc.) are integrated together on a single chip, which is typically called a *microcontroller*. A microcontroller is truly a computer on a chip. Using a microcontroller can greatly ease the hardware architecture design, especially when it has all the necessary peripherals for the system to be developed.

The first commercial microprocessor was the Intel 4004, introduced by Intel Corporation in 1971 for electronic calculators. Since then, powerful, low-cost microprocessors have been used in myriads of embedded systems found almost everywhere: household appliances, cars, toys, DVD players, AV receivers, cell phones, and high-definition TVs, to mention only a few.

3.1.1 Commonly Used Microprocessors

A few commonly used microprocessors are given in Table 3.1.

Intel 4004 is a 4-bit processor with only 2048 bits of addressable memory. Intel 8080 was introduced in 1974. It is an 8-bit processor with 16 address lines that can access 64 KB memories. The most successful family of processors started with the 16-bit Intel 8086, which sets the basis for the Intel x86 architecture (also referred to as IA-32).

Introduced in 1985, Intel 80386 features three operating modes: real mode, protected mode, and virtual mode. The *protected mode* allows the processor to address up to 4 GB of memory. The *virtual mode* makes it possible to run one or more programs in a protected environment.

The Pentium processor was the first x86 processor with superscalar architecture.[1] The Pentium processor also features a 64-bit external data bus, which doubles the amount of information it is possible to read or write on each memory access.

Pentium Pro was introduced in 1995. It can not only access a bigger memory space, but also implements a new architecture called microarchitecture—a decoupled, 12-stage superpipelined architecture. In 2001, Intel introduced Itanium, a family of 64-bit

---

[1] A superscalar processor executes more than one instruction during a clock cycle by simultaneously dispatching multiple instructions to redundant functional units on the processor. A superscalar processor can be envisioned as having multiple parallel pipelines, each of which processes instructions from a single instruction thread.

Table 3.1 The characteristics of some commonly used microprocessors

Architecture	Year	Family	Processing Width (bits)[a]	Address Width (bits)	Addressable Memory Space (1 byte = 8 bits)	External Memory Bus: Data (address)
Intel	1971	4004	4	12	$2^{12} = 4\,K(\frac{1}{2}B)$	4
	1974	8080	8	16	$2^{16} = 64\,KB^b$	8
	1978	8086	16	20	$2^{20} = 1\,MB$	16
	1985	80386	32	32	$2^{32} = 4\,GB$	32
	1993	Pentium	32	32	$2^{32} = 4\,GB$	64
	1995	Pentium Pro	32	36	$2^{36} = 64\,GB$	64
	1999	Pentium III	32	36	$2^{36} = 64\,GB$	64
	2001	Itanium	64	64	$2^{64}\,B$	128
Motorola	1979	68000	32 (16)	24	$2^{24} = 16\,MB$	16
Microchip	2000	PIC18F2XK20	16	C^c: 21	$2^{21} = 2\,MB$	No
				D^d: 12	$2^{12} = 4\,KB$	
	2000	PIC18F8X20	16	C: 21	$2^{21} = 2\,MB$	16 (20)
		PIC18F97J60		D: 12	$2^{12} = 4\,KB$	
	2007	PIC32MX	32	32	$2^{32} = 4\,GB^e$	Not yet. 16 (16) through PMP
ARM	2001	ARM9E	32	C: 18	$2^{18} = 256\,KB$	32
				D: 18	$2^{18} = 256\,KB$	32
	2006	Cortex-M3	32	32	$2^{32} = 4\,GB^e$	32

PMP, parallel master port.
[a]The processing width of a processor is the same as the width of the general-purpose registers and the arithmetic logic unit of the processor. It is also called the word width of the processor.
[b]When the address width is larger than the processing width, memory segmentation is done by the processor internally.
[c]Program code space.
[d]Data (random-access memory) space arranged in banks of size 256 bytes.
[e]Program code space and data space are still separated, but they are mapped into a unified virtual memory space.

microprocessors that implement the IA-64 architecture. Itanium processors are targeted for enterprise servers and high-performance computing systems.

Motorola 68000 (also called 68K) is a 16-/32-bit microprocessor. Although introduced in 1979, it is still in use in embedded applications. For example, it has been used in low-end printers such as HP's LaserJet introduced in 1984, and Apple's LaserWriter introduced in 1985, in game consoles such as Sega's System 16, Mega Drive (Genesis) console, and Saturn console.

The PIC family of microcontrollers is made by Microchip Technology. PIC microcontrollers are popular in both industry and education because of their low cost, large user base, and wide availability of development tools. PIC microcontrollers feature on-chip program and data memory as well as many peripheral components. Some PIC microcontrollers (e.g., PIC18F8X20 and PIC18F97J60) even have an external memory interface to expand the internal memory space.

ARM processors have been used extensively in consumer electronics, including personal digital assistants, tablets, cell phones, music players, handheld game consoles, and computer peripherals such as hard drives and routers. As of 2009, ARM processors accounted for approximately 90% of all embedded 32-bit reduced instruction set computing (RISC) processors [35]. In 2010, over 6.1 billion ARM-based chips were sold, representing over 95% of the smartphone market, 10% of the mobile computer market, and 35% of the digital TV and set-top box market [53].

As integrated circuit technology advances, it becomes feasible to manufacture more and more complex processors on a single chip. It is clear that high-end applications demand more address space and computing power than those offered by 32-bit processors, and the transition from 32-bit computing to 64-bit computing has become the trend.

3.1.2 Microprocessor Characteristics

3.1.2.1 Architectures

As far as memory access is concerned, a microprocessor may belong to either von Neumann architecture or Harvard architecture.

The *von Neumann architecture* is named after the mathematician and early computer scientist John von Neumann. The von Neumann architecture has two features:

- Data and instructions (executable code) are stored in the same address space. On the one hand, this allows for self-modifying code. On the other hand, this opens security holes. For instance, a program may attempt to run instructions from memory that contains data, or a program might write data into memory containing instructions.
- The processor interfaces with memory through a single set of address/data buses. Since an instruction fetch and a data operation cannot occur at the same time because they share a common bus, this often limits the performance of the system. This is referred to as the *von Neumann bottleneck*.

The *Harvard architecture* is named after the Harvard Mark I computer. The Harvard architecture has two features:

- Data and instructions (executable code) are stored in separate address spaces. For instance, the instruction space may be accessed by 20 address lines, while addresses in the data space may only have 16 bits. In addition, program memory is typically read-only memory (ROM). It is impossible for program contents to be modified by the program itself.
- There are two sets of address/data buses between the processor and the memory. Since instruction fetches and data operations are carried on separate buses, it is possible to access program memory and data memory simultaneously.

Modern microprocessor designs (e.g., ARM) incorporate aspects of both Harvard and von Neumann architectures. For instance, a modified Harvard architecture has separate instruction and data caches that are backed by a common address space. While the processor executes from cache, it acts as a pure Harvard architecture. When accessing backing memory, it acts like a von Neumann architecture (where code can be moved around like data).

As far as the instruction set is concerned, a microprocessor may be one of two types: a complex instruction set computing (CISC) processor or an RISC processor.

A *CISC* processor has multiple addressing modes and runs "complex instructions" where a single instruction may execute several low-level operations (such as a load from memory, an arithmetic operation, and a memory store). While this leads to high code density, it often requires manual optimization of assembly code for embedded systems.

An *RISC* processor runs compact, uniform instructions where the amount of work any single instruction accomplishes is reduced (e.g., separate instructions for I/O and data processing). Such compact instructions allow effective compiler optimization and facilitate pipelining, typically leading to higher performance than CISC. As an overhead, however, RISC inevitably produces more lines of code (larger memory footprint) than CISC.

Today, CISC processors dominate the personal computer market and are used in a significant fraction of the low- and mid-range servers and workstations. RISC processors have been successfully used across a wide range of platforms, from mobile devices such as cellphones and tablets to some of the world's fastest supercomputers.

Figure 3.1 gives some example microprocessors for each category, where SHARC (for "*super Harvard architecture*") is a processor family that offers many features required by digital signal processor applications: exceptional core and memory performance and outstanding I/O throughput.

	Von Neumann	Harvard
CISC	X86 (8086, Pentium) Motorola 68000	SHARC (DSP)
RISC	ARM7, SPARC, MIPS, PowerPC	ARM9, PIC

Figure 3.1

Microprocessor architecture types.

3.1.2.2 Processing width

The *processing width* is an important characteristic of a processor design. For example, the Intel 8086 is a 16-bit processor, where the number 16 is its processing width. The processing width of a processor is exactly the width of its working memory (registers).

The term "word" also refers to the largest processing unit that can be transferred to and from the working memory in a single operation of a processor. In other words, the number of bits in a word (the word size or word width) is the same as the processing width (register width). For instance, PIC18F8720 is a 16-bit processor; its word size is 16 bits and a word is composed of 2 bytes. ARM926EJ-S is a 32-bit processor; its word size is 32 bits and a word is composed of 4 bytes.

Modern processors usually have a word size of 16, 32, or 64 bits.

3.1.2.3 I/O addressing

A microprocessor typically accesses I/O devices in two ways.

I/O devices can be placed in a microprocessor's memory address space. This approach is called *memory-mapped I/O*, where I/O devices and memory components are indistinguishable to the processor. Memory-mapped I/O simplifies coding and testing. Any instruction that references memory may be used to access an I/O port located in the memory space. For example, the Intel 80386 processor can use the MOV instruction to transfer data between any register and a memory-mapped I/O port. The Motorola 68000 uses memory-mapped I/O. PIC family processors also have all peripherals memory mapped (through ports and port registers).

Some processors may feature a special signal pin (e.g., M/$\overline{\text{IO}}$ in the Intel 8086 and Pentium processors) to indicate whether a memory device or an I/O device is accessed. In such a case, the designer can choose whether to map I/O devices into the memory space or use a separate *I/O address space*. In the latter case, all I/O devices are treated differently from normal memory locations. Typically special I/O instructions are needed to access I/O ports.

3.1.2.4 Reset vector

The *reset vector* of a processor is the default location where, upon a reset, the processor will go to find the first instruction to execute. In other words, the reset vector is a pointer or address where the processor should always begin its execution. This first instruction typically branches to the system initialization code.

Most microprocessors have reset vectors fixed at the two ends of their address spaces. Here are a few examples:

- The Intel 8086 processor has its reset vector at FFFF0h, the high end of its address space.
- PIC18 processors have the reset vector located at the low end, 0000h.

- The Motorola 68000 processor has a 1024-byte vector table beginning at 000000h. The first entry is the reset vector, which is at 000000h. This vector table also contains pointers to routines used by the processor, operating system, and users.
- ARM family processors have the address 0x00000000 reserved for the vector table (including reset vector, undefined instruction vector, software interrupt vector, interrupt request vector, fast interrupt vector), and the reset vector is at 0x00000000. On some processors (e.g., ARM926EJ-S) the vector table can be optionally located at a higher address in memory, 0xffff0000.

3.1.2.5 Endianness

The term "endian" or "*endianness*" refers to the ordering of individually addressable units (e.g., bytes) within a larger data item (e.g., word) as stored in external memory (or, sometimes, as sent on a serial connection).

As far as microprocessors are concerned, a *big-endian* processor stores the most significant byte (MSB) first (i.e., at the lowest byte address), while a *little-endian* processor stores the least significant byte (LSB) first (see Figure 3.2). When we write the hex value 0x0a0b0c0d from left to right, we are implicitly writing in big-endian style.

Different processors order their multibyte data (i.e., 16-, 32-, or 64-bit words) in different ways. For example, the Intel X86 family and PIC processors use little-endian mode, and the Motorola 68000 family uses big-endian mode. Some processors, such as ARM processors, can be set up to be in either little-endian or big-endian mode.

Endianness is typically transparent to a user of a computer. However, difficulty arises when different types of computers attempt to communicate with one another over a network. Internet Protocol (IP) defines *big endian* as the standard network byte order used by IP and many higher-level protocols over IP for all numeric values. When communicating over a

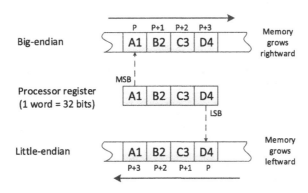

Figure 3.2
Microprocessor endianness.

network composed of both big-endian and little-endian machines, *high-level software should format packets of data in network byte order*. The Berkeley sockets application programming interface defines a set of functions[2] for little-endian host machines to convert 16-bit and 32-bit integers to and from network byte order.

A software system is called "endian clean" when it always reads and writes a multibyte data item as a whole rather than byte-by-byte. An "endian-clean" software system can be recompiled with no changes for big-endian or little-endian machines. It is encouraged to write "endian-clean" software, whenever possible, for networked embedded systems.

3.2 Microchip PIC18F8720

Microchip PIC18F8720 is a family of high-performance RISC microprocessors [4]. PIC18F8720 features a hardware stack for storing return addresses and a wide range of hardware interfaces including 10-bit A/D converters with 16 input channels, five timers, an external memory interface, and nine general-purpose I/O ports (one of which can be reconfigured as an 8-bit parallel slave port for direct processor-to-processor communications). Figure 3.3 shows the pin diagram of an 80-pin PIC18F8720 microprocessor, where many pins are physically multiplexed. For instance, the external memory interface (A19:A16, AD15:AD0) is multiplexed with I/O ports D, E, and H.

The PIC18F8720 device runs from a clock (the OSC1 pin) that is four times faster than its instruction cycle. The four clock pulses are a quarter of the *instruction cycle* in length and are referred to as Q1, Q2, Q3, and Q4. Generally speaking, instruction fetching and execution are pipelined: while a 2-byte instruction is executed in an instruction cycle, the next 2-byte instruction can be fetched during the same cycle.

3.2.1 Memory Organization

PIC18F8720 devices have separate spaces for data memory and program memory, which is illustrated in Figure 3.4.

Data space in electrically erasable programmable ROM (EEPROM). PIC18F8720 devices have 1024 bytes of data EEPROM, which is a byte-addressable memory space that has been optimized for the storage of frequently changing values (e.g., program variables that are updated often). Variables that change infrequently (such as constants, IDs, etc.) should be stored in flash program memory.

---

[2] Functions htons() and htonl(), respectively, convert a short and a long integer from a small-endian host to the network format. Functions ntohs() and ntohl() are for the opposite direction.

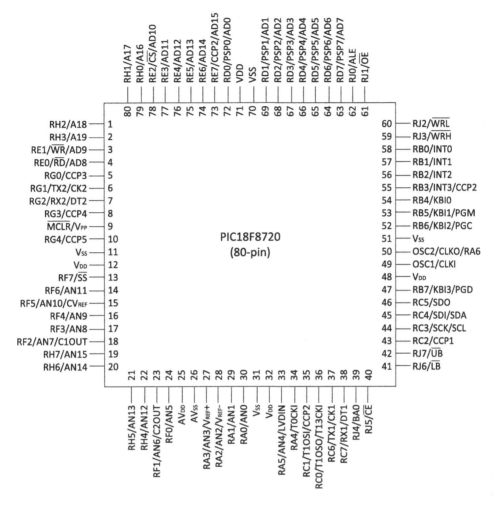

Figure 3.3
Microchip 16-bit 80-pin PIC microprocessor.

When interfacing with the data EEPROM block, the register EEDATA holds the 8-bit data for read/write, while the register EEADR and the two least significant bits (LSbs) of EEADRH hold the address of the EEPROM location being accessed.

Data space in random-access memory (RAM). As shown in Figure 3.4, PIC18F8720 devices also have a data RAM space that contains both special-function registers (SFR) and general-purpose registers (GPR). The SFRs are used for control and status of the microcontroller and peripheral functions, while GPRs are used for storing application data such as intermediate computational values, local variables of subroutines, and task contexts.

Each register in the RAM space has a 12-bit address. To enable rapid access to those registers, a *banking scheme* is implemented where the whole space is partitioned into 16 banks of

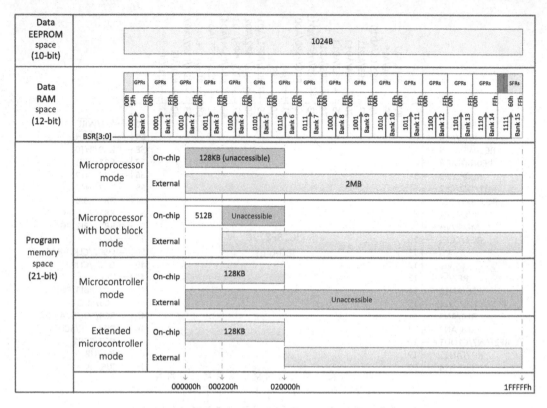

Figure 3.4
PIC18F8720 memory organization.

256 bytes each. To *directly access*[3] a register, the lower 8 bits of its address are embedded in the instruction, and the upper 4 bits of its address, which specify the bank to be accessed, are taken from the lower 4 bits of the bank select register (BSR[3:0]).

GPRs start at the first location of bank 0 and grow upward up to the last location of bank 14. Bank 15 contains 160 SFRs with addresses ranging from F60h to FFFh; its lower portion (from F00h to F5Fh) is unused.

The first segment (96 bytes) of bank 0 and the SFR segment (160 bytes) of bank 15 also form a so-called *access bank*, which allows the commonly used registers (SFRs and some GPRs) to be accessed in a single instruction cycle. A reserved bit in the instruction word is used to specify if the operation is to occur in the bank specified by the bank select register or in the access bank.

---

[3]   There is also an *indirect addressing mechanism*, where any instruction using an INDF register actually accesses the data memory location pointed to by the 12-bit file select register that is associated with that INDF register.

Program memory space. PIC18F8720 has 128 KB of on-chip flash memory. Moreover, it implements a 21-bit program counter, which is capable of addressing a 2 MB program memory space through the external memory interface. Depending on the two LSbs of the CONFIG3L register[4], PIC18F8720 may operate in one of four modes:

(i) Microcontroller mode. This is the default mode. Under this mode, the processor can access only the on-chip flash memory. Attempts to read above the physical limit of the on-chip flash memory return 0's.

(ii) Microprocessor: Under this mode, the processor can access only the external program memory; the contents of the on-chip flash memory are ignored. The 21-bit program counter permits access to a 2 MB linear program memory space.

(iii) Microprocessor with boot block. Under this mode, the processor accesses the boot block (from addresses 000000h to 0001FFh)[5] of the on-chip flash memory. Above this, external program memory is accessed all the way up to the 2 MB limit. Program execution automatically switches between the two memories, as required.

(iv) Extended microcontroller mode. Under this mode, the processor can access its entire on-chip flash memory; above this, the device accesses external program memory up to the 2 MB program space limit. The execution automatically switches between the two memories, as required.

The program memory space is typically used for storing instructions. The address of the instruction to fetch for execution is specified by the 21-bit program counter, where the low byte, the high byte, and the upper 5 bits are contained in registers PCL, PCH, and PCU, respectively.

The program memory is addressed by bytes. However, since PIC18F8720 uses a 16-bit instruction set, the LSb of the PCL register is fixed to a value of 0 to prevent the program counter from becoming misaligned with word instructions. In other words, the program counter increments by 2 to address sequential instructions in the program memory.

The program memory space can also be used for storing data. A block containing data is not required to be word aligned. There are two instructions that allow a processor to move bytes between the program memory space and the data RAM: (a) table read TBLRD, which retrieves data from program memory (or data EEPROM[6]) and places it into the data RAM space, and

---

[4] The configuration bits can be programmed (set to "0") to select various device configurations. These bits are mapped to the *configuration memory space*, starting from 300000h through 3FFFFFh, which is beyond the user program memory space (2 MB: 000000h to 1FFFFFh), and can be accessed using only table reads and table writes.

[5] By use of a bootloader routine located in the protected boot block at the top of program memory, it becomes possible to create an application that can update itself in the field.

[6] Depending on the settings of the EECON1 register.

(b) table write TBLWT, which stores data from the data RAM space in program memory (or data EEPROM). Table read and table write operations move data between these memory spaces through an 8-bit table latch register (TABLAT).

A table pointer TBLPTR is used in reads, writes, and erases of the program memory. TBLPTR comprises three SFRs—TBLPTRU, TBLPTRH, and TBLPTRL—which jointly form a 21-bit-wide address pointer.

To read a byte from the flash program memory, simply set TBLPTR to point to its address in the program space, then execute the TBLRD instruction, which will place the byte into TABLAT.

Writing to the flash program memory is a little complicated because the minimum programming block is 8 bytes (four words). In particular, prior to a flash programming operation, the TBLWT instruction has to be executed eight times, with each essentially writing 1 byte contained in TABLAT to one of the eight holding registers. At the end of updating the holding registers, instruction execution is halted and a long write cycle is started to write the contents of the eight holding registers to the flash program memory.

In all modes, a PIC18F8720 processor has complete access to the data RAM and data EEPROM spaces. In addition, the processor, except for the microcontroller mode, can access external memory devices (such as flash memory, erasable programmable ROM (EPROM), and static RAM (SRAM)) as program or data memory through the external memory interface. Depending on the values of the two LSbs of the MEMCON register, there are three operating modes for the *external memory interface*: word write, byte select, and byte write.

3.2.2 Word Write Mode

This mode allows instruction fetches and table reads from, and table writes to, all forms of 16-bit (word-wide) external memories (EPROM, flash memory, or SRAM). An example configuration of word write mode is shown in Figure 3.5, where, in contrast to Figure 3.3, only those pins that are relevant to the external memory interface are shown.

Whenever the microprocessor accesses external memory, the chip enable signal $\overline{CE}$ is active (asserted low); it is thus connected to the $\overline{CE}$ pin of the flash memory device.

To allow the microprocessor to fetch data from external memory, the corresponding output enable signals $\overline{OE}$ (active low) are connected. $\overline{OE}$ is inactive during a cycle where a table write is executed. During a cycle where an instruction is fetched or a table read is executed, $\overline{OE}$ is asserted at the beginning of Q3 and deasserted at the end of Q4. Data (16-bit word) are fetched from external memory at the low-to-high transition edge of $\overline{OE}$.

The write high control signal $\overline{WRH}$ is active (asserted low) whenever the microprocessor writes to an odd address (or the high byte of a word). It is connected to $\overline{WR}$ of the memory device so

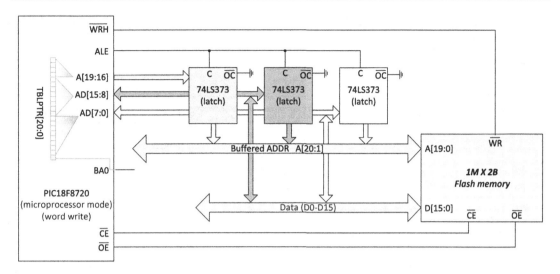

Figure 3.5
PIC18F8720 word write mode.

that the device is ready for writing whenever the microprocessor attempts to write the high byte of a word.

The external memory interface has 20 address lines: A[19:16] and AD[15:0]. Internally, the 21-bit table pointer TBLPTR (similarly the program counter) is mapped to the external memory interface as follows:

- the least significant bit TBLPTR[0] is copied to BA0;
- TBLPTR[20:17] appear on address pins A[19:16];
- TBLPTR[16:9] appear on pins AD[15:8]; and
- TBLPTR[8:1] appear on pins AD[7:0].

Consequently, A[19:16] together with AD[15:0] allow memory access based on a word boundary. If necessary, BA0 can be used to access memory by byte.

Because the pins AD[15:0] are multiplexed for both data and addresses, the control pin ALE (for "address latch enable") is used to enable the address latch devices (74LS373), so that the address on the external address bus is still valid while data are being transferred. Note that to be consistent with the table pointer, A[20:1] are used to name the buffered address lines.

Figure 3.6 gives a timing diagram illustrating the signal interactions for two consecutive table write operations while the microprocessor accesses the external memory in word write mode.

First, during Q1 of an instruction cycle, ALE is enabled while the address information is placed on pins AD[15:0] and A[19:16]. On the falling edge of ALE, the address is latched and available on the external bus.

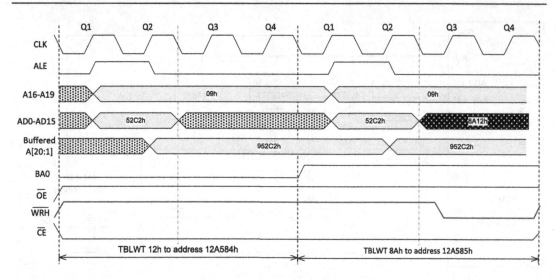

Figure 3.6
PIC18F8720 word write mode timing diagram.

Second, this mode makes a distinction between TBLWT instruction cycles for even or odd addresses. During a TBLWT cycle to an even address, that is, TBLPTR[0] or BA0 is 0, the TABLAT data are transferred to a holding latch and AD[15:0] is tristated[7] for the data portion of the bus cycle. The $\overline{WRH}$ signal is not activated. During a TBLWT cycle to an odd address i.e., TBLPTR[0] or BA0 is 1, the TABLAT data are presented on the upper byte of the AD[15:0] bus. At the same time, the contents of the holding latch are presented on the lower byte of the AD[15:0] bus. The $\overline{WRH}$ signal is activated and the 16-bit data are written to the corresponding word location.

Last but not least, it is worth noting that the address 12A584h used by the table pointer becomes 952C2h; this buffered address does not reflect the LSb of 12A584h. As an exercise, check that 952C2h×2 XOR BA0 equals 12A584h in the first cycle, and equals 12A585h in the second cycle. This is because the LSB of a word location is accessed in the first cycle (BA0 = 0), while the MSB of the word location is accessed in the second cycle (BA0 = 1).

3.2.3 Byte Select Mode

This mode allows instruction fetches and table reads from, and table writes to, 16-bit external memories with byte selection capability, as well as 8-bit external memories. In this mode, the

---

[7] A tristate logic allows an output port to assume a high impedance state in addition to the logic high and low levels. When an output is tristated, its influence on the rest of the circuit is removed. A pull-up or pull-down resistor is typically used to try to pull the node to high or low voltage levels. If the node is not in a high-impedance state, extra current from the resistor will not significantly affect its voltage level.

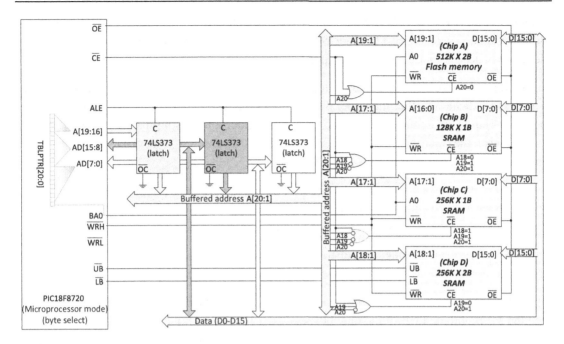

Figure 3.7
PIC18F8720 byte select mode.

write high signal $\overline{\text{WRH}}$ is asserted whenever the microprocessor writes to a memory location, regardless of an odd or even address. An example configuration of byte select mode is shown in Figure 3.7, where $\overline{\text{WRH}}$ is connected to the $\overline{\text{WR}}$ pin of each memory chip.

First, notice that there are four memory chips: chip A and chip D are word-wide memory devices, and chip B and chip C are byte-wide memory devices. At any given time, only one of the four devices can be enabled for reading or writing. This is achieved by the OR gates. Specifically, the chip enable signal $\overline{\text{CE}}$ together with some of the address lines is used to select memory chips:

- Chip A is selected (active) only when $\overline{\text{CE}}$ is asserted and A[20]=0.
- Chip B is selected only when $\overline{\text{CE}}$ is asserted, A[20]=1, A[19]=1, and A[18]=0.
- Chip C is selected only when $\overline{\text{CE}}$ is asserted, A[20]=1, A[19]=1, and A[18]=1.
- Chip D is selected only when $\overline{\text{CE}}$ is asserted, A[20]=1, and A[19]=0.

Second, BA0 is connected to the A0 pin of chip A. While A[19:1] allows the microprocessor to address a word location on chip A, A0 will allow the microprocessor to individually access the MSB or LSB of the word. The A[0] pin of chip C is also connected to BA0. This ensures that chip C gets a contiguous block of addresses. Chip B has almost the same wiring as chip C except that its A[0] is connected to bus line A[1] rather than BA0. Because of this, the block of addresses allocated to chip B is not contiguous: each location on chip B gets one odd address and one even address! In other words, the microprocessor can use two different addresses to access each location on chip B.

Third, chip D is a word-wide SRAM, which, according to the JEDEC memory standards, uses $\overline{UB}$ or $\overline{LB}$ to indicate whether the upper byte or the lower byte is to be selected. Thus, pins $\overline{UB}$ and $\overline{LB}$ of chip D are connected to the corresponding microprocessor pins, respectively. $\overline{UB}$ is asserted when the microprocessor accesses an odd address, and $\overline{LB}$ is asserted when the microprocessor accesses an even address.

The memory mappings of the memory devices are given in Table 3.2. These four chips together form a space with 2 MB addresses, but remember that although chip B has 128 KB, it is allocated with 256K addresses: two for each byte location.

Figure 3.8 gives a timing diagram illustrating the signal interactions for two consecutive table write operations while the microprocessor is in byte select mode.

It is worth noting that $\overline{WRH}$ is asserted in each write cycle, and the TABLAT data are copied on both the MSB and the LSB of the data bus (e.g., 1212h) This is not a problem for word-wide memory devices (say, chip A or chip D), because either BA0 or $\overline{UB}/\overline{LB}$ (they originate from the LSb of the TBLPTR register) can be used to select the right byte to be written.

Also note that the microprocessor writes to an even address in the first cycle, so BA0 = 0 and $\overline{LB}$ is asserted. It then writes to an odd address in the second cycle, so BA0 = 1 and $\overline{UB}$ is asserted. As an exercise, check on which chip the address 12A584h appears.

Table 3.2 The memory mapping of Figure 3.7

Chip	Memory width	CE	Chip Address	A0	Address Range	Size
			Address Bus: A[20] —— A[1]			
A	Word	0	++++++++++++++++++++	+[a]	[0] 0000 0000 0000 0000 0000~ [0] 1111 1111 1111 1111 1111	1 MB
B	Byte	110	+++++++++++++++++		[110]00 0000 0000 0000 000x~ [110]11 1111 1111 1111 111x[b]	128 KB
C	Byte	111	+++++++++++++++++	+	[111]00 0000 0000 0000 0000~ [111]11 1111 1111 1111 1111	256 KB
D	Word	10	+++++++++++++++++	+	[10]000 0000 0000 0000 0000~ [10]111 1111 1111 1111 1111	512 KB

[a]A bit marked by + indicates that it is used by the device; different value combinations of those "+" bits refer to different locations on the device.

[b]A bit marked by x indicates that it is unused by the device. When there are unused bits, each location of the device may be accessed by more than one address. In this case, the last bit can be 0 or 1, implying that each location on chip B virtually has two addresses.

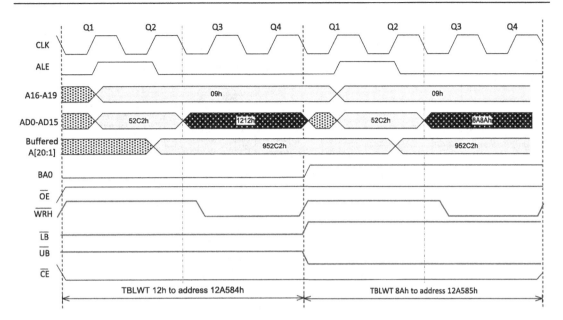

Figure 3.8
PIC18F8720 byte select mode timing diagram.

3.2.4 Byte Write Mode

This mode allows table reads from, and table writes to, 8-bit-wide external memories. In this mode, during a TBLWT instruction cycle, the TABLAT data are presented on both the upper and the lower bytes of the AD[15:0] bus. The appropriate $\overline{WRH}$ or $\overline{WRL}$ control line is asserted on the LSb of TBLPTR.

An example configuration of byte write mode is shown in Figure 3.9. Like the other modes, AD<15:0> from the microprocessor are mapped through the latch to A<16:1> of the buffered address bus, and A<19:16> are mapped directly to A<20:17>.

Let us first look at the wiring of chip A and chip B. They are of the same type, of the same size (512 KB), and even have the same chip selection signal (A[20]=0). They differ in two ways: (a) the data pins of chip A are connected with the upper byte (D[15:8]) of the data bus, while the data pins of chip B are connected with the lower byte (D[7:0]) of the data bus; (b) the write enable pin $\overline{WR}$ of chip A is connected to $\overline{WRH}$, while the write enable pin $\overline{WR}$ of chip B is connected to $\overline{WRL}$. This implies that data are written to chip A when the address is odd ($\overline{WRH}$ is asserted as the LSb of TBLPTR is 1), written to chip B when the address is even ($\overline{WRL}$ is asserted as the LSb of TBLPTR is 0).

The wiring of chip C is almost the same as that of chip B, except that it is selected when A[20] is 1. This ensures that chip C will get a different block of addresses. Note that chip C

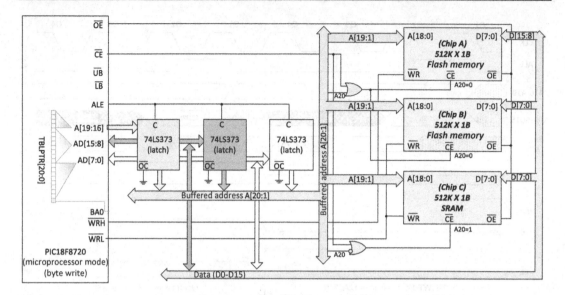

Figure 3.9
PIC18F8720 byte write mode.

permits table operations to only even addresses. Particularly, a table write to an even address of chip C will copy the contents of TABLAT to both bytes of the AD<15:0> bus and activate the $\overline{\text{WRL}}$ signal. The lower byte will then be written to chip C. A table write to an odd address of chip C will also copy the contents of TABLAT to both bytes of the AD<15:0> bus. However, this time, the $\overline{\text{WRH}}$ signal is activated instead. In this case, nothing will be written to chip C. This is similar to the case for chip B in Figure 3.7: although the usage of the memory chips is 100%, a noncontiguous block of addresses is assigned to them. One difference is that each location on chip C gets an even address, while each location on chip B in Figure 3.7 gets an even address and an odd address. The reason is that in byte select mode, $\overline{\text{WRH}}$ is asserted regardless of whether the address is even or odd.

Figure 3.10 gives a timing diagram illustrating the signal interactions for two consecutive table write operations while the microprocessor is in byte write mode. Note that $\overline{\text{WRL}}$ is asserted in the first cycle because an even address is accessed, while $\overline{\text{WRH}}$ is asserted in the second cycle because an odd address is accessed. In each write cycle, the TABLAT data are copied on both the MSB and the LSB of the data bus (e.g., 1212h).

3.3 Intel 8086

Intel 8086 is a 16-bit microprocessor that represents the first design of the x86-family architecture [1]. The Intel 8086 has a von Neumann architecture and belongs to the CISC category. Its pin diagram is shown in Figure 3.11.

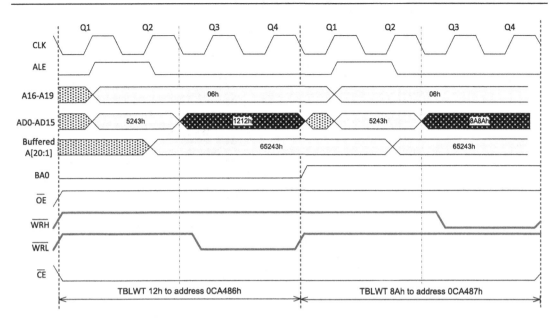

Figure 3.10
PIC18F8720 byte write mode timing diagram.

Intel 8086 supports two hardware modes (hard wired into the circuit and unchangeable by software): maximum mode and minimum mode. In maximum mode pin 33 (MN/$\overline{\text{MX}}$) is wired to ground, and in minimum mode pin 33 is wired to voltage. Changing the state of pin 33 also changes the function of some other pins as indicated in Figure 3.11. The minimum mode is intended for small single processor systems, while the maximum mode is for medium-sized or large systems that use more than one processor (e.g., 8087 coprocessor in an IBM PC).

CLK provides the basic timing for the processor and bus controller. Each *bus cycle* (read, write, or interrupt acknowledge) consists of at least four clock cycles, which are referred to as T1, T2, T3, and T4. The 8086 processor has its address lines and data lines multiplexed (AD0-AD15). This means that the same pins are used to carry both address and data information at different times: the address is emitted from the processor during T1 and data transfer occurs on the bus during T3 and T4. In the minimum mode, the address latch enable signal ALE can be used to enable chips for latching addresses.

The bus high enable signal $\overline{\text{BHE}}$ is asserted (active low) during T1 of a bus cycle when a byte is to be transferred on the high portion (the most significant half) of the data bus, that is, pins D15:D8.

The interrupt request signal INTR is a level-triggered input which is sampled during the last clock cycle of each instruction to determine if the processor should enter into an interrupt acknowledge operation. It can be internally masked by software resetting the interrupt enable

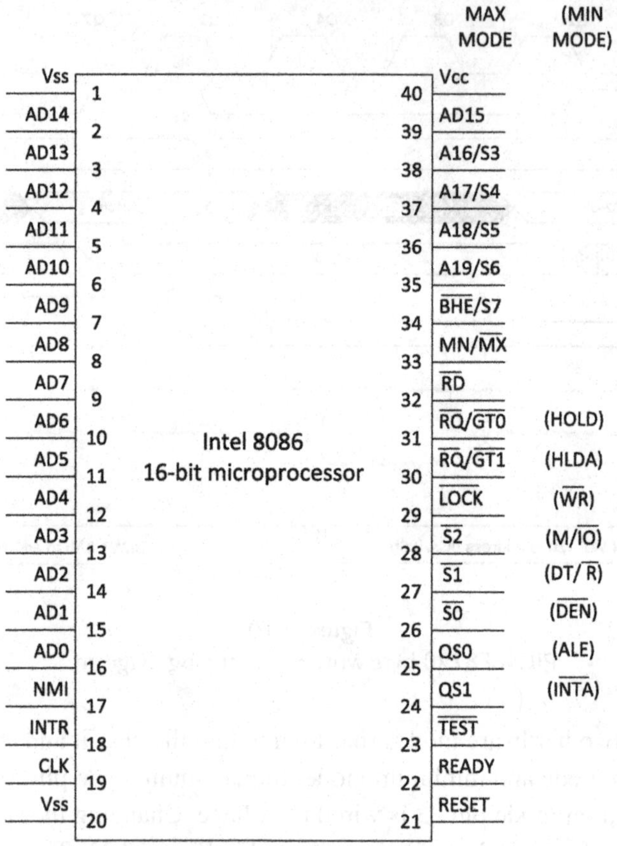

		MAX MODE	(MIN MODE)

```
              Vss  ┌─────────────────┐  Vcc
             ──────┤ 1             40 ├──────
             AD14  │                 │  AD15
             ──────┤ 2             39 ├──────
             AD13  │                 │  A16/S3
             ──────┤ 3             38 ├──────
             AD12  │                 │  A17/S4
             ──────┤ 4             37 ├──────
             AD11  │                 │  A18/S5
             ──────┤ 5             36 ├──────
             AD10  │                 │  A19/S6
             ──────┤ 6             35 ├──────
              AD9  │                 │  BHE/S7
             ──────┤ 7             34 ├──────
              AD8  │                 │  MN/MX
             ──────┤ 8             33 ├──────
              AD7  │                 │  RD
             ──────┤ 9             32 ├──────
              AD6  │                 │  RQ/GT0   (HOLD)
             ──────┤ 10            31 ├──────
              AD5  │  Intel 8086     │  RQ/GT1   (HLDA)
             ──────┤ 11            30 ├──────
              AD4  │ 16-bit          │  LOCK     (WR)
             ──────┤ 12   microprocessor 29 ├──────
              AD3  │                 │  S2       (M/IO)
             ──────┤ 13            28 ├──────
              AD2  │                 │  S1       (DT/R)
             ──────┤ 14            27 ├──────
              AD1  │                 │  S0       (DEN)
             ──────┤ 15            26 ├──────
              AD0  │                 │  QS0      (ALE)
             ──────┤ 16            25 ├──────
              NMI  │                 │  QS1      (INTA)
             ──────┤ 17            24 ├──────
             INTR  │                 │  TEST
             ──────┤ 18            23 ├──────
              CLK  │                 │  READY
             ──────┤ 19            22 ├──────
              Vss  │                 │  RESET
             ──────┤ 20            21 ├──────
                   └─────────────────┘
```

Figure 3.11
Intel 8086 16-bit microprocessor.

bit. The nonmaskable interrupt signal NMI is an edge-triggered input,[8] which has higher priority than the maskable interrupt request signal INTR. The interrupt acknowledge signal INTA is active (low) during T2 and T3 of each interrupt acknowledge cycle.

3.3.1 Memory Organization

The 8086 processor has a 20-bit address bus, which gives a physical address space of up to 1 MB (2^{20}), addressed as 00000h to FFFFFh. However, the maximum linear address space was limited to 64 KB, simply because the internal registers are only 16 bits wide. A technique called "internal segmentation" has to be used for applications that are above the 64 KB boundary. There are four 16-bit segment registers (CS for "code segment," DS for "data

---

[8] A transition from low to high initiates the interrupt at the end of the current instruction.

segment," ES for "extra data segment," and SS for "stack segment"). All memory references are of the form [segment:offset], relative to the base address contained in the corresponding segment register.

In particular, given a [segment:offset] pair, a 20-bit external (or physical) address is produced by segment $\times 2^4$ + offset, where segment $\times 2^4$ is called the *segment address*, which has its 16 most significant bits from the 16-bit segment register, and its four LSbs are all zeros. A 16-bit offset is always added to the 20-bit segment address to yield an external address. As an example, consider the pair [000Ah:000Ch] (i.e., 10:12). The segment value of 000Ah (10) would give a segment address at 000A0h (160) in the linear address space. The address offset 000Ch can then be added to the segment address: 000A0h + 000Ch = 000ACh (160 + 12 = 172). Such address translations are carried out automatically by the segmentation unit of the processor.

Owing to the 4-bit shift, each segment virtually begins at a multiple of 16 bytes, from the beginning of the 1 MB address space. Particularly, the last segment, FFFFh, begins at linear address FFFF0h, 16 bytes before the end of the 20-bit address space. Since all segments are 64 KB long (with wraparound at the high end) and they increment by 16 bytes, there are huge overlaps among segments, and each location in the linear memory address space can be accessed by $2^{16}/2^4 = 2^{12} = 4096$ different [segment:offset] pairs. For example, the memory location 00100h can be referred to by [0000h:0100h], [0001h:00F0h], [0002h:00E0h], [0003h:00D0h], etc. Although complicated, this scheme has its advantages. For instance, a small program (less than 64 KB) can be loaded starting at a fixed offset in its own segment, avoiding the need for relocation, with at most 15 bytes of alignment waste.

Locations from address FFFF0h to address FFFFFh are reserved. Following reset, the CPU will always begin execution at location FFFF0h, which is a jump to the initial program loading routine. The block from 00000h to 003FFh is reserved for the interrupt table, where each of the 256 possible interrupt types has its service routine pointed to by a 4-byte pointer element consisting of a 16-bit segment address and a 16-bit offset address.

The memory is physically organized as a high bank (D15-D8) and a low bank (D7-D0) of 512 KB addressed in parallel by the processor's address lines A19-A1. The processor provides two enable signals, $\overline{BHE}$ and A0, to selectively allow reading from or writing into either an odd byte location or an even byte location, or both. Byte data with an even address (A0 is low) are transferred on the D7-D0 bus lines, and byte data with an odd address (A0 is high) are transferred on the D15-D8 bus lines.

3.3.2 Separate I/O Address Space

An I/O device may be enabled, and the ports or registers on the device may be accessed, merely by the address lines (A19-A0). In such a case, the I/O devices are placed in the 8086 memory address space. As long as an I/O device responds like a memory chip, they are indistinguishable to the processor.

The 8086 processor also allows I/O devices to be placed in a separate I/O address space. This is achieved by the $\overline{S2}$ signal in maximum mode, and by the $M/\overline{IO}$ signal in minimum mode. $M/\overline{IO}$ becomes valid in the T4 preceding a bus cycle and remains valid until the final T4 of the cycle. The I/O address space is accessed by AD15-AD0 (the address lines A19-A16 are zero in I/O operations), which can maximally accommodate 64 KB registers or 32K word registers on I/O devices.

Figure 3.12 shows the wiring of a universal asynchronous receiver/transmitter (UART) which is placed in the I/O address space. The 8086 processor operates in minimum mode, because the $MN/\overline{MX}$ pin is driven high.

3.3.2.1 Timing clock

The processor is connected to an external clock generator, 8284A. A crystal is attached to pins X1 and X2 of 8284A; CLK has an output frequency which is one third of the crystal frequency.

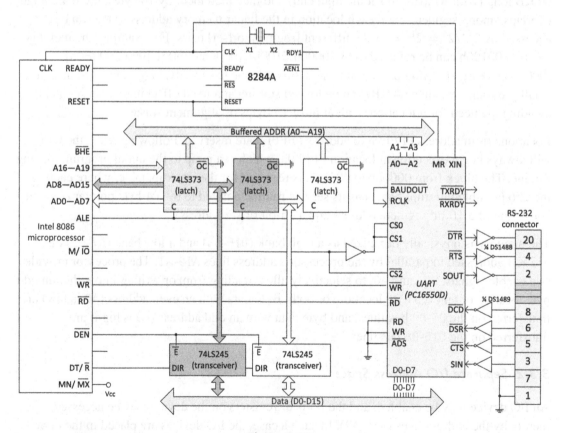

Figure 3.12
Intel 8086 memory I/O mapping.

The processor gets its clock by connecting its CLK input to the 8284A's CLK output. 8284A also has an input pin $\overline{\text{RES}}$, which is used to generate a reset signal through the output pin RESET to reset the 8086 processor. The input pin RDY1 is a bus-ready signal; it is typically driven by memory or I/O devices, acknowledging that data have been received or are available on the data bus. RDY1 is validated by $\overline{\text{AEN1}}$, which is connected to the ground when RDY1 is in use. The input signal RDY1 is synchronized by the 8284A clock generator to generate an output signal READY, which in turn is connected to the READY input of the 8086 processor, indicating the completion of data transfer.

3.3.2.2 External bus

Since AD15-AD0 are multiplexed to carry both data and address information, three 8-bit 74LS373 chips are used to implement an address latch. The address latch enable signal ALE is active (high) during T1 of any bus cycle. ALE is used to enable the three 74LS373 chips so that the address information is captured by the address latch during T1 and remains available on the buffered bus (A19-A0) until the next time ALE transits to high.

AD15-AD0 and the external data bus D15-D0 are connected through a bus transmitter/receiver implemented by two 8-bit 74LS245 chips. The direction input DIR controls transmission of data: from AD15-AD0 to D15-D0 when it is high (say, in a write operation), and from D15-D0 to AD15-AD0 when it is low (say, in a read operation). DIR is driven by the data transmit/receive signal DT/$\overline{\text{R}}$,[9] which is valid in the T4 preceding a bus cycle and remains valid until the final T4 of the bus cycle.

The 74LS245 chip has an enable input $\overline{\text{E}}$; the two buses are isolated when $\overline{\text{E}}$ is high. $\overline{\text{E}}$ is driven by the data enable output signal $\overline{\text{DEN}}$, which, for a read or interrupt acknowledge cycle, is active from the middle of T2 until the middle of T4, while for a write cycle it is active from the beginning of T2 until the middle of T4.

3.3.2.3 I/O device: UART

The processor is wired with a PC16550D UART chip for serial transmission of digital information (bits) through a single wire. The UART performs serial-to-parallel conversion on data characters received from a peripheral device or a modem, and parallel-to-serial conversion on data characters received from the processor.

PC16550D is mainly composed of a programmable baud rate generator, a transmitter, a receiver, and several registers (e.g., 16-bit divisor register) that control UART operations, including transmission and reception of data. Note that the UART chip has three address signals (A0, A1, A2) and eight data signals (D7-D0). Refer to its data sheet [9] for other pin descriptions.

---

[9] $\overline{\text{S1}}$ is used instead in the maximum mode.

In Figure 3.12, pins $\overline{RD}$ and $\overline{WR}$ of the UART are connected to the corresponding pins of the processor. This allows the processor to transmit data to, and receive data from, the UART. Also, the UART chip is selected only when the $M/\overline{IO}$ signal is low; this indicates that the registers of the UART are mapped into the I/O address space rather than the memory space.

3.3.3 Memory Address Space

Figure 3.13 shows a configuration where the Intel 2142 memory chips are mapped into the memory address space. This is because the memory chips are readable or writable only when $M/\overline{IO}$ is high. Specifically, the write enable signal $\overline{WE}$ of the memory chips is from $\overline{MEMW}$, which is asserted when $M/\overline{IO}$ is high and $\overline{WR}$ is low, where $\overline{WR}$ indicates that the processor is performing a write cycle. Similarly, the output disable signal OD of the memory chips is from $\overline{MEMR}$, which is asserted low when $M/\overline{IO}$ is high and $\overline{RD}$ is low, where $\overline{RD}$ indicates that the processor is performing a read cycle.

A 74LS138 1-of-8 decoder/demultiplexer is used to produce chip selection signals. Its truth table is given in Table 3.3. Since the output signal $\overline{O_0}$ is used in Figure 3.13, the memory chips are enabled only when $[A18A17A16] = [000]$.

The 8086 processor is wired with a high memory bank and a low memory bank, each composed of two Intel 2142 chips. A 2142 chip is a 1024×4-bit SRAM. The chip select signal $\overline{CS}$ of the high bank is asserted (active low) whenever $\overline{O_0}$ and $\overline{BHE}$ are low. In such a case, data are written to or read from the high bank through bus lines D8-D15. The chip select signal $\overline{CS}$ of the low bank is asserted (active low) whenever $\overline{O_0}$ and A0 are low. In such a case, data are written to or read from the low bank through bus lines D0-D7. In other words, byte data with an even address (A0 = 0) are read from or written to the low memory bank through

Table 3.3 74LS138 truth table

A	B	C	$\overline{O_0}$	$\overline{O_1}$	$\overline{O_2}$	$\overline{O_3}$	$\overline{O_4}$	$\overline{O_5}$	$\overline{O_6}$	$\overline{O_7}$
0	0	0	L	H	H	H	H	H	H	H
0	0	1	H	L	H	H	H	H	H	H
0	1	0	H	H	L	H	H	H	H	H
0	1	1	H	H	H	L	H	H	H	H
1	0	0	H	H	H	H	L	H	H	H
1	0	1	H	H	H	H	H	L	H	H
1	1	0	H	H	H	H	H	H	L	H
1	1	1	H	H	H	H	H	H	H	L

D0-D7, while byte data with an odd address ($\overline{BHE}$ is asserted as A0 = 1) are read from or written to the high memory bank through D8-D15.

A0 is also called a bank selection signal since it is reserved for that purpose. Care must be taken to ensure that each register within an 8-bit peripheral device located on the lower portion of the data bus be addressed as even. Now, you should understand why in Figure 3.12 it is address lines A1-A3, rather than A0-A2, that are connected to pins A0-A2 of the UART. If A0 of the address bus were used, some of the UART registers would become inaccessible because the UART is wired only to the low portion of the data bus.

3.3.4 Wait States

Normally, a bus cycle (e.g., accessing a memory or I/O port) on an 8086 processor is composed of four clock cycles—T1 to T4. Wait states, called Tw, can be inserted in a bus cycle to help the processor to interface with slow memory or I/O devices. The READY input signal on the 8086 is used to request the processor to stretch out its read or write cycle, to accommodate slow devices. In particular, the 8086 processor samples the READY line at the rising edge of T3. If READY is low, a Tw state is inserted after T3. A "wait" state is of the same duration as a clock cycle. During the Tw state, the READY is sampled again at the next rising edge of the clock, and another Tw is inserted if READY is still low. Whenever READY is sampled high at the rising edge of T3 or Tw, the T4 clock cycle follows.

In Figure 3.13, the processor's READY pin is driven by the READY output of the 8284A clock generator, which, in turn, is synchronized with its RDY1 input. A memory or I/O device can initiate WAIT state generation by bringing RDY1 low.

74LS164 used in Figure 3.13 is an 8-bit shift register, where the two serial inputs A and B are connected to voltage (steady high). When the clear signal $\overline{CLR}$ is asserted, all the outputs (Q_A, ..., Q_H) become low. When $\overline{CLR}$ is not active, Q_A becomes high on the low-to-high transition of the clock input. The outputs Q_B, ..., Q_H, respectively, take the level of Q_A, ..., Q_G before the most recent low-to-high transition of the clock, indicating a 1-bit shift every clock.

Now, let us examine the time diagram in Figure 3.14 to see how the 74LS164 configuration given in Figure 3.13 works.

First, note that the bus cycle begins in T1 with the assertion of the address latch enable signal ALE. The trailing edge of this signal is used to latch the address information, which is valid on the buffered address bus at this time. At T2 the address is removed from the local bus, which goes to a high impedance state.

The 74LS164 shift register is cleared whenever $\overline{INTA}$, $\overline{RD}$, and $\overline{WR}$ are all high. In Figure 3.14 this is indicated by the two shaded areas bounded by the level changes of $\overline{RD}$.

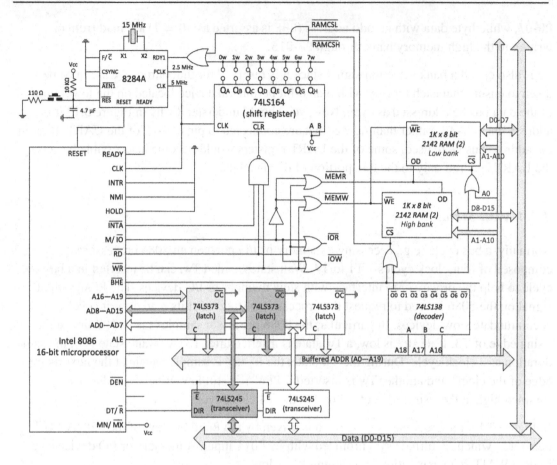

Figure 3.13
Intel 8086 memory mapping.

The read control signal $\overline{\text{RD}}$ is asserted at T2.[10] Whenever $\overline{\text{INTA}}$, $\overline{\text{RD}}$, or $\overline{\text{WR}}$ is asserted, the $\overline{\text{CLR}}$ input of the 74LS164 shift register is driven high, and the 74LS164 shift register becomes ready to drive its outputs. The 74LS164 shift register drives Q_A high when it comes to the rising edge of the next clock, which is the T2 cycle. Q_A remains high until the 74LS164 shift register is cleared. One clock cycle later, Q_B is driven high, taking the level of Q_A; one more clock cycle later, Q_C is driven high, taking the level of Q_B.

To ensure that the 8086 timing requirements are met, the device needs to bring RDY1 low prior to the rising edge of the 8086's T2 clock. As indicated in Figure 3.14, both $\overline{\text{RAMCSH}}$ and $\overline{\text{RAMCSL}}$, which are simply the select signals of the memory chips, are low in clock cycle T1.

---

[10] $\overline{\text{WR}}$ and $\overline{\text{RD}}$ are active for T2, T3, and Tw of any bus cycle.

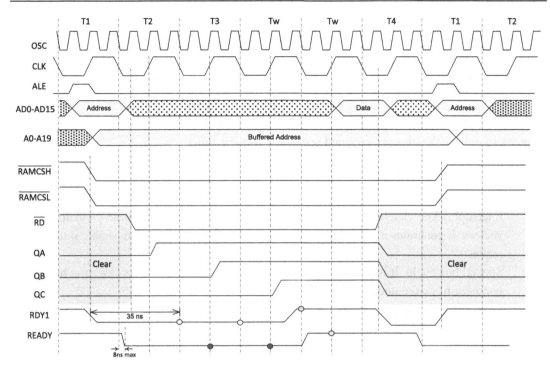

Figure 3.14
Intel 8086 timing diagram.

Since Q_C of the 74LS164 shift register is also low in cycle T1, RDY1 is driven low, which indicates the "not ready" status of the memory devices.

The 8284A clock generator drives the 8086's READY signal low at the falling edge of T1. The 8086 processor samples the READY line at the rising edge of T3. In our case, it finds that READY is low, and it inserts a Tw clock cycle after T3. The 8086 processor samples the READY line again at the rising edge of Tw, and inserts another Tw clock cycle after the first Tw.

Another timing constraint is that whenever a new Tw cycle is no longer needed, the memory device has to bring RDY1 high early in T3 or Tw so that the 8284A clock generator can bring READY high before the rising edge of T3 or Tw. In order to bring RDY1 high, the 74LS164 shift register has to drive Q_C high because only Q_C is wired to the OR gate. Q_C is driven high at the rising edge of the first Tw, which is two clock cycles later than when Q_A changed to high. The 8284A clock generator then starts to drive the 8086's READY signal high at the falling edge of the first Tw. This happens before the 8086 processor samples the READY line at the rising edge of the second Tw. As a result, no more Tw cycle is added before the T4 clock cycle.

3.4 Intel Pentium

The Intel Pentium series is the fifth generation of the x86 microprocessor family. The Pentium series is positioned as Intel's mid-range processor series, in between the Celeron and Core series.

The Pentium processor is a 32-bit processor with a 64-bit data bus. Its pin diagram is given in Figure 3.15, where the pins are grouped according to their functions.

The Pentium processor can initiate the following bus cycles:

- Single-transfer cycle: the transfer of one data item (up to 8 bytes or 64 bits).
  - Interrupt acknowledge cycle: The Pentium processor has 256 possible interrupt vectors. It generates interrupt acknowledge cycles in response to maskable interrupt requests.
  - Special bus cycle: It is generated in response to certain instructions (e.g., the halt instruction is executed).
  - Read cycle: It is generated by the processor to read data (or code) from memory or I/O devices.
  - Write cycle: It is generated by the processor to write data to memory or I/O devices.
- Burst-transfer cycle: the transfer of four data items (up to 32 bytes or 256 bits).
  - Burst read cycle: 32 bytes are read consecutively from memory into the internal cache in one bus cycle.
  - Burst writeback cycle: Modified lines in the processor's data cache are written back to memory in blocks of 32 bytes.

The pins that are pertinent to the bus operations are described below (interested readers can refer to the data sheet [3] to get more information):

- A31-A3: Address lines. As outputs, A31-A3 along with $\overline{BE7}$-$\overline{BE0}$ define the physical address space (memory or I/O).
- $\overline{ADS}$: Output signal for address status. When asserted, it indicates that a new valid bus cycle is currently being driven by the processor.
- AHOLD: Input signal for address hold. An external system can initiate an *inquire cycle* to a Pentium processor to check whether a particular address location is cached in the processor's internal cache, and if so, what state it is in. To prepare for an inquire cycle, an external system first needs to assert AHOLD to force the Pentium processor to float its address bus,[11] then wait two clocks for the processor to finish housekeeping (so that data can be returned or driven for previously issued bus cycles).

---

[11] A signal is floating if no part on the circuit is driving it.

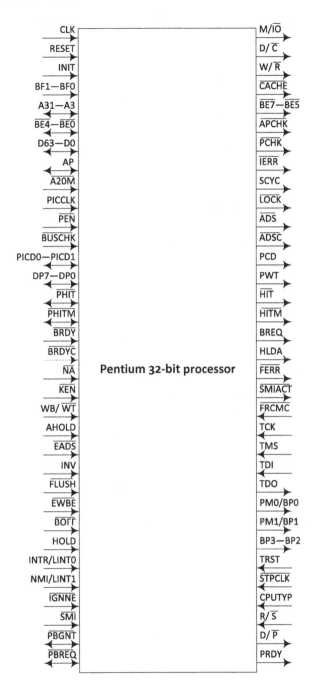

Figure 3.15
Intel Pentium 32-bit processor.

- $\overline{BE7}$-$\overline{BE0}$: Byte enable signals. $\overline{BE7}$-$\overline{BE0}$ are driven in the same clock as address lines A31-A3, used to determine which bytes are transferred (read or written) by the processor for the current bus cycle.
- $\overline{BOFF}$: Input signal for backoff. It is used to abort all outstanding bus cycles that have not yet finished. When it is asserted low, the processor will float all pins in the next clock and maintain its bus in this state until $\overline{BOFF}$ is deasserted, at which time the processor will restart the aborted bus cycle(s).
- $\overline{BRDY}$: Input signal for burst ready. It indicates that the external system has presented valid data on the data pins in response to a read request or that the external system has accepted the processor's data in response to a write request.
- $\overline{CACHE}$: Output signal indicating the internal cacheability of a bus cycle.
- $\overline{KEN}$: Input signal for cache enable. When the processor generates a read cycle that can be cached ($\overline{CACHE}$ asserted) and $\overline{KEN}$ is active, the cycle will be transformed into a burst read cycle.
- D/$\overline{C}$: Output signal for data/code indication. It is driven valid in the same clock as the $\overline{ADS}$ signal is asserted. The current cycle is for data when D/$\overline{C}$ is high, and is for code or special bus cycles when D/$\overline{C}$ is low.
- D63-D0: Data lines. Lines D7-D0 define the LSB of the data bus; lines D63-D56 define the MSB of the data bus.
- $\overline{EADS}$: Input signal for external address status. It indicates that an external system has initiated an *inquire cycle* to the Pentium processor, and a valid inquire address has been driven onto the processor's address pins by the external system.
- $\overline{HITM}$: Output signal for hit to a modified line. It is asserted to indicate to the external system that the cache of the processor contains the most current copy of the data and any device needing to read that data should wait for the processor to write it back. $\overline{HITM}$ remains asserted until two clocks after the last $\overline{BRDY}$ of the writeback cycle is asserted.
- HOLD: Input signal for bus hold request. When it is asserted by another local bus master, the processor will float most of its output pins. After completing all the outstanding bus cycles, the processor will also assert HLDA (bus hold acknowledge) to relinquish the bus to the requester. The processor will maintain its bus in this "bus-hold" state until HOLD is deasserted. In such a case, the processor will deassert HLDA and resume driving the bus.
- $\overline{NA}$: Input signal for next address. When $\overline{NA}$ is asserted, it indicates that the external memory system is ready to accept a new bus cycle although all data transfers for the current cycle have not yet finished. The processor will issue $\overline{ADS}$ for a pending cycle two clocks after $\overline{NA}$ is asserted.
- RESET: It forces the Pentium processor to restart execution at a known state.

3.4.1 Bus State Transition

We are now ready to examine the state transitions of Pentium bus cycles. A *bus cycle* begins with the processor driving an address and status and asserting $\overline{ADS}$, and ends when the last $\overline{BRDY}$ is returned.

The Pentium processor supports up to two outstanding bus cycles. Each bus cycle is composed of several states (clock cycles). Below is a list of valid bus states for the Pentium processor:

- Ti: This is the *bus idle* state. An asserted $\overline{BOFF}$ or RESET will always force the bus back to this state. HLDA will be driven only in this state.
- T1: there is only one outstanding bus cycle, and this is the *first clock* of that bus cycle. Valid address and status are driven out and $\overline{ADS}$ is asserted.
- T2: There is only one outstanding bus cycle, and this is the *second and subsequent clock* of the bus cycle. In state T2, data are driven out (if the cycle is a write cycle), or data are expected (if the cycle is a read cycle), and the $\overline{BRDY}$ pin is sampled. A bus cycle may have multiple T2 states.
- T12: There are two outstanding bus cycles. While the second (newer one) bus cycle is in its T1 state the first bus cycle is in its T2 state. This implies that (a) for the second outstanding bus cycle, the processor drives the address and status and asserts $\overline{ADS}$, and (b) for the first outstanding cycle, data are transferred and $\overline{BRDY}$ is sampled.
- T2P: There are two outstanding bus cycles, and both are in their respective T2 states. In T2P, data are being transferred and $\overline{BRDY}$ is sampled for the first outstanding cycle.
- TD: There is one outstanding bus cycle. Write cycles can be pipelined into read cycles and read cycles can be pipelined into write cycles, but one dead clock is required between consecutive read and write cycles to allow bus turnover. The processor enters TD if (a) a dead clock is needed, and (b) in the previous clock there were two outstanding cycles, and $\overline{BRDY}$ was sampled active (low) for the first bus cycle.

The possible bus state transitions are described in detail in the UML state diagram given in Figure 3.16 (state diagram notations are covered in Chapter 9).

First, the notion of request pending, *ReqPend*, is defined as a Boolean expression:

$$ReqPend \triangleq P \wedge \neg HOLD \wedge \overline{BOFF} \wedge (\neg AHOLD \vee \neg \overline{HITM}),$$

where the proposition P states that the processor has generated a new bus cycle internally. Note that the term $\overline{BOFF}$ says that $\overline{BOFF}$ is *not* asserted (i.e., $\overline{BOFF}$=1).

A composite state named "Active" is used to encapsulate the nonidle states. The bus stays within the idle state Ti if there is no request pending. Whenever the processor has generated a new bus cycle (*ReqPend* is true), in the next clock (i.e., as the current clock cycle expires) the bus transits from Ti to T1. The bus will transit back to Ti upon the next clock if:

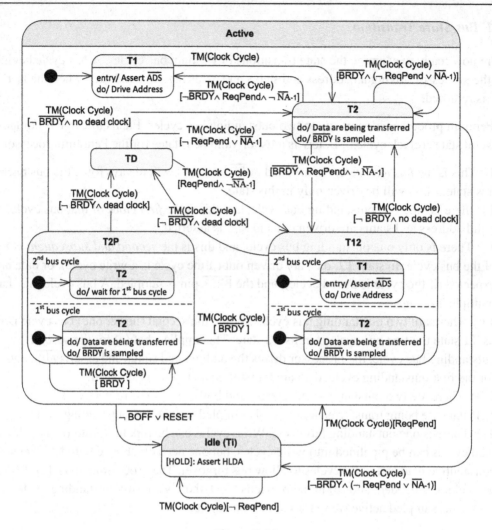

Figure 3.16
Pentium bus control state machine.

- in the current clock, $\overline{BOFF}$ is asserted (low) or RESET is asserted (high); or
- the current clock is T2, the data transfer has finished ($\overline{BRDY}$ was sampled low), and there is no request pending or $\overline{NA}_{-1}$ is high (not asserted). Note here that $\overline{NA}_{-1}$ represents the logical level of $\overline{NA}$ in the clock before this T2.

In a well-designed state diagram, if a state has multiple outgoing transitions triggered by the same event, their guard conditions—say, $C_1, C_2, \ldots, C_k$—ought to be able to form a complete Boolean expression; that is, $C_1 \vee C_2 \vee \cdots \vee C_k = 1$. The state diagram in Figure 3.16 is well designed for the following reasons:

- The guard conditions of the four transitions leaving state T2 are complete. That is, $[\overline{BRDY} \wedge (\neg ReqPend \vee \overline{NA}_{-1})] \vee [\overline{BRDY} \wedge ReqPend \wedge \neg \overline{NA}_{-1})] \vee [\neg \overline{BRDY} \wedge (\neg ReqPend \vee \overline{NA}_{-1})] \vee [\neg \overline{BRDY} \wedge ReqPend \wedge \neg \overline{NA}_{-1})] = 1$.
- The guard conditions of the three transitions leaving state T12 are complete.
- The guard conditions of the four transitions leaving state T2P are complete.
- The guard conditions of the two transitions leaving state TD are complete.

Also note that both T12 and T2P contain two parallel sessions, owing to the existence of two outstanding bus cycles. In T2P, the second bus cycle is waiting for the first bus cycle to finish its data transfer.

Let us use some timing diagrams to examine how the bus changes its states as the timing clock goes on.

Figure 3.17 shows a read cycle followed by an idle state, then a write cycle. Notice that $\overline{ADS}$ and $W/\overline{R}$ are asserted in clock T1; address lines are also driven in clock T1. $\overline{BRDY}$ is sampled in clock T2 to check whether data transfer has finished or not. The transition from T2 to Ti happens because the guard condition $\neg \overline{BRDY} \wedge (\neg ReqPend \vee \overline{NA}_{-1})$ is true.

Figure 3.18 shows a read cycle followed by an idle state, then a write cycle. Notice that the read cycle takes two T2 clocks and the write cycle takes three T2 clocks. This can happen when the processor interfaces with a slow memory device.

Figure 3.19 shows a burst read cycle. Notice that $\overline{CACHE}$ is driven active (low) in clock T1. The read cycle is transformed into a multiple-transfer burst cycle by $\overline{KEN}$ being asserted active on the clock on which the first active $\overline{BRDY}$ is sampled. Consequently, the processor is able to read one data item (i.e., 8 bytes or 64 bits) in each T2 clock. It may take more than one T2 clock to read one data item when the processor interfaces with slow memory.

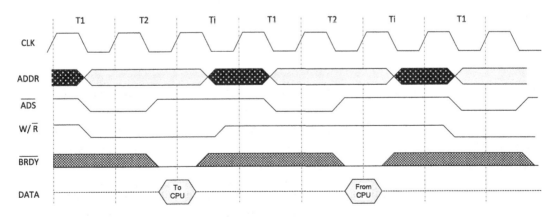

Figure 3.17
Pentium timing diagram example.

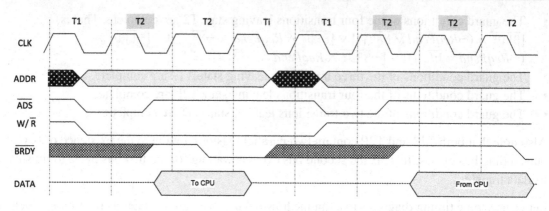

Figure 3.18
Pentium timing diagram: wait states.

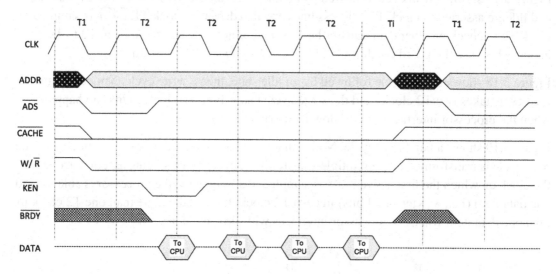

Figure 3.19
Pentium timing diagram: burst read.

Figure 3.20 shows the *pipelining* of a burst read cycle and a write cycle. The first bus cycle starts as a read cycle. $\overline{BRDY}$ is sampled high in the first T2 clock, indicating that another T2 clock is needed to finish the data transfer. $\overline{BRDY}$ is sampled active in the second T2 clock, indicating that one data item has been transferred. However, $\overline{KEN}$ is asserted in the second T2 clock, which transforms the read cycle into a burst read cycle.

Also notice that $\overline{NA}$ is asserted in the second T2 clock, indicating that the external memory system is ready to accept a new bus cycle although all data transfers for the current read cycle have not yet finished. The processor will issue $\overline{ADS}$ for a pending cycle two clocks later.

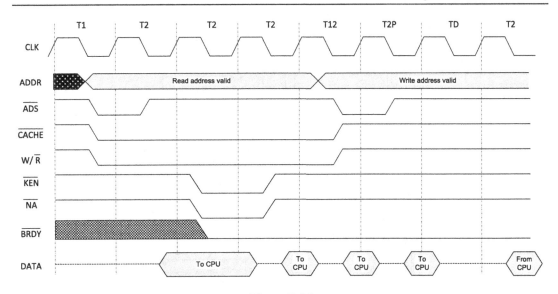

Figure 3.20
Pentium timing diagram: pipelining.

$\overline{BRDY}$ is sampled low in the third T2 clock, indicating that another data item has been transferred. Also, the assertion of $\overline{NA}$ in the second T2 clock makes $\neg \overline{NA}_{-1}$ true in the third T2 clock, which enables the transition from state T2 to state T12.

In the T12 clock, $\overline{BRDY}$ is sampled low, indicating that another data item has been transferred for the burst read cycle. During the same clock, the processor initiates a new write cycle, driving address, status, and $\overline{ADS}$ for this second outstanding bus cycle.

In the T2P clock, $\overline{BRDY}$ is sampled low, indicating that the fourth data item has been transferred for the burst read cycle. Up to this point, the burst read cycle has finished. Upon the next clock, the bus transits from the T2P state to the TD state, because in pipelining a dead clock is needed between read and write cycles. After the TD clock, the bus is ready for transferring data for the write cycle.

3.4.2 Memory Organization

The Pentium processor has a memory space of 4 GB (2^{32} bytes) and a separate I/O space with 64 KB of addressable locations. The memory space is organized as a sequence of 64-bit quantities. Each 64-bit location has eight individually addressable bytes at consecutive memory addresses. The I/O space is organized as a sequence of 32-bit quantities. Each 32-bit quantity has four individually addressable bytes at consecutive memory addresses.

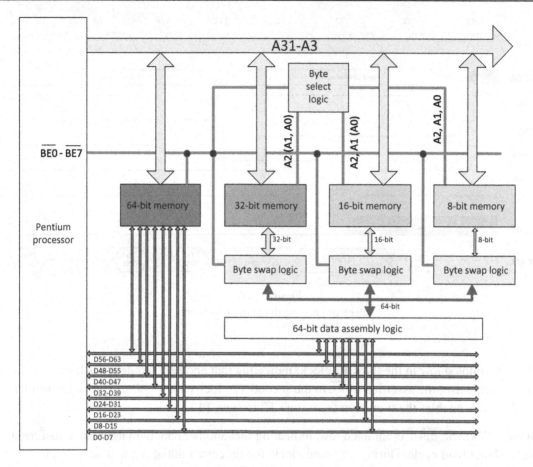

Figure 3.21
Pentium memory mapping.

Figure 3.21 illustrates how the Pentium processor interfaces with memory of various width:

- 64-bit memories are organized as arrays of quadwords (a quadword is a block of 8 bytes). Quadwords begin at addresses evenly divisible by 8. Address lines A31-A3 are used to access quadwords, and $\overline{BE7}$-$\overline{BE0}$ are used to access individual bytes within a quadword.
- 32-bit memories are organized as arrays of dwords (a dword is a block of 4 bytes). Dwords begin at addresses evenly divisible by 4. A dword can be accessed by address lines A31-A3 together with A2. A2 is not a physical address line, but is decoded from $\overline{BE7}$-$\overline{BE0}$ according to Table 3.4:

$$A2 = \overline{BE0} \wedge \overline{BE1} \wedge \overline{BE2} \wedge \overline{BE3} \wedge (\neg\overline{BE4} \vee \neg\overline{BE5} \vee \neg\overline{BE6} \vee \neg\overline{BE7}).$$

Individual bytes within a dword can be accessed by A1 and A0, which are also decoded from $\overline{BE7}$-$\overline{BE0}$ according to Table 3.4.

Table 3.4 Byte access enabled by BE7-BE0

A2	A1	A0	$\overline{BE7}$	$\overline{BE6}$	$\overline{BE5}$	$\overline{BE4}$	$\overline{BE3}$	$\overline{BE2}$	$\overline{BE1}$	$\overline{BE0}$
0	0	0	X	X	X	X	X	X	X	L
0	0	1	X	X	X	X	X	X	L	H
0	1	0	X	X	X	X	X	L	H	H
0	1	1	X	X	X	X	L	H	H	H
1	0	0	X	X	X	L	H	H	H	H
1	0	1	X	X	L	H	H	H	H	H
1	1	0	X	L	H	H	H	H	H	H
1	1	1	L	H	H	H	H	H	H	H

- 16-bit memories are organized as arrays of words (a word is a block of 2 bytes). Words begin at addresses evenly divisible by 2. A word can be accessed by address lines A31-A3 together with A2 and A1. A0 is used to access individual bytes within a word.
- Eight-bit memories are organized as array bytes. A byte can be accessed by address lines A31-A3 together with A2, A1, and A0.

Note that in Figure 3.21 external byte swapping logic and data assembly logic are needed for memory widths smaller than 64 bits so that data can be supplied to and received from the Pentium processor on the correct data pins.

3.5 ARM926EJ-S

The ARM926EJ-S processor [12], a member of the ARM9 family, is a 32-bit RISC microprocessor capable of supporting a full operating system such as Linux, Windows CE, and QNX. The ARM926EJ-S processor has the following key features:

- It supports the 32-bit ARM instruction set and 16-bit Thumb instruction set, enabling a designer to trade off between high performance and high code density.
- It has hardware support for Java bytecode execution, providing Java performance similar to that of a just-in-time compiler.
- It has logic to assist in both hardware and software debugging.
- It has memory-mapped I/O (all peripherals are memory mapped).
- It has a Harvard cached architecture with
  - a memory management unit to support virtual addressing and memory protection;
  - separate instruction and data interfaces to its level 1 memory system, which is known as tightly coupled memory (TCM).

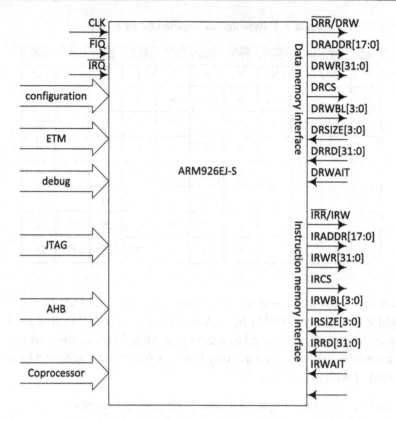

Figure 3.22
ARM ARM926EJ-S processor: pin diagram.

- an advanced microcontroller bus architecture with separate instruction and data interfaces to its level 2 memory system, where the bus is called advanced high performance bus (AHB).

Figure 3.22 gives the pin diagram of the ARM926EJ-S processor, where the pins are grouped according to their functions. It is worth noting that the ARM926EJ-S processor provides two separate interfaces to data memory and instruction memory.

3.5.1 TCM Interface

For real-time systems it is paramount that code execution is deterministic—the time taken for loading and storing instruction or data must be predictable. This is achieved in the ARM926EJ-S processor through the so-called tightly coupled memory (TCM) interfaces.

The ARM926EJ-S processor supports two TCM regions, one for instructions (ITCM) and one for data (DTCM). The physical size of the TCM regions is defined by external inputs (IRSIZE, DRSIZE), and ranges from 4 KB to 1 MB.

Some DTCM signals are described below:

- `DRADDR[17:0]`: DTCM address lines, addressing by words (a word in the ARM926EJ-S processor is 4 bytes).
- `DRCS`: chip select, indicating if an access will take place in the following cycle.
- $\overline{DRR}$/`DRW`: indicates if the access is a read or write.
- `DRRD[31:0]`: input data lines for TCM read.
- `DRSIZE[3:0]`: static configuration input that specifies the physical size of DTCM memory.
- `DRWBL[3:0]` DTCM byte lane indicator, indicating which bytes are to be written. Bits of DRWBL are set only when a write is taking place. In little-endian mode, `DRWBL[0]` indicates the LSB of the word and `DRWBL[3]` indicates the MSB. In big-endian mode, `DRWBL[3]` indicates the LSB of the word and `DRWBL[0]` indicates the MSB.
- `DRWD[31:0]` output data lines for TCM write.

The instruction-side TCM signals are almost identical to the DTCM signals. All the signals on the DTCM have an equivalent on the instruction side.

Figure 3.23 shows a configuration where four banks of byte-wide SRAM are used to form the DTCM region. Note that (a) the 4-byte-wide SRAMs have the same address block (by address

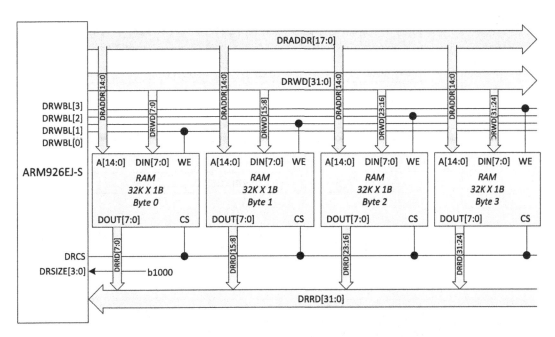

Figure 3.23
ARM926EJ-S processor: 8-bit memory.

lines DRADDR[14:0] and are selected simultaneously by the same chip-select signal;[12] (b) the chip labeled "byte 0" is connected to data input lines DRWD[7:0] and data output lines DRRD[7:0] (The other three chips are connected similarly); (c) the write-enable pin WE of each chip is controlled by a distinct line from DRWBL[3:0], so that each chip can receive only 1 byte of a word on the data bus.

Figure 3.24 shows a configuration where two-word-wide SRAMs are used to form the DTCM region. Note here that the two chips are write enabled by both DRW and address line DRADDR[14]: chip A is accessed when DRADDR[14]=0, and chip B is accessed when DRADDR[14]=1. As an exercise, figure out the address ranges of chip A and chip B.

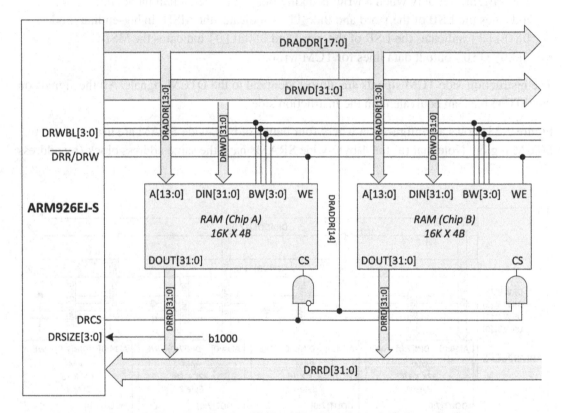

Figure 3.24
ARM926EJ-S processor: 32-bit memory.

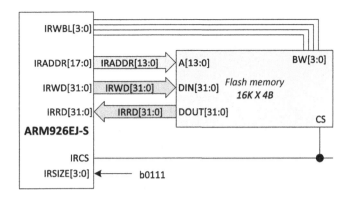

Figure 3.25
ARM926EJ-S processor: instruction.

Figure 3.25 shows a configuration where one-word-wide flash memory is used to form the instruction TCM region. Note here how the instruction-side TCM signals are used.

Problems

3.1 Refer to Figure 3.26, and list the memory address ranges for each of the four chips.

3.2 Refer to Figure 3.27, and list the memory address ranges for each of the four chips. Which chips are memory devices? Which chips are I/O devices?

3.3 Refer to Figure 3.7; which chip would be written to if in the first cycle the instruction is "TBLWT 12h to address 1A6B4Ch" and in the second cycle the instruction is "TBLWT 8Ah to address 1A6B4Dh"? Would the data 12h and 8Ah be stored on the memory chip correctly? Explain why.

3.4 Refer to Figure 3.9; which chip would the data 12h be written to? Which chip would the data 8Ah be written to? Explain why.

3.5 Refer to Figure 3.9, and follow the format of Table 3.2 to work out the memory mappings of chips A, B, and C.

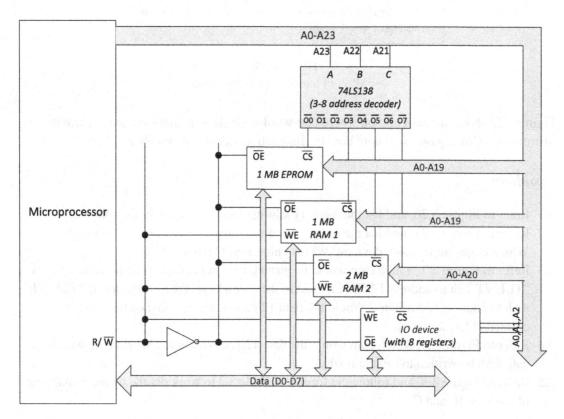

Figure 3.26
Memory-mapped I/O.

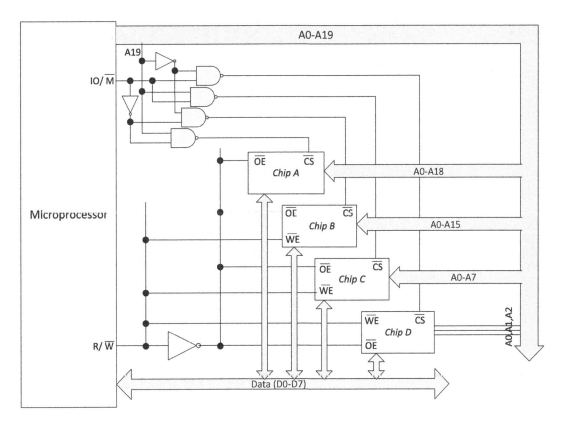

Figure 3.27
Memory mapping configuration.

Interrupts

Contents

Interruptions can be viewed as sources of irritation or opportunities for service, as moments lost or experience gained, as time wasted or horizons widened....Monopolize our minutes or spice our schedules, depending on our attitude toward them.

William Arthur Ward

Real-Time Embedded Systems. http://dx.doi.org/10.1016/B978-0-12-801507-0.00004-3

4.1 Introduction to Interrupts

Interrupt is a mechanism by which a microprocessor can alter its flow of execution to handle asynchronous or synchronous events. *Asynchronous events* are those that can occur at any time and typically occur at unanticipated spots of the running program, while *synchronous events* are those that can occur only at planned or anticipated spots of the running program.

Interrupts can be classified into three categories: external interrupts, software interrupts, and internal interrupts:

* External interrupts, also called *hardware interrupts*, are asynchronous events generated by external hardware devices to get the microprocessor's attention.
* Software interrupts, also called traps, are synchronous events generated by special processor instructions placed in a program. Software interrupts are unconditional in the sense that the execution of the special instruction will always generate a software interrupt.
* Internal interrupts, also called *exceptions*, are synchronous events generated by the processor itself whenever some abnormal condition occurs during instruction execution. Internal interrupts are conditional in the sense that the execution of some valid instruction (e.g., the division instruction) may cause an exception (e.g., if the divisor is 0).

4.2 External Interrupts

I/O devices are the liaison between an embedded system and its work environment. There are mainly two approaches for an embedded system to interact with I/O devices: polling and interrupts.

Polling is the simplest approach used in embedded systems to handle I/O activities synchronously. Assume that a peripheral I/O device intermittently receives data, which must be processed by the processor. In polling, the processor needs to continuously (or at least regularly) check if data have arrived. The processor typically does nothing other than check the status register of the I/O device until it is ready, at which point the device is accessed and serviced.

Polling is relatively straightforward in design and programming. In a simple system with only one I/O device, polling is perfectly appropriate because the system can just wait until the I/O device becomes ready for a service. Waiting for a device to get ready is no longer appropriate in systems involving multiple peripherals. In such a case, the polling mechanism could be improved such that the readiness of all the I/O devices is sequentially checked in an endless loop. When a device is not ready, instead of passively waiting until it becomes ready, the processor turns to check (and service, if possible) the next device. (This is actually called round-robin architecture, and is covered in Chapter 12.)

When there are too many I/O devices to check, the time required to poll them can be considerable, and the system might break the deadlines of certain tasks. In such a case, the so-called *interrupt-driven I/O* comes into play.

Interrupt is a commonly used mechanism for computer multitasking, especially in real-time computing. Hardware interrupts are events generated by external hardware devices to get the microprocessor's attention. For example, pressing a key on the keyboard or moving the mouse triggers hardware interrupts that cause the processor to read the keystroke or mouse position. Interrupts allow an embedded system to respond rapidly to multiple real-time events.

Hardware interrupts are triggered when the interrupt request (IRQ) line(s) of a microprocessor is asserted active by the electrical signals sent from hardware devices. Typically, the processor samples its interrupt input request line(s) at predefined times during each bus cycle. An IRQ is detected if the interrupt line is active when the processor samples it.

Hardware interrupts are *asynchronous* in the sense that they can occur at any time and at any place in the running program. When an interrupt is received, the processor automatically suspends the program that is currently running, saves its status, and transfers control to a special program called the *interrupt service routine* (ISR). Once the ISR has run to completion, the control returns back to the original program that was suspended.

There are two approaches for a microprocessor to locate the ISR of an interrupting device: nonvectored interrupting and vectored interrupting.

4.2.1 Nonvectored Interrupting

In nonvectored interrupting, a fixed memory location is used as a common vector (direction) for all interrupt sources.

Figure 4.1 illustrates the process of nonvectored interrupting:

1. While the processor is executing an instruction *j* of a user program, one of the I/O devices raises an IRQ signal to the interrupt input pin (INT).
2. The processor detects the IRQ. Upon the completion of the instruction *j*, the processor starts an interrupt acknowledge cycle. At this point, the value contained by the program counter register (PCR) is the location of the next instruction of the user program.
3. In the interrupt acknowledge cycle, the status register (SR) and the PCR are saved (say, into spare registers) so that the user program can be resumed later.
4. The PCR is loaded with the fixed memory location where the common interrupt vector is saved.
5. This common interrupt vector is typically a jump instruction, pointing to the start location of the ISR.
   5.1. Inside the ISR, the commonly used registers are first saved onto the interrupt stack. In so doing, the processor switches its context from the user task to the ISR.

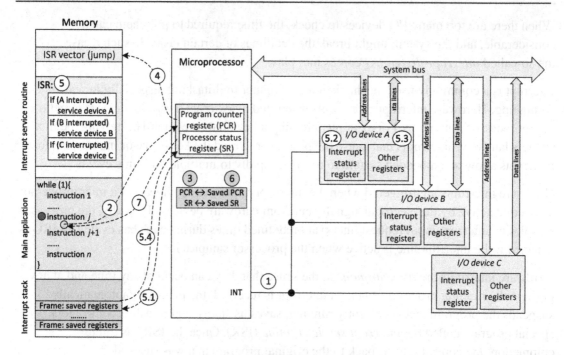

Figure 4.1
Nonvectored interrupting.

5.2.   Next, the processor searches through the devices, from high priority to low priority, to identify the interrupt source (requesting device). An I/O device often has one or more interrupt status registers that latch its IRQ; the processor can check such registers to ensure IRQs are not missed.

5.3.   The portion of code pertinent to the requesting device is executed, which typically entails the access of the ports (registers) on the device.

5.4.   At the end of the ISR, the top frame of the interrupt stack is popped up and the context of the user task is restored.

6.   The original PCR value is restored.

7.   The processor is ready to run the next instruction of the user program.

Nonvectored interrupting does not require extra hardware, but involves serial testing at step 5.2, which can introduce some unacceptably long delays in the response to the interrupting device. It is applicable when there are only a few external interrupt sources.

4.2.2 PIC and Vectored Interrupting

Multiple interrupt sources can fire interrupt requests simultaneously. To respond to multiple requests, it is critical that the microprocessor can rapidly identify each interrupting device and jump to the exact memory location where its own ISR is stored. Here, the identity of a device

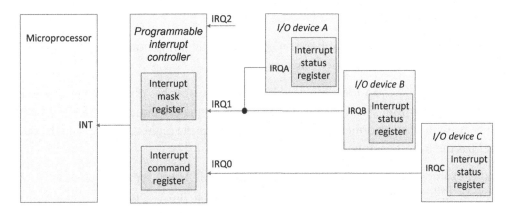

Figure 4.2
General PIC settings.

is called its *interrupt vector number*, which is stored in a programmable register on the device itself, or more typically on a programmable interrupt controller (PIC).

A PIC, as shown in Figure 4.2, typically has an output pin, which is connected to the INT pin of the microprocessor, and several IRQ lines, each of which can take interrupt signals originating from one or multiple I/O devices. In such a sense, a PIC serves as a liaison to link many external devices and the microprocessor together.

A PIC normally has three functions. It can

- *p*rovide a hub for maskable interrupt sources;
- *i*mplement prioritized interrupting at the hardware level;
- *c*ope with different interrupt source modes.

4.2.2.1 Maskable interrupts

Hardware interrupts can be either maskable or nonmaskable. A nonmaskable interrupt can never be ignored, and is used for critical tasks such as system resets and watchdog timers.

A PIC typically has an interrupt mask register (IMR), which allows you to individually enable and disable interrupts from devices on the system. In general, there is one bit in the IMR that corresponds to each IRQ line of the PIC: writing a 0 (or 1) to enable interrupts emanating from devices wired to the corresponding IRQ line; writing a 1 (or 0) to disable interrupts on the corresponding IRQ line.

When a maskable interrupt source is disabled, its IRQs will not be forwarded by the PIC to the processor.

4.2.2.2 Interrupt priorities

Interrupt sources are typically prioritized on the basis of their importance. A PIC can be programmed to offer various priority schemes.

For instance, the Intel 8259A PIC chip used for x86 processors has eight IRQ lines. The BIOS (the name originates from "basic input/output system") typically initializes the 8259A chip to use a scheme with fixed priorities: the device on IRQ0 has the highest priority and the device on IRQ7 has the lowest priority.

As another example, the advanced interrupt controller (AIC) used for ARM processors can handle up to 32 interrupt sources. Each interrupt source can have a programmable priority level of 7-0: level 7 is the highest and level 0 is the lowest priority. The interrupt controller has a source mode register for each interrupt source. A user can assign an appropriate priority level to a device by setting the last 3 bits of the corresponding source mode register. If several interrupt sources of equal priority are pending, the interrupt on the IRQ line with the lowest number is considered first.

A priority level is used to dictate the order in which the interrupts will be serviced. When several IRQs are received simultaneously by the PIC, it serializes these requests according to their priority levels: the request from a device with a higher priority is always forwarded to the processor prior to requests from devices with a lower priority.

What happens if the PIC detects a new request from another interrupt source while the processor is executing an ISR? If the new request has a lower priority, the PIC will not forward it to the processor until after the completion of the existing ISR. If the new request has a higher priority, a mechanism called interrupt nesting comes into play.

Interrupt nesting allows interrupts with an equal or higher priority to interrupt an existing interrupt. This is illustrated in Figure 4.3, where the normal execution is interrupted by ISR *x*,

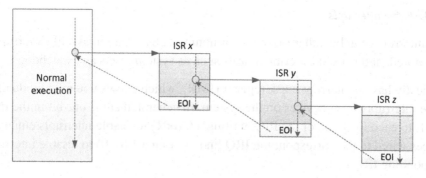

Figure 4.3
Interrupt nesting.

which is interrupted by ISR y, which is further interrupted by ISR z. Once ISR z runs to completion (indicated by end of interrupt, EOI), the control is relinquished back to ISR y, which resumes from the interrupted point and runs to completion, the control is then relinquished back to ISR x, and so on. In other words, the execution of ISR z is nested within ISR y, which is nested within ISR x, which is injected into the normal program execution.

Also notice that interrupt nesting is prohibited at the two ends of an ISR, referred to as the *context switching* sections. This is for good reason: further interrupts are allowed only when enough of the processor context has been saved onto the interrupt stack. Since interrupts are disabled automatically during context switching, to support interrupt nesting, a designer has to explicitly enable interrupts inside an ISR. This can be done immediately after context switching in order to achieve faster response to higher-priority interrupts.

4.2.2.3 Interrupt source mode

The external interrupt sources can operate in one of two modes: level-sensitive mode or edge-triggered mode.

To send an interrupt signal to the PIC, a *level-sensitive* device needs to drive its IRQ line to the active level, and then hold it at that level until it has been cleared (say, upon service completion). For devices operating in this mode, PIC has a better chance of minimizing spurious signals from a noisy interrupt line (a spurious pulse is often too short to be detected by a PIC).

When multiple level-sensitive devices share an IRQ line, the line remains asserted as long as one or more than one of the devices has an outstanding IRQ. After a short hardware delay (typically a few clock cycles), the PIC drives the INT line to its active level to inform the processor about the IRQ. For a level-sensitive device, upon the completion of its ISR, it should be set to stop driving the corresponding IRQ line. If the interrupt source is not cleared in time, when the processor signals the EOI to the PIC, the PIC would mistakenly detect another outstanding request from the same device on the IRQ line. If this IRQ line has the highest priority, the PIC would immediately reissue an interrupt to the processor, which would loop forever, continually calling the ISR.

When multiple devices—say, A and B—share an IRQ line, if device A sends an IRQ while the processor is serving device B, the request from device A might be transparent to the PIC (hidden by device B's request). Device A's request would be detected and honored only if device A is still driving an active signal on the IRQ line after source device B is cleared at the end of ISR-B. Should ISR-B take too long, the processor might not be able to service device A in a timely manner. Even worse, if there is a device that the processor does not know how to service, then any interrupt from that device could permanently block all interrupts from the other devices sharing the same IRQ line.

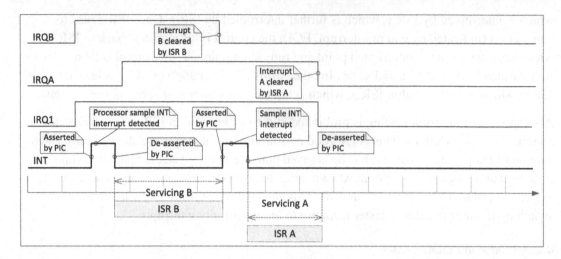

Figure 4.4
Level-sensitive sources.

Figure 4.4 shows a scenario with two level-sensitive sources. Notice when each of the interrupt sources is cleared and when the INT line is asserted and deasserted (there is a hardware delay between the IRQ1 and the INT signals).

To send an interrupt signal to the PIC, an *edge-triggered* device needs to drive a level transition on the interrupt line, either a falling edge (high to low) or a rising edge (low to high). In other words, a PIC detects IRQs from edge-triggered devices by clear-assert transitions.

If the source of the interrupt can be cleared only by the corresponding ISR, there exists a critical issue when multiple edge-triggered devices share an interrupt line. Suppose device A asserts an IRQ before ISR-B clears its source (device B). The IRQ line is still asserted (by device A's request) even after ISR-B has cleared its source. However, the IRQ from device A cannot be detected by the PIC because it is waiting for a clear-assert transition. The code for clearing source device A is in ISR-A, which unfortunately is not running. Consequently, the PIC will never see a clear-assert transition on that IRQ line, and the system will behave abnormally.

Advanced interrupt controllers can clear the interrupt source after the processor has started to handle its IRQ. In such a case, service of one device can be postponed arbitrarily, and IRQs from other devices on the same IRQ line can still be detected and honored. Even though there is a device that the processor does not know how to service, it may be taken as spurious interrupts, and will not interfere with the IRQs from other devices.

Figure 4.5 shows a scenario with two edge-triggered sources. Notice when each of the interrupt sources is cleared by the PIC and when the INT line is asserted and deasserted. Also notice that the service to device B is interrupted by the service to device A.

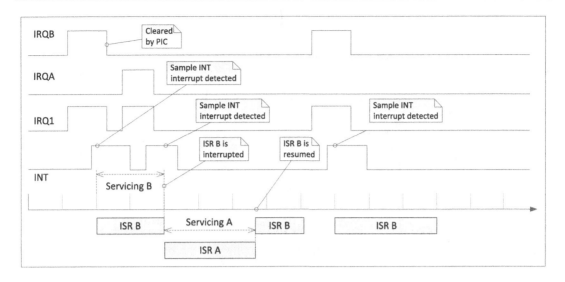

Figure 4.5
Edge-triggered sources.

4.2.2.4 Interrupt vectoring process

In vectored interrupting, a special block of memory is typically reserved to accommodate a data structure called an *interrupt vector table*—a table with a sequence of entries called interrupt vectors. Each *interrupt vector* contains information that points the processor to the start address of the corresponding ISR.

Figure 4.6 illustrates the process of vectored interrupting:

1. Device A drives an IRQ on the IRQ0 line of the PIC.
2. The PIC detects the IRQ on IRQ0. It first converts the request into a vector number corresponding to device A, stores it in a register, and then asserts the INT line to inform the processor.
3. While the instruction *j* of the current program is being executed, the processor samples INT and detects an asserted line. After the execution of the instruction *j* is completed, the current program is suspended. At this point, the value contained by the PCR is the location of the next instruction of the current program.
4. In the interrupt acknowledge cycle, the processor asserts the INTA (interrupt acknowledge) line to the PIC, expecting an interrupt vector number.
5. The PIC drives the interrupt vector number associated with device A to the system bus.
6. The PIC then deasserts the INT line (so that a new IRQ can be asserted);
7. The status register and the PCR are saved (say, into spare registers or pushed to the interrupt stack). This will allow the processor to resume the execution of the original program later.

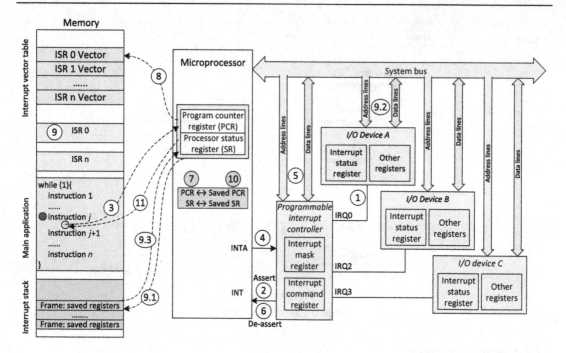

Figure 4.6
Vectored interrupting.

8. To protect context switching, further interrupts are disabled. The PCR is loaded with the address of the appropriate interrupt vector (an entry in the interrupt vector table).

9. The interrupt vector is typically a jump instruction, pointing to the start location of the ISR for the requesting device.

   9.1. Inside the ISR, the commonly used registers are first saved onto the interrupt stack. At this point, the processor has switched its context from the last task to the current ISR. Further interrupts can now be enabled, if interrupt nesting is desired.

   9.2. The portion of code pertinent to the requesting device is executed, which typically entails the access of the ports (registers) on the device.

   9.3. At the end of the ISR, interrupts are again disabled during context switching. Sometimes a special instruction is needed to inform the PIC about the end of the current interrupt. The top frame of the interrupt stack is popped up and the context of the original task is restored.

10. The PCR and the status register are restored to their respective values before the interrupt;

11. The processor is ready to run the next instruction of the original program.

For a level-sensitive device, it can continuously assert its IRQ line until it is serviced. By deasserting the INT in step 6, the PIC is virtually isolating the interrupt source being serviced from the processor. Whenever there is a new IRQ with a higher priority, the PIC can assert the INT line again so that the processor can suspend the execution of the current ISR and switch its context to service the higher-priority interrupt.

4.3 Software Interrupts

Software interrupts are *synchronous* events generated by special processor instructions placed in a program.

For example, pressing a key on the keyboard triggers hardware interrupts (the interrupt vector number is 09h on x86 systems), and the corresponding ISR will typically place the code of the key into a keyboard buffer—a circular queue in a protected memory block. However, those keyboard inputs are not useful unless they are processed by a user application, and only programmers know when and where an input is needed. We all know that functions such as getChar() can be explicitly used in a program to read characters typed through a keyboard. Under the veil, however, it is software interrupts that carry the ball: indeed, the function getChar() uses some assembly instruction to trigger a special ISR to read a character from the protected circular queue.

When a PC is turned on, a firmware called the BIOS takes control and populates the processor's interrupt vector table with the addresses of default ISRs. Later, an operating system, if installed, can take advantage of the services (i.e., ISRs) offered by the BIOS to talk to the installed hardware. In so doing, the operating system is actually using software interrupts to request services from the BIOS.

An operating system can also extend the interrupt vector table, if allowable, by adding more vectors to the table to offer extensive services. Many modern operating systems (e.g., QNX) even choose to install their own ISRs to directly control hardware, completely bypassing the built-in BIOS interrupt facility. The interrupt services offered by an operating system are available to the users in the name of "kernel functions." Next time your code invokes a system call, you will know that it is indirectly raising a software interrupt to the processor (through the operating system).

For example, in order to request a hardware service (e.g., read from the keyboard buffer), a user program can use a software interrupt to get the operating system's attention, causing control to be passed back to the operating system's kernel. The kernel will then process the request, and the execution of the user program is resumed once the service has been done.

On x86 systems, software interrupts are initiated by executing the int instruction :

int vector_number;

where *vector_number* is an integer in the range 0-255.

For example, the BIOS offers application-level interfaces to several hardware-related services [32, 40]. An interface with the keyboard services is under the vector number 16H. By int 16H, a program can invoke the ISR addressed by the entry 16H. This ISR actually provides several functions. The AH register is used to indicate the desired function number. For example, the following two instructions,

mov AH, 00H;

int 16H;

together are used to read a character from the keyboard buffer. If the keyboard buffer is empty, it waits until a character is entered.[1] The ASCII code of the key is returned in the AL register.

An ARM processor has a single vector at 0x08, which stores the address of the ISR for software interrupts. On systems built upon ARM processors, software interrupts are initiated by executing the swi instruction:

swi sn;

where *sn* is a 24-bit number used to indicate a specific service type. The service number *sn* is ignored by the processor, but is utilized by the ISR to branch to a second-level ISR corresponding to the number *sn*. In other words, by varying *sn*, an operating system can implement a collection of privileged system functions which can be invoked by applications running in user mode.

4.4 Internal Interrupts

Internal interrupts, also called exceptions, are synchronous events generated by the processor itself whenever some abnormal condition occurs during instruction execution. Internal interrupts are coerced rather than requested.

A processor may assign a fixed interrupt number to an exception type. Take the x86 processor as an example. The vector number 0x00 is reserved for divide error exception; this exception occurs whenever a value is divided by zero. The vector number 0x01 is reserved for single-step exception, which occurs after each instruction if the "trace" bit of the flags register is set to 1 (say, by a debugger).

For ARM processors, the vector at 0x04 is reserved for "undefined instruction" exception, which is raised when an instruction is not recognized after decoding. The vector at 0x0C is reserved for "pre-fetch abort" exception, which occurs when an attempt to load an instruction results in a memory fault. The vector at 0x10 is reserved for "data abort" exception, which

---

[1] Note that this service is a consumer of the keyboard buffer. The provider of the keyboard buffer is the keyboard hardware interrupt (with the vector number 09H).

occurs when the memory controller indicates that an invalid memory address has been accessed (say, if there is no physical memory for an address, or if the processor does not currently have permission to access a memory region).

It is typically the responsibility of an operating system to install an appropriate ISR for each exception type.

To summarize, let us use an example to clarify the differences among the three types of interrupts. Every human being can be viewed as a smart processor. If, as a student, you are programmed (get used) to sleep during the third 10 min of a class, it is a software interrupt when you fall asleep; everyone in the class is well prepared to see that happen. While you are sleeping, the instructor or your neighbor awakens you. This is an external interrupt. If the topic is very interesting, or the instructor does a good job, or you simply feel guilty about sleeping, then you have fortunately revived this is an internal interrupt.

4.5 Design Patterns for ISRs

Now, we discuss a few design patterns for writing ISRs. These patterns are applicable to external interrupts, software interrupts, and internal interrupts.

4.5.1 General ISR Design Pattern

A simple ISR design pattern is given by ISR-PATTERN-MIN().

ISR-PATTERN-MIN()
1   Save the processor context (registers) onto interrupt stack;
        // some processor/compiler does this automatically.
2   Clear the interrupt source;
3   Service section (may access hardware ports for hardware interrupts);
4   Switching contexts: restore context from interrupt stack;
        // some processor/compiler does this automatically.

The first step of ISR-PATTERN-MIN() is to push the current context onto the interrupt stack so that the context can be restored upon interrupt return. To protect context switching, this step has to be performed with the interrupts disabled.

The second step is to clear the interrupt source. This is necessary especially for level-sensitive devices. At this time, interrupts can also be enabled, if interrupt nesting is desirable.

The third step is the main body of the ISR: service the interrupt source. For a hardware interrupt, this is the place to access the ports associated with the hardware to read inputs from external devices or write outputs to external devices. For example, in the ISR for a serial device, this is the place to transfer data from/to the universal asynchronous receiver/transmitter.

After the interrupt service has finished, at the fourth step the original context is restored and the processor is ready to execute the next instruction prior to the interruption. During context switching, interrupts have to be disabled.

ISR-PATTERN-MIN() can be used to write ISRs for those interrupt sources that have to be *fully serviced* rapidly (e.g., clock).

4.5.2 ISR with a Server Task

Not all interrupt sources have to be *fully serviced* at the third step of ISR-PATTERN-MIN(). In such a case, it is critical to optimize the third step to keep it as short as possible.

This gives us another pattern called ISR-PATTERN-SERVER(), where it is assumed that corresponding to the interrupt source there is a dedicated "server task," or device driver, which may be part of the operating system's kernel and offer services to user programs. The "server task" is normally waiting (blocked) for service requests, and becomes active whenever (a) it receives a signal (say, a semaphore) from an ISR—this is the time it needs to finish the bulk of the work deferred by the ISR, or (b) it receives a service call from a user program—this is the time it needs to relay relevant information to the user program.

ISR-PATTERN-SERVER()

1   Save the processor context (registers) onto interrupt stack;
        // some processor/compiler does this automatically.
2   Clear the interrupt source;
3   Service section (minimal processing only);
4   Signal the corresponding server task (driver task) for further processing;
5   Switching contexts: restore context from interrupt stack;
        // some processor/compiler does this automatically.

The *separation of concerns* as implemented in ISR-PATTERN-SERVER() has the following benefits:

- It allows the system to respond to a device faster, leading to higher hardware concurrency. The ISR may contain only the part that has to be done immediately in order to keep the device working correctly. For example, in the ISR of a network device, it is necessary to copy the incoming packet off hardware. The processing of the packet, however, can be deferred.
- It can improve the performance of the whole system. The execution of an ISR blocks the highest-priority task from running. Lengthy processing can defer the execution (thus break the deadlines) of other ISRs and user tasks.

- Only nonblocking functions can be used in an ISR. Complex operations can now be moved out of the interrupt context.
- It offers flexibility for service management. Some task can be more important than servicing interrupts. In such a case, the task can be designed with a higher priority than the server tasks for interrupts. This is not possible in ISR-PATTERN-MIN().

4.5.3 ISR Chaining

Interrupt chaining allows multiple ISRs to share a single interrupt vector. For example, the BIOS may install an ISR for timer interrupts (or real-time clock interrupts, or keyboard interrupts, etc.). An operating system may decide to install its own timer ISR to offer add-on services (say, for real-time scheduling). Later, a user program may also wish to install a timer ISR to get a time-out event every 100 ms. When an operating system or a user installs a new ISR to a vector that is currently in use, it is normal to copy the original ISR to a new location and call the original ISR immediately before the end of the new ISR.

This chaining process is illustrated in Figure 4.7. Here, the ISR installed by a user is at the beginning of the chain: its location is referenced by the vector **x**. Unless the user-installed ISR could get all the jobs done and done well, it typically will invoke the previous ISR to hand over the rest of the task. This is how interrupt chaining gets its name. It is worth noting that only the last ISR on the chain should report EOI.

Next time you use the system call InterruptAttach() to install an ISR, you will know that this ISR is actually being placed in the front of a chain with other ISRs sharing the same vector number. Figure 4.8 gives an example, where the 8254 programmable interval timer (PIT) chip runs at roughly 1.193182 MHz. The PIT chip has a 16-bit register used as a frequency divider, which can take values from 0 to 65,535 (0 represents 65,536).

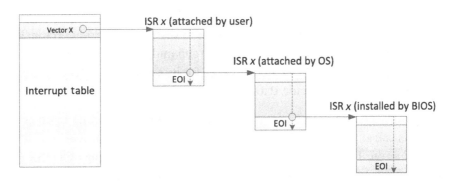

Figure 4.7
Interrupt chaining.

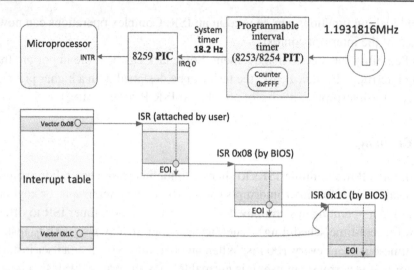

Figure 4.8
Chaining the timer interrupt.

During the boot process, the BIOS typically sets the register to 0xFFFF, which gives an output frequency of 18.2065 Hz (or one output every 54.9255 ms). This output is connected to the IRQ0 line of the PIC. Consequently, the PIC generates interrupts every 54.9255 ms, and the corresponding ISR (with the vector number 0x08 by default) is triggered to execute every 54.9255 ms. In addition, the ISR installed by the BIOS at 0x08 also invokes the ISR at 0x1C, which handles timer ticks—say, updates the time of day. By the system call InterruptAttach(), a user timer ISR is installed prior to the default timer ISR. This user timer ISR can be used to generate customized periodic time-out events demanded by other tasks.

Chaining can also be used by a debug monitor to gather system footprints.

4.5.4 ISR Cascading

Interrupt cascading allows multiple interrupt sources to share one interrupt vector. This pattern is typically used by an operating system to group multiple relevant services together under one entry point. For example, DOS uses 0x21 for its services; all software interrupts on ARM processors share a single vector, 0x0008.

Figure 4.9 illustrates the idea of interrupt cascading, where the ISR directly referenced by the interrupt vector is called a level 1 ISR, and those ISRs referenced by the level 1 ISR are called level 2 ISRs, and so on. The branching typically happens at the beginning of an ISR (switching points): an appropriate next-level ISR is invoked according to the specified function number (software interrupts) or by checking the status registers of hardware devices (hardware

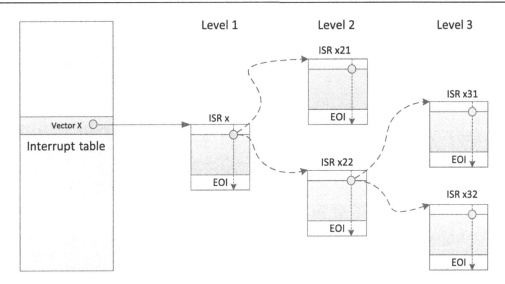

Figure 4.9
Interrupt cascading.

interrupts). It is also worth noting that for each IRQ there is only one ISR to be executed to completion, and after that the control is immediately returned to the interrupted program.

4.5.5 Data Sharing with ISRs

Problems may occur when there are shared data (or resource) in the system that need to be accessed simultaneously by two or more processes. It is even more critical when a user process shares data with an ISR.

We use a classic example [71] to explain this issue. Suppose the time-of-day information is stored separately in four shared variables: *ms, sec, min,* and *hr.* Every 50 ms these four variables are updated by a real-time clock ISR. In addition, there is a user task that refreshes its time-of-day display on a GUI twice a second.

Now, suppose the current situation is 06:59:59:950, and the real-time clock ISR is triggered when the user task has just refreshed the minute display. This is shown in Figure 4.10.

Obviously, the ISR would interrupt the user task. Within the ISR, *ms* is increased by 50 ms, which will in turn trigger the variables *sec*, *min*, and *hr* to be updated to their new values. Upon the completion of the ISR, the new time recorded in the variables is 07:00:00:00.

The execution of the user task is now resumed, and 07:59:59 is displayed on the GUI. This is almost 1 h ahead of the true time, which is annoying to people with sharp eyes (although this could be corrected within 0.5 s).

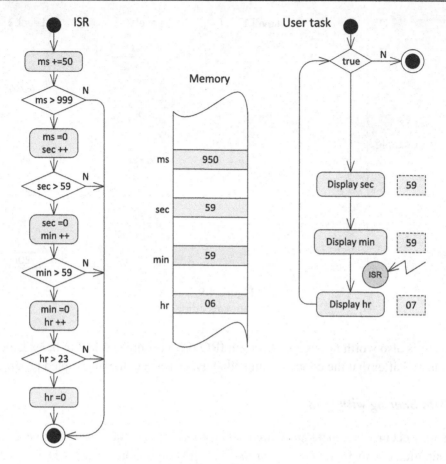

Figure 4.10
Shared data issue.

Figure 4.11 gives a solution, which requires the user program to mark the display section as a *critical section*. The critical section has to be protected by disabling interrupts before it is entered and enabling interrupts afterward.

While interrupts are disabled, the critical section is protected because the current task has exclusive use of the processor (no other task or interrupt can take control). Pending interrupts or higher-priority tasks, if any, are not able to take control until after interrupts have been enabled.

This solution does come with drawbacks. First, disabling interrupts affects the timely update of the time-of-day information contained by the shared variables. Over time, the system time could drift considerably away from the real time. Second, disabling interrupts will also defer

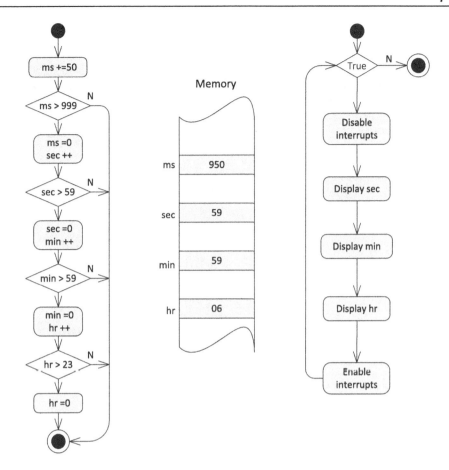

Figure 4.11
Interrupt disabling.

the execution of other ISRs, causing other interrupt-driven tasks to miss their deadlines. When such a solution is used, it is advised to ensure that a critical section encloses necessary code only so that the time for executing the critical section can be minimized.

Another solution, as shown in Figure 4.12, involves the use of *double buffers*. In particular, two sets of variables are used, and the ISR is coded such that it modifies only the set of variables that is not currently being used by the user task.

This solution can avoid the drawbacks of the interrupt-disabling solution. However, it does demand more memory for shared data. The displayed time can drift slightly owing to the use of two variable sets.

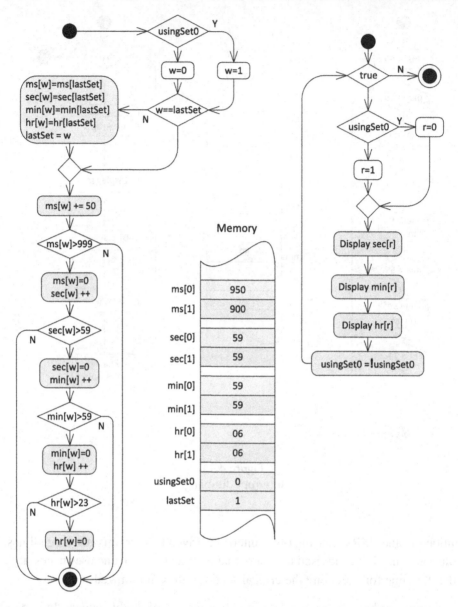

Figure 4.12
Double buffering.

4.6 Interrupt Response Time

Interrupt latency refers to the interval of time from an external interrupt request signal being raised to the *first ISR instruction* being fetched and executed. The interrupt response time refers to the interval of time from an external IRQ signal being raised to the *completion* of the

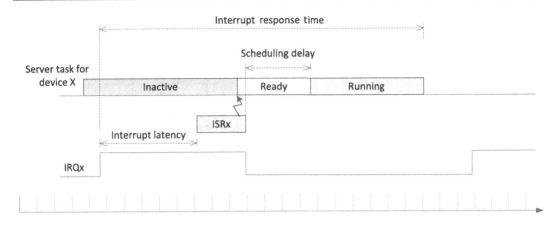

Figure 4.13
Interrupt latency and interrupt response time.

service (including the execution of both the ISR and the server or driver task, if any). In other words, the interrupt response time is the sum of the interrupt latency and the total service time. Low interrupt latency (or a shorter interrupt response) is critical for real-time applications.

Figure 4.13 illustrates the difference between interrupt latency and interrupt response time.

There are two major causes for interrupt latency. First, when a device X raises an IRQ, the processor might be executing an ISR at a higher priority level. Second, interrupts might have just been temporarily disabled by a user task. In embedded systems and many real-time operating systems, it is not uncommon for a user application to disable/enable interrupts in order to obtain greater control of system resources. However, this ought to be done carefully. If a user program occupied the processor for too long, the interrupts would not be handled in a timely manner, and the whole system would suffer owing to a ripple effect.

As shown in Figure 4.13, although the server task for device X has been wakened by the signal from the ISR, there is a *scheduling delay* before this task starts to run. The scheduling delay also has two causes. First, the processor might have been executing an ISR: the execution of a lowest-priority ISR blocks the highest-priority task from running. Second, a higher-priority task may be running. It is thus very important to assign priorities to user tasks appropriately or the whole system may suffer from a slow response. Note that the context switching time is not shown in Figure 4.13 because it is of a lower magnitude relative to the factors considered here.

4.7 Case Study: x86

The interrupt vectoring procedure of x86 processors is illustrated in Figure 4.14. Shown in the top-left corner is the interrupt vector table, which starts at 0x0, the very beginning of the

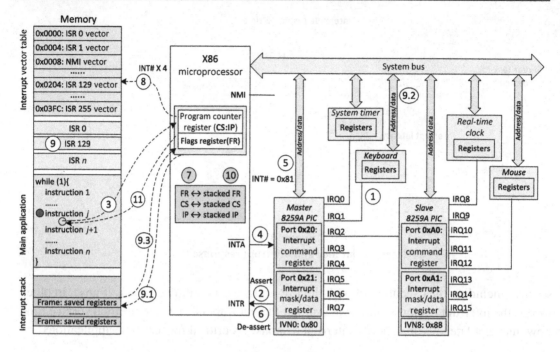

Figure 4.14

x86 interrupt vectoring.

memory space, and ends at 0x03FF. The table has 256 interrupt vectors. Each interrupt vector has 4 bytes, containing 2 bytes for the IP register followed by 2 bytes for the CS register. The 4 bytes together, CS:IP, form an address, pointing to the location of an ISR.

Each interrupt is associated with an *interrupt vector number*:

- For a hardware interrupt, its interrupt vector number is provided by the hardware device or PIC.
- For a software interrupt, its interrupt vector number is provided by the int instruction.
- For an internal interrupt, its interrupt vector number is fixed and is known by the processor.

Given an interrupt vector number, the corresponding interrupt vector is located at the memory address corresponding to four times the interrupt vector number.

The default x86 processor interrupt vector table is given on the left in Figure 4.15. The first 32 interrupt vectors (0x0-0x1F) are either reserved or defined by the x86 processor for internal interrupts (except 0x02, which is a vector for nonmaskable hardware interrupts). All the rest can be used for software interrupts and hardware interrupts.

IVN	Default processor IVT	Remapped by BIOS	Remapped by an OS (Honoring the default IVT)
0x00h	Divide by zero	Divide by zero	
0x01h	Single step trace	Single step trace	
0x02h	NMI, power failure	NMI, power failure	NMI
0x03h	Breakpoint	Breakpoint	
0x04h	Numeric overflow	Numeric overflow	
0x05h	Screen dump to printer	Screen dump to printer	
0x06h	Invalid instruction	Invalid instruction	
0x07h	No coprocessor	No coprocessor	
0x08h	Double fault	IRQ0: system timer	Internal interrupts
0x09h	Coprocessor segment overrun	IRQ1: keyboard	
0x0Ah	Invalid task state segment (TSS)	IRQ2: cascade	
0x0Bh	Segment not present	IRQ3: Serial COM2/4	
0x0Ch	Stack segment overrun	IRQ4: Serial COM1/3	
0x0Dh	General protection fault (GPF)	IRQ5: Sound card	
0x0Eh	Page fault	IRQ6: floppy disk ctr	
0x0Fh	Reserved	IRQ7: Parallel port	
0x10h	Coprocessor error	BIOS: video services	
0x11h	Alignment check (486+only)	BIOS: check	
0x12h	Machine check (Pentium+only)	BIOS: memory size	
0x13h		BIOS: disk services	
0x14h		BIOS: serial ports	
0x15h		BIOS: system services	
0x16h		BIOS: keyboard services	
0x17h	Reserved	BIOS: Printer services	Reserved
0x18h		BIOS: BASIC	
0x19h		BIOS: load OS	
0x1Ah			
....		BIOS: other services	
0x1Fh			
0x20h			
0x21h		DOS services	
....			Software interrupts
0x70h		IRQ8 : real-time clock	
0x71h		IRQ9 . PCI device	
0x72h		IRQ10: Reserved	
0x73h	User definable for	IRQ11: Reserved	
0x74h	Hardware interrupts or	IRQ12: PS/2	
0x75h	Software interrupts	IRQ13: 8087	OS Services
0x76h		IRQ14: Hard Disk 1	
0x77h		IRQ15: Hard Disk 2	
....			
0x80h			Master PIC
....			
0x88h		Software interrupts	Slave PIC
....			
0xFFh			Software interrupts

Figure 4.15
x86 interrupt vector table.

4.7.1 Hardware Interrupts

The 8259A PIC is normally used on x86 systems to manage hardware interrupts. Each PIC can support eight IRQ lines. It is very common to cascade a master PIC with up to eight slave PICs to support 64 IRQ lines. In Figure 4.14, two 8259A PIC chips are used to support 15 interrupt sources.

An 8259A PIC has an interrupt request register (IRR) and an in-service register. The IRR and the in-service register are internal registers that are not directly accessible. The IRR is used to specify which interrupts are pending acknowledgment. The in-service register is used to indicate which interrupts have already been acknowledged and are being serviced by the processor.

Each PIC also has an interrupt command/status register (ICR) and an interrupt mask/data register (IMR). They are accessible to the processor. The ICR and IMR of the master PIC are respectively mapped at ports 0x20 and 0x21 in the I/O address space, while the ICR and IMR of the slave PIC are, respectively, mapped at 0xA0 and 0xA1.

The ICR consists of a command register and a status register, which are two different registers sharing the same port number. The command register is write only, while the status register is read only. The PIC determines what register to access depending on whether the write or read line is asserted. The command register on a PIC is used to (a) initialize the PIC and (b) clear the in-service register upon receipt of an EOI command.

The IMR consists of a mask register and a data register, which share the same port number. The mask register is obviously used to disable/enable interrupt sources. The data register is used in the initialization phase to (a) configure PIC cascading and (b) set up a base interrupt vector number for IRQ0.

By manipulating the PIC's data register, the default x86 interrupt vector table can be remapped to accommodate hardware interrupts. In remapping, some operating systems may override entries in the default x86 interrupt vector table. For example, as shown in the middle of Figure 4.15, the BIOS (and DOS) on a PC traditionally maps the master 8259A's IRQ0 to interrupt vector number 0x08 and the slave 8259A's IRQ8 to interrupt vector number 0x70.

Some operating systems have chosen to honor the processor's default vectors by remapping at least the master PIC to an unused vector as its base offset. An example is given on the right in Figure 4.15. This mapping is adopted in Figure 4.14, where the base interrupt vector numbers set up for the master PIC and the slave PIC are 0x80 and 0x88, respectively.

Once the base vector number for IRQ0 has been set, all the other IRQ lines get their vector numbers by adding the offset. For example, the vector number for the keyboard in Figure 4.14 is 0x81 (it is 0x09 in DOS for the reason mentioned above).

The interrupting process shown in Figure 4.14 is explained below:

1. A user presses a key on the keyboard, which drives an IRQ on the IRQ1 line of the 8259 master PIC.

2. The PIC detects the IRQ on IRQ1. It changes the IRR by setting the bit corresponding to IRQ1. The PIC then examines the IMR to see if the interrupt source is disabled. If not, the PIC then determines if there is any higher-priority interrupt waiting to be serviced. If there is, this new IRQ has to wait until the higher-priority interrupt is serviced. If there is no higher-priority interrupt waiting to be serviced, the PIC stores the vector number 0x81 (the base 0x80 plus one offset) in an internal register, and asserts the INTR line to inform the processor.

3. While the instruction *j* of the current program is being executed, the processor samples INT and detects an asserted line. After the execution of the instruction *j* has been completed, the current program is suspended. At this point, the value contained by CS:IP is the location of the next instruction of the current program.

4. The processor examines the interrupt flag within the flags register. If the interrupt flag is set, the processor acknowledges the IRQ by asserting the INTA line to the PIC, expecting an interrupt vector number from the PIC.

5. The PIC drives the interrupt vector number 0x81 to the system bus. It also sets the corresponding bit inside the in-service register, indicating that interrupt source 1 is currently being serviced.

6. The PIC then deasserts the INTR line (so that a new IRQ can be asserted).

7. The value of the flags register is pushed onto the interrupt stack. The values of CS and IP are also pushed onto the interrupt stack, so that the original program can be resumed later.

8. To protect context switching, the "interrupt enable" flag of the flags register is cleared to disable further interrupts. The processor comes to the keyboard interrupt vector, which has 4 bytes, located at $4 \times 0x81 = 0x0204$.

9. The keyboard interrupt vector contains the address of the keyboard ISR. By loading IP with the 16-bit data at 0x0204 and loading CS with the 16-bit data at 0x0206, the processor jumps to the start location of the keyboard ISR.

   9.1. Inside the ISR, the commonly used registers are first saved onto the interrupt stack. At this point, the processor has switched its context from the last task to the current ISR. The "interrupt enable" flag of the flags register is set to enable further interrupts, if interrupt nesting is desired.

   9.2. The portion of code pertinent to the requesting device is executed. In this case, the code of the key being pressed is stored in a memory area called the keyboard buffer.

9.3. At the end of the ISR, interrupts are again disabled for context switching. The ISR first sends an EOI command to the master PIC's command register. Upon receiving this command, the PIC can clear the appropriate bit in the in-service register, getting ready for new interrupts. The ISR then performs an instruction to pop up the top frame of the interrupt stack.

10. The context of the original task is restored. Especially, CS:IP now refers to the next instruction of the original program.

11. The processor is ready to resume the execution of the original program.

4.7.2 Put It All Together

Figure 4.16 summarizes the process of all three types of interrupts. They differ in how the interrupt vector number is obtained:

- For a hardware interrupt, the vector number is typically read from the PIC or the hardware device itself.
- For a nonmaskable hardware interrupt, the vector number is fixed to 0x02.

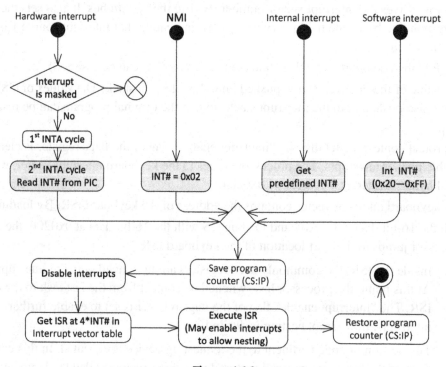

Figure 4.16

x86 interrupting process.

- For an internal interrupt (exception), depending on the type of the exception that occurred, the vector number is also fixed. In the case of an error, an x86 processor may raise an exception prior to completing the current instruction (which is re-executed upon exception return), or may raise an exception after completing the current instruction (which is not re-executed upon exception return). For serious errors, the processor might abort the misbehaving program.
- For a software interrupt, the vector number is simply part of the instruction.

Once the vector number has been determined, the rest of process is the same for all: the return address is saved, interrupts are disabled during the context switching, then the corresponding ISR is executed in the interrupt context (or kernel mode), and the original program execution is resumed upon the completion of the ISR.

4.8 Case Study: ARM Processor

Now, let us look at the interrupt vector table of a 32-bit ARM processor [12, 16]. As shown in the top-right corner in Figure 4.17, the vector table starts at 0x00000000, the very beginning of the memory space, and ends at 0x0000001F. Quite often, the read-only memory (ROM) is located at 0x00000000. Since static random-access memory (SRAM) runs much faster than ROM, the system initialization code typically remaps SRAM to the location 0x00000000. Prior to the remapping, the vector table has to be copied to SRAM at its default address. This also allows interrupt vectors to be dynamically updated as requirements change during program execution.

The interrupt vector table has eight entries. Each vector has 4 bytes, containing a branching instruction in one of the following forms:

- B adr: Upon encountering a B instruction, the ARM processor will jump immediately to the address given by adr, and will resume execution from there. The adr in the branch instruction is an offset from the current value of the program counter (PC) register.
- MOV PC, #immediate: By loading the value #immediate into the PC register, the ARM processor will jump immediately to the address given by #immediate.
- LDR PC, [PC, #offset]: Like PC=PC + #offset. The processor will jump immediately to the address given by PC + #offset.
- LDR PC, [PC, #-0xF20]: Like PC=PC - 0xF20. Suppose the instruction LDR PC, [PC, #-0xF20] is stored at address 0x00000018, which is the vector address of hardware interrupts. Note that the value of the PC register always refers to the address of the instruction being fetched by the processor. Owing to the ARM pipelining mechanism, while the processor is executing an instruction j, it is also fetching the instruction $j + 2$—the second instruction after j. Since each ARM instruction has 4 bytes, while the processor is executing the instruction LDR PC, [PC, #-0xF20] at address 0x00000018,

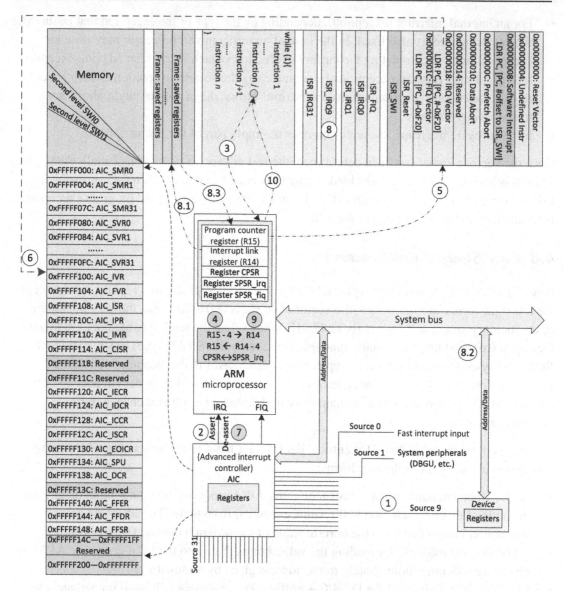

Figure 4.17
Vectored interrupting in ARM processors.

the program counter value is actually $0x00000018 + 2 \times 4 = 0x00000020$. Then, by executing the instruction LDR PC, [PC, #-0xF20], the new program counter value becomes PC - 0xF20 = 0x20 − 0x F20 = 0xFFFFF100 (wrapped around at 0x0), which is the address of the interrupt vector register to be discussed below.

In the interrupt vector table, there are three vectors for internal interrupts (undefined instruction exception, instruction fetch memory abort, and data access memory abort), one vector for software interrupts (at 0x0008), one reset vector (at 0x0000), and two vectors for hardware interrupts (interrupt vector at 0x0018 and fast interrupt vector at 0x001C). The vector entry at 0x0014 is reserved.

4.8.1 Hardware Interrupts

An AIC is used to coordinate hardware interrupts.

ARM processors use memory-mapped I/O. As shown on the left in Figure 4.17, all the registers on the AIC are mapped into the memory space starting from 0xFFFFF000 and ending at 0xFFFFF1FF. Table 4.1 lists all the registers used for interrupt processing, including those registers on the AIC (with the prefix "AIC"). For example, the interrupt vector register AIC_IVR is at 0xFFFFF100, and the fast interrupt vector register AIC_FVR is at 0xFFFFF104.

AIC supports 32 interrupt sources. The ISR address corresponding to the ith ($0 \leq i \leq 31$) interrupt source is stored in the register AIC_SVRi. The registers AIC_IVR and AIC_FVR have a special property. When the processor reads AIC_FVR, the value of AIC_SVR0 corresponding to the fast interrupt source is returned. When the processor reads AIC_IVR, the value of AIC_SVRi ($1 \leq i \leq 31$), where i is the current active interrupt source, is returned. This feature allows the processor to branch in one single instruction to the ISR corresponding to the current interrupt. More specifically, when the processor executes the instruction LDR PC, [PC, #-0xF20], the program counter is loaded with the value read in AIC_IVR (or AIC_FVR), which leads the processor to the corresponding ISR.

The process of ARM interrupting as illustrated in Figure 4.17 is explained below:

1. A device wired on the ninth line (source 9) of the AIC drives the line to raise an IRQ.
2. The AIC detects the IRQ from source 9. It asserts the $\overline{IRQ}$ line to inform the processor.
3. While the processor is about to execute the instruction j of the current program, it samples $\overline{IRQ}$ and detects an asserted line. The execution of the instruction j is abandoned. Owing to pipelining, the value of the program counter at this point is the location of the instruction $j + 2$ (which is 8 bytes ahead of the instruction j because each instruction has 4 bytes).
4. To prepare for return, the value of the program counter (register R15) is saved in the link register R14, which is further decremented by 4 by the ARM core. Consequently, the value of R14 is the location of the instruction $j + 1$. The current program status register (CPSR) is also saved to the banked register SPSR_irq.
5. The CPSR is set up such that (a) the processor enters into the interrupt mode, and (b) further hardware interrupts are disabled (to protect context switching). The program counter is loaded with 0x18, the vector address for hardware interrupts.

Table 4.1 Registers used for interrupt processing

Register	Full Name
SP	Stack pointer register (R13)[a]
LR	Link register (R14)[b]
PC register	Program counter register (R15)
CPSR	Current program status register[c]
SPSR	Saved program status register[d]
AIC_SMR	Source mode register (one for each of the 32 sources)
AIC_SVR	Source vector register (one for each of the 32 sources)[e]
AIC_IVR	Interrupt vector register[f]
AIC_FVR	FIQ interrupt vector register[g]
AIC_ISR	Interrupt status register
AIC_IPR	Interrupt pending register
AIC_IMR	Interrupt mask register
AIC_CISR	Core interrupt status register
AIC_IECR	Interrupt enable command register
AIC_IDCR	Interrupt disable command register
AIC_ICCR	Interrupt clear command register
AIC_ISCR	Interrupt set command register
AIC_EOICR	End of interrupt command register
AIC_SPU	Spurious interrupt vector register
AIC_DCR	Debug control register
AIC_FFER	Fast forcing enable register
AIC_FFDR	Fast forcing disable register
AIC_FFSR	Fast forcing status register

[a]R13 is a banked register that can be switched in to support IRQ, FIQ, supervisor, abort, and undefined mode processing, referred to as R13_irq, R13_fiq, R13_svc, R13_abt, and R13_und, respectively.

[b]R14 is a banked register that can be switched in to support IRQ, FIQ, supervisor, abort, and undefined mode processing, referred to as R14_irq, R14_fiq, R14_svc, R14_abt, and R14_und, respectively. R14 and its banked variants are used to hold the return values of R15 when interrupts and exceptions arise, or when branch and link instructions are executed.

[c]The I and F bits in CPSR are the interrupt disable bits. The I bit disables IRQ interrupts when it is set and the F bit disables FIQ interrupts when it is set. The mode bits in CPSR determine the mode in which the processor operates. CPSR may also be changed as a result of arithmetic and logical operations in the processor.

[d]SPSR is a banked register that can be switched in to support IRQ, FIQ, supervisor, abort, and undefined mode processing, referred to as SPSR_irq, SPSR_fiq, SPSR_svc, SPSR_abt, and SPSR_und, respectively. SPSR is loaded with the CPSR when an exception occurs. There is one SPSR for each privileged mode.

[e]The registers AIC_SVR0 to AIC_SVR31 each store the ISR address of the corresponding interrupt source.

[f]When the processor reads (e.g., executes a load program counter instruction) AIC_IVR, it virtually reads from the register (in the range from AIC_SVR1 to AIC_SVR31) that corresponds to the current interrupt. This allows the processor to branch to the ISR corresponding to the current interrupt efficiently in a single instruction.

[g]When the processor reads AIC_FVR, it virtually reads from AIC_SVR0. This allows the processor to branch to the ISR corresponding to FIQ in a single instruction.

6. The vector at 0x18 is a branching instruction LDR PC, [PC, #-0xF20]. When it is executed, the program counter is loaded with the value read in register AIC_IVR, which is the address of the ISR corresponding to interrupt source 9.

7. The AIC deasserts the $\overline{\text{IRQ}}$ line, and clears the interrupt source. The AIC also pushes the current interrupt priority level and interrupt number onto its internal stack.

8. The processor starts to execute the ISR.

   8.1. Inside the ISR, the commonly used registers are first saved onto the interrupt stack. At this point, the processor has switched its context from the last task to the current ISR. Further interrupts can now be enabled (by clearing the I bit in the CPSR), if interrupt nesting is desired.

   8.2. The portion of code pertinent to the requesting device is executed, which typically entails the access of the ports (registers) on the device. During this phase, an interrupt with priority higher than that of the current level will restart the sequence from step 1.

   8.3. At the end of the ISR, interrupts are disabled during context switching. The EOI command register (AIC_EOICR) is written to indicate to the AIC that the current interrupt has finished. This causes the AIC to pop the current level from its stack, restoring the previous level (nested interrupt) if one exists on the stack. Finally, the top frame of the interrupt stack is popped up and the context of the original task is restored.

9. The CPSR is restored. The link register R14 is decremented by 4 before it is loaded to the program counter. Consequently, the value of the program counter is the location of the instruction *j*, the instruction that was interrupted.

10. The processor is ready to resume the execution of the original program.

4.8.2 Put It All Together

Figure 4.18 summarizes the process of all three types of interrupts:

- For a hardware interrupt, the adjusted value of the PC register (i.e., PC-4) is first saved to the banked link register (R14). After interrupts have been disabled, the program counter is set to 0x01C if it is a fast IRQ, or to 0x018 if it is a normal IRQ. Upon completion of the ISR, the program counter is set to (R14-4), which allows the processor to return to the interrupted instruction.

- For a nonmaskable hardware interrupt (RESET), after interrupts have been disabled, the program counter is set to 0x0. The corresponding ISR allows the processor to initialize the system (say, caches, interrupt stacks, and external interrupt sources). This ISR has to be carefully implemented to avoid any interrupts taking place during its execution. In particular, this ISR may not use swi instructions, undefined instructions, or memory access.

- For an internal interrupt:
  - If it is an "undefined instruction" exception, the program counter is decremented by 4 before it is saved to the banked link register (R14). After interrupts have been

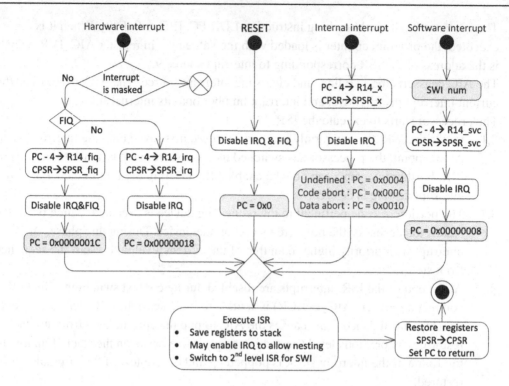

Figure 4.18
ARM interrupting process.

disabled, the program counter is set to 0x004. Upon completion of the ISR, the program counter is set to R14, which allows the processor to return to the immediate next instruction after the undefined instruction.

- If it is a "data abort" exception, the program counter is decremented by 4 before it is saved to the banked link register (R14). After interrupts have been disabled, the program counter is set to 0x010. Upon completion of the ISR, the program counter is set to R14-4, which allows the processor to return to the interrupted instruction.

- If it is a "prefetch abort" exception, the program counter is decremented by 4 before it is saved to the banked link register (R14). After interrupts have been disabled, the program counter is set to 0x00C. Upon completion of the ISR, the program counter is set to R14-4, which allows the processor to return to the interrupted instruction.

- For a software interrupt, the program counter is decremented by 4 before it is saved to the banked link register (R14). After interrupts have been disabled, the program counter is set to 0x008. Upon completion of the ISR, the program counter is set to R14, which allows the processor to return to the immediate next instruction after the swi instruction.

As usual, if interrupt nesting is desired, interrupts can be re-enabled in an ISR after context switching. Also, for a software interrupt, the ISR can trigger other ISRs by switching on the value of the function number specified in the swi instruction.

When multiple interrupts/exceptions arise at the same time, they are handled by the processor in the following order: reset, data abort, FIQ, IRQ (eight levels of priority), prefetch abort, undefined instruction, and software interrupt. Specifically, hardware reset has the highest priority, while software interrupts have the lowest priority.

Problems

4.1   Why are hardware interrupts called asynchronous events? Why are software interrupts called synchronous events?

4.2   Compared with polling, what are the advantages offered by interrupts?

4.3   Describe the process of nonvectored interrupting.

4.4   Why is it important to push the current context onto the interrupt stack?

4.5   How can one disable interrupts from a certain interrupt source?

4.6   What is interrupt nesting?

4.7   How does a level-sensitive device raise its IRQs? How does an edge-triggered device raise its IRQs?

4.8   How does a processor get the vector number of the current interrupt?

4.9   What is a software interrupt? Why is the software interrupt mechanism important to an operating system?

4.10 Why is it important to move the lengthy processing code from an ISR to a separate server task?

4.11 What is interrupt chaining? As a laboratory exercise, use operating system calls to attach a user-defined ISR to the timer interrupt.

4.12 What is interrupt cascading? How is it used by DOS to offer its services?

4.13 Describe how the shared data issue is resolved by interrupt disabling and double buffering.

4.14 What factors may contribute to interrupt latency? What factors may contribute to scheduling delay?

4.15 How many vectors are supported by the interrupt table on an x86 system?

4.16 Where does the BIOS map the master PIC and the slave PIC to the interrupt vector table? Can this map be changed by an operating system?

4.17 Explain the sequence of events after a person presses a key on the keyboard of an x86 system.

4.18 Consider the AIC in Figure 4.17; does the ARM processor use memory-mapped I/O or does it use a separate I/O address space? Why?

4.19 Explain why the instruction LDR PC, [PC, #-0xF20] stored at 0x018 allows the processor to read the interrupt vector register.

4.20 How is interrupt cascading exploited by ARM processors to support software interrupts?

Embedded System Boot Process

Contents

Chaos reigns within. Reflect, repent, and reboot. Order shall return.

Suzie Wagner

5.1 System Bootloader

Given that an embedded application—an object in Executable and Linking Format (ELF)—is ready, our focus now is on how to run the application on its target platform. This job is performed by a special piece of software called the system bootloader or bootstrap code,[1] which is typically programmed in the nonvolatile memory (NVM) at a location where the reset vector points.

The bootloader's main job is to initialize some hardware components to prepare for loading an ELF object. In particular, it

---

[1] It is called a bootloader in reference to a legend about Baron Müanchhausen, who made an impossible task possible—to pull himself up by his bootstraps.

Real-Time Embedded Systems. http://dx.doi.org/10.1016/B978-0-12-801507-0.00005-5

- prepares a "loading path" for the ELF object by initializing peripheral devices such as the system timers and the serial/network interfaces;
- prepares a "storage room" for the ELF object by initializing the memory controller and memory chips;
- prepares a "running environment" by initializing the interrupt controller and installing default interrupt service routines.

Some specialized bootloaders are primarily used in the development phase. For instance, a monitor is a bootloader that further allows a developer to debug the target system at run time. A monitor typically has a well-defined command interface. Via a terminal connected to the debug serial port of the target board, a developer can issue commands to debug a running application, including activities such as

- resetting the embedded system;
- accessing system registers or memory locations;
- setting/clearing breakpoints; and
- stepping through instructions.

Some specialized bootloaders are primarily used in released products. For instance, some bootloaders, upon system reset, are directed to load software (user application) from an attached electrically erasable programmable read-only memory (EEPROM) accessible through an Inter-Integrated Circuit (I2C) interface or a serial peripheral interface (SPI). Such an I2C/SPI bootloader is designed to allow easy upgrade of software in the field: changing the firmware is as simple as replacing a memory chip in a socket. As another example, the controller area network (CAN) bootloader, extensively used in the automotive industry, allows in-field upgrades of software/firmware over the CAN bus to fix bugs on cars that have been sold.

5.2 System Boot Process

Figure 5.1 illustrates the system boot (or bootstrapping) process. Upon reset of an embedded system, the microprocessor, directed by the instruction at its reset vector, starts to run the bootloader.

After minimum hardware initialization, the system bootloader comes to a decision point, where it may follow a specific scanning sequence to locate an ELF object.

5.2.1 Load Embedded Software

A target board, especially an evaluation board used in the development phase, may have jumper pins that allow a developer to use jumpers to control which memory chip the bootloader uses to detect and download the embedded software upon system reset. During the

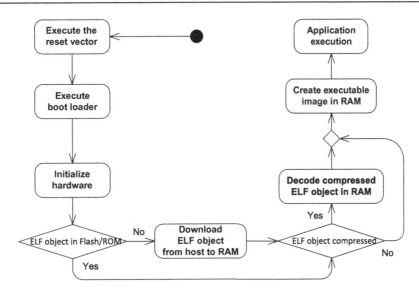

Figure 5.1
Embedded system boot process.

development phase, in order to avoid repeatedly programming the target board, it is preferable to load an ELF object from the host platform via a serial or network connection. Once the embedded software is finalized and ready for performance testing or delivery, the bootloader can be set up to load an ELF object directly from the NVM on board.

5.2.1.1 From host node

In this case, the system is configured such that the bootloader expects to receive an ELF object from a host node.

The bootloader first needs to initialize a hardware interface prior to "listening" for incoming bytes via that interface. Commonly used interfaces include a serial interface and a network interface:

- *Serial interface:* The bootloader cooperates with a utility program running on the host platform to make an agreement on communication parameters such as baud rate and packet size. Afterward, commands and binary data (i.e., the ELF object) can be transferred over the serial connection.
- *Network interface:* The bootloader needs to work with a utility program, such as a File Transfer Protocol server or a Trivial File Transfer Protocol server, running on the host platform. Commands and binary data can be transferred over the network connection after communication parameters have been set up.

If the bootloader detects data arriving, the data are saved at a temporary location in random access memory (RAM). The bootloader is responsible for checking the integrity of the transferred object once all the data have been received.

If no data come through the intended interface in a given amount of time, the bootloader will time-out, then execute a jump instruction to NVM and start executing instructions there, if applicable.

5.2.1.2 From NVM on board

In this case, the bootloader is instructed to get the embedded software from some specific NVM on the target board itself. Hence, the assumption is that the software has been directly "burned" into the NVM. If so, the boot process simply moves to the next step.

5.2.2 Prepare Embedded Software for Execution

5.2.2.1 Decode compressed ELF object

An ELF object, regardless of whether it is programmed in NVM or loaded temporarily in RAM, might be in a compressed form to reduce demand for storage memory or transmission time.

A compressed ELF object is not directly executable until it has been decompressed. Decompression can be performed by the system bootloader, but often it is performed by a secondary bootloader to which the system bootloader transfers the execution control. The uncompressed ELF object is typically relocated to another location in RAM, and the memory space holding the compressed object can be recycled afterward.

5.2.2.2 Create executable image

At this point, the (secondary) bootloader can create an executable image for the ELF object—mapping file segments into memory segments. In this way, the loadable segments of the ELF object are relocated to their respective run addresses:

- The .text section (segment) is copied to its run address specified by vaddr (see Section 2.4.2.2).
- The .data section (segment), which contains initialized data, is copied to its run address.
- The .rodata section (segment), which contains constant data, is copied to its run address.
- The .bss section (segment) contains uninitialized data; its filesz is zero. The loader reserves memory space for the .bss section according to the memory requirement specified by vaddr and memsz.

- In addition, the bootloader also reserves a stack space in RAM for the newly created "image" (recall that the stack space is used for holding local values in function calls). The processor's stack register is set to point to the beginning of this stack.

Lastly, the bootloader executes a processor-specific instruction to jump to the beginning of the .text section of the newly created image. At this point, the boot process completes. The processor continues to execute the application code in the .text section until it runs to completion or the system is powered off.

5.3 Case Study: AT91SAM9G45 Boot Process

The AT91SAM9G45 evaluation board embeds an internal read-only memory (ROM), which is mapped at address 0x0040 0000 and contains the bootloader and the program SAM-BA [13, 16]. The bootloader integrates different boot programs that interface with different memory components on board. The embedded bootloader is used upon reset if the boot mode select pin is detected at high level (logical 1); otherwise, the memory connected on the Chip Select 0 of the External Bus Interface is used.

When the embedded bootloader is used, it will perform the following actions upon system reset:

1. It performs minimum hardware initialization—say, starts the on-chip RC oscillator and enables the 32,768 Hz slow clock oscillator.
2. It attempts to boot from NVM devices. The sequence below is used to download an application from an external storage medium into internal static RAM (SRAM):
   a. The *SD Card Boot program* is executed first. It looks for a boot.bin file in the root directory of a FAT12/16/32 formatted SD card. If such a file is found, code is downloaded into the internal SRAM. Go to step 4.
   b. If the SD card is not well formatted or if a boot.bin file is not found, the *NAND Flash Boot program* is then executed to search for a valid application in the NAND flash memory. If a valid application (described by a sequence of seven valid ARM exception vectors[2]) is found, this application is downloaded into the internal SRAM. Go to step 4.
   c. If no valid application is found, the *Data-Flash Boot program* is then executed to search for a valid application in a data-flash memory connected to the SPI. If a valid application is found, it is downloaded into the internal SRAM. Go to step 4.

---

[2] This is somehow relevant to a technique called Secure Boot: When the system starts, the firmware checks the signature of each piece of boot software, including firmware drivers and the operating system (OS); the system boots only if the signatures are good.

3. If no valid application is found in NVM devices, *SAM-BA Boot* is then executed. It initializes the USB High Speed device port and the debug unit serial port. It then starts automatic baud rate detection, waiting for transactions either on the USB High Speed device port or on the debug unit serial port.
4. It remaps memory, which lays out the internal SRAM bank to 0x0.
5. It jumps to the first instruction of the application in SRAM and run it.

In practice, a developer may want to use the option given in step 3 to program embedded software to the NAND flash memory device. Once an ELF object is on the board, the option given in step 2b could be used to load and run the application.

5.4 Load ELF Objects Embedded Within an OS Image

Embedded software can be built upon an OS. In such a case, a user program (i.e., a binary ELF object) may have one INTERP program header element that specifies the location of a program interpreter (see Section 2.4.2.2). A program interpreter is able to initialize itself without relying on other programs to relocate its process image. When exec, a system call for creating a new process, is invoked on the user program, the OS retrieves the path name of the prespecified program interpreter and creates an *initial process image* from the interpreter's file segments. In other words, instead of directly using the user program's segments, the OS composes a process image for the interpreter first; this initial image is then expanded to form a complete image for the user program.

As we explained in Section 2.4.1.3, there are two types of program interpreters: a program loader[3] and a dynamic linker. A program loader simply loads a program and creates a process image for it. A dynamic linker is more than a program loader; it can also resolve symbolic references dynamically by loading and binding external shared libraries. We now focus on the roles played by a dynamic linker.

Once execution control has been obtained from the OS, a dynamic linker is responsible for creating a process image for the user program, which entails the following actions:

1. *Program loading:* It must expand the initial process image by adding the user program's loadable segments into the process image.
2. *Dynamic binding:* After the program segments have been loaded into memory, it must complete the process image by resolving symbolic references. A dynamic linker may be configured to operate in one of two modes:

---

[3] Note that a program loader is different from a system bootloader, the former is an OS utility, while the latter is well beyond the scope of an OS.

- *Immediate binding:* It first evaluates the procedure linkage table entries of the user program; it then adds segments of the referenced shared objects to the process image. On the one hand, the segments of a shared object may be relocated to virtual memory addresses that are different from the addresses recorded in the object file's program header table. This means that the dynamic linker needs to calculate and update the affected address references (say, the global offset table) to ensure that the absolute addresses are available during execution. On the other hand, a shared object itself may reference other shared objects. This means that the binding/linking process is recursive and it stops until after all the symbolic references have been resolved. Hence, by repeatedly connecting referenced shared objects and their dependencies, the dynamic linker builds a complete process image.
- *Lazy binding:* In this mode, the dynamic linker evaluates procedure linkage table entries lazily: segments of a referenced shared object are loaded only when necessary. This may slightly affect the system's run-time performance, but it can avoid the symbol resolution and relocation overhead for functions that are not called.

3. *Control transferring:* Once a complete process image has been formed, the linker will transfer the execution control to the newly created process. If applicable, before transferring its control, the linker will also close the file descriptor that was used to read the user program. From a user's perspective, it seems like the user program has received control directly via the invocation of exec.

5.5 Case Study: Boot Process of QNX-based Embedded Systems

Figure 5.2 shows the boot process of an example QNX-based embedded system [6]: a general process illustrating the transfer of execution control is shown on the left, and the transfer of objects among storage media is shown on the right. The boot procedure is as follows:

1. When the system is powered on, the processor starts execution from the reset vector, which typically is a jump instruction to a *first-level bootloader* (say, a monitor or a BIOS bootloader) available on the target board.
2. The role of the first-level bootloader is to perform minimum device initialization (say, set up the processor, oscillator, and serial ports), and try to detect and boot from external NVMs.
3. If the NVM (say, NAND flash memory) contains valid code, it is copied from the external NVM to internal RAM. In Figure 5.2, this code is an initial program loader (IPL)—a *secondary bootloader*. The control is then transferred to the IPL. Sometimes a system may not have a BIOS or ROM monitor program (say, in a cost-reduced embedded system). In such a case, the IPL must be located at the reset vector and it needs to play the role of the first-level bootloader as well. This is called a "cold start" IPL.[4]

---

[4] In warm start, the IPL gets control immediately from the BIOS or ROM monitor.

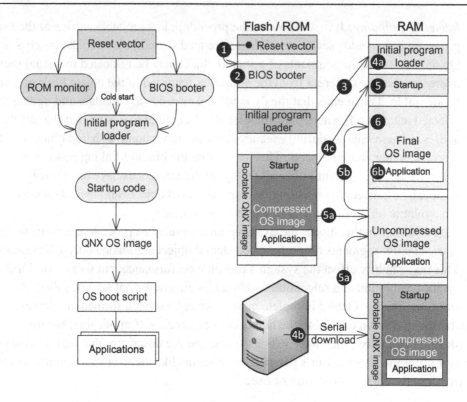

Figure 5.2
The boot process of QNX-based embedded systems.

4. The IPL performs the following action steps to transfer a QNX image to linearly addressable memory:
   a. Hardware initialization: create an environment that allows the startup program to run (e.g., configure clocks, memory controller, memory chips, and set up a stack).
   b. Image download: skip this step if a bootable QNX image is available in ROM/flash memory; otherwise, it tries to receive data via the serial interface from a host and copy it to a temporary location in RAM, and then scan the downloaded object to validate its integrity.
   c. Image setup: copy the startup code from its original location (ROM/flash memory or RAM) to a final location in RAM, then transfer control to the startup code.
5. The startup code performs the following action steps:
   a. Image decompression: skip this step if the QNX image is uncompressed; otherwise, it tries to decompress the QNX image (in ROM or RAM) into a new location in RAM.
   b. Image finalization: copy the QNX image to its final location in RAM.
   c. Hardware initialization: create an environment that allows the OS kernel to run (e.g., set up the debug interface and interrupt controller interface), then transfer control to the OS.

6. The OS kernel may perform the following action steps:
    a. Software initialization: initialize the OS kernel and OS services (file systems, interrupts, memory management, etc.).
    b. User application execution.

Problems

5.1 What is a system bootloader? What is a monitor? What is their major difference?

5.2 Given that a target board has a serial port reserved for loading executable objects from a host platform, explain the process of loading a binary object via the serial interface.

5.3 Given that an ELF file object is in RAM of a target board, explain how an executable image is created from the ELF object.

5.4 What is immediate binding? What is lazy binding?

5.5 In the case study given in Section 5.5, the IPL is called a secondary bootloader. Describe what the IPL can do.

5.6 In the case study given in Section 5.5, the startup code is the first portion of a bootable QNX image. Explain the roles played by the startup code.

PART II

Real-Time System Modeling

CHAPTER 6

Fundamental UML Structural Modeling

Contents

You can define a complex thing with a fairly simple model.

Gordon Hammond

Real-Time Embedded Systems. http://dx.doi.org/10.1016/B978-0-12-801507-0.00006-7

6.1 Unified Modeling Language

Unified Modeling Language (UML) is a general-purpose modeling language for software-intensive systems. UML 1.1, the first version officially adopted by the Object Management Group (OMG)[1] in November 1997, integrated James Rumbaugh's object-modeling technique, Grady Booch's component notation, Ivar Jacobson's use case notation, Archie Bowen's timing analysis, and David Harel's statecharts. UML 2.0, a major revision, was adopted by the OMG in 2005 [10], and UML 2.4.1, the current version, was formally released in August 2011 [10].

For thousands of years in human history, engineers, artists, and craftsmen have built various models to try out designs before building the real things [21]; examples include airplane scale models, blueprints, and pencil sketches. The benefit of such a model-first approach is manifold:

- *Models reduce complexity:* To deal with a system that is too complex to understand, human beings often use abstraction to omit nonessential details. By abstraction, those parts that are important for the goal concerned can be isolated from those unimportant details.

- *Models exhibit diverse views:* Separation of concerns has been a key principle in engineering disciplines. For instance, in software engineering, separation of concerns has been widely applied in system architectures, design patterns, algorithm designs, testing, and software evolution.[2] In general, more than one model can be built for an entity, each capturing one crucial aspect. This allows people to thoroughly study the entity from multiple perspectives (angles).

- *Models allow early correction of flaws:* A model, be it a physical model (i.e., scale model) or a computer model (i.e., software simulation), is much cheaper than building the real system itself. In addition, it is very convenient for people to repeatedly test and adjust models, and quickly correct design flaws or mistakes whenever detected, which could be very costly or impossible once the real system has been finished.

- *Models offer concrete contexts for communication:* A model is a mock-up that imitates certain desired features of the system to be developed. Thus, a model is a very important artifact that can be used to share ideas and concerns among members of a design team or between the designers and customers.

UML is an industry standard for modeling systems [18, 21, 37, 59]. It is a graphical language for specifying, visualizing, constructing, and documenting system artifacts, and is useful in a variety of engineering problems, from single-process, embedded systems and stand-alone user applications to concurrent, distributed systems.

---

[1] The OMG is a consortium which is focussed on modeling (programs, systems, and business processes) and model-based standards.

[2] Particularly, a new programming paradigm called aspect-oriented programming [44] offers language-level support for the separation of cross-cutting concerns.

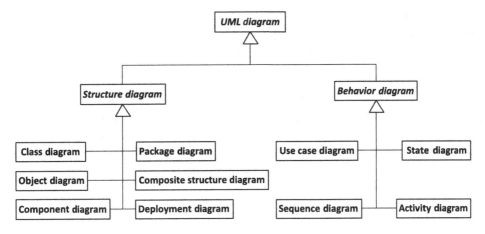

Figure 6.1
Commonly used UML diagrams.

As a modeling language, UML specifies a wide spectrum of notations for representing different aspects of a system. Some commonly used UML diagrams are given in Figure 6.1, which itself is an example of a UML class diagram.[3] UML diagrams represent two different views of a system model:

- *Static view*, which captures the *static structure* of the system using objects, attributes, operations, and relationships.
- *Dynamic view*, which captures the *dynamic behavior* of the system by visualizing collaborations among objects and internal changes of object states.

In this chapter we focus on UML class diagrams and package diagrams.

6.2 Class Diagram and Class Modeling

Prior to the rise of *object-oriented programming* (OOP), modular (procedural) programming had been the dominant paradigm for many years. It focussed on the use of linked modules (subroutines), which raised great concern in software maintainability and manageability as systems became increasingly complex.

In the 1960s, researchers started to investigate ways to maintain software quality and developed the OOP paradigm, where the notions of objects, classes, and inheritance were introduced. By the mid-1990s, OOP became the dominant programming method as OOP languages (such as Simula 67, Smalltalk, C++, and Visual Basic) became widely available.

---

[3] For the treatment of the UML standard [10, 11], sometimes I try my best to honor the original term definitions and descriptions, sometimes I use my own discretion to make adjustments in order to maintain a consistent context in this book.

OOP continues to be the mainstream programming paradigm and will be in the foreseeable future, in part owing to the ever-increasing importance of GUIs and event-driven programming, and the widely acceptance of Java and the .NET framework in industry.

The term *"object-oriented modeling"* (OOM) refers to a collection of modeling concepts used to model a problem domain as interacting objects. Although the modeling concepts in OOM can be directly supported in an OOP language, using OOM does not necessitate the use of OOP languages in implementation. Owing to their efficiency and accessibility to hardware, C and its variants are still the dominant programming languages for embedded systems, but at the same time, OOM is well adopted in the design of embedded systems.

The first OOM concept is the "object." The notion of the object is used to model either physical or conceptual things in the problem domain under consideration. Some have real-world counterparts (e.g., George Washington, Mount Rushmore), while others are purely conceptual entities (e.g., a specific computing thread). As far as modeling is concerned, customers, products, workers, invoices, and jobs are objects in a typical business system. Radar transmitter, receiver, antenna, and signal processor are objects in a radar control system. Reader accounts, borrowing records, and various kinds of publications are objects in an e-library system. A software engineer would also treat stacks, queues, lists, and interface items as objects.

An object has one or more attributes (aka. properties). For example, an e-banking system may manage many "bank account" objects, each of which has three attributes: the owner's social security number, account number, and balance.[4] The current values of an object's attributes are together called the object's state. The state of an object may change over time.

An object also has behavior, which is a list of operations that can be performed on the object. For example, two operations, deposit and withdraw, can be performed on a bank account object; the execution of these operations may change the state of the bank account object (say, balance increased by $100). Treating an object as a wrapper of both data and logic is called *encapsulation*.

Objects of the same kind have the same attributes and exhibit the same behavior, which is true at least in the same modeling context. For instance, the e-banking system mentioned above may have thousands of personal bank account objects. They have the same attributes and operations but they may differ in the attribute values (different owners and balance

---

[4] For complex objects such as an airplane, it is formidable, if not possible, to get a complete description of the object state. We typically restrict our consideration to only those attributes (properties) that are relevant to the modeling task at hand.

information).[5] The notion of *class* is used to capture the common features (attributes and behavior) of objects belonging to the same kind or category.

Software engineers have a lifetime job in working with abstract data types, such as queues, stacks, and trees, which are basic data structures bound together with appropriate manipulation operations. At a low level of abstraction, classes are merely abstract data types.

A class describes a set of objects with the same attributes (properties), behavior (operations), kinds of relationships, and semantics.[6] A class can be viewed as a pattern, template, or blueprint for a category of structurally identical objects (or instances). By grouping objects into classes, which is often called the *abstraction* process, we can generalize (factor out) common features (including attribute names and operations), store them once per class, and reuse them in all similar cases. For example, from the bank account objects we could abstract a class, and let us call it BankAccount.

Sometimes we need to distinguish class attributes and object (instance) attributes. *Object attributes* are owned by each individual object of the class; each object has its own copy of the attribute, and its value is not visible to other objects of the same class. So far, all the attributes of the BankAccount class are object attributes.

In contrast, a *class attribute* holds a value that is shared by all the objects of that class. In particular, a class attribute is owned by the class; whenever its value is changed by any one object, the change is reflected in all the other objects as well. Let us assume that as a designer of the banking system you want to add a new attribute called "counter" to the BankAccount class and use it to keep track of how many BankAccount objects have been created. The design rationale is that this counter information could be used as a seed in the generation of the next bank account number. This counter attribute is not related to any individual BankAccount object, but is related to the BankAccount class as a whole; it is thus a class attribute.

Inheritance is a core concept of object orientation. New classes can be defined by inheriting attributes and behavior from pre-existing classes called superclasses or parent classes. The resulting classes are known as subclasses or child classes. Conceptually, a superclass establishes a common part for its subclasses: a subclass object has all the attributes and

---

[5] Inherently, each and every object also has a unique identity. Twins are two distinct people although they are identical in many ways. Objects with the same attribute values are possible, but they should be distinguished from each other by their identities.

[6] Above and beyond the requirement of common attributes and behavior, the objects in a class also share a common semantics, the interpretation of which depends on the purpose of each model and application. For example, in a drawing system, although both square shapes and rectangular shapes have the same attributes (position, color, width, height) and operations, they are semantically different two-dimensional shapes (say, a square has a class invariant property where its width equals its height). Hence, they should be modeled as two different classes.

behavior declared in the superclass. This factoring of commonality is one way to improve reusability.

A subclass may polymorphically modify (redefine) an operation inherited from its superclass (this is called specialization), or may add new attributes and behavior (this is called extension). Subclassing should be avoided unless it is appropriate to apply specialization or extension to capture semantically related domain concepts. In our banking example, we can model a checking account and a savings account as subclasses of a bank account.

6.2.1 Class

A UML class diagram models the structure of a system by showing classes and their relationships. Class diagrams are useful both for conceptual modeling and for designing actual systems.

The graphical notation for a class is a box with three sections or compartments: the upper section holds the name of the class; the middle section contains the attributes of the class; and the bottom section gives the operations the class can take. A few example UML classes are given in Figure 6.2. The class name should be centered, with the first letter of the class name capitalized. Class attributes and operations are underlined (the employeeCounter in the Employee class is a class attribute). Depending on the purposes of modeling, the attribute section and/or the operation section can be suppressed.

In class diagrams, visibility defines whether an attribute or an operation of a class can be seen and used by other classes. UML defines the following visibility keywords and symbols:

- Public: A public element is visible to all that can access the contents of the enclosing (or owning) namespace. In the class diagrams hereinafter, the term "element" refers to an attribute or operation, and the enclosing namespace refers to a class.[7] A public element is proceeded by a "+" mark.

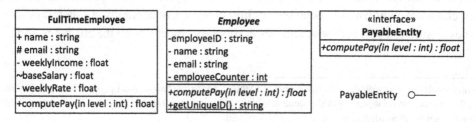

Figure 6.2
UML class notation.

---

[7] In the context of package diagrams (see Section 6.5), the term "element" can also refer to a class, and the enclosing namespace can refer to a package.

- Private: A private element defined in a class, proceeded by a "−" mark, is visible only inside the class itself.
- Protected: A protected element defined in a class, proceeded by a "#" mark, is visible in the class's subclasses.
- Package: A package feature (attribute or operation) defined in a class, proceeded by a "~" mark, is visible to all classes that are in the nearest enclosing package of the class that owns the feature. Outside the nearest enclosing package, an element with package visibility is not visible.

Information hiding is a modeling principle where those design decisions that are most likely to change are segregated, thus protecting other parts of the system from extensive modification if a design decision is changed. Specific to class modeling, information hiding is the ability to compel a client to access the object information through well-defined operations. Hence, it is good practice in OOP to define private attributes only, forcing potential clients to access object properties by getters and setters. In Figure 6.2, information hiding is applied to the Employee class, but not the FullTimeEmployee class.

A class may have special attributes (properties) called *derived* properties. The value of a derived property can be derived (computed) from other properties, and the derivation is typically specified by a constraint. A derived property is often specified to be read only. For example, a Rectangle class may have attributes height, width, and area, where area is a derived property because its value can be computed from the values of width and height (here the constraint is that area equals width$\times$ height). In class diagrams, a derived property is indicated by putting a slash in front of the property name.

So far, we know that the behavioral features of a class are given by operations. The term "operation" should be distinguished from "method." An operation (signature) is simply a specification or operational contract, describing the number, order, and types of arguments and the type of return value. A method, on the other hand, has a body of code that implements an operation.

When a class is degenerated such that it defines only a collection of public operations (i.e., no method is to be provided in implementation), it becomes a named *service contract* and is called an *interface*. Interfaces may be shown in two forms. One looks like a class except for the keyword «interface» preceding the name. The other form is a circle or ball labeled with the name of the interface. An interface called PayableEntity is given in Figure 6.2.

Sometimes the features defined in a class are coherent but incomplete for good reasons. For instance, it is decided to not provide a method for a certain operation because it is either awkward or impossible to implement the operation at this abstraction level (better to do it in subclasses, each of which offers a specialized context with more properties to consider). Such an "incomplete" class is called an *abstract* class, and the features declared by it can be reused by inheritance. The name of an abstract class is shown in italics. Employee in Figure 6.2 is an abstract class; the computePay operation is not implementable because there is no rate of pay information at this abstraction level.

6.2.1.1 OCL constraints on classes

Object Constraint Language (OCL) is a specification language mainly used to annotate the UML models of a system with various constraints [15].

OCL is a typed language. To be well formed, an OCL expression must conform to the type conformance rules of the language. OCL predefines a few basic types (Integer, Real, Boolean, String, UnlimitedNatural, and Collection types) and some operations on the basic types. For instance, the Boolean type has two values: true and false. The operations defined for the Boolean type include "or," "xor," "and," "not," and "implies." For example, p.implies(q) is a Boolean expression, which is evaluated to true if p is false, or if both p and q are true at the same time. In addition to the predefined types, all the class types defined in a UML model can be used in OCL expressions that are to be applied to the UML model.

Constraints in class diagrams are typically described as OCL Boolean expressions. A *class invariant* is a Boolean expression that must be true for all objects of that class at any time. An operation can have a *precondition*, which is a Boolean expression that should evaluate to true whenever the operation starts executing. An operation can also have a *postcondition*, which is a Boolean expression that should be true at the moment when the execution of the operation has just finished.

Let us look at a few OCL expressions defined for the class *FullTimeEmployee* given in Figure 6.2.

```
context e: FullTimeEmployee inv baseIncome:
         e.baseSalary > 300.
```

This example shows the typical structure of a constraint specification:

- The constraint is specified by the OCL expression e.baseSalary > 300, where *e* is the object on which the Boolean expression "baseSalary > 300" is to be evaluated.
- The object *e* has a type, which is declared after the "context" keyword. It says that the class *FullTimeEmployee* is the context for evaluating the OCL expression.
- The label "inv" declares that the specified constraint is a class invariant, which can be referenced by the name "baseIncome."
- The value of an object property is specified in an OCL expression by placing the dot operator between the object and the property name.
- In sum, this example specifies a class invariant, declaring that any full-time employee must have a weekly payment of more than $300.

The constraint name is optional. Also, if no contextual object is explicitly declared, the reserved word "self" can always be used as an object of the contextual class. The class invariant below is equivalent to the previous one:

```
context FullTimeEmployee inv:
        self.baseSalary > 300.
```

In OCL, you can use the "::" operator to invoke class operations. See the example below:

```
context Employee::employeeID : String init:
    Employee::getUniqueID().
```

Here, the label "init" indicates that the constraint body is an expression representing an initial value. It says that a client can invoke the *Employee* class's static operation "getUniqueID" to generate a unique ID, which can then be used in the initialization of the *employeeID* attribute.

OCL has a pre-defined operation called *allInstances*. The following constraint specifies that all full-time employees have unique e-mail addresses:

```
context FullTimeEmployee inv:
    FullTimeEmployee.allInstances()->forAll(e1, e2 |
        e1 <> e2 implies e1.email <> e2.email)
```

where the *allInstances* operation returns a collection of *FullTimeEmployee* objects. To access a property of a collection, the infix operator "->" is placed between the collection and the name of the property. In this example, the function "forAll" is invoked. See Table 6.1 for a list of operations defined for collections.

OCL expressions can also be used to describe preconditions and postconditions on operations. A postcondition expression can refer to the attribute values upon completion of the operation, and the values at the moment when the object just starts the operation (postfix the property name with the keyword "@pre" to distinguish the pre-value from the value upon completion). An example is given below:

```
context FullTimeEmployee:: computePay(level: integer): float
pre priorPay: (level<>0) and (self.weeklyRate > 0)
post: result = baseSalary + 40*weeklyRate/level
```

where

(a) The context is the operation signature confined by the class name.
(b) The precondition has a name "priorPay," whereas the postcondition is anonymous. Although optional, the name of a precondition or postcondition allows the constraint to be referenced by name in other places.
(c) "Self" is used in the precondition to get the *weeklyPay* property. There is no difference if the keyword "self" is dropped.
(d) The reserved word "result" denotes the result of the operation. In this case, the result equals an amount computed from the parameter value and the object attributes.

In a class diagram, constraints, such as class invariants, operation preconditions, and post conditions, can be attached to a class as UML notes.

6.2.1.2 Template class

A class with template parameters is often called a *template class*[8] or a *parameterized class*. Template classes are useful for working with collections or design patterns involving parameterized collaborations. This concept is widely supported in OOP languages, say, template in C++ and generics in Java.

A template class is not instantiable—no objects can be created—unless the list of formal template parameters are bound to a list of actual elements. Each template binding produces a new instantiable class, where the original template class is reified such that the template parameters are replaced by the corresponding actual elements. The new classes derived from a template class by binding are called *bound classes*.

In Figure 6.3, *Set* is defined as a template class, where a dashed rectangle containing a list of the formal template parameters is superimposed on the upper right-hand corner of the class box.

A *template binding* is a directed relationship, specifying the substitutions of actual parameters for the formal parameters of the template class. A template binding is shown as a dashed line with the tail on the bound class and the arrowhead on the template class. The line is generally labeled by the stereotype «bind» together with the binding information—a comma-separated list of actual parameters. There is another form where the binding information is attached to the name of the bound class. Examples of both forms are given in Figure 6.3.

Semantically, the binding relationship implies that the contents of the template class are copied into the bound class, substituting any elements exposed as formal template parameters by the corresponding elements specified as actual parameters in the binding.

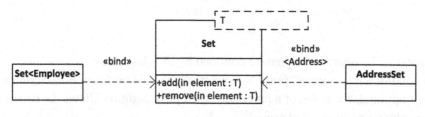

Figure 6.3
UML template class notations.

---

[8] "Template classifier" is used in UML where a classifier can be any data type, including classes and interfaces.

6.2.2 Instance-Level Relationships

A class diagram may have three types of relationships: association, generalization, and dependency. Association describes instance-level relationships, while both generalization and dependency describe class-level relationships. Only instance-level relationships may appear (as links) in object diagrams. From another perspective, association is concerned with structural features, dependency in class diagrams is concerned with behavioral features, and generalization is concerned with both.

6.2.2.1 Association

An *association* is a type of logical connection between classes, specifying a semantic relationship shared among the objects (instances) of the corresponding classes.

An association end is the end of an association. A binary association has two ends; a ternary association has three ends; and so on. In general, associations with three or more ends are called n-ary associations.

Binary association

A binary association by default is bidirectional, and it is normally drawn as a solid line connecting two classes. In the case of a unidirectional association, an open arrowhead drawn on an end indicates that end is navigable. For example, in Figure 6.4, there is a bidirectional association between the *Company* and *Employee* classes, and a unidirectional association from the *Employee* class to the *Office* class.

An association can have a name (label), which is used to explain why the association exists between the two classes. The association name is usually read in a particular direction, which

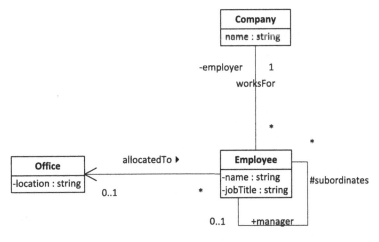

Figure 6.4
UML association relations.

is marked by a solid triangular arrowhead next to the association name. The label direction is unrelated to the association direction. For example, in Figure 6.4, *worksFor* is the name of the association between the *Company* and *Employee* classes; *allocatedTo* is the name of the association between the *Employee* and *Office* classes, where the label direction is the opposite of the association direction.

When there are multiple associations among the same classes, we need ways to resolving the ambiguity. Labeling associations by names is one way. In addition to this, each end of an association can have its own name (also called a role name because it typically indicates the role to be played by the objects of the class at that end). An end name is treated as a *property of the other end*; it thus can be adorned by a visibility. For example, In Figure 6.4, the *worksFor* association has an end name *employer* close to the class *Company*, and it has a "private" visibility.

In Figure 6.4, the line connecting the *Employee* class to itself is called a self-association (reflexive relationship). A self-association can be better described by association end names. For example, from the end names, it is clear that the self-association of class *Employee* describes a management relationship where some employees are managers of the others. The end name *manager* has a "public" visibility; this allows, for example, a company object to access an employee's manager directly, as if it were a normal public attribute defined in the class *Employee*.

By default, an association end represents a set of objects of the end class. The notion of *multiplicity* can be applied to an association end to specify the number of objects of the end class that may be related to a single object of the class on the other end. UML specifies multiplicity with an interval, such as "1" (exactly one), "0…1" (zero or one),"*" (zero or more), or "1..*" (one or more). Specific ranges of non-negative integers—say, "0…7" (zero to seven, inclusive)—can also be used.

It is good practice to read a bidirectional association in two sentences, one for each direction. For instance, the "worksFor" association in Figure 6.4 can be read like this: "an employee can work for exactly one company," and "a company hires many employees." The self-association can be read like this: "an employee as a manager can have zero or many subordinates," and "an employee may report to zero managers or one manager." It is unclear if you just say "zero or one manager has zero or more subordinates."

In addition to role names, multiplicity, and visibility, an association end can also be adorned with a property string enclosed in curly braces. A property string can have several expressions of the following forms:

• *readOnly*: to indicate that this end property is read only (unchangeable after being initialized);

- *subsets <property-name>*: to show that this end property is a subset of the property identified by <property-name>;
- *redefines <property-name>*: to show that the property redefines an inherited property identified by <property-name>;
- *union*: to show that the end is derived by being the union of its subsets;
- *unique*: to show that the end represents a set (no duplicates);
- *ordered*: to show that the end represents an ordered set;
- *bag*: to show that the end represents a collection that permits the same element to appear more than once (nonunique);
- *sequence* or *seq*: to show that the end represents a sequence (an ordered bag).

Figure 6.5 shows some examples of property strings. The property string at the *ownedMember* end says that a package can own many *NamedElement* objects, and the derived property *ownedMember* is a subset of the property called *member*, whose definition is not shown in Figure 6.5.[9] Also, the property *ownedMember* is unchangeable, and it is a derived union of its subsets, one of which is defined by the association end called *packagedElement* (close to the class *PackageableElement*). In turn, the *packagedElement* property has two subsets:

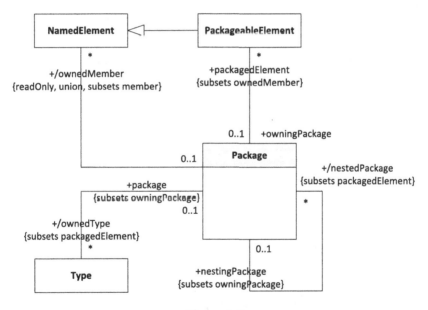

Figure 6.5
UML association end properties (package concepts as defined in [10]).

---

[9] For example, in addition to those owned members, the property *member* may also have named elements imported from other packages.

nestedPackage and *ownedType*. Indeed, a package, as an *owningPackage* object, could contain many *Type* objects and nested subpackages.

Now, let us look more deeply at the semantics of the association relationship. A reference is an attribute in one object that refers to another object. Semantically, a one-to-one association[10] between two classes *A* and *B* establishes a mutual reference between them. A mutual reference means two things.

First, it implies a pair of references, one from an *A* object to a *B* object, and one from a *B* object to an *A* object. By the one-to-one association, class *A* inherently has one reference attribute of type *B*, and at the same time, class *B* inherently has one reference of type *A*. Moreover, if an association has end names, it is common practice to use the name on the opposite end of the association as the name of the "implicit" reference attribute.

Second, a mutual reference carries a stronger semantics than merely a pair of references. The forward reference (say, from *A* to *B*) and the inverse reference (say, from *B* to *A*) depend on each other. For instance, when an *A* object *a*1 switches its relation from a *B* object *b*1 to another *B* object *b*2, the object *a*1's reference is changed from *b*1 to *b*2. Moreover, such a change would entail changes in both *b*1 and *b*2: *b*1 lost its reference to *a*1 and *b*2 acquired a new reference to *a*1.

At this point, you may be wondering how to implement association relations in programming languages. According to the above-mentioned semantics, an association is normally implemented as a pair of references.[11] In particular,

- for a unidirectional association from source class *A* to target class *B* where the multiplicity at the target end is "zero" or "one," declare in *A* a reference variable of type *B*.

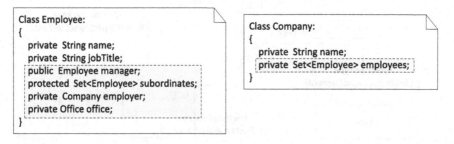

Figure 6.6
Implement association relations.

---

[10] For associations with other end multiplicities, it can be analyzed similarly—say, using the notion of arrays or lists.

[11] However, you should be aware that the mutual dependency semantics is lost. If needed, the mutual dependency has to be enforced explicitly by code.

- for a unidirectional association from source class *A* to target class *B* where the multiplicity at the target end is more than one (say, "*"), declare in *A* a reference variable of type *Set⟨B⟩*, which is a *Set* type with objects of type *B* as elements.
- for a unidirectional association from source class *A* to target class *B* where the target end is annotated with *{bag}* or *{sequence}* or *{ordered}*, declare in *A* a reference variable of type *List⟨B⟩*, which is a *List* type with objects of type *B* as elements. A *List* preserves orders and allows the same element to occur multiple times. For example, a college student may take many courses, and the same course can be taken more than once (say, retaking a course the student failed).
- the name of the target end, if it exists, is used as the variable name in the source class. The visibility of the target end also becomes the visibility of the reference variable in the source class.
- for a bidirectional association between class *A* and class *B*, consider it as two unidirectional associations (one from *A* to *B* and one from *B* to *A*) and implement each accordingly.

As an example, a possible implementation of the classes *Company* and *Employee* in a Java-like programming language is sketched in Figure 6.6. It is interesting to see that the self-association on the class *Employee* leads to two reference variables: one for the *manager* end and one for the *subordinates* end.

Association class

An *association class* is an association which is also a class containing attributes and operations. The attributes of an association class are joint properties that cannot be ascribed to either end class without losing information.

Figure 6.7a gives a situation where an association can have its own properties. A student can register for as many listed courses as permitted. Each of the registrations can have information such as registration date and grade. The *{sequence}* annotation at the *ListedCourse* class end indicates that a student is allowed to take the *same course* multiple times—retake a course the student failed in a previous semester.

Normally, association classes are not directly supported in programming languages. In such a case, an association class needs to be "promoted" to a normal class. Figure 6.7b shows how the association class in Figure 6.7a is promoted to a normal *Registration* class, where the original many-to-many association becomes two one-to-many associations. The *{sequence}* annotation in Figure 6.7a implies that there may exist multiple *Registration* occurrences (objects) for each object pair of *Student* and *ListedCourse* classes. The *{sequence}* annotation is no longer needed in Figure 6.7b because it is automatically enforced by the two one-to-many associations: each student and each course can be related by multiple registrations.

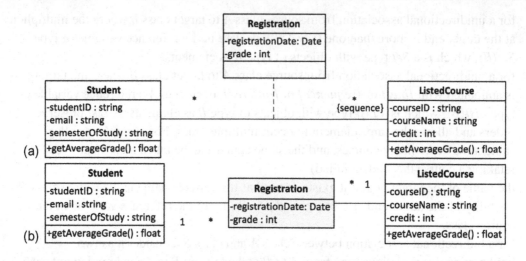

Figure 6.7
(a) The notation of an association class; (b) promote an association class to a normal class.

However, when there is no "sequence" or "bag" annotation at an association end, the association generally enforces there being *at most one link* for each object pair of the end classes. This semantic constraint can be lost in the promotion process [21]. For example, the association class *AccessibleBy* in Figure 6.8 is used to model the accessibility relations between files and file users. The model perfectly reflects reality: there can be at most one access permission defined for each pair of users and files.[12] In general, there are at most

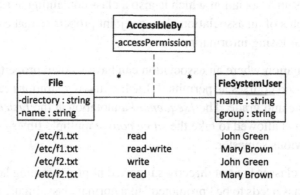

Figure 6.8
An association class: FileSystemUser.

---

[12] Here, a composite permission such as "read-write" is viewed as a single access permission; it is one object rather than a set of objects of simple permissions. If composite permissions are not allowed, we would have to use *{bag}* to annotate the *FileSystemUser* end of the association, specifying that each file and a user can be related by several simple permissions.

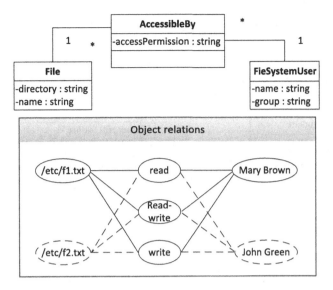

Figure 6.9
Constraint on relation uniqueness is lost in promotion.

$M \times N$ accessibility relations between M number of users and N number of files. However, this constraint cannot be captured by the model in Figure 6.9, where the association class is promoted to a normal class. The model in Figure 6.9 allows up to $M \times K \times N$ accessibility relations defined for M number of users and N number of files, where K is the number of access types. In implementation, special application code would have to be written to enforce the uniqueness of object combinations.

It is bad practice to fold the attributes of an association into an end class. To simplify our discussion, let us ignore the *{sequence}* annotation of the association in Figure 6.7a. By folding association attributes, we could get a model as shown in either Figure 6.10a or b.

This "folding" approach is discouraged for good reasons. First, the attributes of an association do not belong in either of the end classes. Folding breaks the semantic integrity of class modeling, resulting in monolithic or dirty classes with loosely related properties. For example, information about course grades is not an integral part of a student object.

Second, the folding approach inevitably complicates application logic. Consider Figure 6.10a for an example. Suppose a professor, who has a reference to a listed course he taught, would like to know the average student grade for that course. The association allows a course object to navigate to all the student objects corresponding to students who took that course. However, each of those students might have taken many other courses (that is why map structures are used). Hence, each student object needs to search the *courseGrades* attribute by *courseID*.

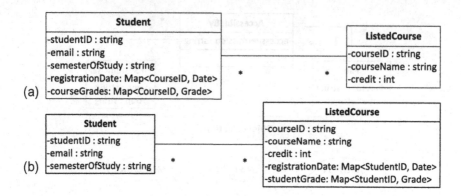

Figure 6.10
(a) Place association properties in Student; (b) place association properties in ListedCourse.

Consequently, given a reference to the listed course, it takes a two-level nested loop to get the average grade.

In contrast, for the model in Figure 6.7b, given a reference to the listed course, it takes only a one-level loop to get the average grade from those related *Registration* objects. It is unnecessary to navigate to the *Student* objects. A similar analysis can be conducted for the model shown in Figure 6.10b.

In other words, when applied to a many-to-many association, the folding approach undesirably adds another layer of barriers, preventing a client from the other end of the association from accessing the association attributes directly.

N-ary association

You may occasionally encounter n-ary associations connecting three or more classes. Just like binary associations, an n-ary association can also have its own attributes captured by the notion of an association class, and which, being a class, may participate in other associations. Figure 6.11 shows such a complex situation [21].

An n-ary association is an atomic unit, and it can be described by N sentences each starting with a unique combination of $N - 1$ ends. For example, the ternary association in Figure 6.11 says that (a) a specific professor in a particular semester can deliver zero or more listed courses, with each delivered course having a classroom and a schedule; (b) a specific professor can deliver a particular listed course in zero or more semesters, with each delivered course having a classroom and a schedule; and (c) a particular listed course in a specific semester can be delivered by zero or more professors, with each delivered course having a classroom and a schedule.

When it comes to implementation, you can promote an n-ary association, regardless of whether it has an association class or not, to a normal class and reduce the original n-ary association to *N* one-to-many binary associations, each of which has the "many" multiplicity at the end of the introduced association class. Figure 6.12 shows a promoted model of the ternary association given in Figure 6.11.

Similarly to binary associations, an n-ary association by default also enforces there being at most one link for each object combination of the end classes. In contrast, a promoted class

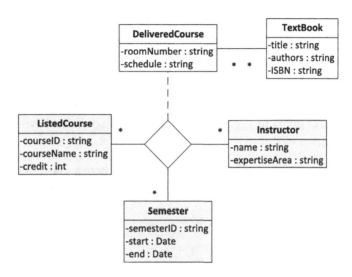

Figure 6.11
UML ternary association.

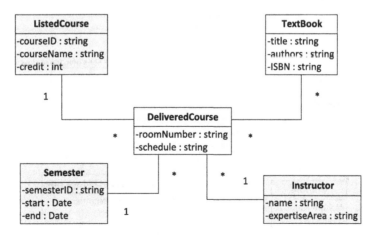

Figure 6.12
Promoting a ternary association.

permits any number of relations for each object combination of the end classes. This might reflect the real case in Figure 6.12. For example, at some colleges, a specific professor in a particular semester can deliver a specific listed course multiple times (we call each occurrence a session). However, we should be aware that (a) this "extra" semantics is not captured by the model in Figure 6.11, and (b) if relation uniqueness is desirable, it has to be enforced by application logic.

6.2.2.2 Aggregation

Aggregation is a strong form of binary association, where the class at one end plays a role called "assembly" and the class at the other end plays the role of "constituent part." Aggregation captures a "whole-part" or "has-a" relationship.

Since an aggregation is a special association, it is drawn like an association, but the assembly end is adorned with a hollow diamond. In addition, all of the properties and adornments that apply to associations also apply to aggregations, including association labels, multiplicity, visibility, and role names.

An example of using aggregations is given in Figure 6.13, where a magazine may have many (zero or more) volumes; a volume may have many issues, and each issue of the magazine may contain many articles.

6.2.2.3 Composition

There is a strong form of aggregation called "composition," where the class at the assembly end is called "composite." The notation for composition is a small solid diamond next to the composite class. Since composition is a special kind of association, it can likewise have all the adornments defined for an ordinary association.

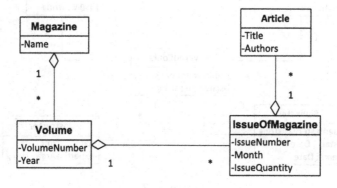

Figure 6.13
UML aggregation relations.

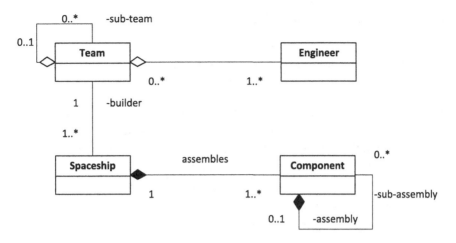

Figure 6.14
UML composition relations.

Figure 6.14 shows an example, where a spaceship is composed of many components, and each component can be composed of many smaller components.

While both aggregation and composition capture "whole-part" relationships, composition is a relatively stronger form in the sense that composition implies ownership of the parts by the whole. This notation of ownership, in particular, means two things: (a) once the "part" objects have been assembled to form a composite object, deletion of the composite object triggers deletion of all the constituent parts; (b) a part object can belong to at most one composite object at any one time, although it may participate in other associations or even belong to different composite objects at different times (say, a part object was removed from one composite object then assigned to another).

The composition relationship is often called "physical" containment to distinguish it from aggregation's "catalog" containment. The composition relationship is most appropriate for representing real-world whole-part relationships—for example, an engine is a part of a car. In class diagrams, the part end of a composition may have any multiplicity, but the whole end must have a multiplicity of 0...1 or 1, indicating that a part must belong to only one whole at a time. See Figure 6.14 for examples.

The aggregation relationship is appropriate when there is no lifetime dependency between a whole and its parts. For example, a certain processor chip is part of a circuit board design (the same processor chip can be used in many deigns at the same time). Different from compositions, in class diagrams, the whole end of an aggregation can have any multiplicity. In Figure 6.14, an engineer can be a member of multiple teams simultaneously. Composition is not appropriate here, because after a team has been dissolved, the members can still be involved in other teams or activities.

6.2.2.4 OCL expressions for association navigation

Navigating class models is important because it allows you to scrutinize the class models and uncover hidden flaws and omissions so that you can fix the problems at an early stage of development.

As mentioned before, by default an association end represents a set of objects. Navigation in a class diagram means we have to work with collection types. In line with the possible adornments at an association end, OCL defines an abstract type *Collection* and four concrete subtypes: *Set*, *OrderedSet*, *Bag*, and *Sequence*. A *Set* does not contain duplicate elements. An *OrderedSet* is like a *Set* in which the elements are ordered. A *Bag* is a collection that may contain duplicates (i.e., the same element may be in a bag twice or more times). A *Sequence* is like a *Bag* in which the elements are ordered.

Now, let us focus on the class model in Figure 6.7a and look at the OCL constructs for traversing class models:

- Access the attribute of an object: *object.attributeName*. For example, suppose *self* refers to a specific student, then *self.email* returns the student's e-mail address.
- Access the attribute of each object in a collection: *collection.attributeName*. For example, suppose *self* refers to a set of specific students, then *self.email* returns a set of e-mail addresses, one for each student.
- Invoke an operation on an object: *object.operation(para1,...)*. For example, suppose *self* refers to a specific student, then *self.getAverageGrade( )* returns his/her average grade for all the courses.
- Invoke an operation on each object in a collection: *collection.operation(para1,...)*. For example, suppose *self* refers to a bag of specific students, then the *self.getAverageGrade( )* returns a bag of grades.
- Invoke an operation on a collection: *collection->operation(para1,...)*. For example, suppose *self* refers to a bag of grades, then *self->select(g: float | g>3.5)* return a new bag of grades that are greater than 3.5. The more complicated expression *self->count(self->max())* returns the number of students who have earned the highest grade. Tables 6.1 and 6.2 list a few collection operations predefined in OCL.
- Access the other end of a binary association: *object.associationEndName*. By default, the value of this expression is a set of objects which are instances of the class on the other side of the association. If the association end is adorned with *ordered* (or *sequence* or *bag*), the navigation results in an *OrderedSet* (or *Sequence* or *Bag*). If no name is specified for an association end, an implicit name is taken from the name of the class to which the end is attached. For example, suppose *self* refers to a specific student, then *self.ListedCourse* returns a *Sequence* object containing a list of courses the student has taken (the same course may appear several times). If the multiplicity of the association end is "0...1" or

Table 6.1 OCL collection operations

Form	Description
collection->select(boolean-exp), or collection->select(v | boolean-exp), or collection->select(v: Type | boolean-exp)	It results in a collection that contains all the elements from collection for which the Boolean expression evaluates to true. The iterator v, if used, iterates over the collection.
collection->reject(boolean-exp), or collection->reject(v | boolean-exp), or collection->reject(v: Type | boolean-exp)	It results in a collection that contains all the elements from collection for which the Boolean expression evaluates to false. If used, v iterates over the collection.
collection.propertyname, or collection->collect(propertyname), or collection->collect(v | v.propertyname), or collection->collect(v: Type | v.propertyname)	It results in a collection that contains all the values of the named property from those elements in the collection. The iterator v, if used, iterates over the collection.
collection->forAll(boolean-exp), or collection->forAll(v | boolean-exp), or collection->forAll(v1, v2,... | boolean-exp), or collection->forAll(v: Type | boolean-exp)	The result is true if the Boolean expression is true for all elements of collection. If the Boolean expression is false for one or more v in collection, then the complete expression evaluates to false.
collection->exists(boolean-exp), or collection->exists(v | boolean-exp), or collection->exists(v1, v2,... | boolean-exp), or collection->exists(v: Type | boolean-exp)	The result is true if the Boolean expression is true for at least one element of collection. If the Boolean expression is false for all v in collection, then the complete expression evaluates to false.
source>closure(expression), or source>closure(v | expression-with-v), or source>closure(v : Type | expression-with-v)	The result is a set or ordered set populated by recursive invocation of the expression. The recursion starts from each distinct element of the source collection and terminates when the expression returns empty collections or collections containing only already accumulated elements.
source->iterate(element : Type; accumulator : Type – <initial-value> | evaluation-expression)	The result is denoted by accumulator, which is built up by iterating over the elements of the source collection After each evaluation of the expression, its value is used to update the accumulator. The result is a collection if the accumulator is defined as a collection.

"1," then the value of this expression is an object. For example, in Figure 6.4, suppose *self* refers to an employee, then *self.employer* returns the company for which the employee works.

- Access a simple association class: *object.associationClassName*. For example, in Figure 6.7a, suppose *self* refers to a specific student, then *self.Registration* returns a set of *Registration* objects the student made.

Table 6.2 An incomplete list of collection operations from OCL Standard Library ("self" refers to the collection object on which an operation is invoked)

Operation	Description of Return
Collection<T>	
=(c:Collection<T>): Boolean	True if c is a collection of the same kind as self and contains the same elements in the same quantities and in the same order, in the case of an ordered collection
<> (c : Collection<T>) : Boolean	True if c is not equal to self
size() : Integer	The number of elements in the collection self
includes(object : T) : Boolean	True if object is an element of self
excludes(object : T) : Boolean	True if object is not an element of self
count(object : T) : Integer	The number of times that object occurs in self
includesAll(c: Collection<T>) : Boolean	True if self contain all the elements of c
excludesAll(c: Collection<T>) : Boolean	True if self contain none of the elements of c
isEmpty() : Boolean	True if self is empty
notEmpty() : Boolean	True if self is not empty
max() : T	The element with the maximum value of all in self
min() : T	The element with the minimum value of all in self
sum() : T	The addition of all elements in self
asSet() : Set<T>	The Set containing all the elements from self, with duplicates removed
asOrderedSet() : OrderedSet<T>	An OrderedSet that contains all the elements from self, with duplicates removed, in an order dependent on the particular concrete collection type
asSequence() : Sequence<T>	A Sequence that contains all the elements from self
asBag() : Bag<T>	The Bag that contains all the elements from self
flatten() : Collection<T>	If the element type is a collection type, the result is a collection containing all the elements of all the recursively flattened elements of self
Set<T>	
union(s : Set<T>) : Set<T>	The union of self and s
intersection(s : Set<T>) : Set<T>	The intersection of self and s
symmetricDifference(s:Set<T>):Set<T>	The sets containing all the elements that are in self or s, but not in both
- (s : Set<T>) : Set<T>	The elements of self, which are not in s
including(object : T) : Set<T>	The set containing all elements of self plus object
OrderedSet<T>	
append (object: T) : OrderedSet<T>	The set of elements, consisting of all elements of self, followed by object
reverse() : OrderedSet<T>	The ordered set of elements with same elements but with the opposite order
at(i : Integer) : T	The ith element of self
first() : T	The first element in self
last() : T	The last element in self

- Access an association class on a reflexive association: we have a more convolved notation *object.associationClassName[endName]*. For a reflexive association, the association class name alone is not enough to distinguish the two possible navigation directions. The rolename (end name) of the direction in which we want to navigate is added to the association class name, enclosed in square brackets. Let us introduce an association class to the self-association in Figure 6.4. We name this association class *PerformanceRating*, which describes the mutual rating between each worker and his/her manager (say, two scores, each evaluated by one for the other). Now, suppose *self* refers to a specific employee, then *self.PerformanceRating[subordinates]* returns a set of *PerformanceRating* objects, each of which is between this manager and one of his/her subordinates. In contrast, *self.PerformanceRating[manager]* returns just a single *PerformanceRating* object, which is between this employee and his/her manager.
- Navigate from an association class: *object.associationEndName*. Navigation from an association class to one of the association ends will always produce exactly one object (or a singleton set). If no name is specified for an association end, an implicit name is taken from the name of the class to which the end is attached. For example, in Figure 6.7a, suppose *self* refers to a registration object, then *self.Student* returns the specific student who made that registration.

Below are a few examples that use the collection operations given in Table 6.1. Note that the expressions are all equivalent, saying that the collection of all registrations before the deadline is not empty (see Figure 6.7a):

```
context Student inv:
  self.Registration->select(registrationDate < deadline)->notEmpty()

context Student inv:
  self.Registration->select(r |r.registrationDate < deadline)->notEmpty()

context Student inv:
  self.Registration->reject(registrationDate >= deadline)->notEmpty()

context Student inv:
  self.Registration->reject(r |registrationDate >= deadline)->notEmpty()
```

Now, let us use Figure 6.13 to work on some navigation examples. Suppose *Times* is a *Magazine* object. The expression *Times.Volume* returns a set of *Volume* objects; *Times.Volume.IssueOfMagazine* returns a set of *IssueOfMagazine* objects (each of which belongs to a specific *Volume* object); and *Times.Volume.IssueOfMagazine.Article* returns a set of *Article* objects that have ever been published by the *Times* magazine (each article appears in a specific issue of a specific volume). To get the number of authors who have contributed to the *Times* magazine, we write *Times.Volume.IssueOfMagazine.Article.Authors->size()*.

6.2.3 Dependency Relationships

Dependency is a relation that can be used in class diagrams, package diagrams, and deployment diagrams.

In class diagrams, a dependency is a directed relation from *A* to *B*, where *A*, a class or interface, is called the depending or client element, and *B*, a class or interface, is called the supplier element. A dependency signifies that the behavioral features of *A* (operations or methods) depend on the supplier *B*. In other words, a dependency implies that the operational semantics of the client is not complete without the supplier.

A dependency is shown as a dashed line with an open arrowhead, where the client element is at the tail and the arrowhead points to the supplier. A dependency is usually labeled with a stereotype, conveying special semantics.

6.2.3.1 Abstraction

An *abstraction* is a special kind of dependency that represents the same concept at different levels of abstraction. Abstraction has predefined stereotypes, such as «derive», «refine», and «trace». An abstraction relationship is shown as a dependency with an «abstraction» or a specific predefined stereotype name attached to it.

A «derive» dependency specifies that the client can always be computed from the supplier. The client and the supplier are usually, but not necessarily, of the same type. The "derived" client contains logically redundant information; it is modeled/implemented merely for convenience or efficiency. The downside is that this redundancy adds a point of structural weakness to the model because if a change is made to the supplier, it is often necessary to make a change to the client.

Figure 6.15 shows an example where the class *InternationalTransaction* depends on the class *ExchangeRateManager* by «derive». Each transaction keeps a record of the transaction amount both in the foreign currency and in the local currency. Only one of the two is needed because the other one could always be calculated by requesting the exchange rate for a particular date from the *ExchangeRateManager* class. However, the redundancy could increase the system performance.

A «trace» dependency is mainly used for tracking requirements and changes across models, indicating that one or more elements in one model represent the same or an analogous concept in another model.

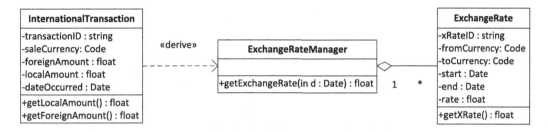

Figure 6.15
The «derive» dependency.

In Figure 6.16, some class names of the domain model are changed in the design model. Also note that in the design model, the class *Employee* is used to replace the classes *Manager* and *Worker* used in the "preliminary" domain model.

A «refine» dependency is similar to a «trace» dependency in that it connects two elements that are related by evolution. The difference is that the evolution in a trace usually indicates different stages in a development, and therefore cuts across models (from analysis to design to implementation), whereas the refinement relationship represents different evolutionary stages of the same concept at different semantic (or abstraction) levels. Elements connected by the

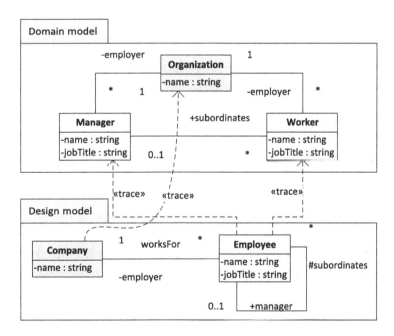

Figure 6.16
The «trace» dependency.

«refine» dependency are usually in the same model. The refinement relationship is especially useful for capturing model/concept transformations.

6.2.3.2 Interface realization

UML defines realization as a special kind of abstraction dependency. In general, realization dependency can be used to model stepwise refinement, optimization, and transformations. Realization has two forms: interface realization and substitution.[13]

An interface realization relation connects a class to an interface, where the interface specifies a *contract* (set of behavioral features) which must be fulfilled by the class. To fulfill a service contract, the class (or its concrete subclasses if it is an abstract class) must define a method (implementation) for every operation specified by the interface (and its parent interfaces, if any). When a class *C* realizes an interface *I*, the interface *I* actually offers a "window" that allows a client (or user) of *C* to selectively view and use a subset of the features encapsulated in the class *C*.

A realization dependency may be shown in two forms. If the interface concerned is shown as a class stereotyped by «interface», a realization dependency is shown as a dashed line with a triangular arrowhead pointing to the interface. If that interface is shown as a named circle (operations are suppressed), a realization dependency is shown by a solid line connecting the circle to the class that realizes this interface. Figure 6.17 shows an example of both forms.

6.2.3.3 Comparison: interface and abstract class

Interface and abstract class are two similar modeling notations that allow us to model a set of common features as a contract to be completed or fulfilled by other classes. Since they are incomplete, neither can be used to create objects directly.

Interface differs from abstract class in many ways:

- Semantic connection: It is highly recommended to use abstract classes to declare common features for semantically related classes, and to use interfaces to declare common features for classes belonging to different semantic categories. For example, *Employee* and *Invoice* in Figure 6.17 implement the same interface; they are concepts in two different semantic categories.
- Scope: An abstract class covers both attributes and behavioral features, whereas an interface covers behavioral features only.

---

[13] The substitution relationship is instrumental to specify run-time substitutability for domains that do not support inheritance. It requires that the substituting class implement a more specialized interface type than (or the interface type same as) that implemented by the contract class. Substitution implies compliance of contracts only, and it is shown as a dependency with «substitute» attached to it.

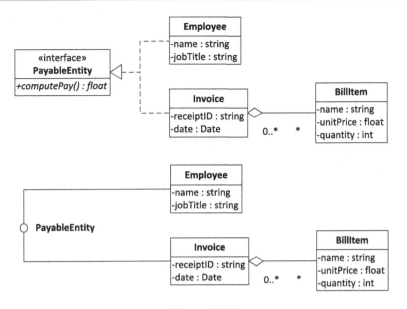

Figure 6.17
The interface realization dependency.

- Reuse of default implementation: An interface is merely a list of operation signatures, whereas an abstract class could be partially implemented (some operations are provided with default methods).
- Stability of contract: It is preferable to use an abstract class to capture a contract if the contract is expected to change. When an abstract class is used, any change to it (e.g., adding new operations or changing signatures) will be automatically reflected in the subclasses. However, when an interface is used, any change to the interface would entail many changes in the implementing classes, because the change inevitably breaks the "realization" relation.

Some other differences are language dependent. For instance, in certain OOP languages (e.g., Java), a class may implement an unlimited number of interfaces, but may inherit from only one abstract class. A class that inherits from an abstract class may implement interfaces at the same time.

6.2.3.4 Usage

A usage dependency from a class (or interface) *A* to another class (or interface) *B* indicates that *A* uses *B* for its operation or full implementation. *A* is called the client, and *B* is called the supplier.

Typically a usage dependency is used when one class (or interface) temporarily uses another class (or interface) in its operation signatures (i.e., as a parameter or return type).

A usage dependency from a client to a supplier is generally shown as a dependency labeled by the stereotype «use». If the supplier is an interface and it is shown as a half circle labeled with the name of the interface, the usage dependency is denoted by a solid line connecting the client to the half circle.

UML also provides a few predefined stereotypes for usage dependency:

- «call»: A call dependency specifies at least one of the operations in the client class invokes ("depends" on) one or more operations specified in the supplier.
- «create» or «instantiate»: A create dependency signifies that at least one of the operations in the client class creates instances of the supplier class.
- «send»: A send dependency exists between an operation of a client and a signal. Signals (e.g., exceptions) are asynchronous messages that can be sent by a sender to receivers. A send dependency specifies that the client operation can send the signal. However, the receiver of the signal is beyond the scope of the send dependency.

Figure 6.18 shows examples of the usage dependency, where the *Starter* class can create *TimeSensitiveTask* and *TimerManager* objects. A *TimeSensitiveTask* object needs to use a *TimerManager* object to manage its timers, this service is available to *TimeSensitiveTask* through the interface *TimerInterface*; *TimerManager* can send *TimeoutSignal* objects, and the *TimeSensitiveTask* itself uses the *Timer* interface.

One key difference between association and usage dependency is that an association implies a much stronger bond between the end classes. Indeed, a bidirectional association entails that

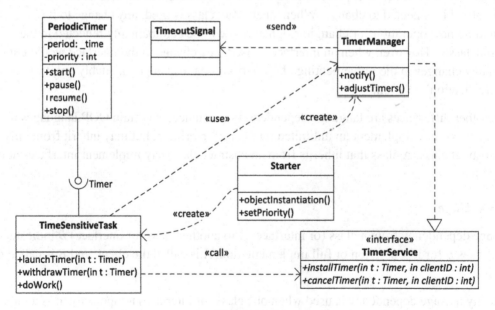

Figure 6.18
The usage dependency.

each end is part of the other end's structural features (reference variables), and such a bond typically exists for a long time. In contrast, the bond in usage dependency is very weak, and it usually needs a stereotype to confine the precision of the usage. Usage dependency has no impact on the structures of the client and the supplier, and it exists only temporarily (used not as instance variables, merely as parameter or local variables). For example, by «create» usage, a client may temporarily use a supplier class as a local variable for one of its operations; a temporary utility object may create all the other objects to carry out the services of the system.

6.2.4 Generalization

Given a class *A*, the reasoning process of introducing subclasses for *A* is called *specialization*, while the process of introducing superclasses for *A* is called generalization. Generalization is a directed relationship from a subclass *S* to a superclass *B*, indicating that *S* is considered to be a specialized form of *B*.

In class diagrams, a generalization is shown as a line between a subclass and its superclass; the line has a hollow triangle as its arrowhead pointing to the superclass. Figure 6.19 gives some abstract examples, where *T3* is a subclass of *T2*, which in turn is a subclass of *T1*. As is shown here, subclassing often forms a hierarchical structure, which is called class hierarchy or *inheritance hierarchy*. A class (e.g., *T1*) is the root of a hierarchy if it is not a subclass of any. There is another class hierarchy, where *ClassC* is a subclass of *ClassB*, which in turn is a subclass of *ClassA*. *ClassD* depends on *ClassA*, which is indicated by the dependency relationship labeled by the «use» stereotype.

A generalization implies many things. First, a subclass as a class can define its own features (attributes and/or operations). This is called *extension*. Indeed, it is not necessary to introduce a new class if the same semantics (features) have been captured in an existing class. In Figure 6.19, classes *T1*, *T2*, and *T3* define different operations; classes *ClassA*, *ClassB*, and *ClassC* differ from each other in both attributes and operations.

Second, a subclass does not merely have the features declared by itself; it also inherits from its direct superclass and indirect superclasses along the class hierarchy up to the root class. This is summarized below:

- *Inherit attributes declared in superclasses*: When you request a system to create an object of a subclass, the system will create just a single object of that subclass. It is *not* the case that an object is created for each of its superclasses along the inheritance hierarchy. But do not worry. The object is really a "super ball," containing a slot for each attribute declared by the superclasses! It even has a slot for the attributes declared as "private" by one of the superclasses; however, this object can access those private attributes only through public/protected getter and setter operations. For example, in Figure 6.19, when a *ClassC* object is created, it has three attributes. Code in *ClassC* can access *attr2* and *attr3* directly,

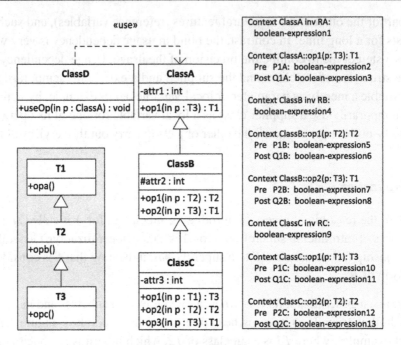

Figure 6.19
Class inheritance hierarchy.

and can access *attr1* by using the corresponding getter and setter operations. Note that a class diagram can be cluttered with getter and setter operations; they exist but are typically suppressed to highlight more important features.

- *Inherit public/protected operations declared in superclasses*: Since a subclass object can be viewed and used as a superclass object, the behavior (public operations) of a subclass object cannot surprise its users. Hence, a subclass will implicitly inherit all the public operations declared in superclasses. Protected operations are defined to be inherited by subclasses; they are essential to the behavioral integrity of the subclass. For example, the class *T3* has three operations, one defined by itself, one from class *T2*, and one from class *T1*.

A subclass can also modify an inherited operation (this is called *overriding*) in two ways. First, the signature of an inherited operation can be "slightly" adjusted. Here, the word "slightly" is used because you cannot add more operation parameters or remove operation parameters. If you were to do this, you would not be revising the inherited operation, but would actually be defining a completely new operation with the same operation name (this is called *overloading*). To adjust the signature of an inherited operation, you can change only the return type and/or the parameter types. Moreover, to avoid any surprise to the potential users, (a) the return object cannot be beyond what is expected from the original operation, and thus you cannot redefine a new return type with a wider range (superclass), and (b) any object acceptable by the original operation ought to be acceptable by the inherited operation, and thus you cannot limit the user by redefining a more restricted type (subclass) for a parameter. For example, *ClassA* defines an operation *op1(p:T3):T1* which has a return type *T1* and a parameter of type *T3*. This operation is overridden in *ClassB* with a new return type *T2*, which is more restricted than *T1*, and a new parameter type *T2*, by which the revised operation can accept more possible inputs than what is allowed by *T3*. You can conduct the same analysis for the other operation overridings in Figure 6.19.

Second, a subclass can modify the method implementation of an inherited operation. A superclass at its best can have its operations implemented for general cases. A subclass typically has more concrete information that allows you to implement the inherited operations more efficiently (better quality of service) and even achieve more (add-on benefits) than what is promised in superclasses.

- *Inherit associations and dependencies of superclasses*: A generalization is often called an "is-a" relationship, meaning that a subclass object is also an object of its superclass. Hence, a subclass object can be related to other objects by the associations and dependencies attached to its superclasses. In particular, the mutual reference implied by a binary association between class *A* and class *B* will be inherited by their respective subclasses, if any. For example, in Figure 6.19, *ClassD* depends on *ClassA* because its operation *useOp(...)* uses *ClassA* as its parameter type. Obviously, this «use» dependency applies to the subclasses of *ClassA* as well—say, a *ClassC* object could be passed to the operation *useOp*.

- *Inherit constraints declared in superclasses*: All the constraints (class invariants, preconditions and postconditions of public operations) declared in a superclass also apply to its subclasses. This can be better understood by using the "contract" concept as introduced next.

6.2.4.1 Inheritance as subcontracting

The preconditions and the postconditions of the public operations defined in a supplier class together form a contract for its potential clients (users). Particularly, the precondition of an operation states the prerequisites for calling the operation. It is the responsibility of the client to fulfill the precondition (otherwise the request is unwarrantable). The postcondition of an operation states the effect of the operation, which is the responsibility of the supplier itself. Basically, the supplier declares to its client that it can deliver the requested service (specified by the postconditions) as long as the client conforms to its assumption (specified by the preconditions).

By inheritance, a subclass forms a new contract. This new contract, however, has to be a *subcontract* of the superclass's contract in the sense that the client should not experience any difference when it needs a service from a superclass object but the service is actually delivered by a subclass object.

In order to realize this notion of *subcontracting*, we have to enforce certain requirements on the preconditions, postconditions, and class invariants across the class hierarchy. First, the *postcondition* of a redefined operation in a subclass must be *equal to or stronger than* the postcondition of the corresponding public operation in the superclass; otherwise, the redefined operation would not be able to deliver the service as promised by the contract in the superclass.

Second, the *precondition* of a redefined operation in a subclass must be *equal to or weaker than* the precondition of the corresponding public operation in the superclass; otherwise, the service would fail if the client could not fulfill the stronger preconditions.

Third, the *invariants* of a subclass must be *equal to or stronger than* the invariants in the superclass. A class invariant describes a requirement on the integrity of the object state (represented by instance variables). Since a subclass inherits all the instance variables declared in the superclass, it should sustain the invariants specified in the superclass.

So how are good inheritance decisions made? Substitutability is the key: subclass objects must behave in a manner consistent with the promises made in the superclass's contract. For instance, if you specify in the *Ellipse* class that an ellipse can be resized asymmetrically, it is then inappropriate to design *Circle* as a subclass of *Ellipse*. The reason is that the resizing behavior in *Ellipse* is a promise that the *Circle* class cannot satisfy.

The notion of subcontracting is closely related to the Liskov substitution principle, which is discussed next.

6.2.4.2 Liskov substitution principle

In OOP, each class also represents a data type. In such a sense, the type represented by a subclass is subsumed by the type represented by its superclass. From the set theory perspective, the value range (i.e., individual objects) declared by a subclass is a subset of the value range declared by its superclass. In other words, a superclass represents a relatively wider type (supertype) and its subclass is a narrower type (subtype).

Since subclassing introduces new subtypes, it is desirable to preserve type conformance in the sense that a subclass object can always substitute for an object of its superclass. This is known as the *Liskov substitution principle*, which is formally given below.

Let $B \trianglerighteq A$ denote that B is a subclass of class A, and let $X \sqsubseteq Y$ denote that X is a subtype of type Y.[14] We use $A\{R : p(S)\}$ to denote that class A has a public operation p which takes a parameter of type S[15] and returns a value of type R. Given $A\{R : p(S)\}$, let pre$[A, p(S)]$ denote the precondition of operation $p(S)$ in class A, and let post$[A, p(S)]$ denote the postcondition of operation $p(S)$ in class A. We also use inv(A) to denote the set of class invariants defined in A.

The class generalization principle implies that any subclass object must be at least able to do what the superclass objects can. This imposes some standard requirements on subclassing:

1. Sustain public operations: $A\{R_A : p(S_A)\} \wedge B \trianglerighteq A \rightarrow \exists R_B, S_B \cdot B\{R_B : p(S_B)\}$.
2. Contravariance of method arguments:
   $A\{R_A : p(S_A)\} \wedge B \trianglerighteq A \wedge B\{R_B : p(S_B)\} \rightarrow S_A \sqsubseteq S_B$. That is, parameters in the subclass have wider types.
3. Covariance of return types: $A\{R_A : p(S_A)\} \wedge B \trianglerighteq A \wedge B\{R_B : p(S_B)\} \rightarrow R_B \sqsubseteq R_A$. That is, return values in the subclass have narrower types (stricter).
4. Preconditions cannot be strengthened in a subclass:
   $A\{R_A : p(S_A)\} \wedge B \trianglerighteq A \wedge B\{R_B : p(S_B)\} \rightarrow \text{pre}[A, p(S_A)] \rightarrow \text{pre}[B, p(S_B)]$.
5. Postconditions cannot be weakened in a subclass:
   $A\{R_A : p(S_A)\} \wedge B \trianglerighteq A \wedge B\{R_B : p(S_B)\} \rightarrow \text{post}[B, p(S_B)] \rightarrow \text{post}[A, p(S_A)]$.
6. Invariants of the superclass must be preserved in a subclass: $\text{inv}(A) \subseteq \text{inv}(B)$.

In sum, to be substitutable, a subclass is *not* allowed to remove public operations defined in the superclass, and is allowed (not required) to (a) add new attributes and/or operations, (b) weaken preconditions and/or strengthen postconditions for each inherited public operation, and (c) have stronger class invariants.

For the abstract example in Figure 6.19, proper inheritance implies the following:

- $RA \subseteq RB \subseteq RC$.[16]
- $P1A \rightarrow P1B \rightarrow P1C$.
- $P2B \rightarrow P2C$.
- $Q1C \rightarrow Q1B \rightarrow Q1A$.
- $Q2C \rightarrow Q2B$.

In short, proper inheritance means "is substitutable for" [25]. It is a good thing to choose appropriate names for classes. However, class names are merely labels. Whether a class is a subclass of another is based not on their names, but on the conformance of their behavioral contracts.

---

[14] Obviously $B \trianglerighteq A$ implies $B \sqsubseteq A$, but not vice versa.

[15] It is simplified for clarity. In general, an operation can have a list of parameters.

[16] Here we use names to refer to the corresponding Boolean expressions. In OCL, a subclass object can access the properties of its superclasses by using the *oclAsType()* operation. For example, suppose *self* refers to a *ClassC* object, then *self.oclAsType(ClassB).op1* accesses the operation op1 defined in *ClassB*.

6.3 Class Modeling Principles

6.3.1 Model Evolution

Software development is an iterative process, where key activities such as requirements analysis, design, implementation, and testing are conducted iteratively: the models/artifacts produced in one iteration typically refine those produced in a previous iteration.

Requirements analysis starts with a requirements statement, which describes the desired services (functional requirements) and the desired quality of services (nonfunctional requirements) from the user's perspective. However, what are described in a requirements statement are often ambiguous, incomplete, or inconsistent. The goal of requirements analysis is to determine a scope for the system to be developed, refining the real-world requirements and domain concepts into a so-called domain model. A domain model may be captured in many forms, such as use cases and UML class diagrams.

The left portion of Figure 6.20 illustrates the refinement process of a domain model, where each oval represents one of its footprints. The domain model of a system typically starts with a *big* footprint, which may contain as many domain concepts as possibly elicited from the problem statement. As far as a UML class diagram is concerned, in this first footprint of the domain model, each domain concept is modeled as a class. The domain model can be deflated in successive footprints

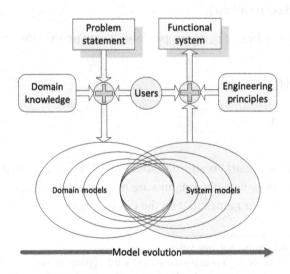

Figure 6.20
Model evolution.

- by removing those classes that are beyond the system scope;
- by relating classes via associations and dependencies, removing synonymous classes;
- by relating classes via generalizations, capturing common features at the highest possible abstraction level (so that redundant features/associations once at a lower level are removed).

With the help of domain experts and users, ambiguities and inconsistencies are gradually clarified and resolved. This refining process continues until you feel the domain model, although not perfect yet, can be a good starting point for design.

The evolution of a system model is another refinement process, which is illustrated on right in Figure 6.20. The final domain model is the first footprint of the system model. Starting from there, one can incrementally expand the system model. For example, in the second footprint, you may want to introduce abstract superclasses or interfaces or generics into a class hierarchy to improve feature reuse; in the third footprint, you may want to apply certain design patterns to improve the software architecture; yet in the next footprint, you may want to organize the class model into different packages to address system complexity. This refining process continues until you feel the system model, although not perfect yet, can be a good starting point for implementation.

In sum, domain/system modeling is a refinement process, and you can rarely get the model right the first time. As far as class modeling is concerned, as time goes on, the set of classes is shrunk or condensed in the evolution of the domain class model, and the set of classes is expanded in the evolution of the system class model. We next discuss some generic strategies that can be used for evolving system class models: subclassing, refactoring, and minimum information redundancy.

6.3.2 Subclassing

Subclassing is a powerful way for refining a class model. By subclassing, one can expand a class diagram with new classes related by inheritance (generalization). However, if not used carefully, subclassing may lead to bad designs.

The following is a short list of good practices regarding the use of subclassing:[17]

- *Symmetry may suggest missing subclasses*: For example, suppose initially we only have the left part of the class model as shown in Figure 6.21. That is, the problem statement involves only publications such as journals, newspapers, and magazines. These are periodicals.

---

[17] Readers may refer to [36] for more object-oriented design heuristics.

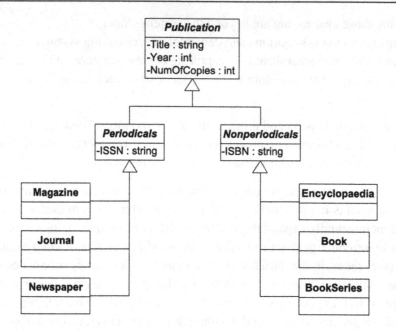

Figure 6.21
A class model of publications.

As an experienced software engineer, you know how important it is to design for future needs. So, by symmetry, you would like to consider nonperiodical publications such as books, book series, and encyclopedias.

- *Avoid singleton subclasses*: You may have singleton classes (a class from which at most one instance is to be created) in your model for good reasons. For example, in a gaming system you can use a single *Counter* object to keep track of how many enemy entities have been destroyed. However, be careful when you try to introduce a singleton subclass for an existing class. Most likely, the singleton subclass ought to be an instance of the existing class. For example, it is problematic (overengineered) to define *President George Washington* as a subclass of *USPresident*. Instead, it is just an instance of *USPresident*.
- *Good subclassing by enumeration*: You may run into a situation where objects with different attribute values have different behaviors. As a principle, it is highly discouraged to define an operation by explicit case analysis on attribute values. It is good practice to introduce subclasses, and often a subclass can be derived from each value of the attribute.

For example, you may have a *Shape* class with an operation *draw()* to display itself at a designated position. However, drawing a circle is different from drawing a rectangle; gluing them together would end up with a monolithic method composed of many blocks, each dealing with a specific kind of shape. For situations like this, it is better to introduce a subclass for each case—say, classes *Rectangle* and *Circle*.

- *Bad subclassing by enumeration*: You should not introduce subclasses if they have features identical to those of an existing class (with the same attributes, operations, and associations) and your sole purpose is to use them to indicate an enumeration. For situations such as this, what you need is not subclassing but an attribute in the existing class to tag the distinction.

For example, suppose you already have a class *RecordingCategory* to model a category of media recordings. It is not a hard step for you to consider modeling some specific categories of recordings such as jazz, blues, and rock'n'roll. At first look, you might want to use subclassing to derive three subclasses. However, on reflection, you may realize that as media recordings, they really share the same attributes and behaviors. You can simply use an attribute—say, *type*—in *RecordingCategory* to mark the specific categories.

- *Avoid subclassing from multiple parents*: Multiple inheritance can degrade maintainability and understandability of the system. It is preferable to develop class models where each class has only a single path leading to its "root" superclass along the inheritance hierarchy.
- *Avoid subclassing by evolving roles*: You may run into a situation where an object may change its roles over time. For example, suppose we have a *CollegeStudent* class and would like to define freshman, sophomore, junior, and senior as its subclasses. This, however, might cause problems in object management because as students change their levels, the system needs to create a new object for each student, copy all the relevant state information from the old object, and finally destroy that old object. A better solution is to simply use an attribute to indicate the current level of study. If necessary (say, different levels have different behavioral privileges), you can design a *StudentRole* class with four singleton subclasses—*FreshmanRole*, *SophomoreRole*, *JuniorRole*, and *SeniorRole*—and associate the *CollegeStudent* class with the *StudentRole* class. As a student moves into a new level, the system simply switches the corresponding *CollegeStudent* object to the new role object he/she is to play.
- *Avoid subclassing by annihilating public operations.* When you introduce a new subclass *B* for an existing class *A*, it is bad practice to nullify a public operation (i.e., empty its method body) inherited from class *A*. In so doing, you are semantically removing the inherited public operation, which breaks type conformance. When such a situation occurs, it usually suggests an opportunity to adjust the inheritance hierarchy. A solution is to introduce an abstract class *S* and declare both *A* and *B* as subclasses of *S*. This way, class *S* encapsulates the features common to both *A* and *B*, and it is no longer necessary for *B* to nullify those operations applicable to *A* only.

There is a principle called the *open-closed principle* [52] stating that software entities (classes, modules, operations, etc.) should be open for extension but closed for modification. It encourages subclassing from well-designed classes. For instance, given a third-party package (with a well-established class model), subclassing is a preferable way to accommodate new features as required by the system under development, especially when the source code of the third-party package is not released to the public.

6.3.3 Minimum Information Redundancy

The principle of *minimum information redundancy* is very useful in the abstraction process where new classes are introduced to capture the common features factored from concrete objects. The consequence of applying this modeling approach is that the class diagram is expanded with a few new classes related by association.

Let us consider the issue of information redundancy in a general setting. Suppose we need to model a collection of objects which (a) have a set of common features, (b) share common values (information) in some features, but (c) have different values in some other features. Obviously, information redundancy exists at two levels: common features and common instance values. How do we model such a collection of objects in a class diagram?

To understand the problem better, let us now consider a concrete example. Assume that a library would like to switch to online services. You were asked to develop a library information system to manage all the library books electronically. To simplify the problem, let us suppose the library has only two kinds of books: 50 copies of book *Real-Time UML* and 80 copies of book *Real-Time Embedded Systems*. Some book properties are shown in Figure 6.22: (a) they share a set of common features (e.g., title, author, and publisher); (b) all copies of the same book share common values (e.g., information about the title, author, and publisher); (c) each specific book copy has some unique information (e.g., library index number, notes, and reservation information). How do we model this collection of book objects?

Our first response might be to simply introduce a class *Book*, as shown in Figure 6.23a, to capture the set of common features. However, this seemly straightforward answer could lead to a good system with poor performance. In particular, the system would have the same set of information duplicated many times (50 times for the book *Real-Time UML* and 80 times for the book *Real-Time Embedded Systems*). Of course, by separation of concerns, we can refine the model in Figure 6.23a to that in Figure 6.23b by subclassing, using the subclass

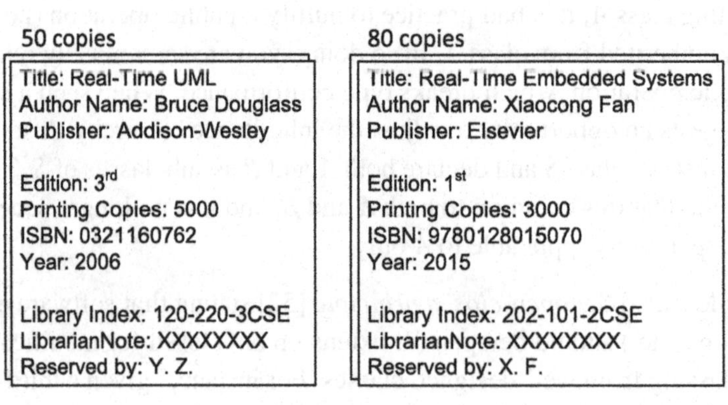

50 copies	80 copies
Title: Real-Time UML Author Name: Bruce Douglass Publisher: Addison-Wesley Edition: 3rd Printing Copies: 5000 ISBN: 0321160762 Year: 2006 Library Index: 120-220-3CSE LibrarianNote: XXXXXXXX Reserved by: Y. Z.	Title: Real-Time Embedded Systems Author Name: Xiaocong Fan Publisher: Elsevier Edition: 1st Printing Copies: 3000 ISBN: 9780128015070 Year: 2015 Library Index: 202-101-2CSE LibrarianNote: XXXXXXXX Reserved by: X. F.

Figure 6.22
A collection of books.

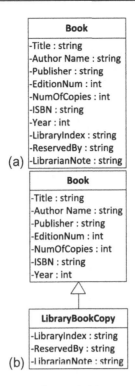

Figure 6.23
(a) A wholistic model; (b) a model improved by subclassing.

LibraryBookCopy to capture copy-specific features. However, this improvement will not address the redundancy issue: each *LibraryBookCopy* object still maintains a copy of all the attribute values specified by the superclass *Book*.

You may be wondering whether class attributes (static variables) can be used to attack the redundancy issue. The answer is no, because a class attribute can hold only one value for all its objects. In this example, if you define the book attributes—say, title—as a class attribute, the system would allow you to model only one kind of book!

At this point, you might be tempted to introduce a separate class for each kind of book, because in so doing, you could use the technique of class attributes to reduce redundancy without causing any interference. Figure 6.24 shows such a model.

It looks like the problem has been solved, but it has not been solved well. Our assumption that the library has only two kinds of books is not realistic. What if after system delivery, the library purchases new books? You would have to keep changing your design and implementation, and before long, both you and the library would suffer from the pain or lose business.

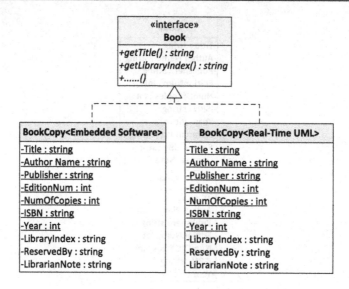

Figure 6.24
Another model of books.

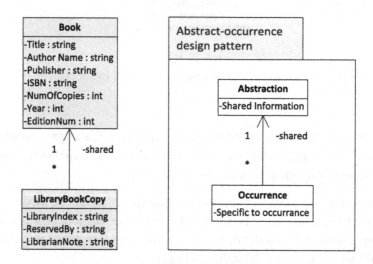

Figure 6.25
A good model of books and the abstraction-occurrence pattern.

Obviously, the model in Figure 6.24 is an overengineered model: subclassing has been applied too much.

We can gracefully address the problem by simply using an association, as shown on the left in Figure 6.25. This model indicates that the system simply needs to create as many *Book* objects as the number of kinds of books, one for each kind, to manage the information common to all the copies. Of course, one *LibraryBookCopy* object needs to be created for each book copy to manage copy-specific information. For the "mini-library" example, the system can create two *Book* objects. Since there are 50 copies of *Real-Time UML* and 80 copies of *Real-Time Embedded Systems*, the system can create 50 *LibraryBookCopy* objects, all of which refer to the same *Book* object for *Real-Time UML*, and 80 *LibraryBookCopy* objects, all of which refer to the same *Book* object for *Real-Time Embedded Systems*.

This solution, in general, is an application of the so-called abstraction-occurrence pattern, which is shown on the right in Figure 6.25, where each *Abstraction* object is shared by many *Occurrence* objects.

6.3.4 Refactoring

The term *"refactoring"* originally referred exclusively to modifying the structure of software code without affecting its functionality [36]. Here, by refactoring a class model, we refer to a series of model transformations that lead to a set of concise, well-named, single-purpose classes connected by semantic relations (inheritance, association, dependency) in a consistent way.

Let us continue our library example to examine a few refactoring techniques.

The first principle is that *a class should capture one and only one key abstraction*. For the *Book* class, we can factor out the information about the book edition because this attribute is not essential to books: some have it, while others do not. We know that different editions of a book are closely related in many ways, but there also exist significant changes. Since each edition of book is also a book, we can introduce a new subclass *EditionOfBook* for *Book* and place the edition information into *EditionOfBook*. In so doing, we have separated the concerns captured by the original *Book* class into two classes.

Actually, there may exist several solutions to a design problem. Here, instead of subclassing, we could also apply the "abstraction-occurrence" pattern, with the *Book* class playing the abstraction role and the *EditionOfBook* class playing the occurrence role. The beauty of this solution is that all the *EditionOfBook* objects of the same book can share a single *Book* object. Undoubtedly, up to this point, the "pattern" solution is better than the solution using subclassing.

However, do remember that the "abstraction-occurrence" pattern has a constraint: each *Abstraction* object must be shared by all the corresponding *Occurrence* objects. This is not true for our book example. Let us expand the model and consider how to model the chapters of a book. It is straightforward to introduce a class *BookChapter* and relate it to the *Book* class by an aggregation. Now we are facing a new issue: different editions of a book may have different numbers of chapters and different chapter arrangements (chapter titles, section

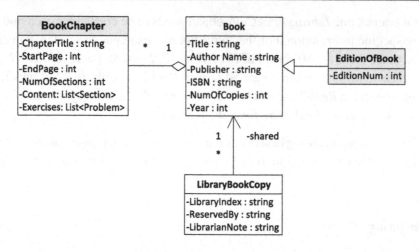

Figure 6.26
A refined class model of books.

numbers, problems, etc.). This is a dilemma for the "pattern" solution because the information about book chapters is not common among all the *EditionOfBook* objects! You may argue that you could resolve the dilemma by pushing down the aggregation relation to the *EditionOfBook* class. This does help, but as we continue evolving the model, we have to push down all the relations from *Book* to *EditionOfBook*, which is no longer a graceful solution. This analysis brings us back to the "subclassing" solution, which is shown in Figure 6.26. After all, information redundancy is tolerable here because unlike book copies, it is rare to see a book with tens of editions.

The second refactoring principle is to *introduce abstract superclasses to capture common features*. What is common among *Book* objects and *BookChapter* objects? They are semantically related because both are creative-writing materials. Among other things, both have a "title" attribute. So, a new abstract class, let us call it *BookItem*, can be introduced to be the superclass of both *Book* and *BookChapter*.

Have you ever seen a book consisting of many "smaller" books? The Bible is one example, and the Chronicles of Narnia is another. To cover these cases, we can add a self-aggregation relation to the *Book* class. Now, we have a model capturing (a) a book is composed of many "inner" books, and (b) a book is composed of many chapters. Since books and book chapters are *BookItem* objects, we can actually say "a book is composed of many *BookItem* objects." This gives us the model shown on the left in Figure 6.27, where one aggregation relation between *Book* and *BookItem* is used instead of the *Book-Book* aggregation and the *Book-BookChapter* aggregation.

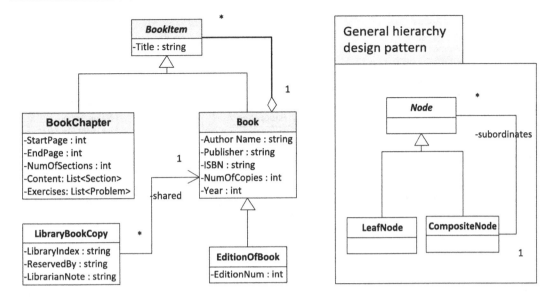

Figure 6.27
A further refined class model of books.

This concise solution, in general, is an application of the so-called general hierarchy design pattern, which is shown on the right in Figure 6.27, where a *CompositeNode* object is composed of many other *Node* objects, each of which can be either a *CompositeNode* object or a *LeafNode* object. Such a recursive aggregation relation terminates at *LeafNode* objects. As an exercise, you can think about how to apply the "general hierarchy" design pattern to model operating system files and directories.

The third refactoring principle is to *adjust features to the appropriate generalization level.* As the refinement process goes on, you need to shift common features, including attributes, operations, and associations, up into superclasses, as well as to shift specialized features down into subclasses. The class model ultimately becomes concise and neat as each of the features is placed into the most general class for which it is appropriate.

Now, let us consider different types of publications as shown in Figure 6.21 to derive a more general class model. Refer to the model in Figure 6.28, and notice the following changes:

- Abstract classes *CollectedWorks* and *SingletonWorks* are added to model collection-like and indecomposable publications. Now, *Book* is simply one kind of *CollectedWorks*, and *BookChapter* is merely one kind of *SingletonWorks*.
- The class *BookItem* is replaced by *Publication*. The aggregation relation is retained but now is between *CollectedWorks* and *Publication*. Let us examine whether the aggregation relation makes sense in this general context. A journal (magazine, newspaper) is a collection of volumes, a volume is a collection of issues, an issue is a collection of

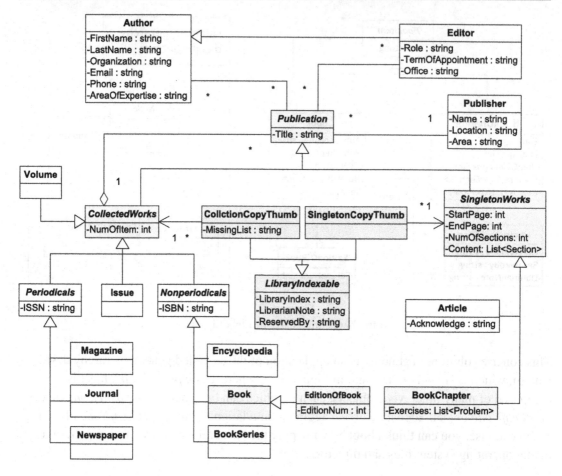

Figure 6.28
A class model of publications.

articles. An encyclopedia is a collection of volumes, a volume is a collection of articles. A book series is a collection of books. A book (or edition of a book) is a collection of book chapters.

- The *LibraryBookCopy* class is renamed *CollectionCopyThumb*, which is a subclass of *LibraryIndexable*, to model copy-specific information. By symmetry, *SingletonCopyThumb* is introduced for a similar purpose.

- The original *Book* attribute *NumOfCopies* is pushed up to the class *CollectedWorks* and renamed *NumOfItem*. *NumOfItem* has different meanings for different types of *CollectedWorks* objects. For instance, for a book object, it still tracks the number of printed book copies. For a journal, it records the number of volumes that have ever been published; for a volume object, it records the number of issues included; and for an issue object, it records the number of printed copies of that issue.

- The attribute *ISBN* is pushed up to the abstract class *Nonperiodicals* because it applies to all nonperiodicals. By symmetry, *ISSN* applies to all periodicals.
- The author information and publisher information originally captured by the *Book* class are now pushed up to the class *Publication* and are further modeled as separate classes by the principle of separation of concerns.
- Some publications have editors instead of authors, and some have both. The class *Editor* is modeled as a subclass of *Author*.

6.4 Object Diagram

Classes exist at design time, whereas objects exist at run time. A UML *object diagram* shows individual objects and their relationships. A class diagram models an infinite set of object diagrams.

Only instance-level relationships can appear in an object diagram. This means, you will not see generalization and dependency relations. An association relation in a class diagram becomes "links" between objects of the corresponding classes.

Figure 6.29 shows an example object diagram. There is a *BookSeries* object linked to three *Editor* objects and one *Publisher* object. The *BookSeries* object also links to two *Book* objects and two *EditionOfBook* objects. The *EditionOfBook* object with the title *Real-Time UML* (third edition) is linked to one *Author* object, one *BookChapter* object for each chapter, and one *CollectionCopyThumb* object for each book copy. As an exercise, you can examine which link corresponds to which association as captured by the class diagram in Figure 6.28.

6.5 Package Diagram

The class diagrams of a nontrivial system often have a tendency to grow very fast as the design process goes on. Packages are UML constructs that enable you to organize model elements into groups, making your UML diagrams simpler and easier to understand.

A package is depicted as a tabbed file folder (a large rectangle with a small "tab" rectangle). The name of the package can be placed within the large rectangle. Alternatively, when the members of the package are shown within the large rectangle, the name of the package is placed within the tab. A *UML package* provides a namespace for its members, each of which may have a public or private visibility (preceding the element name by a "+" for public and "−" for private) that determines whether they are available outside. By default, a member of a package has a public visibility. The public members are always accessible outside the package through the use of qualified names.

Figure 6.30 shows an example of packaging a class diagram. The package *Course Management* has seven members, including five classes and two packages—*CourseRegistrar*

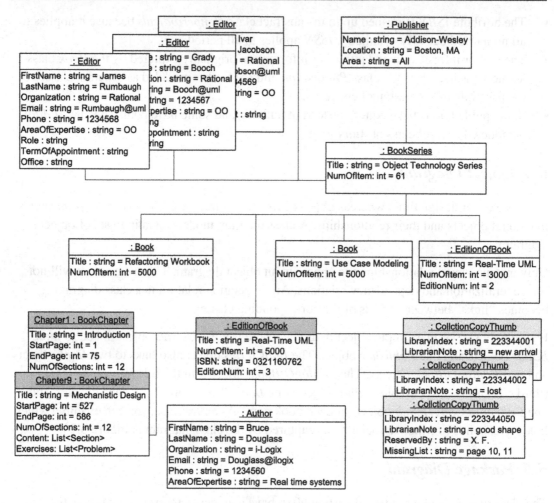

Figure 6.29
An object diagram of library books.

and *CourseMentoring*. Outside the package *CourseManagement*, the *Semester* class can be referred to by *CourseManagement::Semester*, and the *Registration* class can be referred to by *CourseManagement::CourseRegistrar:: Registration*. Notice that the classes *ListedCourse* and *DeliveredCourse* have a "private" visibility; they are not visible outside *CourseManagement*. By default, classes *TextBook*, *Semester*, and *Instructor* have a "public" visibility. Also, the *schedule* attribute of the class *DeliveredCourse* has a "package" visibility; it is visible to all the other classes defined in *CourseManagement*.

There are a few rules of thumb in packaging class diagrams. For instance, classes in the same inheritance hierarchy typically belong in the same package. Classes related to one another via association (aggregation or composition) often belong in the same package.

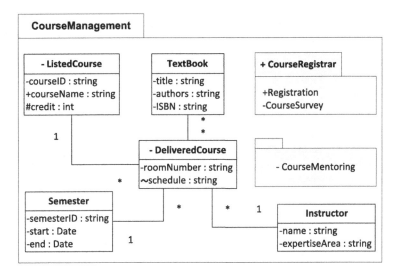

Figure 6.30
A class package diagram.

6.5.1 Package Import

As a namespace, a package can import either individual members of other packages or all the members of other packages. A *package import* is a directed relationship between an importing package and an imported package, indicating that the *names* of the *public* elements in the imported package are all added to the namespace of the importing package. Each public element of the imported package can be referenced in the importing package by its name without a qualifier.

A package import itself has a visibility that can be either public or private. For a *public package import* from the importing package *A* to the imported package *B*, *A* keeps the "public" visibility of all the imported elements originally defined in *B*; for a *private package import*, *A* changes the visibility of all the imported elements to private (i.e., they are no longer visible outside *A*).

A package import is shown using a dashed arrow with an open arrowhead from the importing package to the imported package. A keyword is shown near the dashed arrow to identify which kind of package import is intended. The predefined keywords are «import» for a public package import, and «access» for a private package import.

Figure 6.31 shows the notation for package import. *PackageA* is connected to *CourseManagement* by a public package import relation, which implies that the namespace of *PackageA* contains the names of all the public members of *CourseManagement*, including *Semester*, *TextBook*, *Instructor*, and *CourseRegistrar*, all of which can be referenced within

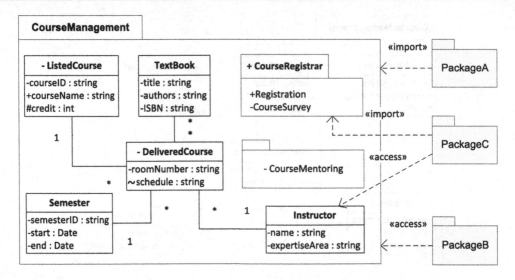

Figure 6.31
Package import.

PackageA by names without using the qualifier *CourseManagement*. Moreover, all these imported elements are visible outside *PackageA* and can be further imported by other packages.

PackageB is connected to *CourseManagement* by a private package import relation, which expands the namespace of *PackageB* in the same way as when «import» is used, except that all the imported elements have a private visibility in *PackageB*.

An *element import* is defined as a directed relationship between an importing package and an element defined in another package, allowing the external element to be imported selectively. By element import, the name of the external element (or its alias, if available) is added to the namespace of the importing package. Similarly to *package import*, «import» is used to annotate a public element import, and «access» is used for a private element import. In Figure 6.31, *PackageC* imports the package *CourseRegistrar* and keeps its public visibility, and imports the class *Instructor* but sets its package visibility to "private."

6.5.2 Package Merge

A package merge is a directed relationship between two packages, specifying an operation that takes the contents of two packages and produces a new package that combines the contents of the packages involved in the merge. In practice, a domain concept might be modeled in various ways (different definitions by different modelers to capture different aspects). Starting from a "base" definition, a concept can be extended in increments, with each increment defined in a separate package. Since the elements with the same name defined in different packages are intended to model the same concept, package merge can be used to derive a combined definition of the concept for a specific need.[18]

Let us clarify a few terms before we come to the details of package merge:

- Merged package: one operand of the merge. It is the package that is to be merged into the receiving package.
- Merging package: another operand of the merge. It is the package that will contain the results of the merge.
- Receiving package: this is merging package before the merge happens.
- Resulting package: this is the merging package after the merge has been performed.
- Matching element pair: this refers to two elements which have the same name and type and refer to the same concept. The two are combined to produce an element representing a new evolution of the concept.
- Merged element:this is a model element that exists in the merged package.
- Merging element:this is a model element in the merging package.
- Receiving element: this is the merging element before the merge happens.
- Resulting element: this is the merging element after the merge has been performed.

A package merge is shown using a dashed line with an open arrowhead pointing from the merging package to the merged package. The keyword «merge» is shown near the dashed line. Figure 6.32 gives an example.

Let us first consider the merge relation between the packages P and R. P is the merged package which defines a class A and a subclass B, while R is the merging package which defines a class A and a class D with a self-association. Notice that P and R define the same concept A differently: in P, class A has an attribute $a1$ of type $T1$ and an operation accepting a parameter of type $T1$; in R, class A also has the same attribute $a1$ but has a different operation $pa2$ accepting a parameter of type $T2$. Before the merge, R is also called the receiving package. After the merge has been performed, the resulting package R is the combination of P and the receiving package R. The resulting package R has three classes:

- A class A with
  - an attribute $a1$ of type $T1$;
  - an operation $pa1$ accepting a parameter of type $T1$; and
  - an operation $pa2$ accepting a parameter of type $T2$.
- A class B that inherits from class A.
- A class D with a self-association.

---

[18] To some extent, the package merge concept seems to implement a version control mechanism at the model level.

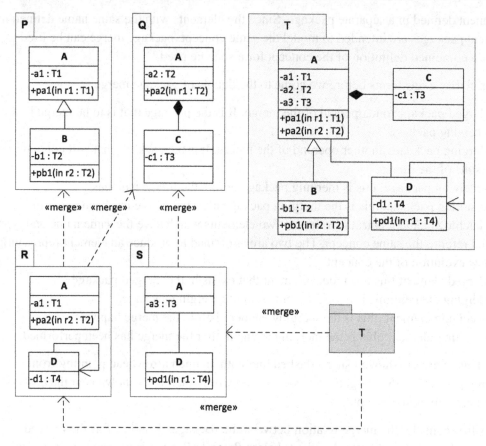

Figure 6.32
Package merge.

If we further consider the merge relation from *R* to *Q*, the resulting package *R* will have four classes:

- A class *A* with
  - an attribute *a1* of type *T1*;
  - an attribute *a2* of type *T2*;
  - an operation *pa1* accepting a parameter of type *T1*; and
  - an operation *pa2* accepting a parameter of type *T2* (the two implementations are also combined into one).
- A class *C* that is a part of class *A*.
- A class *B* that inherits from class *A*.
- A class *D* with a self-association.

By doing a similar analysis, you will find that the resulting package *S* will contain three classes—*A*, *C*, and *D*. In addition, the two package merges from the empty package *T* will eventually transform *T* into a package that combines all the definitions in packages *P*, *Q*, *R*, and *S*.

Problems

6.1 Identify some class invariants for the model given in Figure 6.4, and write OCL expressions for those invariants.

6.2 When do we use associations? When do we use aggregations? When do we use compositions?

6.3 What is the Liskov substitution principle? Why is it important?

6.4 What is the abstraction-occurrence pattern? When is it useful?

6.5 What is the general hierarchy pattern? When is it useful?

6.6 Figure 6.33 gives a class diagram model of a train control system [34]. Explain each of the associations used in the design. Can you tell the design pattern used in the model?

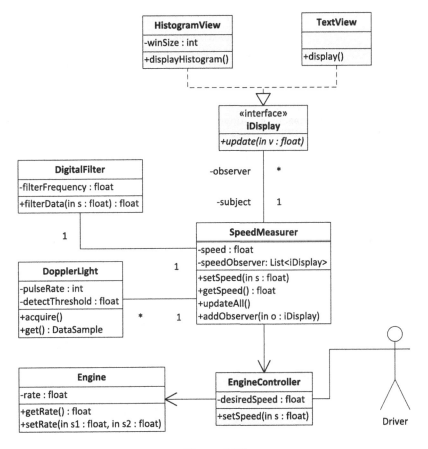

Figure 6.33
Train control system.

6.7 In embedded systems, resources can be classified in many ways. As far as the purpose of use is concerned, there are resources used for communication (e.g., shared memory, pipes), for processing tasks (e.g., processor), and for specific services (e.g., hardware devices). In terms of activeness, there are active resources (e.g., threads, hardware devices) and passive resources (e.g., shared memory). Resources can also be classified into protected resources (e.g., memory blocks, files) and unprotected resources. Figure 6.34 illustrates a concept called "power type." A power type is a class with its subclasses grouped into several collections. Each collection is referred to as a *generalization set*. Each generalization set represents an *orthogonal dimension* of specialization of the general class. Each generalization set of a class defines a particular set of generalization relationships that describe the way in which the class may be subclassed. In Figure 6.34, *Resource* is a power type, which is subclassed along three orthogonal dimensions named *activenessKind*, *protectionKind*, and *purposeKind*. Work out a class diagram to show that *InsurancePolicy* is a power type.

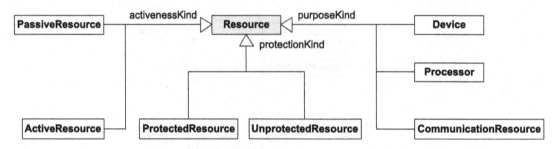

Figure 6.34
Notation of power type.

Architecture Modeling in UML

Contents

> *There are two ways of constructing a software design: One way is to make it so simple that there are obviously no deficiencies and the other way is to make it so complicated that there are no obvious deficiencies.*
>
> *C.A.R. Hoare*

7.1 Levels of Architectural Abstraction

At the systems engineering level, a system is the integration of many designs/implementations, each from a unique angle [33]. For example, an embedded system may involve engineering activities from several disciplines: mechanical engineers see the mechanical parts such as engines and heat transfers, electrical engineers see the electrical parts such as power supplies and motors, computer engineers see the circuit parts such as processors and wirings, and software engineers "see" the intangible parts such as task scheduling and collaboration.

From each disciplinary perspective, a system can be further organized at several abstraction levels. Let us now take the software engineering perspective.

At the system level, a software system has a boundary which represents its scope of responsibility or functionality. Through its interface, the system provides services to or request services from human users or other systems.

At the *subsystem level*, the system is treated as the integration of many interacting subsystems. The term *"software architecture"* comes into play at this level. It is used to refer to the high-level structure of a software system, concerning key software elements (or modules) and their relationships [34]. For example, a software system may be composed of several subsystems: one is designated for data management, one is responsible for security control, and yet another supports graphical user interfaces. All the subsystems collaborate together, directly or indirectly, to form a complete functional system.

Below the subsystem level is the *component level*, where a subsystem is formed by many components that collaborate with each other through their respective ports and interfaces. For example, a home intrusion detection system may have a subsystem running at the security center and a subsystem deployed within each house. The subsystem at the security center may be composed of a monitoring component, a communications component, and an image processing component. The subsystem in a private dwelling may be composed of a control panel component and a sensing component with several sensors.

The lowest abstraction level is the object level, where each component is composed of many role parts to be played by objects. A component can be passive in the sense that the contained objects execute only in response to an external service invocation. An *active* component, on the other hand, has at least one active object; each active object has its own thread of execution.

The question now is how to document the design of components and subsystems of a complex system. We already know that a UML package diagram can be used to organize classes and data types belonging to certain software elements (or modules). However, UML package diagrams are not appropriate for capturing intermodule relations other than "import" and "merge." Indeed, a package merely serves as a namespace to group semantically related concepts (classes) together; it does not dictate how objects will be organized and deployed at run time.

We next introduce the notion of a structured class, which is the basis for modeling tasks, components, and subsystems.

Throughout the rest of this chapter, we take the computer engineering perspective, showing how to model components, subsystems, and the complete system of the AT91SAM9G45 evaluation board [13].[1]

7.2 UML Structure Diagram

A *structured class* is a class whose behavior can be completely or partially described through interactions between parts.

---

[1] UML structured classes are mainly used to model software architectures. In computer engineering, block diagrams are typically used in the design of circuit boards. Here, we adopt the notion of a structured class to demonstrate that it can be useful in modeling hardware architectures as well.

A structured class can have properties, each of which is an element with a name, a type, and a multiplicity. When an instance of the structured class is created, depending on its multiplicity, a set of instances can be created for each property. The term "part" refers to a special kind of property that is *strongly owned* by the containing class.

A *port* is a property of a structured class that specifies a distinct interaction point between the class and its environment (public port) or between the class and its internal parts (protected port). A port may be associated with *provided interfaces*, specifying the services the class offers to its environment, as well as *required interfaces*, specifying the services that this class expects from its environment. A simple port has just a single required or provided interface; a complex port can have as many provided and/or required interfaces as needed. The provided and required interfaces at a port completely characterize the interactions that may occur between the class and its environment through that port. Multiple ports can be defined for a class, enabling different types of interactions to be distinguished on the basis of the port through which they occur.

By decoupling the internals of a class from its environment, ports allow a class to be defined independently of its environment, making the class reusable in any environment that conforms to the interaction constraints imposed by its ports.

The properties, parts, and ports of a structured class are related by connectors (sometimes also called contextual associations). A *connector* specifies a link that enables communication between two or more property instances. The link may be realized by something as simple as a pointer or by something as complex as a network connection. In contrast to associations, which specify links between any instances of the associated class, connectors specify links between those property instances involved merely in a specific instance of the structured class.

Besides parts and ports, a structured class may also have properties that are "weakly" contained by reference. Such a property is called a "referenced property." Whenever an instance of a structured class is deleted, excepts for referenced properties, all the instances corresponding to its parts, connector links, and ports are destroyed recursively.

A structured class can be shown in a UML internal structure diagram, a special kind of composite structure diagram.[2] As an example, Figure 7.1(a) shows a structured class *Interrupt Controller*. The structured class itself is shown by a rectangle separated into a name compartment and a body compartment. Inside the body compartment, a part is shown by a box with a solid outline. The box symbol may contain a string of the form *name:type [int]*, specifying the property name, type, and multiplicity. For example, *d: EdgeSignalDetector[1]* denotes a property *d* which is an object of type *EdgeSignalDetector*. A referenced property is

---

[2] A collaboration use diagram is another type of composite structure diagram, and is generally used to explain how a collection of cooperating objects achieve a joint task or set of tasks [10]. A collaboration is often defined in terms of the roles typed by interfaces, and is especially useful for describing design patterns.

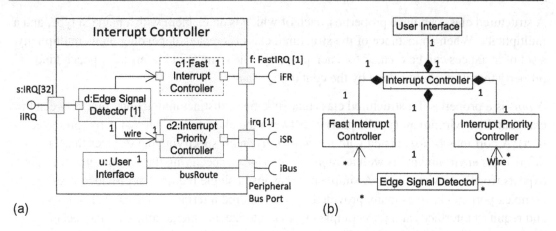

Figure 7.1
An example UML internal structure diagram.

shown by a box with a dashed outline (e.g., *c1: FastInterruptController*[3]). The type of property can be elided, implying that the property has a type which is an anonymous class nested within the namespace of the structured class. The multiplicity for a property may also be shown in the top-right corner of the box symbol (e.g., the *UserInterface* part). When elided, the default multiplicity is one.

A port of a class is shown as a small square, placed either overlapping the boundary of the class box (public port) or inside the class box (protected port). All the ports in Figure 7.1(a) are public ports. A port can be optionally annotated by a string of the form *name: type [int]*, specifying the port name, port type, and multiplicity. A provided interface may be shown using the "ball" notation attached to the port. A required interface may be shown by the "socket" notation attached to the port. If there are multiple interfaces associated with a port, the list of interfaces is separated by commas. The class *Interrupt Controller* has four ports declared:

(1) *s:IRQ[32]:* a port named *s* of type *IRQ* with a provided interface *iIRQ*. At run time, an object of *Interrupt Controller* has 32 interrupt request port instances, each of which can be individually referenced. Through these ports, an *Interrupt Controller* object supports up to 32 interrupt request lines from peripheral devices.

(2) *f:FastIRQ[1]:* a port named *f* of type *FastIRQ* with a required interface *iFR*. Through this port, an *Interrupt Controller* object can request a fast interrupt service from a connected processor.

---

[3] There is no reference property in this example. We purposely treat *c1: FastInterruptController* as a reference property to merely show its notation.

(3) *irq[1]:* a port named *irq* of anonymous type with a required interface *iSR*. Through this port, an *Interrupt Controller* object can request a standard interrupt service from a connected processor.

(4) *Peripheral Bus Port:* a port of an anonymous type with a required interface *iBus*. Through this bus port, an *Interrupt Controller* object can access peripheral devices.

The required and provided interfaces of a port specify everything that is necessary for interactions through that interaction point. If all interactions of a class with its environment are achieved through ports, then the internals of the class are fully isolated from the environment [10]. This allows the class to be used in any context that satisfies the constraints specified by its ports.[4] In Figure 7.1(a), the classes *Interrupt Controller, FastInterruptController, and InterruptPriorityController* are completely isolated from their respective environments. Specifically, *Interrupt Controller* can be designed and implemented without any knowledge of the environment where it will be embedded. Afterward, an *Interrupt Controller* object can be connected to any microcontroller and still function properly as long as the two parties obey the constraints specified by the provided and required interfaces.

A connector is drawn using the notation for association; it thus can have a name and end names, and can be bidirectional or unidirectional. The multiplicity on a connector end indicates the number of instances that may be connected to each instance of the role on the other end. Connectors need not necessarily attach to parts via ports. For instance, the *wire* connector attaches to a port of the *InterruptPriorityController* object but attaches directly to the *EdgeSignalDetector* object. Recall that when a port and/or an interface is used, this implies that the corresponding service is completely encapsulated and the port is the only way to invoke the service.[5] Since *wire* is unidirectional, the object *c2* would never invoke a service from the object *d*, and it thus makes sense to attach *wire* directly to *d* without a port or an interface. The connector *busRoute*, although bidirectional, is also attached directly to the *UserInterface* object *u*. In general it is preferable to use ports, but the choice of whether to use a port or not really depends on judgment and the nature of the problem.

The whole-part composition relations of a structured class captured by an internal structure diagram can also be modeled by using a class diagram. However, an internal structure diagram is semantically rich. Let us compare Figure 7.1(b) and Figure 7.1(a) to see the differences:

(1) An internal structure diagram shows not only the structure of a complex class, but also captures run-time information such as objects and object roles, which is beyond the capability of a class diagram.

---

[4] In so doing, you are actually applying the so-called *facade pattern*, which is commonly used to reduce intercomponent dependencies and to improve reusability and maintainability.

[5] A port by default is a service port. Sometimes in an internal structure diagram you may also see state-like symbols (rounded rectangles) with names, each of which represents a behavior. A port of a class is called its "behavior port" when it is linked by a connector to one of its behaviors.

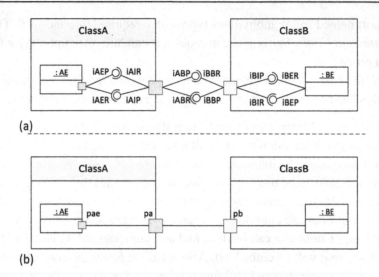

Figure 7.2
Two forms of port connections: (a) interface agreements are shown explicitly; (b) interfaces are suppressed.

(2) An internal structure diagram gives a specific context (the enclosing class) to make sense of the contained relations, whereas a class diagram, owing to lack of context, can represent relations in only a general sense. For example, Figure 7.1(b) specifies that any instance of class *EdgeSignalDetector* can be linked to an arbitrary number of instances of class *InterruptPriorityController*. This might be true before board production because a designer is free to use one *EdgeSignalDetector* to test many interrupt priority controllers. In contrast, Figure 7.1(a) specifies that within the context of class *Interrupt Controller*, the *EdgeSignalDetector* object playing the role of *d* may be connected to only one *InterruptPriorityController* object playing the role of *c1*. This exactly captures the postproduction board composition. Of course, instead of modeling the general case, you could modify Figure 7.1(b) such that the *wire* association becomes a one-to-one relation. This, however, is still not as good as the model in Figure 7.1(a), which also asserts additional constraints on the linked instances: the two objects are not related by chance; they are related (linked) merely because they were owned by the same *Interrupt Controller* object. The model in Figure 7.1(b) would not prevent such a situation where an *EdgeSignalDetector* object is linked to one *InterruptPriorityController* object but they belong to two different *Interrupt Controller* objects.

(3) An internal structure diagram exposes services by ports only. As will be shown in the sections below, the notation of a structured class can be carried over to model large-scale concepts such as components and (sub)systems.

Connectors between ports can be shown in two ways. Figure 7.2(a) shows interface agreements explicitly—say, the service provided by interface *iABP* meets the needs specified by interface *iBBR*. Figure 7.2(a) also indicates that the ports on the boundary of a structured class serve as a service relay: the service specified by interface *iABP* is actually offered by the *AE* object—an internal part of *ClassA*; similarly, the needs specified by interface *iBBR* are exactly derived from *iBER*—the needs of the *BE* object inside *ClassB*.[6] Going deeper, we have the following relations among the interfaces in Figure 7.2(a):

- $iAEP = iABP = iBIP$;
- $iAIR = iBBR = iBER$;
- $iAER = iABR = iBIR$;
- $iAIP = iBBP = iBEP$.

Figure 7.2(b) shows a simplified form where the provided and required interfaces are suppressed. This can be used at or below a certain level of abstraction to prevent balls and sockets from cluttering a diagram. A connector is called a *delegation connector* when it connects a public port of a class to its parts. The connector between ports *pa* and *pae* and the connector between *pb* and the anonymous *BE* object are delegation connectors.

Let us look at a slightly bigger structured class. Figure 7.3 gives a class model of the ARM926EJ-S processor [12]:

- An ARM926EJ-S processor is composed of a processor core, a memory management unit, two memory interface units, a 32 KB data cache, a 32 KB instruction cache, a data bus interface unit, an instruction bus interface unit, and an in-circuit emulator.
- Eight public ports are shown on the boundary of the *ARM926EJ-S Processor* class, where

  - *ice* is a port providing debugging functions such as downloading code and single-stepping through programs. The interface *iJTAG* uses the JTAG protocol.
  - *d* is a port providing debugging information (e.g., to serial port). The interface *iDEBG* uses the Xmodem protocol.
  - *t* is a trace port providing a real-time trace capability for the processor core. The provided interface *iETM* is a trace protocol. An external trace port analyzer can be used to capture the trace information.
  - the two anonymous ports connected to the two bus interface units are used to request bus services for accessing data or instructions in external memories.
  - *c* is a port for clock signals (400 MHz).
  - *i* refers to two ports for servicing interrupt requests.

---

[6] Of course a public port of a structured class can be connected with more than one parts of that class. In such a case the port acts as a service integrator as well as a service relay.

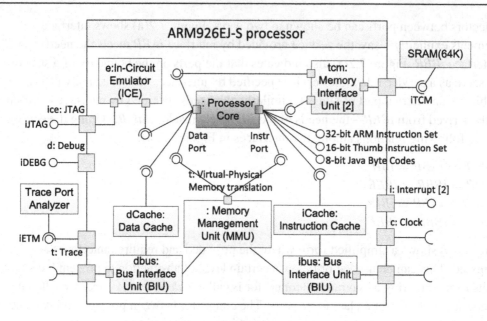

Figure 7.3
The internal structure of the ARM926EJ-S processor.

- the port connected to the two *Memory Interface Unit* objects is used to request access to fast memory devices. The required service is described by the interface *iTCM*, which, in this case, is fulfilled by a 64 KB static random-access memory chip.
- The *Memory Management Unit* object offers a memory address translation service to the processor core.
- The processor core object is encapsulated by two public ports: a data port and an instruction port.
- The data port of the processor core
  - offers a service to *e*—an *In-Circuit Emulator* object;
  - relies on data-access services from *tcm*, *dCache*, and the anonymous *Memory Management Unit* object.
- The instruction port of the processor core
  - has three provided interfaces—accepting a 32-bit ARM instruction set, a 16-bit Thumb instruction set, and eight-bit Java bytecodes;
  - relies on instruction-access services from *tcm*, *iCache*, and the anonymous *Memory Management Unit* object.

7.3 Modeling Components

Component is a *design-time concept* for organizing run-time objects. It is especially useful in component-based system structuring and development.

A *component* is an autonomous unit with its internals hidden and inaccessible other than through the provided and/or required interfaces. The provided interfaces specify a formal contract of the services that it provides to its clients, and the required interfaces specify what it requires from other components or parts of the system. The environment is able to interact with a component as long as the environment complies with the component's contract expressed by the provided and required interfaces. This makes a component substitutable: it can be replaced at run time by any other component that offers/requires equivalent or compatible services [10].

Components in UML are just structured classes. A component is shown as a rectangle with a name, optionally followed by the keyword «component» or a component icon (a component box with two tiny interface boxes on the border). Just like any structured class, a component can contain ports, parts, and connectors, which, respectively, define interfaces, roles, and their connections. Note that objects of the same type (class) may be instantiated to play different roles of a component.

A port can be a simple port—exposing just a single required or provided interface—or a complex port—exposing more than one interface (required and/or provided). Two related parts inside a component can be connected directly by a connector, or indirectly through their ports. Alternatively, the ball-and-socket notation can be used to make the contract explicit: the offered interface (ball) on the port of one part meets the required interface (socket) on the port of the other part. Although visually the ball is connected to the socket, in fact it is the two parts that are connected.

As a side note, a component, being a structured class, can inherit from supertype components and can be extended by subtype components. In such a situation, the Liskov substitution principle (type conformance) applies to the contracts defined on the public ports. The inherited interfaces are indicated on the component diagram by preceding the name of the interface by a forward slash.

Figure 7.4 gives a diagram depicting the *Peripheral Composite* component of the AT91SAM9G45 chip [16]. A *Peripheral Composite* has 13 parts; there is one instance for each part except for *Media Card Controller* and *USART* (according to the multiplicity, a *Peripheral Composite* has two media card controllers and four universal synchronous/asynchronous receiver/transmitter units). The *Peripheral DMA Controller* object manages four direct memory access devices: an LCD controller, an Ethernet controller, a USB controller for programming the AT91SAM9G45 chip, and a controller for normal USB connections.

A *Peripheral Composite* object also has 12 public ports, each of which is connected to an internal part through its port and offers an interface to the external users.

Figure 7.5 gives a more complicated diagram depicting the *System Controller* component of the AT91SAM9G45 chip. A *System Controller* contains six parts: an instance of a structured

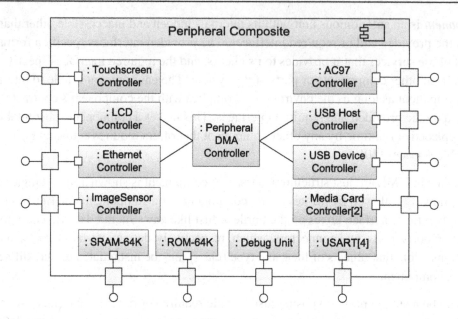

Figure 7.4
A UML Component Diagram: Peripheral Composite of the AT91SAM9G45 chip.

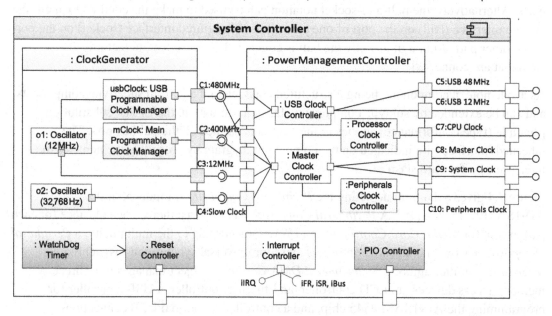

Figure 7.5
A UML Component Diagram: System Controller.

class named *ClockGenerator*, an instance of a structured class named *Power Management Controller*, an instance of a structured class named *InterruptController* (see Figure 7.1), a watchdog timer, a reset controller, and a set of parallel I/O controllers. The internal structures of *WatchDog Timer*, *Reset Controller*, and *PIO Controller* are not our concern here.

A *ClockGenerator* object contains four parts: an *Oscillator* object *o1* operating at a frequency of 12 MHz, an *Oscillator* object *o2* operating at a frequency of 32,768 Hz, a *USB Programmable Clock Manager* object *usbClock*, and a *Main Programmable Clock Manager* object *mClock*. Through the four public ports, a *ClockGenerator* object offers clock signals at 480 MHz (port C1), 400 MHz (port C2), 12 MHz (port C3), and 32,768 Hz (port C4).

A *PowerManagementController* object contains four parts: a USB clock controller, a master clock controller, a clock controller for the processor, and a clock controller for peripherals. A *PowerManagementController* object has 12 public ports. On the left border there are six ports, each of which requires clock signals of a certain frequency offered by the clock generator. On the right border there are also six ports, offering clock signals to the system, each of which is linked by a connector to a public port of the *System Controller* object.

The *Interrupt Controller* object has a port with one provided interface *iIRQ* and three required interfaces *IFR, iSR,* and *iBus*. Since this port is connected to a public port of the *System Controller* object, the provided/required interfaces are thus available to the environment of the *System Controller* object.

7.4 Modeling Subsystems

A *subsystem* is simply a large-scale component, formed by assembling multiple related components. Thus, a subsystem diagram is visually the same as a component diagram, except that it can be optionally annotated by the stereotype «subsystem» or a subsystem icon (a fork symbol).

Figure 7.6 gives a subsystem diagram, depicting the composition of the AT91SAM-9G45 chip [16]. It has four components: a system controller, a peripheral composite, an ARM926EJ-S processor, and a bus matrix. As labeled, the processor has two bus ports connected to the bus matrix, three system ports (one for clock signals and two for interrupts) connected to the system controller, and a port with four provided/required interfaces connected to the peripheral composite.

As another example, Figure 7.7 shows the Internet Protocol stack as a subsystem composed of subsystems. Internally, the subsystems inside the Internet Protocol stack form a layered architecture, with each offering a service to its immediate upper layer. At the same time, the service offered at each layer can be used by the clients through the corresponding public port. For example, a Web browser relies on application-layer protocols such as HTTP and SMTP; the OSPF protocol functions above the network layer.

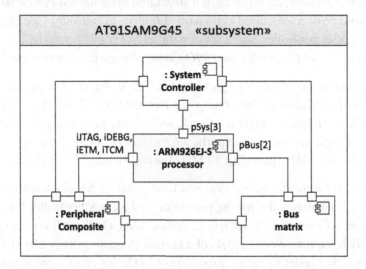

Figure 7.6
The AT91SAM9G45 subsystem.

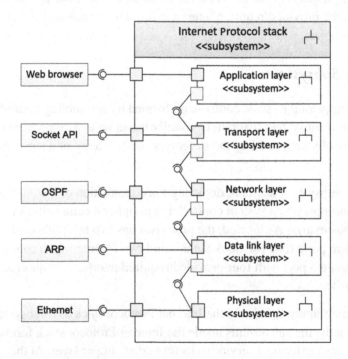

Figure 7.7
The Internet Protocol stack in UML.

7.4.1 Class Diagram Versus Subsystem Diagram

Let us use another example to see the differences between class diagrams and subsystem diagrams.

In Figure 7.8 we use a class diagram to model the key concepts in a data processing subsystem. Note that here the term "subsystem" is used as a synonym of "package," quite differently from the subsystem concept defined in UML. Figure 7.9 shows the subsystem diagram representing a run-time configuration instantiated from classes defined in the data processing package.

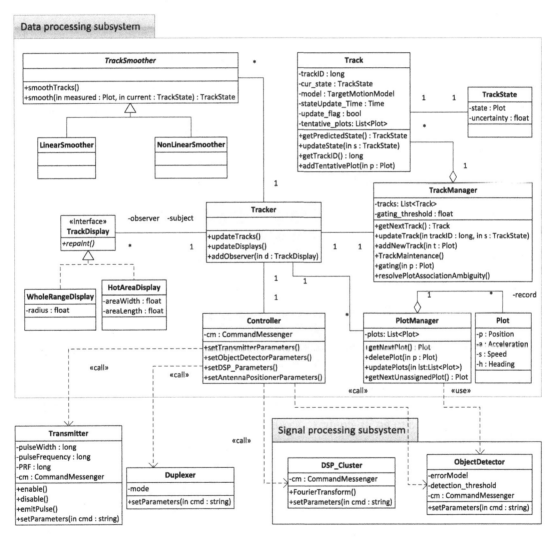

Figure 7.8
Radar data-processing subsystem in a class diagram.

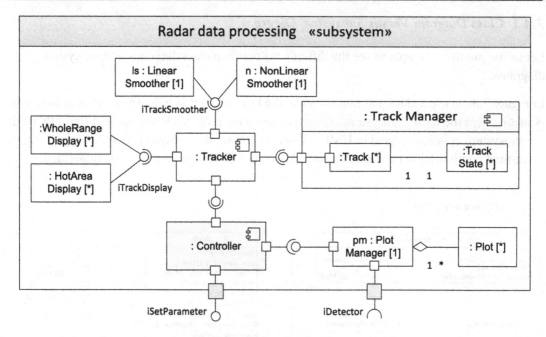

Figure 7.9
Radar data processing subsystem diagram.

Obviously, the class diagram is a static model, while the subsystem diagram is a dynamic model. Particularly,

- a class diagram defines a template for infinitely many object diagrams, specifying object relationships in a general sense. In a class diagram, the objects of one class can play different roles specified by *different associations*. As a comparison, a subsystem diagram defines a template for many subsystem instantiations, specifying object relationships in terms of roles and role fulfillment under a specific context. In a subsystem diagram, multiple objects of the same class can be referenced by one part to play the same role or by *different parts* to play different roles.
- a class is typically defined in a single place (package). Objects of a class can exist in many places of a system. Neither class diagrams nor object diagrams can specify where the objects are supposed to go. In contrast, a subsystem diagram (and its kinds such as structured class diagram and component diagram) dictates a specific environment (context) where the contained objects will reside.
- a subsystem diagram (and its kinds) hides the internal structure and exposes its services only through the public ports. For example, in Figure 7.9, the subsystem offers a service specified by the provided interface *iSetParameter*, and needs a service specified by the required interface *iDetector*. Interaction points cannot be explicitly specified in class diagrams. In contrast, a class diagram uses visibility to hide or expose attributes and behaviors.

- in a subsystem diagram, an interface can be used to define a part type or to label the ball or socket associated with a port. The interface detail is seldom shown. In a class diagram, the detail of an interface is typically shown, especially when it is further extended. For example, the *TrackDisplay* interface in Figure 7.8 plays the "observer" role of the *Observer* design pattern. Its detail is suppressed in Figure 7.9, and the ball-and-socket notation is used instead.
- generalization is a class-level relation, and it is typically not used in subsystem diagrams. It is better to capture generalization in class diagrams.
- in a subsystem diagram, the detail of a class (as a type of some part) is suppressed. It is better to capture class features in a class diagram.

7.5 Modeling a Complete System

A system is composed of subsystems. Figure 7.10 gives a "simplified" system diagram for the AT91SAM9G45 evaluation board [13].

The evaluation board contains the AT91SAM9G45 chip and many other objects, such as an LCD, a touchscreen, USB ports, debug ports, an Ethernet port, on-board buttons, joysticks, and memory chips. In particular, the *Test Program* part has a dashed outline, indicating that it is not physically owned by the board; a user can load a different test program to the NAND memory chip.

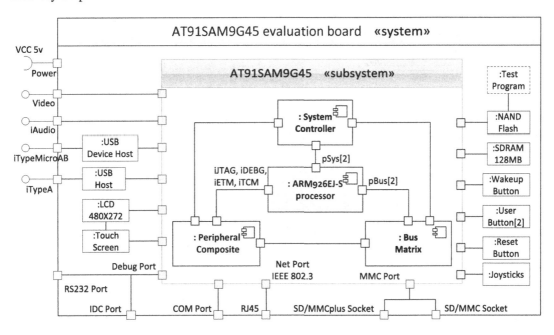

Figure 7.10
UML system diagram.

This concludes the modeling of system architectures at various abstraction levels.

7.6 Deployment Diagram

A software artifact is a physical piece of information that is used or produced by a software development process, or by the execution of a system. Examples of artifacts include model files, source files, scripts, and binary executable files.

A node is a computational resource upon which artifacts may be deployed for execution. A node can be a device (e.g., a computer) or an execution environment (e.g., a Web services container). In UML deployment diagrams, a node is shown as a figure that looks like a three-dimensional view of a cube. Nodes can be connected to each other by associations or links, which indicate communication paths for the nodes to exchange signals and messages.

A deployment is a dependency relationship from one or more artifacts to a node. Figure 7.11 gives an example deployment diagram, representing the execution architecture of a radar data processing subsystem. A deployment can be shown by placing an artifact within the node to

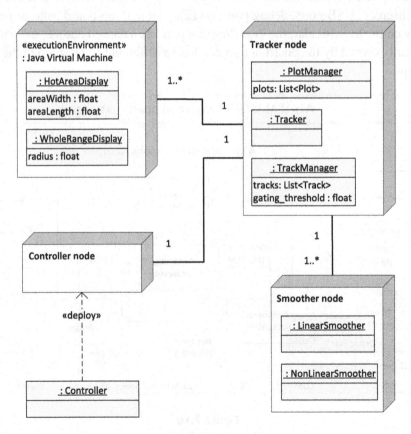

Figure 7.11
A UML deployment diagram.

be deployed. For example, the Tracker object is contained in the Tracker node. Alternatively, it can be shown by a dependency labeled «deploy» that is drawn from the artifact to the node. In Figure 7.11, the Controller object is deployed in the Controller node.

Problems

7.1 For a complex system, why is it necessary to represent its system architecture at multiple abstraction levels?

7.2 What is a structured class? How does a structured class expose its services to the external users?

7.3 How does a component interact with its environment?

7.4 Find an example system to practice component modeling and subsystem modeling.

CHAPTER 8

Fundamental UML Behavioral Modeling

Contents

> *To map out a course of action and follow it to an end requires courage.*
>
> **Ralph W. Emerson**

8.1 Use Case Diagram and Use Case Modeling

In software engineering, the term "actor" is used to refer to a role that a user can play when she/he/it interacts with a system. Here, a user can be a person, a company or organization, another system (hardware or software), or a physical environment [26].

Potential actors can be identified from the stakeholders of the system under consideration. However, many stakeholders are not actors if they never interact directly with the system [26]. Moreover, while interacting with a system, a user may play multiple roles, thus acting as different actors.

The concept "use case" was originally introduced by Jacobson [41] and was later developed by others [19, 26, 46]. A use case of a system describes a main sequence of interactions through which one or more actors work with the system to accomplish a specific goal.

8.1.1 Use Case Diagram

In system requirements analysis, the relationships between use cases and actors are represented in one or more UML *use case diagrams*. A use case diagram modeling of an online social game system is shown in Figure 8.1.

Real-Time Embedded Systems. http://dx.doi.org/10.1016/B978-0-12-801507-0.00008-0

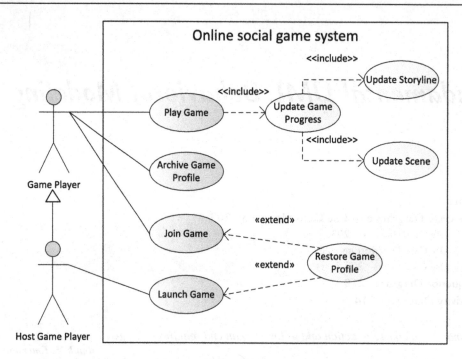

Figure 8.1
A use case model of a game system.

As shown in Figure 8.1, the system under consideration is isolated from the rest by a system boundary (box). Stick figures are used to represent system actors. In this example, we have two actors: Game Player and Host Game Player, which are connected by a generalization relation (solid line with triangle head). In Chapter 6 we have already seen the generalization relation used for class inheritance; it is used here to state that the Host Game Player is a special Game Player—it inherits all the actions that the Game Player can do with the system.

A use case is denoted by an oval labeled by an active verb phrase, representing a sequence of actions (activities or tasks) to be performed by a user and/or the system. An actor can be related to a use case by a communication link, over which information or control flows between the actor and the system. In Figure 8.1, there are three communication links between the Game Player and the system: the Game Player can join a game, play a game, and archive personal game profiles. The system has only one use case "Launch Game" connected to the Host Game Player, meaning that only the Host Game Player has the privilege to initiate a game. However, owing to the generalization relation, the Host Game Player can also conduct the three use cases linked to the Game Player, even though they are not explicitly linked to the

Host Game Player. Generally, the set of use cases directly linked to external actors represent the main system functionality (or "functional requirements" in software engineering jargon).

Some use cases are *not* directly linked to an external actor. In such a case, they are typically connected with other use cases by two types of internal links: include and extend. In UML, *stereotyping* is an extensibility mechanism that allows designers to extend the vocabulary of UML in order to create new model elements from existing ones. A stereotype is rendered as a name enclosed by guillemets «». The UML standard predefines a few built-in stereotypes, including «include» and «extend». Both are used to label the dependency relations (dashed line with a stick arrowhead) between use cases.

X «include» Y means that X has to include Y in the flow of events: in the process of completing the tasks in use case X, tasks in use case Y will be completed at least once. An analogy in programming would be sub-procedure calls. The main reason of separating Y from X is for factoring commonality to achieve potential reusability.

X «extend» Y means that the base logic Y may invoke the extra logic X under certain conditions: X contains a few extra tasks that go above and beyond the tasks that are performed in Y. It is a dependency relation where the extending use case X defines logic that may be *conditionally* required during the execution of the base use case Y. An analogy in programming would be exception handling.

In Figure 8.1, the Play Game use case includes the Update Game Progress use case, which in turn includes the Update Storyline use case and the Update Scene use case. This means that the logic defined in the Update Game Progress use case will be triggered at least once for every input activity when a user is playing a game.

The Restore Game Profile use case extends both the Join Game use case and the Launch Game use case. Here «extend» is used because the Restore Game Profile use case is triggered conditionally by the system:

```
As a player is launching/joining a game, if his/her game profile exists (say, this
is not the player's first time to play the game and the player archived his/her
profile before), the system would automatically load the player's profile.
```

If the Restore Game Profile use case is designed as a main functionality of the system—a task that a game player can directly initiate, then it could be connected to the Game Player via a communication link.

As another example, Figure 8.2 gives a use case diagram for a complex system that consists of two subsystems.

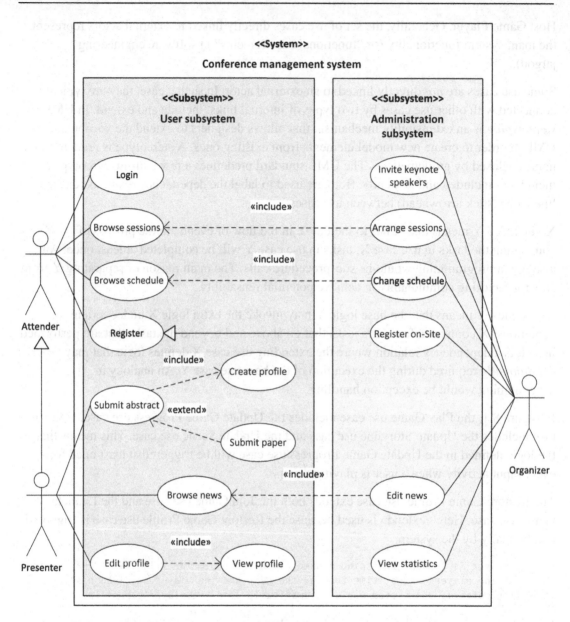

Figure 8.2
A use case model of a system with subsystems.

8.1.2 Use Case Descriptions

Use cases capture the scope of functional requirements of a system from the users' perspective,[1] which is critical to the entire software development process.

- In requirements analysis, use cases allow stakeholders to virtually go through all the critical scenarios related to the system functionality. This can help lay out the business workflow, reduce ambiguity in requirements, elicit implicit requirements,[2] and clarify the primary capabilities of the system.
- In the design phase, use cases suggest a potential system partition—objects closely related to a use case can be organized into one package.
- In system implementation, use cases can be used to plan development iterations—use cases with higher business values are implemented and evaluated with higher priorities.
- In user acceptance testing, use cases are important sources for generating test plans or test cases. Test cases can be created by enumerating all the potential sequences of user-system interactions.

However, as we can see from Figures 8.1 and 8.2, a use case diagram is nothing but a list of use case names offered to the potential users of the system. It captures nothing about the detailed course of actions that an actor or the system may perform. Software practitioners typically use a table-like template to document use case details. One such template is given in Table 8.1.

In this template, a use case is structured into three sections: (a) a section describing meta-level information such as the use case name, a unique ID, goal, scope, level, primary actor, trigger action, preconditions, and postconditions. Their meanings are given in Table 8.1; (b) a success scenario section describing the sequence of actions when everything goes well; and (c) an extension section containing a few branching blocks describing what to do when certain condition happens (i.e., when it is necessary to do special processing, or when something goes wrong).

The extension section of a use case is really a collection of stripped-down use cases [27].[3] The greatest value of a use case lies not in the main success scenario, but in those alternative behaviors captured in the extension section, which need to be carefully considered in system design, testing, and validation.

Each entry in the extension section has

- an index referencing the relevant action step given in the success scenario section;

---

[1] Use cases are not well suited for capturing computing logics (such as algorithm or mathematical requirements) or nonfunctional requirements (such as platform, performance, timing, or safety-critical aspects) [27]. For these cases, UML activity or sequence diagrams are more appropriate.

[2] Sometimes use cases can be supported by prototyping to elicit requirements—say, for user interfaces, storyboarding can be used with hand-drawn designs.

[3] In such a sense, a use case actually describes a set of scenarios with a common user goal.

Table 8.1 A template for documenting use cases

Use Case	#num: < the use case name as a short active verb phrase>
Goal in Context	< what the users want to accomplish by this use case>
Scope	< the system (or one of its subsystems) in which this use case belongs>
Level	< one of summary, primary task, subfunction>
Primary Actor	< the primary role who participates in this use case>
Preconditions	< the relevant situation description prior to the start of this use case>
Minimal Guarantee	< the relevant situation description upon an unsuccessful pass of the use case>
Success Guarantee	< the relevant situation description upon a successful pass of this use case>
Trigger	< the action upon the system that starts this use case>
Success Scenario	Action step
1	< the first step of the success scenario>
2	< the second step of the success scenario>
...	......
i	< a reference to another use case related by «include»>
...	......
n	< the last step of the success scenario>
Extension Step	Branching action (a step may have several branches)
2a	< the condition causing this branch>
	2a1: < a reference to another use case related by «extend»>
2b	< the condition causing this branch>
	2b1: < the first step of this branch>
	2b2: < the second step of this branch>
	......
	2bj: < the last step of this branch>

- a branching condition describing the trigger condition under which the system takes a different behavior; and
- a sequence of action steps to be performed (by the user or the system) in response to the branching condition.

The goal of an extension is either to complete the use case goal or to recover from the exception that caused this extension. An extension may rejoin the success scenario at a specified step or simply at the step where the branching happened, or may end with success or failure.

A common pitfall in writing use cases is the inclusion of design specifics [26]. Use cases are black-box requirements of system behaviors. A use case describes very well what the system should do, without saying how the system is to do it, which is a design issue. For instance, being a veteran in computing, a developer often has the tendency to use specific user interface terms (such as buttons, text fields, menus, mouse clicks, and list selections) to describe use case steps. Design specifics such as these should be avoided because the readers of use cases include not only the development team, but also other stakeholders who might be unfamiliar with the user interface terms. Moreover, including design details in the requirements document will later either limit your design choices or cause unnecessary revisions to use cases when the actual design changes.

8.1.3 Use Case Levels

Use cases are typically specified at three levels: summary, primary task, and subfunction [26]. The three use case levels are compared in Table 8.2.

First, they are used for different purposes. A summary-level use case is generally used as an index table, just like the table of contents of a book. A primary task (aka user goal) level use case is used to lay out user-system interactions in detail, while a subfunction-level use case is used to describe the reusable interaction steps that may be shared by several primary-task-level use cases.

Second, use cases at different levels differ in their completion times. It may take a user a few hours or days to finish a summary-level use case. A primary-task-level use case can typically

Table 8.2 Comparisons of use case levels

Aspect	Summary Level	Primary Task Level (User Goal Level)	Subfunction Level
Purpose	Used as "Table of contents." One for each (sub)system	Used to describe base (backbone) logic	Used to describe small reusable logic
Level test	Takes a long time to finish, maybe a few hours or days	Takes one-sitting time to finish, maybe 2–20 min	Takes a very short time, say, less than 1 min
Diagram indicators	Covers nearly all the use cases inside the system or subsystem boundary	May have direct links to actor(s); May «include», or be «extended» by, other use cases	Has no direct link to the actor; its primary actor is the one linked to the base use case (by «extend» or «include»)
Scenario steps	Generally has "user steps" only; often contains references to primary task level use cases	Generally lists "user-system" steps in an interleaving manner; often contains references to other use cases (via «include» or «extend»	Generally has only a few "user-system" steps in an interleaving manner

be finished in a few minutes or less than 1 h, which is often referred as people's one-sitting time. A subfunction-level use case can be completed in a very short time—say less than 1 min.

Third, use cases at different levels may be revealed differently by their relationships with each other in the use case diagram. Each (sub)system boundary suggests one or more summary-level use cases. For example, an actor may follow a certain order to perform most or all of the use cases inside the (sub)system boundary. A use case inside the (sub)system boundary must be at the primary task level if it links directly to one or more actors. A primary-task-level use case may include or be extended by other use cases. Subfunction-level use cases usually have no direct links to actors, and they should be included by or extend several other use cases.

Fourth, the scenario of a summary-level use case normally consists of user action steps only, providing a sequence of references to several primary-task-level use cases. A primary-task-level use case contains "user-system" action steps in an interleaving manner; sometimes it may include or be extended by other use cases. A subfunction-level use case contains only a few "user-system" action steps in an interleaving manner; it can rarely include or be extended by other use cases.

Table 8.3 gives an example use case at the summary level. An example use case at the primary task level is provided in Section 11.1.

8.2 Sequence Diagram

A system is composed of objects. A sequence diagram shows object interactions arranged in a time sequence.

A UML *sequence diagram* can be used together with a use case, depicting the objects involved in a scenario of the use case and the sequence of messages exchanged between the participating objects as they collaborate to fulfill the goal of the use case. This allows a developer to virtually step through the sequence of events to flesh out the high-level business process and validate the completeness of a use case.

UML sequence diagrams can also be used within the design context to explore a system or algorithm design, allowing a designer to trace the thread(s) of execution that may involve many system objects. Let us use Figure 8.3[4] to explain the sequence diagram notation.

First, a sequence diagram may be enclosed by an *interaction frame*, which is a solid-outline rectangle with a pentagonal area in its upper-left corner. The pentagonal area is used to display the keyword **sd** followed by the interaction name (and parameters, if applicable). Although the interaction frame is optional, sometimes it can be useful to cross-reference a sequence diagram by its name (say, composing simpler diagrams into a complex sequence diagram by references).

---

[4] This example is adapted from [5].

Table 8.3 A summary-level use case

Use case	#1: play the online social game
Goal in Context	A host player playing the game
Scope	The game system
Level	Summary
Primary Actor	Host game player
Preconditions	The game is not launched yet
Minimal Guarantee	The game terminated
Success Guarantee	The game profile updated
Trigger	The host player starts the system
Description Step	Action
1	The host player launches the game «Launch Game»
2	The game system initializes the environment
3	A player Bob joins the game «Join Game»
4	The host player plays the game «Play Game»
5	Bob plays the game «Play Game»
6	Bob saves his profile and exits «Archive Game Profile»
7	The host player saves her profile and exits «Archive Game Profile»
Extension Step	Branching action
2a	Environment performance is too low
	a1: The game terminated

Inside the interaction frame there are five objects: a *WebBrowser* object named *w*, an anonymous *WebServer* object, an anonymous *VideoServer* object, a *VideoPlayer* object named *aPlayer*, and an anonymous *VideoViewer* object. Notice that three stereotypes—«Model», «View», and «Controller»—are used to specify the use of the *Model-View-Controller* design pattern explicitly.

Vertically, each object has a lifeline that represents the lifetime of its participation in the interaction. The order of events/actions along a lifeline denotes the order in which the events/actions will occur. In the example, there are five lifelines, two of which are created in the middle of the interaction.

Horizontally, objects can exchange messages. A message reflects either an operation call and start of execution or the sending and reception of a signal. Each message has a sending event, which occurs at the sender object, and a receiving event, which occurs at the receiver object. Messages of the following forms can be used:

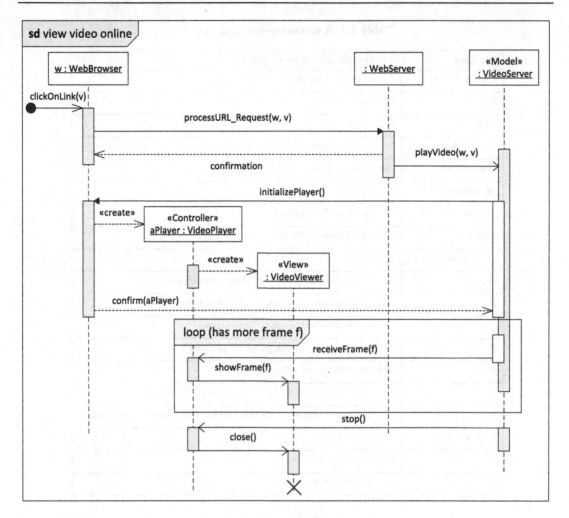

Figure 8.3
A UML sequence diagram for an online video player.

- Asynchronous message: this represents a signal or asynchronous operation call and is shown as a line with an open arrowhead.
- Synchronous message: this typically represents a synchronous operation call and is shown as a line with a filled arrowhead.
- Reply message: acknowledgment to a previously received message. It is shown as a dashed line with a stick arrowhead.
- Object creation message: a message indicating the creation of a new object. It is shown as a dashed line with an open arrow, usually labeled by «create».
- Object deletion message: a message indicating the destruction of the receiver object. It is shown as a cross at the bottom of a lifeline.

- Lost message: a message where the sending event is known, but there is no receiving event. It is shown as a small black circle at the arrow end of the message.
- Found message: a message where the receiving event is known, but the sending event is unknown or outside the scope of the current context. It is shown as a small black circle at the starting end of the message.

In the example, *clickOnLink(v)* is a found message; the objects *VideoPlayer* and *VideoViewer* are created by object creation messages; the object *VideoViewer* sends to itself an object deletion message; *processURL_Request(w,v)* and *initializePlayer()* are synchronous messages; *confirmation* and *confirm(aPlayer)* are return messages; all others are asynchronous messages. Now, you may realize why some of the objects have a name, while others are anonymous: a named object can be referenced by its name. For example, the *VideoServer* object needs to know who is the receiver object of its video frames.

In response to a message, the receiver object performs an appropriate operation, which is indicated by a thin rectangle drawn on top of the receiver's lifeline. If an object sends a message to itself, an execution rectangle can be overlapped on top of another to indicate a further level of processing. In the example, the *VideoServer* object has two operations performed in the execution of *playVideo(w,v)*.

Sometimes a block of interactions can be grouped together by the notion of interaction fragments. A few kinds of interaction fragments are defined in UML, including **alt** and **opt** for choice of behavior depending on the truth value of the associated guard expression, **par** for parallel behavior, **seq** for sequential execution, **loop** for iterative execution, **critical** for critical regions, and **ref** for cross-reference. In Figure 8.3, there is a 'loop' fragment, saying that the *VideoServer* object keeps sending frames to *aPlayer* until there is no more frame left.

Figure 8.4 shows another sequence diagram named *plot_gating*, which has three nested interaction fragments. A more complicated example is given in Figure 8.5, which has a **ref** fragment referencing the *plot_gating* sequence diagram.

Figure 8.5 depicts the sequence of operations for data processing in radar control systems. There are three parallel regions. It is worth noting that in the second parallel region there are five named rectangles with rounded corners on the *Tracker* lifeline. The named rectangle is called a *state symbol*. As the sequence of execution reaches a state symbol, the object enters that specific state and the object's behavior is relatively stable and predictable while it is in that state. A state could be the internal state of the corresponding lifeline object (say, a thread has states such as idle, running, and blocked), or it could be an imaginary system state from an external view (say, we can externally ascribe states to a traffic light system by the colors of the lights). In the former case, the object is called a *stateful* object, and its behavior can be further modeled by a state diagram (this is covered in Chapter 9).

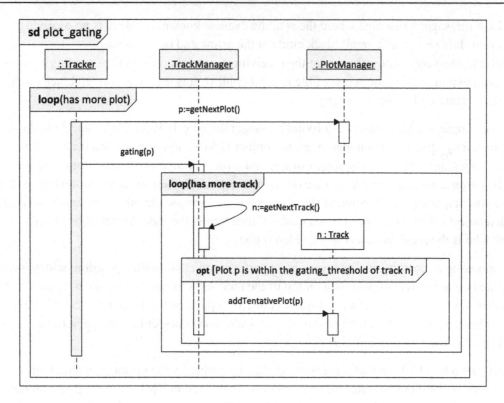

Figure 8.4
A UML sequence diagram with nested fragments.

The "stableness" of an object state can be described by *state invariants*—a concept similar to class invariants. A state invariant represents run-time constraints on the participants of the interaction. The constraints associated with a state should be valid while the object is in that state. In a sequence diagram, a state invariant is shown as text in curly brackets on the lifeline. For example, the invariant *Tracker.smoother != Null* says that the *Tracker* object should have a smoother while it is in the *TrackAssociation* state.

8.3 Activity Diagram

UML activity diagrams are used for modeling the control flows and data flows of a business process, an engineering workflow, or a procedural computation.

An activity represents a behavior that is composed of individual actions. Each action within an activity represents a single-step operation that may or may not be atomic, and a nonatomic action can be arbitrarily complex and even refers to another activity capturing a behavior at a finer level of granularity.

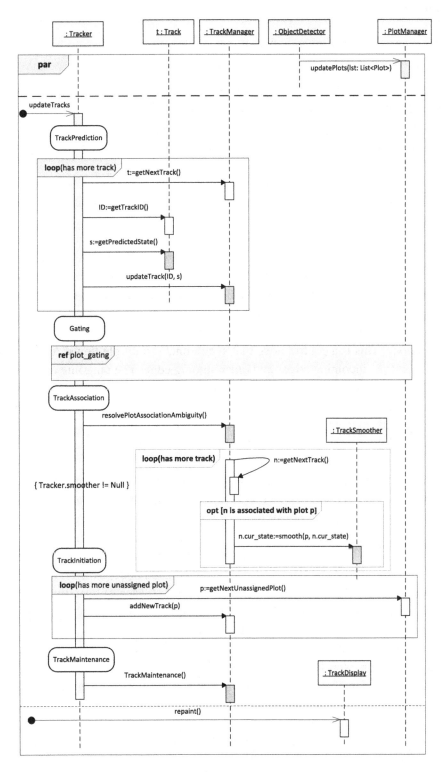

Figure 8.5
A UML sequence diagram for a radar system.

Each activity diagram defines an activity that may contain three kinds of nodes—control nodes, object nodes, and action nodes:

(1) *Control nodes.* A control node is used as a traffic switch to coordinate the flows between other nodes. There are several kinds of control nodes, including

    (a) *initial node.* This is a control node at which a flow starts when the activity is invoked. An initial node, notated as a solid circle, is a starting point for executing an activity.

    (b) *activity final node.* This is a control node that stops all flows in an activity. In particular, it stops all executing actions in the activity. An activity final node is notated as a solid circle within a hollow circle.

    (c) *flow final node.* This is a control node that terminates a single flow. It has no effect on other flows in the activity. It is notated as a circle with an "X" through it.

    (d) *fork node.* This is a control node that splits a flow into multiple concurrent flows. A fork node has one incoming edge and multiple outgoing edges, which are either all object flows or all control flows. A fork node is notated as a thick line segment, with a single edge entering it and two or more edges leaving it.

    (e) *join node.* This is a control node that synchronizes multiple flows. A join node has multiple incoming edges and one outgoing edge. The outgoing edge is a control flow if all the incoming edges are control flows; otherwise, it is an object flow. A join node is notated as a thick line segment, with two or more edges entering it and a single edge leaving it. A join node is an AND-type filter in the sense that the flow passes to the outgoing edge if and only if a flow has traversed each of the incoming edges.

    (f) *decision node.* This is a control node that chooses between outgoing flows. A decision node is shown as a diamond with one incoming edge and several edges leaving it. The edges are either all object flows or all control flows.

        An outgoing edge may have a guard expression. Which of the outgoing edges is actually traversed depends on the run-time evaluation of the guards on the outgoing edges. For a *well-designed* decision node, exactly one outgoing edge is traversed at a time, which is warranted if (i) each of the outgoing edges has a guard expression, (ii) all the guards together form a complete set, and (iii) they are disjoint (no overlapping area).

        A decision node may also have a behavior/operation without any side effects— say a Boolean expression involving some objects being processed by the activity. The output of the behavior is available to the guards on the outgoing edges, which are evaluated afterward. The behavior of a decision node can be specified by a note with the keyword «decisionInput» followed by an appropriate decision condition.

    (g) *merge node.* This is a control node that brings together multiple alternate flows. A merge node has multiple incoming edges and a single outgoing edge, all of which must be either all object flows or all control flows. The notation for a merge node is a diamond-shaped symbol with two or more edges entering it and a single edge leaving it. Unlike the AND-type join node, a merge node is an OR-type filter in the sense that the flow passes to the outgoing edge if and only if a flow has just traversed one of the incoming edges.

(2) *Object nodes.* An object node is notated as a rectangle with a name inside, where the name indicates the type of the object node, or the name and type of the node in the form "name:type."

(3) *Action nodes.* An action node is notated as a round-cornered rectangle with a name inside. An action node represents the execution of a subordinate behavior, such as an invocation of another activity, a call to an operation, or manipulation of an object attribute.

An action node may have sets of incoming and outgoing edges that specify control flow and data flow from and to other nodes. When the set of incoming edges is not a singleton, it is treated as if the incoming edges are connected to the action node through an *implicit join*, implying that the action can be invoked until all the incoming edges have been traversed. If this is not desired, you should use a merge node between the incoming edges and the action node. When the set of outgoing edges is not a singleton, it is treated as if the action node is connected to the outgoing edges through an *implicit fork*, leading to parallel flows.

An action execution may need to process or produce certain objects. This is captured by pins, small rectangles attached to an action node, each specifying a typed parameter expected by the action as an input (input parameter) or offered by the action as an output (output parameter). A parameter can be *stream* or nonstream. While an action must have all the nonstream inputs ready before its execution, it may contiguously accept values to its stream inputs or post values to its stream outputs while it is executing. In other words, values for streaming input parameters may arrive anytime during a single execution of the action, not just at the beginning, and multiple values may be posted to a streaming output parameter during the execution, not just at the end. A streaming parameter is marked by the string "[stream]" near the pin or one end of the corresponding edge if the pin is elided.

Control flow is used to model the sequencing of behaviors that does not involve the flow of objects. A control flow edge is notated by an arrowed line connecting a source action node to a target action node (there may be some control nodes in between), indicating that the target action cannot start until the completion of the source action. The source node is also called the predecessor node of the target.

Object flow is used to model the flow of data and objects in an activity. An object flow edge is typically denoted by an arrowed line connecting a source action node to an object node and another arrowed line connecting the object node to a target action node, indicting that the object flows from the source node to the target node (for further processing or transformation). Alternatively, when the "pin" notation is used, an object flow edge is notated as one arrowed line connecting an output pin and an input pin attached to the source action node and the target action node, respectively.

Figure 8.6 gives an activity diagram depicting the process for software changes. The activity starts with the *Receive Change Request* action, which has an incoming edge from the *initial*

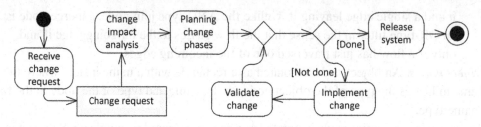

Figure 8.6
UML activity diagram example: software process for changes.

node. In between the actions *Receive Change Request* and *Change Impact Analysis* there is a *Change Request* object, which indicates that the *Change Request* object flows between the two actions. After *Change Impact Analysis*, the control comes to the action *Planning Change Phases*, which is followed by a *merge* node and a *decision* node. After the decision node, the guards of the two outgoing edges are evaluated to decide which edge is to be traversed. The new system is released and the whole activity is finished if all the changes have been implemented; otherwise the next change phase is implemented and validated.

The activity diagram shown on the left in Figure 8.7 depicts how a video server offers services to its clients. The activity is enclosed within a frame with the keyword **ad** followed by the

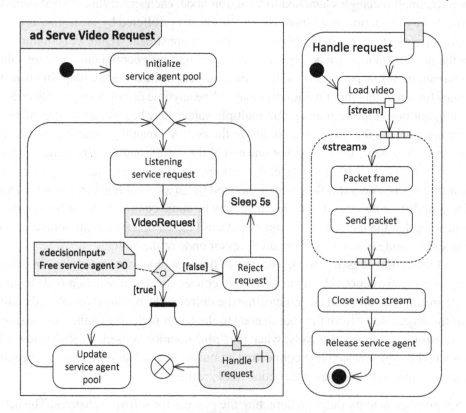

Figure 8.7
UML activity diagram example.

activity name. When a video server starts, it first initializes a pool of service agents. A service agent can be scheduled to serve at most one client at any time, and it is not available to serve a second client until it is released back to the pool. The server has a dedicated "listening" thread for accepting service requests from clients. Whenever a video server receives a service request, it passes the request object to the decision node, which, in this case, has a decision behavior specified by a note. At the decision node the server checks whether there are any service agents available. If there are none available, the request is rejected (maybe ask the client to try later) and the server continues to listen for client requests after a short sleep (some service agents could have been released in the meantime). If a service agent is available, the flow is split into two: one is used to handle the client request, and the other still acts as the "listening" thread after the pool has been updated.

Note that the activity comes to a *flow final* node after the completion of the action *Handle Request*. As explained before, the flow final node stops only this specific service thread. The activity of the video server is still executing, and most likely there are multiple service agents offering video services to their respective clients.

In addition, the *Handle Request* action is marked by a "rake" symbol, indicating that this action is a call to another activity. The activity diagram for *Handle Request* is shown on the right in Figure 8.7, and is enclosed by a rounded rectangle with a pin (small rectangle) on the border. Up to this point, we have shown three styles of activity diagrams, and this one is especially useful to explicitly specify the input and output parameters by pins.

The *Handle Request* activity accepts one parameter, which is a *VideoRequest* object as can be seen from the activity diagram on the left in Figure 8.7. According to the *VidroRequest* object, the *Load Video* action loads the requested video from the storage media. The "[stream]" annotation near the output pin of the action *Load Video* indicates that the action produces a stream of objects (video frames), the target of which is called an *expansion region*—a block enclosed by a rounded box with a dashed outline.

An *expansion* region may have one or more input collections, each containing elements of the same type and shown as a small rectangle with several tiny compartments placed on the boundary of the dashed box. The expansion region is executed once for each element in the input collection. If there are multiple input collections, they must have the same number of elements, and a value is taken from each of them for each execution of the region. An *expansion* region may also have one or more output collections. On each execution of the region, an output value from the region is inserted into each output collection at the same position as the input elements.[5]

---

[5] All the objects contained by the same output collection must be of the same type. An output object can be more complex than the corresponding input object(s), as simple as a flag indicating whether the execution succeeded or not, or a null object indicating no output at all.

An expansion region can be in one of three modes: *iterative, parallel,* and *stream*:

(1) *Iterative mode:* the element executions form a sequence, with one finishing before another can begin. This is the default mode.

(2) *Parallel mode:* the element executions may happen in parallel, or overlapping in time.

(3) *Stream mode:* like pipelining, the element executions form a stream. At any given time, it is likely that multiple or all elements are being processed at different stages by different actions in the region.

Now, let us come back to the expansion region in the activity diagram on the right in Figure 8.7. It is in *stream* mode, which means the service agent may be sending a frame

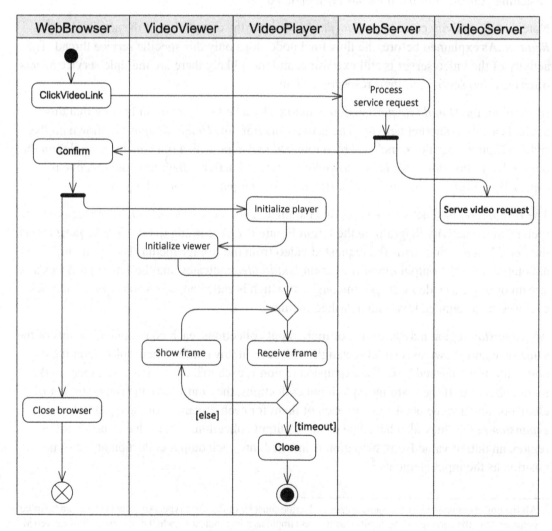

Figure 8.8
UML activity diagram: multiple participants.

packet while it is packeting another frame. Note that the output collection in this example is *not* used to hold the packets to the client, which should have been done by the action *Send Packet*. The output collection in this case could hold flags for load analysis.

Figure 8.8 shows an activity diagram which is partitioned into several "swimlanes." A swimlane is a way to group actions performed by the same actor or in the same thread. In the example, there are five participants, where the *Web browser*, *video player*, and *video viewer* are on the client workstation, while the *Web server* and *video server* are in the cloud. In general, partitions can be hierarchical and multidimensional.

Since the action *Handle Request* performed by a video server is covered in Figure 8.7, Figure 8.8 focuses on the behaviors of the video player and video viewer, which are in the same flow (thread) generated by the Web browser object. In particular, the video player and video viewer execute actions in a loop: keep receiving and showing video frames until time-out happens. Alternatively, the actions *Receive Frame* and *Show Frame* can both be modeled using stream pins, like the action *Load Video* in the activity *Handle Request*.

Problems

8.1 Explain the differences among the three levels of use cases.

8.2 Refer to Section 1.3, and explain the sequence diagrams given in Figures 8.4 and 8.5.

8.3 Start from the initial node and describe the activity modeled by Figure 8.8.

8.4 An expansion region of an activity diagram can operate in three modes. Explain their differences.

8.5 In an activity diagram, how can you tell that an edge represents an object flow rather than a control flow?

8.6 In an activity diagram, the outgoing edges of a decision node should be guarded by a Boolean expression. List a few good practices for specifying guard expressions.

Modeling Stateful Behaviors in UML

Happiness is a state of activity.

Aristotle

9.1 Basics of a State Machine Diagram

A *state* models a stable situation during which some (usually implicit) invariant condition holds. The invariant may represent a static "being" property such as an object being idle while waiting for some external event to occur, or a dynamic "doing" property such as the process of performing some activity that takes a detectible amount of time.

An object is called a *stateful* object if it can be in more than one state and it exhibits different behaviors in different states. For example, a watch is a stateful object because it behaves differently in a measuring-time state and in a setting-time state. A complex system as a whole, especially a real-time system [34], can also be treated as a stateful object.

Real-Time Embedded Systems. http://dx.doi.org/10.1016/B978-0-12-801507-0.00009-2

As a variant of Harel statecharts [39], UML state machine diagrams (also called state-transition diagrams or state diagrams) can be used to model the various states that a single object (or system) may be in and the transitions between those states.

9.1.1 States

A state of a stateful object is in general shown as a rectangle with rounded corners, with the state name shown inside the rectangle [10, 18]. Optionally, a state may have a separate compartment containing a list of internal behaviors (actions or activities) that are performed while the object is in the state. There are three special kinds of behaviors:

(1) *Entry action.* Preceded by the label "entry," it identifies a behavior which is performed upon entry to the state. An entry action is always executed to completion prior to any other actions.
(2) *Exit action.* Preceded by the label "exit," it identifies a behavior that is performed to completion upon exit from the state. It is the final step prior to leaving the state.
(3) *In-state behavior.* Preceded by the label "do," it identifies an ongoing activity that starts immediately after the completion of the entry action, if any, and runs until its completion or as long as the object is in the state.

9.1.2 Transitions and Events

A transition is a directed relationship between a source state and a target state. It is called a *self transition* if the source and the target are the same state. A transition is shown as an arrowed line labeled with a string of the form "triggeringEvent[guard]/ action," which specifies a triggering event, a guard expression, and an action (or action block).

As summarized in Figure 9.1, a triggering event can be one of five types:

(1) *Time event.* A time event specifies an instant in time by a time expression, which might be absolute or might be relative to some other time instant. A time event with a relative time expression is specified with the keyword "after:" followed by an expression that evaluates to a time value, such as "after: 12 seconds." A time event with an absolute time expression is specified with the keyword "at:" followed by an expression that evaluates to a time value, such as "at: Feb. 12, 2012, noon." If the time expression is relative and no explicit starting time is defined, then it is relative to the time of entry into the source state of the transition. In such a case, the time event is generated only if the object is still in the source state when the deadline expires.
(2) *Completion event.* A transition originating from a state is called a *completion transition* if it has no explicit trigger event. A completion transition is implicitly triggered by a completion event, which is generated once the entry actions and the in-state activities have been completed. If multiple completion transitions are defined for a state, then they

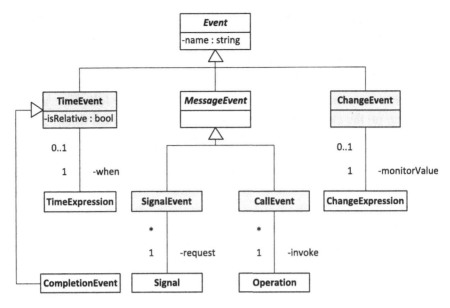

Figure 9.1
Classification of triggering events.

should have mutually exclusive guard conditions. Completion events are handled at a higher priority than any other events.

(3) *Signal event.* A signal event represents the reception of an asynchronous signal. The sender of a signal will not block waiting for a reply but will continue its execution immediately.

(4) *Call event.* A call event represents the reception of a request message to invoke a specific operation on the receiver object. A call event may result in the execution of the requested operation.

(5) *Change event.* A change event models a change in the object (or system) state that makes a condition true. A change event is specified with the keyword "when:" followed by a change expression that takes Boolean values. A change event is generated every time the value of the change expression changes from false to true. In order to do so, the change expression may be continuously evaluated.

A state becomes active when it is entered as a result of firing an incoming transition, and becomes inactive if it is exited as a result of firing an outgoing transition. The firing of a transition relies on the constraint specified by the guard expression (it is always evaluated to true if no guard is specified). When the triggering event on a transition happens, the transition is *enabled* if and only if the source state is active and the corresponding guard is evaluated to true.

All the enabled transitions will fire if they are not conflicting with each other. Two transitions are said to *conflict* if they both exit the same state. In other words, a state cannot be exited more than once simultaneously. For example, consider the case of two transitions originating from the same state, triggered by the same event, but with different guards. If that event occurs and both guard conditions are true, then only one transition can fire, but the selection of which is undefined. The existence of conflicting transitions can cause nondeterministic behavior, which, if not desired, typically indicates a design error.

Once a transition has been enabled and has been selected to fire, the following steps are carried out in order: (a) the exit actions of the source state are executed; (b) the source state is exited; (c) the action (block) of the transition is executed; (d) and the target state is entered [10].

9.1.3 Pseudostates

There is a special kind of state called a *final state*, which has no internal behavior or outgoing transition. A final state typically represents the completion of a successful execution. A final state is shown as a circle surrounding a small filled circle.

UML also defines some pseudostates, which are typically used to connect multiple transitions into more complex state-transition paths:

- *Initial pseudostate.* This is shown as a small solid filled circle; it points to a default active state in a state diagram. The outgoing transition from the initial pseudostate may have a behavior, but not a trigger or guard.
- *Terminate pseudostate.* This is shown as a cross; it typically represents the end of an unsuccessful execution. The stateful object is immediately destroyed when one of its terminate pseudostates is entered.
- *Junction pseudostate.* This is shown as a small black circle; it is merely used to chain together multiple transitions to construct compound transition paths between states. A junction can be used to converge (merge) multiple incoming transitions into a single outgoing transition representing a shared transition path. It can also be used to split an incoming transition into multiple outgoing transition segments with different guard conditions. In so doing, it realizes a *static conditional branch* in the sense that the decision regarding which outgoing branches to take is made statically, before the incoming transition is executed. In other words, the guards on the outgoing branches are evaluated *before* the action (block) on the incoming transition is executed.
- *Choice pseudostate.* This is shown as a diamond; it realizes a *dynamic conditional branch* in the sense that the decision regarding which path to take is made dynamically, after the incoming transition has been executed. In other words, the guards on the outgoing branches are evaluated *after* the action (block) on the incoming transition has been executed. A well-designed choice point should have at least one guard evaluate to true at the evaluation time. An arbitrary one is selected if more than one of the guards evaluates to true.

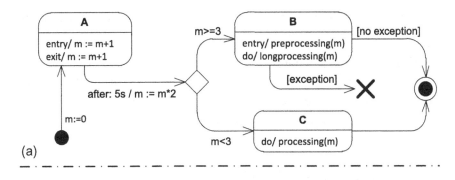

(a)

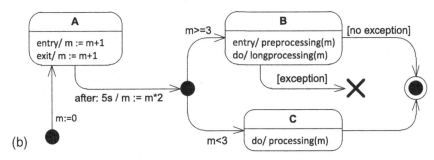

(b)

Figure 9.2
An example state machine diagram.

State machine diagram notation is illustrated in Figure 9.2, where Figure 9.2(a) and Figure 9.2(b) are exactly the same except that the choice pseudostate used in Figure 9.2(a) is replaced by a junction pseudostate in Figure 9.2(b).

In Figure 9.2(a), the transition from the initial node to state *A* has an action which initializes the variable *m* to 0. State *A* has an entry action, an exit action, and a transition triggered by a 5 s time event. Once state *A* is entered, the value of *m* becomes 1, and it becomes 2 when state *A* is exited after 5 s. Since the transition leaving state *A* reaches a choice pseudostate, the action on the transition (m:=m*2) is executed before the guards on the outgoing branches of the choice node are evaluated. In other words, the value of *m* is 4 at the choice point. Consequently, the branch leading to state *B* is fired and state *B* is entered afterward.

In state *B*, the action *preprocessing(m)* runs to completion, followed by the activity *longprocessing(m)*. Note that there are two transitions leaving state *B*, one to the final node and the other to the terminate node. Neither transition has an explicit triggering event, which indicates that both transitions are *completion* transitions, triggered by the implicit completion event raised by state *B* upon the completion of its in-state activity *longprocessing(m)*. The guards of the two completion transitions are exclusive and form a complete set, which guarantees that no ambiguity or transition conflict will occur. Depending on whether there is any exception raised in state B, the execution of the stateful object will terminate either successfully or unsuccessfully.

In Figure 9.2(b), the transition leaving state A reaches a junction pseudostate, which implies that the action on the transition (m:=m*2) is executed after the guards on the outgoing branches of the junction node have been evaluated. Since the value of m is 2 when state A is exited, state C will be the next active state. Of course, the value of m becomes 4 as state C is entered.

9.1.4 A Network Protocol Modeled by State Machines

An object implementing a protocol is inherently a stateful object. A protocol specifies which operations can be called in which state and under which condition, thus specifying the allowed call sequences on the operations of the implementing object. It is thus straightforward to model protocols by state machine diagrams, specifying all allowed transitions for all possible states.

Let us consider a situation involving three computing devices A, B, and C connected to each other via a broadcast channel—a packet sent by one device is carried by the channel to all the other devices. Further, assume that the broadcast channel can independently lose and corrupt packets (say, a packet sent from A might be correctly received by B but not by C). Our problem is to design a stop-and-wait error-control protocol for reliably transferring packets from server A to clients B and C, such that A will not send a new packet until it knows that both B and C have correctly received the current packet.

Figure 9.3 gives a protocol design for server A. The variable *seq* is a state variable. Its value is used to label an outgoing data packet so that the receiver can tell whether it is a new packet or simply a duplicate of an already accepted packet.

Upon start-up, server A enters a state called *Waitfor send request*, where it waits for a call event *rtd_send(data)* from the application layer. When this event happens, the action block on the only outgoing transition is executed, where (a) a new packet is created from the data, its checksum, and the sequence number, (b) the packet is sent out, and (c) a timer is started, which has a relative time expression *tm_wait* that can take a predefined value or be adjusted dynamically. The timer represents a "reasonable" amount of time that the server should wait: the server would assume that the packet it just sent was lost if it does not receive an acknowledgment from the receiver before the timer raises a time event. This is handled by the self transition with a time event trigger *after: tm_wait* in state *Waitfor first ACK*, where the data packet is resent and the timer is restarted.

Assume that in the state *Waitfor first ACK* the server can receive a packet from either B or C in time. The received packet is ignored (no action is defined for that self transition) if it is corrupted or is not acknowledging the last data packet. Otherwise, depending on whom is the

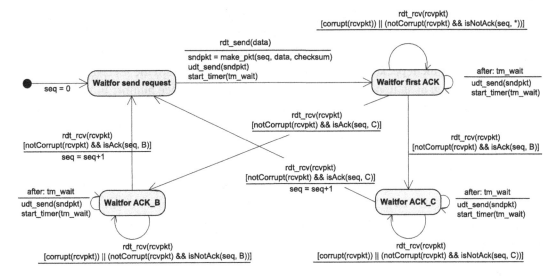

Figure 9.3
Reliable data transmission protocol: sender.

sender of the acknowledgment packet, the server enters either the state *Waitfor ACK_B* or the state *Waitfor ACK_C* to wait for a confirmation from the other receiver. Similarly, both of these states need to handle time-out events and unwanted packets by self transitions. If the server receives an uncorrupt acknowledgment packet from the right receiver, it enters the state *Waitfor send request* again and the sequence number is increased by 1 to make it ready for sending the next data packet.

The protocols for the clients are relatively simpler. The two diagrams shown in Figure 9.4 are exactly the same except that the acknowledgment packets created by different receivers

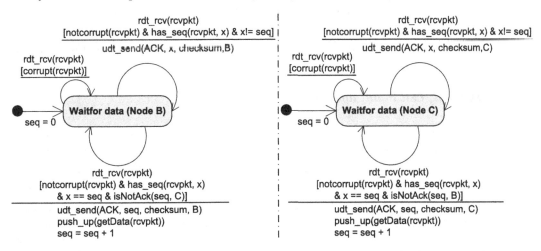

Figure 9.4
Reliable data transmission protocol: receiver.

contain different identifiers. From the single state, each receiver needs to handle corrupted packets and unwanted (duplicate) packets. In response to a packet that is received in the correct order and is not a noise packet from the other receiver, a receiver will (a) create and send an acknowledgment packet to the broadcast channel, (b) forward the received data to the consumer application at an upper layer, and (c) increase the sequence number to indicate the next expected data packet.

9.2 Composite States

UML defines a region as an independent area containing states and transitions. A region as a whole can be treated as a "state," or can be technically called a simple composite state. A composite state, in general, can have more than one orthogonal (or independent) region. Any state enclosed within the region of a composite state is called a substate of that composite state.

Like simple states, a composite state is also shown as a rectangle with rounded corners. In addition to the optional name and internal activity compartments, a composite state may have an additional graph compartment that contains a nested diagram describing its substates and transitions. When a composite state has more than one region, the graph compartment is further divided into regions by dashed lines. Each region may have an optional name and contains the states and transitions defined for it.

Figure 9.5 shows an example state machine diagram for phones, where *Idle* is a simple state and *Active* is a composite state. There are time events, call events, and signal events associated with the transitions.

A composite state is entered through its default initial pseudostate when a transition pointing to the boundary of the composite state is fired. So the phone is in the state *ReadyToDial* as its receiver is lifted. A transition that emanates from the boundary of a composite state is a "group" transition in the sense that it applies to all the substates of the composite state. In the example, whenever a user hangs up the phone, it becomes idle regardless of which substate the phone was in.

9.2.1 Entry Point, Exit Point, and History

A composite state may have one or more *entry* and *exit* points (pseudostates) on its border or in close proximity to its border (inside or outside). It is not necessary to use the entry/exit point notation if a composite state is always entered through its default initial pseudostate or if it is always exited as a result of the completion of the state activities. However, entry points and exit points may give you a lot of flexibility when it is necessary to capture situations and conditions other than the default that would cause your composite state to be entered or exited.

An entry point is shown as a small circle, while an exit point is shown as a small circle with a cross. Figure 9.6 shows the use of entry and exit points. The washing machine is modeled with

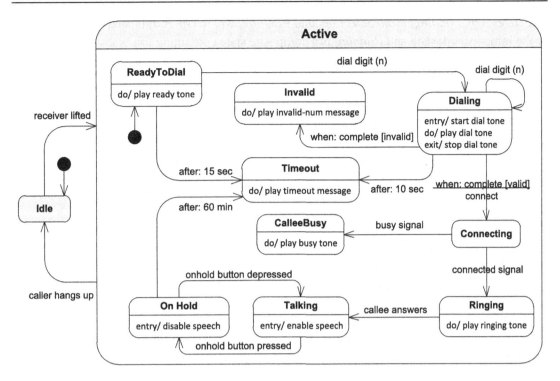

Figure 9.5
A state machine diagram for phones.

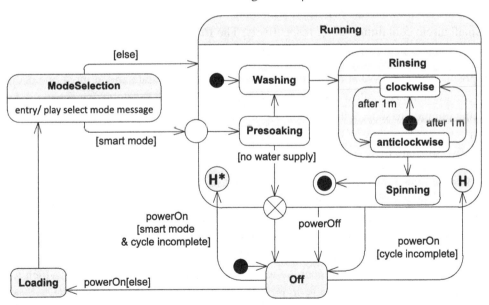

Figure 9.6
A state diagram for a washing machine.

four states: *Off*, *Loading*, *ModeSelection*, and *Running*, where *Off* is indicated as the first state by the initial pseudostate and *Running* is a composite state containing four substates. Unless specified explicitly, all the transitions are triggered by the completion event raised by their respective source states.

There are two transitions leaving the state *ModeSelection*, one directly connected to the border of the *Running* state and the other connected to the *Running* state through an entry point on the border. As indicated by the transition leaving the entry point, the first active substate is *Presoaking* if the smart mode is selected at the state *ModeSelection*. Otherwise, the first active substate is *Washing*, which is the default initial state as marked by the initial pseudostate symbol.

On the border of the *Running* state there is an exit pseudostate, which guides the transition leaving the *Presoaking* state to the *Off* state. As indicated by the transition guard, this happens only under the contingency that there is no water supply. In a normal situation, the *Running* state runs to completion, and upon completion, the washing machine will transit to the *Off* state after the firing of the "group" transition that emanates from the border of the *Running* composite state. Notice that there is another "group" transition leaving the *Running* state, which is triggered by the *powerOff* event, not the completion event. This means that whenever the power is off, the washing machine will transit to the *Off* state regardless of which running substate it was in.

To allow the modeling of "smart machines with memory," UML also defines two pseudostates called *shallowHistory* and *deepHistory*, which are shown as a small circle containing an "H" and a small circle containing "H*," respectively. The *Running* composite state in Figure 9.6 encloses a shallow history node and a deep history node.

A composite state can have at most one shallow history pseudostate and at most one deep history pseudostate. A shallow history node represents the *most recent* active substate of its containing composite state, while a deep history node represents the *innermost* active substate when the composite state was last exited. In other words, a deep history node finds the most recent active substate by recursion, while a shallow history node does not.

For the washing machine example in Figure 9.6, let us first assume that a power failure happens while the washing machine is in the *Washing* state. When the power is on again, since the last washing cycle is not complete yet, if the smart mode was not selected before, the washing machine will come to the shallow history node, which leads to *Washing*—the *most recent* active substate of the *Running* composite state. What if the smart mode was selected before? In this case, the washing machine would come to the deep history node, which would also lead to *Washing* because there is no recursion to take on a simple substate.

Now let us assume that a power failure happens while the washing machine is in the *anticlockwise* substate of the *Rinsing* substate of the *Running* composite state. When the power is on again, since the last washing cycle is not complete yet, if the smart mode was not selected before, the washing machine will come to the shallow history node, which leads to *Rinsing*—the *most recent* active substate of the *Running* composite state. What if the smart mode was selected before? In this case, the washing machine would come to the deep history node, which would lead to the *anticlockwise* substate because it is the innermost active substate by recursion.

9.2.2 Concurrency

When a composite state has concurrent behaviors, they are shown in independent (orthogonal) regions. Each region of a composite state may have an initial pseudostate and a final state.

In the example shown in Figure 9.7, the home intrusion detection system has three states: *Idle*, *Monitoring*, and *Alarming*, where *Idle* is the default initial state, and *Monitoring* and *Alarming* are both composite states with two concurrent regions.

A transition to a composite state represents a transition to the initial pseudostate in each region. In our example, when the system is turned on, it goes to the *Monitoring* state, and the two substates *Sensor Off* and *Waitfor Data* are entered concurrently. Alternatively, you can

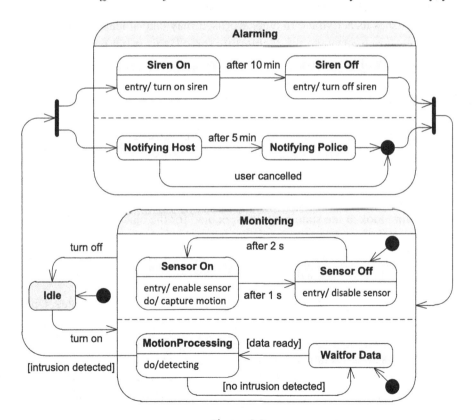

Figure 9.7
A state diagram for a home intrusion detection system.

use a *fork pseudostate*, shown as a short heavy bar, to split an incoming transition into two or more transitions terminating on substates in different regions of a composite state. Notice that the segments going out from a fork node must not have guards or triggers. In our example, a *fork pseudostate* is used before the *Alarming* state. Whenever an intrusion is detected, the system goes to the *Alarming* state, and the two substates, *Siren On* and *Notifying Host*, are entered as indicated by the two transitions leaving the fork node.

There are several ways to leave a composite state with concurrent regions. The first one is shown by the transition triggered by a *turn off* event, which will terminate all the concurrent activities of the *Monitoring* state. The second one is shown by the transition with a guard *[intrusion detected]*, which has the *MotionProcessing* substate as its source. Similarly, this transition will terminate all the concurrent activities of the *Monitoring* state, although it has no connection with the other region. The third way is the use of a *join pseudostate*, which merges several transitions emanating from source states in different orthogonal regions. Note that the transitions entering a join pseudostate cannot have guards or triggers. Just like the join node used in UML activity diagrams, a *join pseudostate* is typically used for synchronization. In our example, the host may cancel the notification in the case of a false alarm. If this happens, this region runs to completion and reaches the join pseudostate. However, it has to wait until the other region finishes its activities. In other words, it may take at least 10 min for the system to transit from the *Alarming* state to the *Monitoring* state. If this is not desired, it may suggest a design error. However, it is a perfect example for us to learn about concurrent regions.

9.3 Inheritance of State Behavior

As discussed in Section 6.3.2, by inheritance, the properties, associations, and constraints defined in a class also apply to its subclasses. Moreover, a subclass can also inherit state behaviors from its superclasses.

We use Figure 9.8 to explain how a subclass inherits the state behavior specified in its superclasses. We first look at the state diagram specified for the class *DigitalWatch*. A digital watch has two states: a simple state *ShowMeasure* and a state *Setup*, which contains a special icon consisting of two tiny state symbols connected together. The special icon indicates that *Setup* is a composite state that has a decomposition shown in a separate diagram.

As shown on the left in Figure 9.8, the *Setup* composite state has three substates. Upon entering *Setup*, the watch first blinks the hour digits. As button *B* is pressed, the hour value is increased by 1 (and wrapped around to 01 after 12). Note that the entry *ButtonB/ incr hour* is special, and represents an *internal transition*. The watch object executes the action *incr hour* in response to the specified event *ButtonB*. An internal transition is similar to a self transition except that the state entry/exit actions are executed for a self transition, but not for an internal transition.

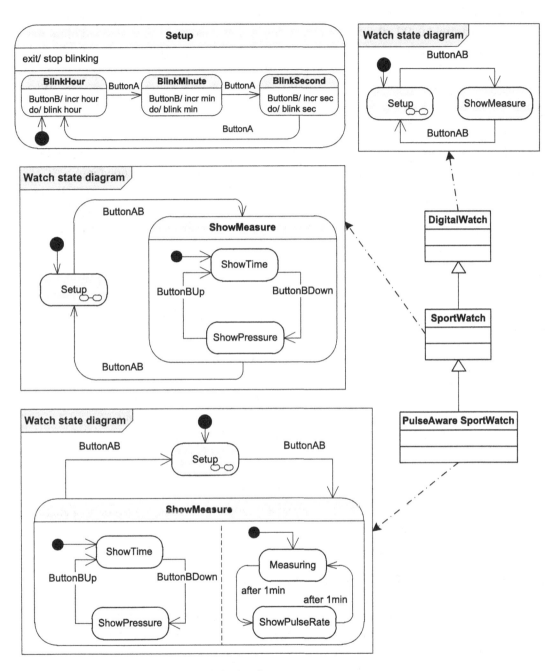

Figure 9.8
Inheritance of state behavior.

When button *A* is pressed, the watch enters the *BlinkMinute* substate, starting to blink the minute digits. Within this substate, as button *B* is pressed, the minute value is increased by 1 (and wrapped around to 00 after 59). When button *A* is pressed, the watch starts to blink the second digits. Within this *BlinkSecond* substate, as button *B* is pressed, the second value is increased by 1 (and wrapped around to 00 after 59). When button *A* is pressed, the watch starts to blink the hour digits again and begins another round of the time adjustment process. When buttons *A* and *B* are pressed together, the watch enters the *ShowMeasure* state. The watch stops blinking upon exiting the *Setup* state.

SportWatch is a subclass of *DigitalWatch*. As a subclass, in addition to inheriting the state behavior specified for *DigitalWatch*, the simple state *ShowMeasure* is extended into a composite state in *SportWatch*. Upon entering the *ShowMeasure* state, a sport watch displays time by default. If button *B* is pressed, the sport watch displays the blood pressure of the person who is wearing it, and it returns to the *ShowTime* substate as button *B* is released.

PulseAware SportWatch is a subclass of *SportWatch*. As a subclass, in addition to inheriting the state behavior specified for *SportWatch*, the state *ShowMeasure* is further extended to contain an additional region. This new region models an extra feature of a pulse-aware sport watch, which can measure and display the pulse rate of the person who is wearing it.

In general, the Liskov substitution principle applies to the inheritance of state behavior. In a subclass,

- states and transitions defined in the superclasses cannot be deleted. A subclass object can be used as a superclass object without causing any surprise to the clients.
- the triggering events on transitions (including internal transitions) cannot be changed. A subclass object, when used as a superclass object, should respond to a sequence of events in the same way as when a superclass object is used. However, the action (block) on a transition can be added, removed, or specialized in the subclass.
- the entry/exit actions and in-state activity of a state can be added, removed, or specialized.
- orthogonal regions may be added.
- new states and transitions may be added.

9.4 Stateful Object Timing Diagrams

An object timing diagram shows the state changes of an object, with the vertical axis marked with various states that the object may go through in its lifetime. When the object is in a certain state, a line is drawn for the duration of time when it is in that state. The state changes of an object are triggered by events (or stimuli) from the other objects or the environment. If it is desirable to show the event causing a state change, it is annotated by an arrow stemming from the time instant when the event occurs. A time-out event indicates the expiry of a time duration; it is typically labeled by *tm(e)*, where *e* can be a time expression or simply a predefined time value, which is evaluated relative to a past time instant.

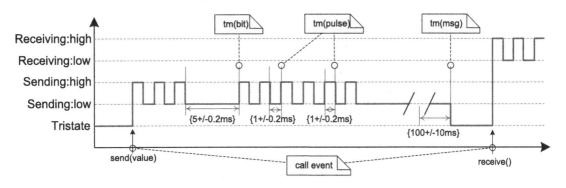

Figure 9.9
An object timing diagram.

Figure 9.9 shows the timing behavior of a low-level system object, which can transform a digital message into electrical pulses and vice versa. A message is composed of a sequence of bytes, which is composed of a sequence of (eight) bits. Each bit has to be represented as a certain number of consecutive pulses. For example, the bit *1* can be represented as six consecutive pulses and the bit *0* can be represented as three consecutive pulses. A protocol could be designed such that

- a pulse low or pulse high is counted only when its width is in the range *(0.8, 1.2)* ms;
- the time interval in between two consecutive bits must be 5 ms with a jitter of 0.2 ms;
- the time interval in between two consecutive bytes must be 10 ms with a jitter of 0.5 ms; and
- the time interval in between two consecutive messages must be at least 100 ms with a jitter of 10 ms.

Some of these constraints are shown in Figure 9.9. Also shown are the request (call event) to send a message, the detection of a pulse indicating the start of an incoming message, and time-out events for pulses, bits, and messages. A portion of the timing diagram is suppressed for simplicity.

We may have an elaborate form of a timing diagram where more than one object is shown. Figure 9.10 gives an example, where each object resides within a partition representing its lifeline. When a user presses the "open" button on the remote door opener, a signal is sent to the receptor of the garage door, which should respond in 2 s (with a jitter of 0.5 s). For safety, the user should wait a few seconds after the door has been opened.

9.5 Example: Modeling Stateful Behavior of a Radar System

Let us look at how to model the state behavior of a radar system as introduced in Section 1.3. Since radar is a very complex system [51], we focus here on the *pulse-level communication* package as shown in Figure 9.11, which specifies how a radar controller communicates with a radar transceiver (transmitter/receiver) at the pulse level.

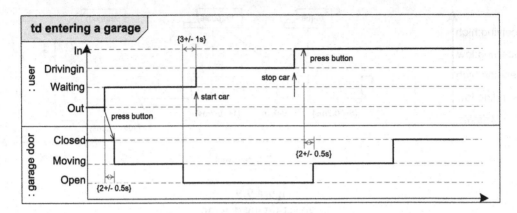

Figure 9.10
An object timing diagram with multiple objects.

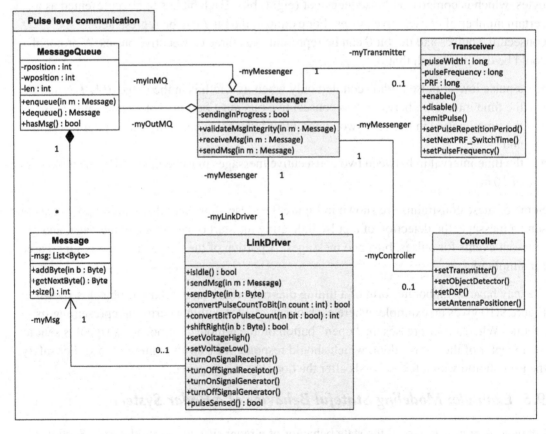

Figure 9.11
Pulse-level communication in a radar system.

A radar transceiver and its controller are physically connected by wires. When it is necessary to adjust some transmitting parameters, the controller needs to send commands over the wire to the transceiver, which may optionally send back confirmations. So both the controller and the transceiver can be a sender and a receiver. As captured in Figure 9.11, a radar transceiver is associated with a command messenger, which has an input message queue and an output message queue. The message queues are, respectively, populated with messages received by the link driver object or messages to be sent out. On the other end of the wire, a controller object is responsible for adjusting parameters for the transceiver, as well as other parts such as the digital signal processor and the antenna positioner. A controller object is also associated with a command messenger, which has an input message queue and an output message queue. The link driver objects are used to transform incoming pulses to messages and to transform outgoing messages to electrical pulses.

The sequence diagram in Figure 9.12 illustrates the sequence of operations as a commander adjusts the transceiver parameters through a controller object. Below, we examine the stateful behavior of the transceiver, link driver, and command messenger, respectively.

9.5.1 Modeling the Transceiver

Figure 9.13 gives the state behavior of the transceiver, which has an *Idle* state and an *Active* composite state.

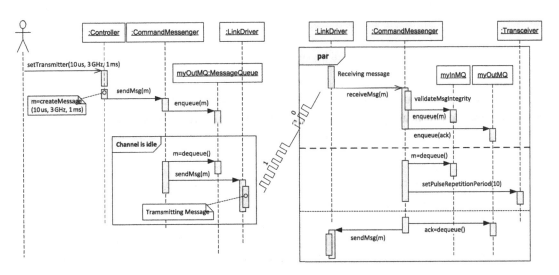

Figure 9.12
Pulse-level communication in a sequence diagram.

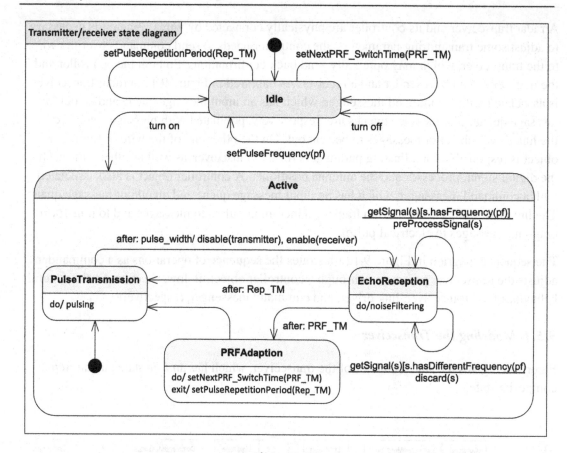

Figure 9.13
Stateful behavior of a transceiver.

The *Idle* state has three self transitions triggered by call events, which allows a user to preset the radar signal transmitting parameters such as the pulse repetition period, the switching rate of the radar's pulse repetition frequency, and the radio pulse frequency. In response to a "turn-on" event, it enters the *Active* composite state.

In the *Active* state, the transceiver first transmits radio pulses at the specified frequency *pf*. After enough signals have been transmitted (specified by *pulse_width*), the transmitter is disabled and the receiver is enabled, and the radar is ready to receive echo signals from the environment. Given a detected signal *s*, it is preprocessed if *s* is of the expected frequency; otherwise the signal is dropped. When a time event with a relative timer *Rep_TM* happens, it is the time for the radar to send another batch of signals (switching back to the transmitter mode). When a time event with a relative timer *PRF_TM* happens, the *PRFAdaption* substate is entered—it is the time for the radar to change the switching rate of the radar's pulse repetition frequency.

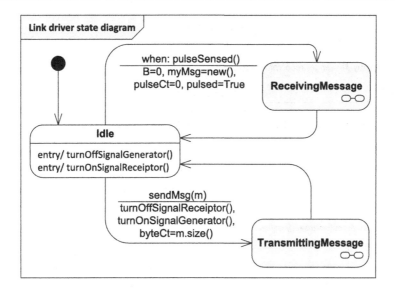

Figure 9.14
Stateful behavior of a link driver.

9.5.2 Modeling the Link Driver

Figure 9.14 gives the state behavior of a link driver object, which has an *Idle* state, a *ReceivingMessage* state, and a *TransmittingMessage* state.

When started, a link driver object is in the *Idle* state with the signal receptor turned on and the signal generator turned off. When it has sensed a pulse through the signal receptor, the transition leading to the *ReceivingMessage* is fired, and the action block is executed to prepare for receiving a message from the wire. The transition leaving the *ReceivingMessage* state implies that the link driver becomes idle once the activity of the *ReceivingMessage* state runs to completion.

When a call event *sendMsg(m)* occurs (as indicated in Figure 9.12 this event is triggered by a call from the command messenger), the link driver enters the state *TransmittingMessage*. The action block on the transition is executed to prepare for transmitting the message *m*. Upon completion, the link driver becomes idle again.

Let us next look at the composite state *ReceivingMessage*. As shown in Figure 9.15[1], it has a simple substate *WaitingforNextByte* and a composite substate *ReceivingByte*, which in turn has two simple substates *WaitingforBit* and *ReceivingBit*. When *ReceivingMessage* is entered, the innermost active state is *ReceivingBit*.

---

[1] This state model is enlightened by a similar treatment of pulse-level communication in [34].

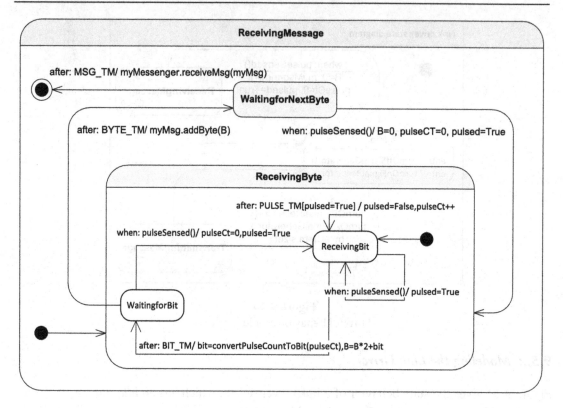

Figure 9.15
How a link driver receives messages.

According to the transition leading to the state *ReceivingMessage* (see Figure 9.14), several state variables are initialized when the transition is fired: *B* is used for the current byte being received, *pulseCt* is used to record the number of pulses sensed by the link driver, *pulsed* is a Boolean variable used to indicate that a pulse has just finished, and *myMsg* is used to store the message string being received.

You may have noticed that there are quite a few transitions triggered by time events. This is the norm for real-time systems [34]. In our example, we take a message as a list of bytes and a byte as a sequence of bits (typically eight), and a bit is encoded by a certain number of pulses (say a binary 1 is represented by six consecutive pulses and 0 is represented by three consecutive pulses). Let us assume there are at least four time-out constants defined for the pulse-level communications: *PULSE_TM* is a time interval for (high voltage level) pulses and in between two pulses, *BIT_TM* is a time interval in between two bits, *BYTE_TM* is a time interval in between two bytes, and *MSG_TM* is a time interval in between two messages. For example, two bytes cannot be distinguished if the no-energy period between them is less than *BYTE_TM*.

We also assume that we have some utility functions ready. In particular, the function *convertPulseCountToBit( )* takes a (pulse) count number as input and decodes it to 0 or 1; *convertBitToPulseCount( )* takes a 0 or 1 and returns a number, which is the number of pulses needed to represent the bit; *shiftRight( )* takes a byte, shifts its binary form to the right by one bit, and returns the last bit; and *pulseSensed( )* returns true when the link driver detects a voltage transition from low to high (edge sensitive). These functions are defined as operations in the class *LinkDriver*.

The substate *ReceivingBit* starts with a pulse being detected. It has three outgoing transitions, two of which are triggered by time events. As explained before, the timer of a time event on a transition is started whenever the source state was recently entered. Hence, after *PULSE_TM* relative to the time *ReceivingBit* was entered, if the variable *pulsed* is true, the pulse counter is increased by 1 and *pulsed* is reset to false afterward. Notice that the self transition causes the timer to be restarted upon re-entry. While in the state *ReceivingBit*, if a pulse is detected before *BIT_TM*, the other self transition is fired, which sets *pulsed* to true to indicate the beginning instant of a new pulse. This new pulse will be counted after another *PULSE_TM* period. This loop terminates when there is a silent time period of length *BIT_TM*, which indicates that a bit has just been received. Whether it is a 0 or 1 depends on the decoding, and it is appended as the last bit of *B*.

In the state *WaitingforBit*, if the link driver is able to sense a pulse before the time-out *BYTE_TM* happens, then the current byte has not been fully received yet, and the *ReceivingBit* is entered again to receive the next bit. If a byte is completely received (triggered by the *BYTE_TM* time event), the byte is added to *myMsg*, and the state *WaitingforNextByte* is entered.

In the state *WaitingforNextByte*, if the link driver is able to sense a pulse before the time-out *MSG_TM* happens, then the current message has not been fully received yet, and the *ReceivingByte* is entered again to receive the next byte. If a message is completely received (triggered by the *MSG_TM* time event), all the activities of the *ReceivingMessage* state have been done. This will raise a completion event, causing the link driver to enter its *Idle* state, as shown in Figure 9.14.

9.5.3 Modeling the Command Messenger

As shown in Figure 9.16, the command messenger has three regions with parallel activities. The region at the top is for handling an incoming message, the region in the middle is for processing messages in the input message queue, and the region at the bottom is for handling an outgoing message. The operation of the command messenger relies on the services offered by the lower-level link driver.

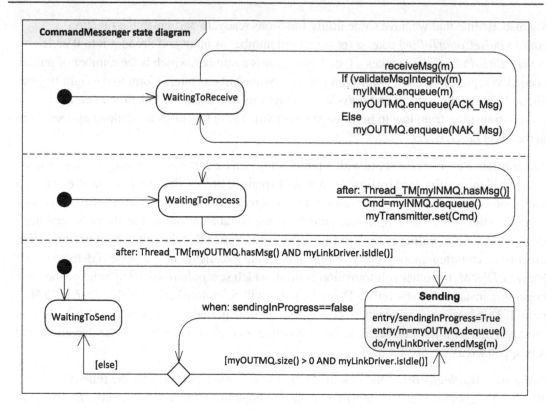

Figure 9.16
Stateful behavior of a communication manager.

The top region contains a single state, which responds to a call event *receiveMsg(m)*. This event is triggered by the lower-level link driver—an action on the final transition of the *ReceivingMessage* state. Whenever the link driver has received a message, the command messenger will check the integrity of the message. If it is valid, the message is put into the messenger's input message queue to be processed later by another thread. The messenger always acknowledges a received message by sending a feedback to the sender.

The middle region also contains a single state, which is periodically scheduled to process the messages, if any, in the input message queue. In particular, it may forward a message to the transmitter if the message is a command from the controller to set up transmitting parameters.

The bottom region contains two states. This thread periodically checks whether there are messages in its output message queue. If there are messages and the link driver is not currently busy, the thread will try to request the link driver to send a message out. When the link driver has done its job, *sendingInProgress* is set to false, which brings the messenger to a choice point. It continues to request the sending of another message if the output message queue is not empty; otherwise, it comes to the state *WaitingToSend*.

Problems

9.1 A simple protocol is given in Figure 9.17. Follow the transitions to explain the state behavior of the protocol object.

9.2 Find an example to practice concurrent state modeling.

9.3 Find an example to specify the inheritance of state behavior.

9.4 Work out the state diagram for the *TrabsmittingMessage* composite state of the link driver object. Note that it is roughly the reverse process of the *ReceivingMessage*.

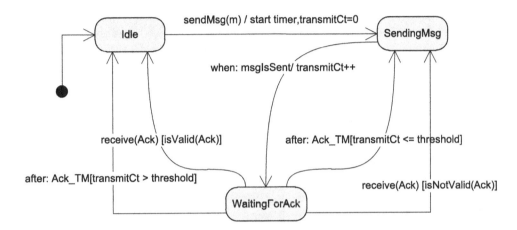

Figure 9.17
Reliable data transmission protocol: sender.

CHAPTER 10

Real-Time UML: General Resource Modeling

Contents

> *Time is the scarcest resource and unless it is managed nothing else can be managed.*
> *Peter F. Drucker*

10.1 Real-Time UML Profile

In response to the needs of the real-time software community, the Object Management Group has adopted a real-time UML (RT-UML) profile, formally known as the UML Profile for Schedulability, Performance, and Time [5].

As shown in Figure 10.1, RT-UML profile is structured into a number of packages, including the General Resource Modeling (GRM) package and the Analysis Models package.

The GRM package is the core, providing a common base for the rest of RT-UML profile. This package is further partitioned into three separate subprofiles:

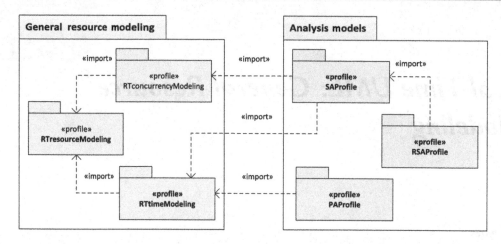

Figure 10.1
Real-time UML profiles as specified in [5].

(1) *RTresourceModeling*, a subprofile for resource modeling;
(2) *RTconcurrencyModeling*, a subprofile for concurrency modeling; and
(3) *RTtimeModeling*, a subprofile for modeling time and time-related mechanisms.

The Analysis Models package defines the analysis subprofiles, including

• *PAprofile* for performance analysis;
• *SAprofile* for schedulability analysis; and
• *RSAprofile* for real-time CORBA schedulability analysis.

The RT-UML profile,[1] if widely accepted, could offer a common vocabulary to the real-time software community, allowing human users and automatic tools to exchange and interpret appropriately-annotated models, and to perform quantitative analysis on them. This would enable designers and developers to predict the salient real-time properties of a system, so that the models could be modified early in the development cycle to improve system schedulability or real-time performance.

This chapter focuses on the GRM package, and the Analysis Models package will be covered in Chapter 11.

10.1.1 Meta Modeling, UML Stereotypes, and Tags

A metamodel is a language used to specify models. UML is a metamodel language used to specify models built for specific systems. Above UML, there is yet another abstract layer called Meta Object Facility (MOF), which is a meta metamodel—a language used to

---

[1] For the treatment of the RT-UML profile [5], in this book I try my best to honor the original term definitions and descriptions, but sometimes I may make adjustments for clarity and consistency.

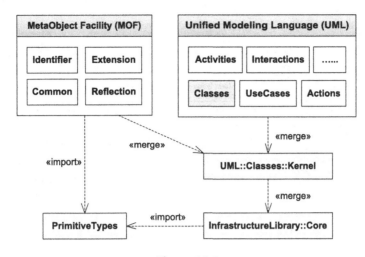

Figure 10.2
MOF, UML, and common core.

specify metamodels such as UML. Indeed, UML is simply one metamodel instance derived from MOF.

If we take a deeper look, MOF and UML actually share a common core, which is illustrated in Figure 10.2. The packages *PrimitiveTypes* and *InfrastructureLibrary* are the cornerstones. The package *PrimitiveTypes* consists of a few predefined primitive types that are commonly used in metamodeling, including *Boolean*, *Integer*, *Real*, *String*, and *UnlimitedNatural*. The package *InfrastructureLibrary* consists of the packages *Core* and *Profiles*, where the latter defines the mechanisms that are used to customize metamodels and the former contains core concepts used in metamodeling [11]. The *Core* package is merged with the *Kernel* subpackage defined in the *Classes* package of UML, which is reused to bootstrap both UML and MOF.

In the four-layer Model Driven Architecture, MOF is at a higher abstract level than UML, which is shown in Figure 10.3. The M3 layer defines elements that can be instantiated to define metamodel elements. For example, the concept "Action" as used in UML activity diagrams is merely an instance of the concept "Class." Note that the "Class" at level M3 is different from the "Class" at level M2, where the former is defined in the core of the infrastructure library and the latter is merely an instance of the former.

The M2 layer defines elements that can be instantiated to define model elements. For example, when a user defines a class such as *X86 Core i7*, she is creating an instance of the UML "Class" concept. All the UML terms we have learned so far—say, *Association*, *Interface*, *Attribute*, and *Dependency*—are concepts at the M2 level.

One interesting thing we can do at the M2 level is to define new stereotypes, which is an extensibility mechanism that allows a designer to extend the UML vocabulary by creating new model elements from existing ones. A *stereotype* is a metaclass with specific attributes that are

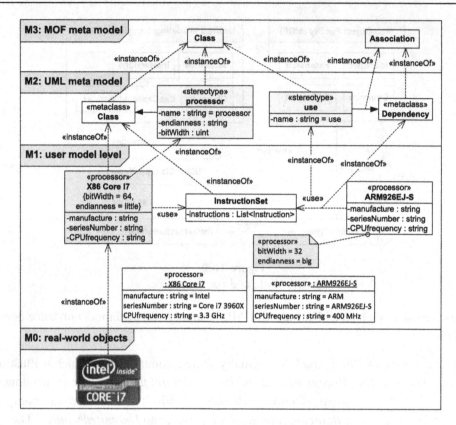

Figure 10.3
Four layers of the Model Driven Architecture.

suitable for a particular problem domain or specialized usage. A stereotype attribute is referred to as a *tag* definition. A stereotype definition is shown within a class box, with the name annotated by the keyword «stereotype». In Figure 10.3 there are two examples given at level M2. The stereotype *processor* has three tag definitions, and *use* has only one tag definition.

A stereotype cannot be used by itself, but must always be used in conjunction with another metaclass, which is usually referred to as the *base metaclass*. A stereotype and its base metaclass is related by an *extension*, which is an instance of the MOF *Association* and is shown as an arrow with a filled triangle pointing from the stereotype to the base metaclass. In our example, the stereotype *processor* extends the UML metaclass concept "Class" and *use* extends the UML metaclass concept "Dependency."

It is exactly at the M2 level that the RT-UML profile defines various stereotypes for supporting real-time modeling and analysis. However, as users of UML and the real-time profile, 99%, if not all, of our activities are within the M1 level. And most likely, some of you never need to jump out of the M1 box in your career as a software engineer.

So what do we do at the M1 level? We build models by introducing classes pertinent to the domain problem at hand. Whenever we do this, we are implicitly creating instances from the M2 level concepts such as "Class," "Association," "Dependency," "State," "Transition," and "UseCase." These instances become our building blocks at level M1, allowing us to model the real-world objects living at level M0.

Now let us take a step back to see how the stereotypes defined at level M2 are used at level M1. A stereotype is rendered as a name enclosed by guillemets and placed above the name of the base element. For example, the stereotype «use» is shown above the dependency relation from the class *ARM926EJ-S* to the class *InstructionSet*. Also, the stereotype «processor» is shown above the class name *X86 Core i7*, with the values of the stereotype attributes (also called tagged values) displayed in the class name compartment. Alternatively, the tagged values of the applied stereotype can be shown in a comment symbol attached to the base element, as used in the «processor» stereotype for the class *ARM926EJ-S*.

The stereotype mechanism should not be confused with subclassing:

- A stereotype is defined at level M2 to extend a base metaclass. When used at level M1, a stereotype applies to the modeling elements, which are instances of the base metaclass. In contrast, a subclass is defined and used at level M1 to extend a superclass at the same modeling level.
- A stereotype is a lightweight collection of tagged values. When a stereotype is applied to a user defined class, the tagged values are information common to all objects of the user class. We already know that information redundancy cannot be addressed by subclassing—say, by introducing a *Processor* superclass (refer to Section 6.3.3 to see why). There are two ways to implementing stereotypes in object-oriented programming without causing information redundancy. The first approach is to treat the tag definitions as *constant class attributes* initialized with the corresponding tagged values. The second approach is to apply the *abstraction-occurrence* design pattern.
- A stereotype can be defined for any base metaclass, not just for "Class." We have seen a stereotype that extends "Dependency." In RT-UML profile, you will see many stereotypes defined for different metaclasses, such as *Message*, *Action*, *Node*, and *Stimulus*.

10.2 Resource Modeling

At the core of the GRM package is the *RTresourceModeling* subprofile, which defines a collection of stereotypes, tagged values, and rules for resource modeling.

In general, a *resource* refers to a run-time entity that offers one or more services. For instance, processing devices (virtual or physical) are resources that offer program execution services; OS objects such as shared memory objects, channels, message queues, and pipes are operating system resources that offer intertask communication services; peripheral devices such as memory chips, I/O ports, and direct-memory-access channels each can offer one or more specific services.

From another perspective, there are active resources, such as hardware devices, operating system processes, and operating system threads, each of which has its own flow of control and can function concurrently or pseudoconcurrently with other entities. There are passive resources, which cannot act by themselves, but can only react to external service requests.

A resource in its nature has limited capacity, which is usually reflected by the effectiveness or quality of services (QoS) offered to its clients. Some example QoS measures in real-time systems include response time, idle time, and worst-case switching time.

10.2.1 UML Core Resource Model

To conduct real-time analysis, we need models that can describe specific instances and their linkages. For this purpose, it is critical to distinguish *descriptor concepts* and *instance-based concepts*. A descriptor concept is a specification or an abstraction of a set of instances, and an instance is a single occurrence of its descriptor. In some cases, an instance may be related to multiple descriptors, each exhibiting a different viewpoint or abstraction of the instance.

Figure 10.4 gives the core resource model described in RT-UML profile [5], where all the elements on the left represent instance-based concepts, and the elements on the right represent their corresponding descriptors. At the top layer, a *resource* describes a collection of *resource instances*. In the middle, a *resource service instance* is a specific incarnation of a *resource service* description. At the bottom, a *QoS characteristic* can have many concrete *QoS values*.

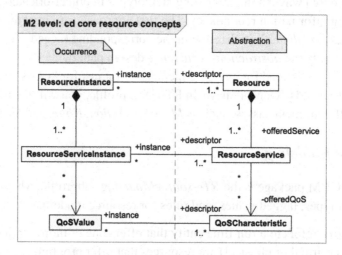

Figure 10.4
RT-UML profile core resource model (at level M2).

Both the right part and the left part of the class diagram can be similarly interpreted, but the right part is at a higher abstraction level. As far as the left part is concerned, a *Resource Instance* object may offer one or more services, each of which is a *ResourceServiceInstance* object. The actual quality measures of each *ResourceServiceInstance* object are given by some *QoSValue* objects, which are often referred to as the offered QoS. An offered QoS value of a resource service instance actually dictates a minimum guarantee, called the worst-case QoS performance, which is a critical measure often considered in schedulability analysis.

Depending on the implementations and the resource usage contexts, different resource service instances may offer services at different quality levels (reflected in the offered QoS values). Also, depending on the nature of the service and the characteristic, QoS values can differ in complexity. Simple QoS characteristics are typically defined as the attributes of a service object. For example, when an action is modeled as a service object, it may have QoS-related attributes such as worst-case execution time. A physical processor offers a "processing service" which may include a throughput characteristic. Some service objects may have QoS characteristics that take complex structured values, such as probability distributions.

In Figure 10.4, each instance-based concept is related to its corresponding descriptor by an association, which is exactly an application of the *Abstraction-Occurrence* design pattern at the M2 level. In particular, each instance of *Resource* (which is a user-defined "abstraction" class at level M1) specifies some common properties among a collection of *ResourceInstance* objects (which are instances of a user-defined "occurrence" class at the M1 level). As an example, Figure 10.5 gives a user model of a resource named *X86 Core i7* and its resource instance concept called *X86 Core i7 Instance*, where the former class models abstract specifications and the latter class models the occurrences.

Note that both the class diagram and the object diagram in Figure 10.5 are level M1 models. The object diagram, however, illustrates the descriptor concept and its instance concept more clearly. In our example, two objects are instantiated from the *X86 Core i7* resource descriptor: they differ in the values of the series number and the CPU frequency. The object diagram also indicates that each resource descriptor object is linked to many (maybe hundreds of thousands of) *X86 Core i7 Instance* objects, each of which corresponds to a real-world processor chip. Indeed, each resource descriptor object captures a collection of common property values for many resource instance objects. The *Abstraction-Occurrence* design pattern has successfully isolated the instance-specific property *chipID* from the descriptor-specific properties *manufacture*, *seriesNumber*, and *CPUfrequency*. Here we purposely treat the *manufacture* attribute as a descriptor-specific property, modeling the possibility of x86 processors being made by a manufacturer other than Intel. If this will never happen, it is better to model the *manufacture* attribute as a tag definition in the stereotype *Processor*.

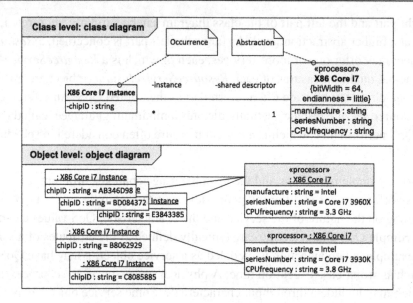

Figure 10.5
An example of the resource model (at level M1).

10.2.2 Action and Action Execution

Action is an abstract concept. In computing disciplines, an action is the specification of an executable entity and is the fundamental unit of processing.[2]

There are atomic actions and nonatomic actions. An atomic action can run to completion without interruption. A nonatomic action, sometimes also called an activity, is a more complex collection of behavior that may run for a long duration. A nonatomic action may be interrupted by events or may be preempted by other actions.

Corresponding to the concept *action*, there is an "occurrence" concept named *action execution*. An action is a specification (or descriptor) of many action executions. Once again, we see the application of the *Abstraction-Occurrence* design pattern. An action execution, the occurrence of an action in a particular thread of execution, can be taken as a resource instance that offers services to its callers or the embedding thread. An action execution may also consume resources of other kinds (e.g., data storage).

Figure 10.6 gives the metamodel concepts related to actions, where the right part is a collection of descriptor concepts, each of which has a corresponding "occurrence" concept in

---

[2] In programming, an action can be a function call or simply a block of statements.

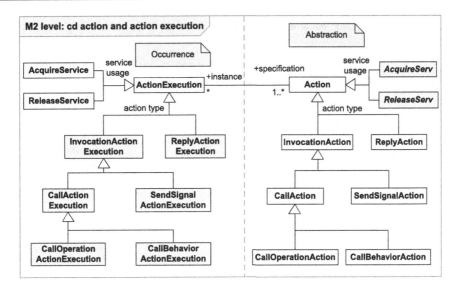

Figure 10.6
Action and action execution.

the left part of the figure. Since real-time analysis is instance based, hereinafter instead of exposing the descriptor concepts in our discussions, we will focus on the "occurrence" concepts only, assuming that there exists such a concept correspondence.

In Figure 10.6, *ActionExecution* is subclassed along two dimensions (refer to Exercise 6.7 for the definition of power type). Along the *action type* dimension, *ActionExecution* is subclassed into many specific kinds of action executions at multiple levels. As far as service usage is concerned, there are *AcquireService* action executions and *ReleaseService* action executions, which are performed to acquire a service instance offered by a resource instance and release a service instance respectively.

The concepts shown in Figure 10.6 are at the M2 metamodel level. We will see some examples at the M1 level in the next two subsections.

10.2.3 UML Stereotypes for Protected Resources

Most resources in real-time embedded systems are *protected resources* in the sense that they can only be accessed (consumed) exclusively by one client at a time. For instance, while a file could have multiple readers at the same time, when a writer is using the file, no other readers or writers can be granted access to the file.

We first look at the M2 level class diagram given in Figure 10.7 (as said, only the "occurrence" concepts are shown). A *ProtectedResource* instance offers one or more *ExclusiveService*

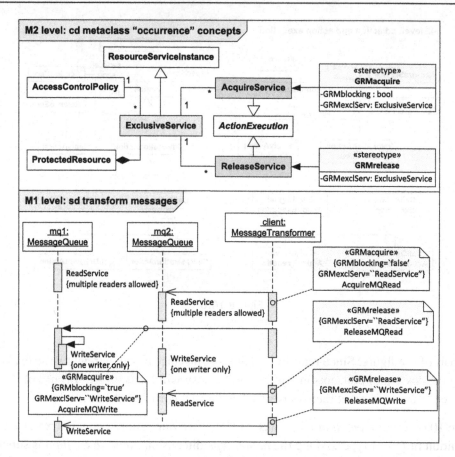

Figure 10.7
Using the RT-UML profile resource model.

instances. The concurrent access to an exclusive service instance is restricted according to some access control policy. To use an exclusive service of a protected resource, it is necessary first to execute an appropriate *AcquireService* action. After each use of a service, the *ReleaseService* action can be executed to release the service.

The RT-UML profile defines two stereotypes «GRMacquire» and «GRMrelease», which are especially useful for annotating accesses to protected resources. The stereotype «GRMacquire» extends the base class *AcquireService*, specifying two tag definitions. *GRMblocking* is of Boolean type, indicating whether the *AcquireService* action execution is blocking or not. The caller of a blocking *AcquireService* action execution is blocked until the exclusive service instance becomes available. The tag *GRMexclServ* refers to the exclusive service instance to which the *AcquireService* action execution is trying to acquire access rights. Similarly, the stereotype «GRMrelease» extends the base class *ReleaseService*.

Remember what we have said about the correspondence between "occurrence" concepts and "descriptor" concepts. Here, you can assume that each of the two stereotypes also extends another base class—the corresponding subclass of *Action*. Since both *Action* and *ActionExecution* have subclasses, the stereotypes «GRMacquire» and «GRMrelease» also apply to other metaclasses such as *Operation*, *Reception*, *Transition*, *Message*, and *Method*.

As an example, the lower part of Figure 10.7 gives a sequence diagram where the two stereotypes are applied to user-defined operations and messages. The *MessageTransformer* object named *client* reads from one message queue named *mq2* to another message queue named *mq1*. The two message queue objects are protected resource instances, each of which offers two exclusive services called *ReadService* and *WriteService*. The access control policy for *ReadService* says that multiple readers are allowed, whereas only a single writer is allowed to use *WriteService*.

The client operation *AcquireMQRead* is an instance of the metaclass concept *AcquireService*. The tagged values indicate that *AcquireMQRead* is nonblocking, and it acquires access to an exclusive service called *ReadService*, which is offered by the *mq2* resource instance as implied by the message from *client* to *mq2*.

AcquireMQRead is followed by an anonymous operation, which sends a message called *AcquireMQWrite* to the resource instance *mq1*. The tagged values of the message indicate that the message sending operation is blocking, and it acquires access to an exclusive service called *WriteService*, which is offered by the message receiver. The message *AcquireMQWrite* is eventually handled by *WriteService* of *mq1*. At the end of the scenario, the *client* releases the service instances to make them available to others.

10.2.4 Resource Usage

We have already seen an example of resource usage in Figure 10.7. In general, a resource usage is built upon a classic client-server model, specifying a pattern about how and when a set of clients use a set of resources and their services. It closely corresponds to a use case instance, describing an ordered series of action execution steps.

The upper part of Figure 10.8 gives the resource usage metaconcepts. A *Scenario* instance consists of an ordered series of *ActionExecution* steps. The ordering of the execution steps follows a predecessor-successor pattern, with the possibility of multiple concurrent successors and predecessors, stemming from concurrent thread forks and joins, respectively.

Each *Scenario* (and its subclass *ActionExecution*) may use one or more resource instances and resource service instances. In each of the service usages, it may specify explicitly the *required QoS* values that it needs in order to meet its obligations. These can be matched against the

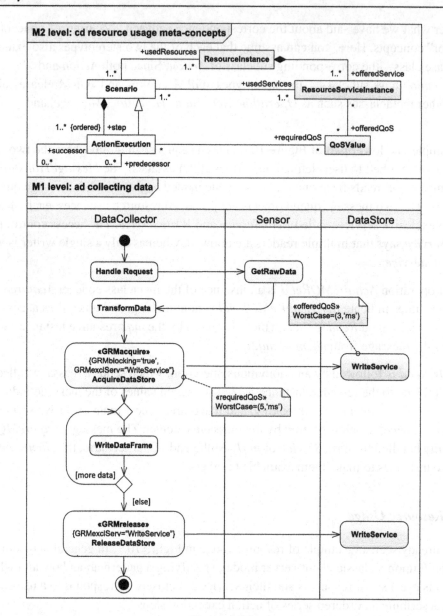

Figure 10.8
Resource usage metaconcepts.

offered QoS values of the corresponding resource (and service) to determine if the requirements can be met. A resource usage request can be satisfied if the offered QoS is at least as good as the required QoS.

The lower part of Figure 10.8 gives a UML activity diagram modeling a scenario instance of collecting data from sensors. *DataStore* is a *ResourceInstance* object offering a *WriteService* to its clients. *WriteService* has a simple QoS characteristic called "WorstCase." The *WriteService* instance used in the scenario has an offered QoS value of 3 ms, implying that its worst-case performance is 3 ms.

The scenario has a sequence of actions, where the action *AcquireDataStore* performed by the *DataCollector* object is an *AcquireService* action as indicated by the stereotype «GRMacquire». The tagged values say that *AcquireService* is a blocking action and its required QoS from *WriteService* is 5 ms. Since the offered QoS value is better than the required QoS value, this resource usage can be satisfied.

10.2.5 Resource-Client Graph

The resource-client relationship can be typically interpreted as a peer collaboration in the sense that the resource and the client are service provider and service consumer modeled at the same level of abstraction. For example, the resource may be a semaphore that protects some shared data and the client could be a software task that uses the semaphore to access the shared data. We next introduce resource-client graphs, which are used in embedded systems to model and analyze resource-client relationships.

In graph theory, a bipartite graph (or bigraph) is a graph whose vertices (or nodes) can be divided into two disjoint sets X and Y such that every edge connects a vertex in X to one in Y. A resource-client graph is a bigraph with a resource set R and a client set C. A directed edge from a resource r to a client c indicates that the client c is currently the owner of the resource r, while a directed edge from a client c to a resource r indicates that the client c is currently requesting for the resource r. In other words, the outgoing edges of a client indicate its outstanding resource requests.

Figure 10.9 gives an example resource-client graph, where the client $C1$ currently owns the resource $R1$, and it requests resources $R2$ and $R3$, which have already been granted to clients $C2$ and $C3$, respectively.

Each edge of a resource-client graph can be labeled by QoS constraints at each of its ends. For instance, in Figure 10.9, there are two labels on the edge from client $C4$ to resource $R1$: the required QoS is that $C4$ has to be served in 4 ms, while the offered QoS is that $R1$ can, in the worst case, provide service in 2 ms.

A resource-client graph can be used to detect potential issues of a system design. For instance, if we assume that each resource client has to obtain all the resources it requests in order to get its job done well, then a circular chain in a resource-client graph would indicate a critical problem. For example, in Figure 10.9 there is a circular chain $C1$-$R2$-$C2$-$R1$-$C1$, with each client holding some resources that are currently requested by the other. This system will inevitably run into a deadlock situation.

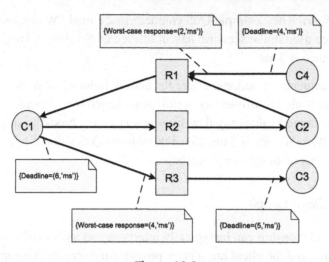

Figure 10.9
A resource-client graph.

10.3 Time Modeling

10.3.1 The Notion of Time

What is time? We value it as money and race against it every day, but few people, except philosophers or theoretical scientists, would be able to give a rigorous definition for it. As engineers, all you might remember is that "time is one of the seven fundamental physical quantities in the International System of Units."

Sean Carroll puts it this way [23]:

> The most mysterious thing about time is that it has a direction: The past is different from the future. That's the arrow of time...The arrow of time is the reason why time seems to flow around us, or why we seem to move through time. It's why we remember the past, but not the future. It's why we evolve and metabolize and eventually die. It's why we believe in cause and effect, and is crucial to our notions of free will.

In the Newtonian view, time is a dimension of the universe in which events occur in sequence and spread out across the time line. In computer science, time is typically modeled as a line structure with a fixed past and future, or a tree structure where there is a linear past but there exist many future possible worlds (states).

According to the International System of Units, a second, the base unit of time, is the duration of 9,192,631,770 periods of the radiation corresponding to the transition between the two hyperfine levels of the ground state of the caesium 133 atom. The basis for scientific time is a continuous count of seconds based on atomic clocks, known as International Atomic Time (TAI). The modern civil time standard, Coordinated Universal Time (UTC), follows

International Atomic Time with an exact offset of an integer number of leap seconds. The Global Positioning System broadcasts worldwide a very precise atomic time signal, which can be converted to UTC with errors at the microsecond level.

Fortunately, to work with real-time systems, you do not have to be a philosopher. We will simply treat time as an ordered series of time instants and use milestones or deadlines to capture the passage of time [34]. A time value is used to denote a specific instant of time. A time interval is the interval between two time instants, and thus is denoted by a pair: a start time instant and an end time instant.

In computing systems, time is measured by clocks. Computer real-time clocks keep time with an electronic oscillator circuit that gives the vibrating quartz crystal tiny "pushes," and generates a series of electrical pulses, one for each vibration of the crystal, which is called the clock signal. The clock signals are added up digitally by integrated circuit counters or dividers to generate periodic events called clock interrupts. In such a sense, a clock interrupt is also called a hardware timer and a computer system may have multiple hardware timers with different interrupt frequencies. A clock interrupt typically triggers a processor to execute an interrupt service routine (ISR), where the passage of time is acknowledged and the system time is adjusted. Periodically—say, every week—computers use Network Time Protocol to synchronize themselves over the Internet with atomic clocks (UTC).

Operating systems and embedded systems often use a single hardware timer to implement an extensible set of software timers. In such a case, the hardware timer's ISR would maintain a countdown variable with a preset value, which is decreased by 1 for each interrupt signal. Once its countdown variable becomes 0, the ISR is responsible for resetting the countdown variable, and informing a utility called a software timer manager. A software timer manager maintains a countdown variable for each software timer installed, and it decreases all the countdown variables by 1 every time it gets a signal from the timer ISR. Whenever a countdown variable becomes 0, which indicates the expiry of the corresponding software timer, the software timer manager will send a time-out event to the owner process. Expired software timers that are continuous would also be reset to a new expiry time based on their timer interval, and one-shot timers would be disabled or removed by the software timer manager (the software timer manager is covered in Chapter 22).

10.3.2 Timing Mechanisms

The UML *General Time Modeling* subprofile [5] describes a conceptual framework for representing time and time-related mechanisms that are appropriate for modeling real-time software systems.

A *timing mechanism* is an abstract concept that captures the common features of resources (timing devices) that is capable of performing time measurement and timing-related functions.

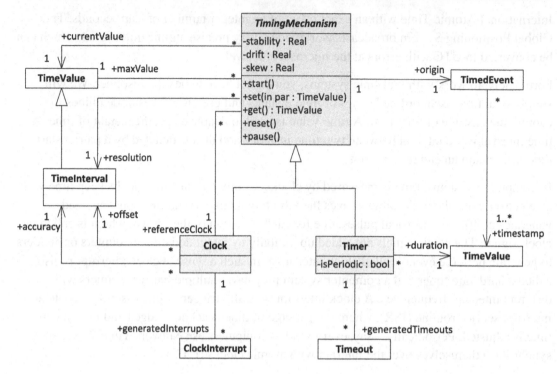

Figure 10.10
The concept of the timing mechanism as defined in [5].

As shown in Figure 10.10, *TimingMechanism* has several operations, which represent its offered resource services. It also has eight attributes (with five reference attributes), which specify the corresponding QoS characteristics:

(1) referenceClock: the reference clock whose QoS characteristics serve as a reference for specifying the QoS characteristics of this timing mechanism. The reference clock of a timing mechanism is typically some kind of near-to-ideal clock, such as a clock maintained by an international standards organization. A clock should be synchronized with its reference clock as closely as possible.

(2) stability: which is the ability of the mechanism to measure the progress of time at a consistent rate.

(3) skew: the rate of change of the offset between the timing mechanism and its reference clock.

(4) drift: the maximum absolute difference between the frequency of the timing mechanism relative to the frequency of its reference clock.

(5) origin: the event occurrence relative to which the timing mechanism starts to measure time. A *TimedEvent* may have multiple timestamps if multiple reference clocks are used.

(6) currentValue: identifies how far in time it has progressed since the origin time instant (with respect to a modulus of the maximum value allowed, and without counting any pauses or resets).

(7) maxValue: the maximum value that the current value of the timing mechanism can take.

(8) resolution: an attribute that identifies the minimal duration that can be distinguished by the timing mechanism. If necessary, we can obtain whatever time resolution we need simply by choosing a sufficiently short cycle time for the reference clock.

Table 10.1 lists some attribute values of an example timing mechanism.

Clocks are mechanisms that periodically generate clock tick events. A clock tick may in turn cause a clock interrupt. In addition to the attributes inherited from the superclass *TimingMechanism*, *Clock* also has two attributes: *offset*—the absolute time difference between the time of its clock tick and the corresponding tick of its reference clock (e.g., time zone adjustment); and *accuracy*—the maximum offset of a clock over time. Specific to a clock, the inherited attribute *resolution* identifies the minimal duration between two successive ticks of the clock.

Clock has two more associations: one indicates that a clock could act as a reference clock for many timing mechanisms, and the other indicates that a clock can generate an ordered set of clock interrupts.

Timers are mechanisms that may generate one or more time-out events. *Timer* defines two more attributes: *duration*—the time interval measured by the timer; and *isPeriodic*—a Boolean value indicating whether the timer is periodic or not (periodic timers keep generating time-out events until they are paused or destroyed). Specific to *Timer*, the inherited attribute *currentValue* represents the duration of time that must expire before the time-out event occurs.

Table 10.1 Some attribute values of an example timing mechanism

Attribute	Value
Frequency	999 ± 1 Hz
Reference clock	1000 Hz (i.e., 1000 ticks per second)
Stability	Precision error less than 2 ms per second
Skew	On average 1 ms per second
Drift	2 Hz
Maximum value	12 h
Current value	3 h 40 min 20 s
Resolution	1 s

The time-out event is generated at the instant when the duration of the timer has expired relative to the instant when the timer was started or to the instant when the preceding time-out was generated (for periodic timers) plus any time that passed while the timer was paused.

10.3.3 Time Modeling Stereotypes

Corresponding to the real-time modeling concepts given in Figure 10.10, RT-UML profile provides a set of stereotypes (and their tag definitions) for annotating UML models with time-related information. A few stereotypes are listed in Table 10.2.

Table 10.2 Some stereotypes defined in the RT-UML timing subprofile

Stereotype	Base Metaclass	Tags (Optional)
«RTtimingMechanism»	Class, Instance, DataType	RTstability: Real
		RTdrift: Real
		RTskew: Real
		RTmaxValue: RTtimeValue
		RTorigin: String /*event name*/
		RTresolution: RTtimeValue
		RTcurrentValue: RTtimeValue
		RTrefClk: Refer to a «RTclock»
«RTclock» (subclass of «RTtimingMechanism»)	Class, Instance, DataType	RTclockId: String /*clock name*/ RTaccuracy: RTtimeValue RToffset: RTtimeValue
«RTtimer» (subclass of «RTtimingMechanism»)	Class, Instance, DataType	RTduration: RTtimeValue RTperiodic: Boolean
«RTstart»	Operation	/*start a timing mechanism*/
«RTaction»	Action, ActionExecution, Message, Method, Transition	RTstart: RTtimeValue RTend: RTtimeValue RTduration: RTtimeValue
«RTdelay» (subclass of «RTaction»)	Action, ActionExecution, Message, Method, Transition	RTduration: RTtimeValue
«RTinterval»	Class, Instance, DataType	RTintStart: RTtimeValue RTintEnd: RTtimeValue RTintDuration: RTtimeValue
«RTevent»	Action, ActionExecution, Message, Method, Transition	RTat: RTtimeValue
«RTtimeout» (subclass of «RTevent»)	Action, ActionExecution, Message, Method, Transition	RTtimeStamp: RTtimeValue
«RTclkInterrupt» (subclass of «RTevent»)	Message	RTtimeStamp: RTtimeValue

Notice that «RTtimingMechanism» is an abstract stereotype, which defines a list of tags that also apply to its subclass stereotypes «RTclock» and «RTtimer». «RTaction» provides a common abstraction for an action that takes time to complete; it can be used for modeling behavior that has a definitive start time and a definitive end time. A special kind of «RTaction» is «RTdelay»—a deliberate delay action which delays execution for some time interval.

In Table 10.2, except for a few tags that have primitive types (e.g., Real, String, Boolean) or reference to other elements, most have a type called *RTtimeValue*.

Table 10.3 Real-time value forms for tag type *RTtimeValue*

<timeVal>	::=	<timeStr> | <dateStr> | <dayStr> | < metricTimeStr>
<timeStr> <hr>	::= ::=	<hr> [":" <min> [":" <sec> [":" <centisec>]]] "00".."23"
<min>	::=	"00".."59"
<sec>	::=	"00".."59"
<centisec>	::=	"00".."99"
<dateStr>	::=	<year> "/" <mon> "/" <dayOfMon>
<year>	::=	"0000".."9999"
<mon>	::=	"01".."12"
<dayOfMon>	::=	"01".."31"
<dayStr>	::=	"Mon" | "Tue" | "Wed" | "Thr" | "Fri" | "Sat" | "Sun"
<metricTimeStr>	::=	"(" [<number> | <PDFstring>] "," <timeUnit>")"
<number>	::=	<Integer> | <Real>
<timeUnit>	::=	" 'ns' " | " 'us' " | " 'ms' " | " 's' " | " 'hr' "
		| " 'days' " | " 'wks' " | " 'mos' " | " 'yrs' "
<PDFstring>	::=	<bernoulliPDF> | <binomialPDF> | <normalPDF>
		| <gammaPDF> | <exponentialPDF> | <poissonPDF>
		| <uniformPDF> | <histogramPDF> | <geometricPDF>
<bernoulliPDF>	::=	" 'bernoulli' ," <Real>
<binomialPDF>	::=	" 'binomial' ," <Real> "," <Integer>
<normalPDF>	::=	" 'normal' ," <Real> " , " <Real>
<gammaPDF>	::=	" 'gamma' ," <Integer> "," <Real>
<histogramPDF>	::=	" 'histogram' ," <Real> "," <Real>* "," <Real>
<exponentialPDF>	::=	" 'exponential' ," <Real>
<poissonPDF>	::=	" 'poisson' , " <Real>
<uniformPDF>	::=	" 'uniform' , " <Real> " , " <Real>

The general format for expressing *RTtimeValue* is described by the extended Backus normal form (or Backus-Naur form) given in Table 10.3. Here are a few simple examples:

- Time string: 11:22 (22 minutes past 11 o'clock); 10:15:35:18 (10 o'clock 15 min 35 sec 18 centisec).
- Date string: 2002/02/12.
- Metric time string: (12, 'ms'); (3.5, 'hr').

A metric time value can also be a standard probability distribution function. Here are a few examples:

- Bernoulli distribution: ('bernoulli', 0.72, 'ms'), which has one parameter specifying a probability (a real value no greater than 1).
- Binomial distribution: ('binomial', 0.5, 12, 'ms'), which has two parameters: a probability and the number of trials (a positive integer).
- Exponential distribution: ('exponential', 1.8, 'ms'), which has one real parameter, the mean value.
- Gamma distribution: ('gamma', 12, 1.8, 'ms'), which has two parameters: a positive integer and a real value specifying the mean.
- Histogram distribution: ('histogram', 1.0, 0.15, 1.2, 0.65, 1.4, 0.20, 1.6, 'ms'), which has an ordered collection of one or more pairs of real values and one end-interval value for the upper boundary of the last interval. Each pair identifies the start of an interval and the probability that applies within that interval.
- Normal (Gaussian) distribution: ('normal', 2.4, 0.8, 'ms'), which has two real parameters: a mean value and a standard deviation (greater than 0).
- Poisson distribution: ('poisson', 960, 'us'), which has a mean value.
- Uniform distribution: ('uniform', 20, 80, 'ms'), which has two real parameters designating the start and end of the sampling interval.

10.4 Concurrency Modeling

In concurrency modeling, a *concurrent unit* represents an *active resource instance* that executes concurrently with others. Following its creation, each concurrent unit starts to execute one main scenario (method execution). During its execution, the main method execution may perform *accept-event* or *accept-call* actions in order to receive incoming requests. In response to a request, the concurrent unit activates an appropriate service instance and its service method. During the execution of the service method, the main method may either be blocked or continue its execution concurrently.

Table 10.4 lists a few stereotypes defined in the RT-UML profile concurrency subprofile [5]. A concurrent unit is indicated by the stereotype «CRconcurrent», which has a tag definition named *CRmain*. The tagged value of *CRmain* refers to the main method of the concurrent unit.

Table 10.4 Some stereotypes defined in the concurrency subprofile

Stereotype	Base Metaclass	Tags (Optional)
«CRconcurrent»	Node, Component, Class, Instance	CRmain: reference to the main method of a concurrent unit
«CRaction»	Action, ActionExecution, Message, Method, Transition	CRatomic: Boolean
«CRasynch»	Action, ActionExecution	
«CRsynch»	Action, ActionExecution	
«CRdeferred»	Operation, Reception, Message	
«CRimmediate»	Operation, Reception, Message	CRthreading: enumeration of {'remote', 'local'}

The stereotype «CRaction» has a Boolean tag definition named *CRatomic*. An atomic action can be annotated by «CRaction» with the tag *CRatomic* set to "true."

A client object can invoke an operation on a service object; such a service request may be either asynchronous or synchronous. If the request is synchronous, then the client execution is blocked until a response is received from the server. If the request is asynchronous, then the client execution proceeds immediately. The stereotype «CRsynch» can be applied to an action execution to indicate that the caller (client execution) is blocked until the service action execution is completed. Likewise, «CRasynch» can be applied to an action execution to indicate that the caller continues with its execution after the call action (request message) is issued.

A service instance may be busy with other processings when a new service request arrives. In such a situation it is necessary for it to defer its response to the new request until the service instance is ready (the deferred requests are typically put into one or more queues, if priority matters). The stereotype «CRdeferred» can be applied to a message to indicate that the receiving service instance handles the request in a deferred fashion. This stereotype can also apply directly to an operation or a reception.

A service instance may also be able to handle requests immediately, and the stereotype «CRimmediate» is defined for this purpose. «CRimmediate» has a tag definition named *CRthreading*, which can take a value of "local" or "remote":

* *CRthreading='local':* the incoming request is handled locally, where the service instance spawns a new local thread to handle the request.
* *CRthreading='remote':* the incoming request is handled remotely, where the requester's thread of execution is utilized for service execution.

Figure 10.11 gives a UML sequence diagram illustrating the usage of some concurrency stereotypes. The stereotype «CRconcurrent» is applied to two active instances: one is a

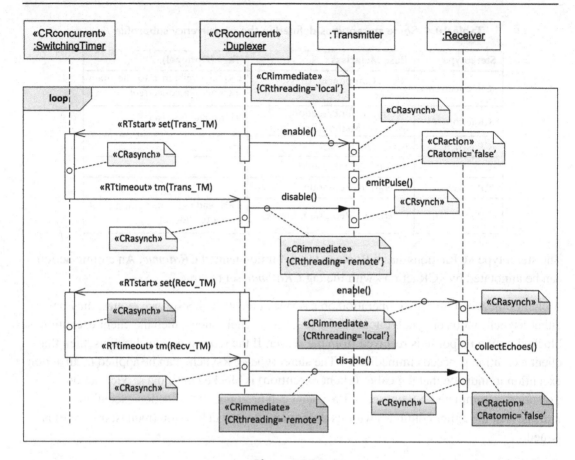

Figure 10.11
Usage of concurrency stereotypes.

SwitchingTimer object and the other is a *Duplexer* object. Although the tagged value for *CRmain* is not shown explicitly, you can assume that each of the two objects has its own main method (and thread of execution).

Both of the action executions of the *SwitchingTimer* object are annotated by «CRasynch», indicating that the *Duplexer* object continues its execution after the calls.

The *enable()* message is an asynchronous message, which is annotated by the stereotype «CRimmediate» with a tagged value "local." This means that the *Duplexer* object continues its own execution after the call, and the *Transmitter* object runs its operations (i.e. the processing of the *enable()* message, and the *emitPulse()* action execution) asynchronously in a new local thread. Indeed, the behavior of the *Duplexer* object is regulated by the timer object; it should not be blocked by the *Transmitter* object because it needs to be responsive to the time-out events from the *SwitchingTimer* object.

In contrast, the *disable()* message is a synchronous message, which is annotated by the stereotype «CRimmediate» with a tagged value "remote." This means that the *Duplexer* object will not resume its own action execution until its request is completely processed. In response to the *disable()* request, the corresponding operation of the *Transmitter* object is invoked and executed in the thread of the *Duplexer* object. Indeed, the *Duplexer* object cannot proceed until after the *Transmitter* object has been disabled; otherwise, the radar would function incorrectly owing to the activeness of both its transmitter and its receiver at the same time.

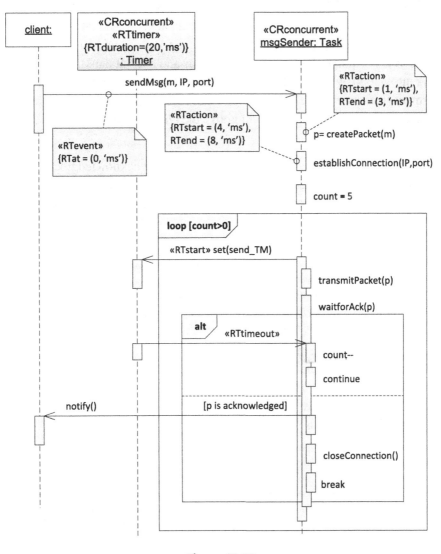

Figure 10.12
A UML sequence diagram annotated with timing stereotypes.

The rest of the sequence diagram deals with the *Receiver* object in a similar fashion. Its interpretation is left as an exercise.

Figure 10.12 gives a UML sequence diagram illustrating the use of timing stereotypes. Notice that the anonymous *Timer* object acts as a timing mechanism as indicated by the «RTtimer» stereotype. The expiry duration of the timer is 20 ms.

The scenario starts with the *client* object sending a message *sendMsg(m, IP, port)* to the *msgSender* object. The message is annotated as a real-time event occurring at 0 ms (relative to the beginning of this scenario).

In response to the *client's* request, *msgSender* creates a packet to encapsulate the message to be sent (this action starts at 1 ms and ends at 3 ms), then tries to establish a connection with the designated receiver (IP, port) (this action starts at 4 ms and ends at 8 ms). Afterward, the sender enters into a loop, where it starts the timer, then transmits the packet and waits for an acknowledgment from the receiver. If the sender can receive an acknowledgment before the timer sends a time-out event, the sender will notify the *client*, the connection is closed, and the message sending task is done.

If a time-out event happens while the sender is waiting for an acknowledgment, the sender will assume the packet was lost and retransmit the packet. As indicated by the diagram, the sender can have five attempts in total, and the timer is restarted at the beginning of each (re)transmission.

Problems

10.1 What is UML metamodeling? What is a UML stereotype?

10.2 Explain the difference between the two concepts of action and action execution.

10.3 Find an example to practice UML concurrency modeling.

10.4 What is a timing mechanism?

10.5 Explain the meanings of the following time expressions:
   (a) ('bernoulli,' 0.58, 'ms');
   (b) ('exponential,' 1.2, 'ms'); and
   (c) ('normal,' 3.6, 0.7, 'ms').

Real-Time UML: Model Analysis

Contents

Essentially, all models are wrong, but some are useful.

George Box

11.1 Elicitation of Timing Constraints

A timing constraint specifies a time value (relative to the onset of an initiating stimulus) by which an action (or a task or a service) must finish. In terms of the consequence of missing a task deadline, there are two types of timing constraints:

(1) Hard timing constraint: a late completion of an action leads to a system or service failure. Hard real-time systems typically come with hard timing constrains. For example, a missile control system has to respond in 200 ms once an unidentified object has been detected approaching a critical region.

(2) Soft timing constraint: a late completion of an action is acceptable or tolerable; it can lead to a degraded quality of service (QoS) instead of a system or service failure. Statistical or probabilistic terms are often used to specify soft timing constraints for a collection of executions rather than a single execution instance. For example, the average processing time of a GPS signal is 10 ms.

Real-Time Embedded Systems. http://dx.doi.org/10.1016/B978-0-12-801507-0.00011-0

As far as system interaction is concerned, the timing constraints of a real-time system can also be categorized as follows [50]:

- S-R timing constraint: the maximum time allowed between the arrival of an external stimulus and the system's response.
- R-S timing constraint: the maximum time allowed between a system's response and the next stimulus.
- S-S timing constraint: the maximum time allowed between the occurrence of two stimuli.
- R-R timing constraint: the maximum time allowed between two system responses.

A critical question is how to obtain the timing constraints in the first place. Identifying timing requirements (or constraints) from the problem domain is challenging because the process might be long and tedious and it relies on the close involvement of domain experts, who may not be available when you need them. Incomplete requirements or requirements with too much unrecognized disambiguation could have serious consequences [42].

Modern software engineering principles encourage the use of use cases and user scenarios for requirements analysis. They can be used for eliciting significant timing constraints of a real-time system as well.

Table 11.1 gives a use case scenario describing how a user plays the Simon game, which is a memory retention game invented by Ralph H. Baer and Howard J. Morrison to measure and challenge a player's memory retention capacity. From the scenario description, it is easy to identify the four types of timing constraints. For example, an R-S timing constraint is given in steps 2-3, where the player has to react within 4 s; an S-R timing constraint is given in steps 3-4, where the system needs to react in 1 s once it has detected the player's input; an R-R timing constraint is given in steps 4-5, where the system needs to flash each button for 1 s; an S-S timing constraint is given in steps 6-7, where the player has to recall the button series in a timely manner (only 4 s of thinking for each flashed button).

As we saw in Chapter 10, the stereotypes and tags defined in the real-time UML (RT-UML) profile are very useful in documenting timing requirements of real-time systems. When used in a multitask context, UML diagrams, especially sequence diagrams, are very powerful in capturing traces of collaboration among multiple objects (or tasks) and the timing relations of their interactions. For example, the sequence diagram in Figure 11.1 shows how concurrency stereotypes and real-time stereotypes may be applied to describe time observation and timing constraints. In particular, the *Duplexer* object sends a message to the *SwitchingTimer* object to set up a time-out duration specified by *Trans_TM*; the execution of the operation *enable()* has to be finished in 5 μs.

Figure 11.2 gives another example showing the transmission of a message at a sender node. At the time instant *0 ms* (relative to the start of the scenario), a user attempts to set transmitter

Table 11.1 Example timing constraints in real-time interaction systems

Use Case	#1: play game
Goal in Context	An experienced user plays the Simon game
Scope	The Simon system
Level	Primary task
Primary Actor	Experienced game player
Preconditions	The Simon game is started and ready to play
Minimal Guarantee	The longest record is unchanged
Success Guarantee	The longest record is updated
Trigger	The player presses the start button
Success Scenario	Action step
1	The Simon game plays welcome sounds for 5 s
2	The Simon game randomly flashes a color button for 1 s
3	Within 4 s, the player presses the flashed button
4	Within 1 s, the Simon game replays the current series (flashes all the buttons one-by-one in the series, each button flashing for 1 s
5	Within 1 s, the Simon game randomly flashes one additional color button for 1 s
6	Within 4 s, the player presses the first flashed button in the series
7	Within 4 s, the player presses the next flashed button in the series
8	Repeat step 7 until the player has pressed all the flashed buttons in the series in the correct order
9	Repeat from step 4
Extension Step	Branching action
3a, 6a, 7a	Upon time-out
	a1: The Simon game flashes the correct button for 2 s
	a2: The Simon game waits for the next round of play
3b, 6b, 7b	The player pressed a wrong button
	b1: The Simon game flashes the wrong button for 2 s
	b2: The Simon game waits for the next round of play

parameters through the controller object. This action starts at 0 ms and ends at 3 ms. Afterward, the controller object creates a message, and requests the command messenger send it. The command messenger has multiple threads, one of which is to enqueue outgoing messages into the outgoing-message queue. The operation enqueue() finishes within 1 ms. Whenever the link driver object is idle, the command messenger retrieves a message from its outgoing-message queue, and requests the link driver physically transmit the message. The transmission of a message may be done within 3 ms, which includes the time to transform the message into pulses and transmit the pulses over wire.

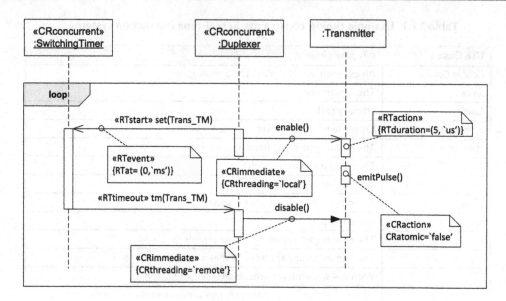

Figure 11.1

An example sequence diagram with real-time constraints.

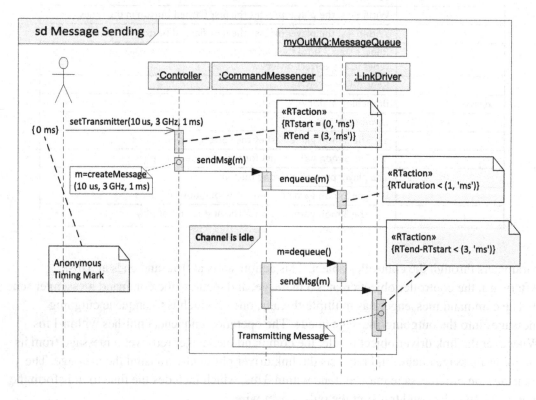

Figure 11.2

An example sequence diagram with real-time messages.

11.2 RT-UML Profile Schedulability Modeling Subprofile

The RT-UML profile schedulability subprofile [5] describes a set of common scheduling annotations for capturing timing *requirements* (time-related QoS characteristics) within the design context, and for documenting schedulability results (computed manually or by analysis tools prior to the real system implementation).

11.2.1 RT-UML Profile Metaconcepts for Schedulability Analysis

The key concepts involved in schedulability analysis include *scheduler*, *scheduling policy*, *scheduling job*, *resources*, and *execution engine*. Figure 11.3 gives a class diagram showing those metaconcepts and their relationships.

11.2.1.1 Concepts for schedulability contexts

We first introduce the concepts for modeling schedulability contexts. A schedulability context is referred to as an *SAsituation*. As shown in Figure 11.3, an *SAsituation* (logically) contains a set of scheduling jobs (processing load), a set of resource instances that are used by the jobs, and a set of execution engines (e.g. processors) that provide the power for processing the jobs.

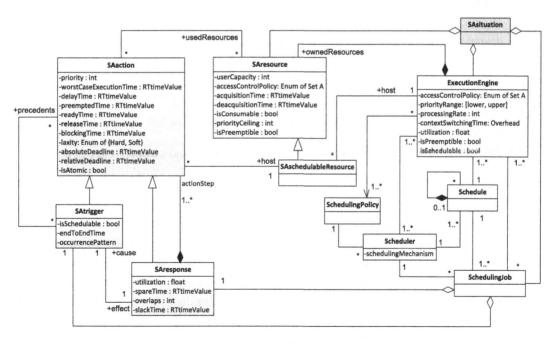

Figure 11.3
Metaclass concepts for schedulability as defined in [5].

The concept *SAresource* is a subclass of *ProtectedResource* defined in the General Resource Modeling framework, representing a physical device or an exclusively accessible object (e.g., a queue or semaphore). It has the following (reference) attributes:

- An *ExecutionEngine*: the execution engine that has exclusive use of this resource.
- A set of *SAaction* objects: an *SAresource* may be shared by multiple actions exclusively and must be protected by a locking mechanism.
- *userCapacity*: the number of permissible concurrent users. For example, a counting semaphore may allow multiple tasks to procure a token.
- *accessControlPolicy*: inherited from *ProtectedResource*, specifying the access control policy for handling resource requests from scheduling jobs. It takes values from a predefined enumeration set of policies, including
  - *FIFO*. This is the simplest first-in first-out (FIFO) protocol, which may or may not support preemption.
  - *NoPreemption*. This protocol allows a task to run to completion.
  - *PriorityInheritance*. When a job blocks one or more higher-priority jobs for some shared resource, it executes its critical section at the highest priority level of all the jobs it blocks. After releasing the resource, the job returns to its original priority level. The priority inheritance protocol prevents unbounded priority inversion (Related topics are covered in Chapter 14). It, however, can cause deadlocks or chain blocking.
  - *HighestLockers*. To each resource is assigned a ceiling priority, which is determined at design time and is equal to the highest priority of all those tasks that are supposed to use that resource. When a task (job) acquires a resource successfully, its priority value is automatically raised to the ceiling priority of that resource. The highest locker protocol prevents unbounded priority inversion and deadlocks. It, however, may incur unnecessary blocking, causing some intermediate-priority tasks to miss their deadlines.
  - *PriorityCeiling*. Similarly to the highest locker protocol, to each resource is assigned a ceiling priority, which is determined at design time and is equal to the highest priority of all those tasks that are supposed to use that resource. In addition, the run-time environment also has a variable ξ, which, at any time, holds the highest ceiling value of all the resources currently in use. A task (job) is granted access to a resource only if the task's priority is higher than ξ; it is blocked on the resource otherwise [64]. As a task is blocked on a resource, the task which is currently using the resource inherits the priority of the blocked task, if it is higher. When a task T releases a resource, T's priority reverts to its original priority, or to the highest priority of all the tasks being blocked on a resource that is still held by T. The priority ceiling protocol prevents deadlocks, chain blocking, and unbounded priority inversion. It also limits inheritance-related priority inversion.

- *acquisitionTime*: a QoS specifying the time delay between the instant when an action is granted access to a resource and the instant when the resource is ready for use.
- *deacquisitionTime*: a QoS specifying the time delay between the instant when an action initiates release of a resource and the instant when the resource becomes available again.
- *isConsumable*: a Boolean value indicating whether the resource is consumed by use.
- *isPreemptible*: a Boolean value indicating whether the resource can be preempted while it is being used.
- *priorityCeiling*: the highest priority of all those tasks that are supposed to use this resource by design.

SAschedulableResource is a subclass of both *SAresource* and *ActiveResource*, representing a unit of concurrent execution, such as a task, a process, or a thread. An *SAschedulableResource* needs to be deployed on an execution engine, and is responsible for hosting (executing) real-time jobs or actions. Often, each schedulable resource (process) is dedicated to executing a single job, but sometimes a multi-threaded process may be used to host a number of jobs.

An *ExecutionEngine* is an active, protected resource that is used to execute schedulable resources. Example execution engines include physical CPUs and virtual machines. An *ExecutionEngine* has the following attributes:

- *accessContolPolicy*. Each policy defines a set of rules that govern when and under what conditions a request for the execution engine can be granted. It takes the same values as defined in *SAresource*.
- *processingRate*. This is a speed factor, expressed as a percentage, for the execution engine as compared with a reference processor.
- *contextSwitchingTime*. This is the length of time (overhead) that it takes to switch from one scheduling job to another.
- *priorityRange*. This is the set of valid priorities permitted on this execution engine. These are used to confine the priorities taken by the jobs (actions) scheduled on this engine.
- *isPreemptible*. This is a Boolean value indicating whether or not the execution engine is preemptible once it begins the execution of an action.
- *isSchedulable*. This is a Boolean value indicating whether or not the execution engine is schedulable under the conditions specified in the design. This is the result of schedulability analysis.
- *utilization*. This is another result of schedulability analysis, indicating the computed utilization of the execution engine expressed as a percentage.
- *SAschedulableResource*. This is a set of schedulable resources that this execution engine is charged with executing.
- *ownedResources*. This is the resources that this execution engine owns exclusively such as pipes, queues, data-stores, etc.

- *SchedulingJob*. It is a set of jobs that this execution engine needs to accommodate in its schedule. A *SchedulingJob* is a unit of work with a defined occurrence pattern and a sequence of actions. Each scheduling job raises a work load to the engine.
- *Scheduler*. This is a set of scheduler instances that is responsible for scheduling jobs on the basis of a scheduling policy. A scheduler is actually a resource broker that is responsible for allocating execution engine resources. A scheduler may schedule jobs for one or more execution engines.
- *Schedule*. This is a particular ordering of execution of a set of scheduling jobs. A composite schedule may be composed of a set of finer-grained schedules, which can be generated for several different execution engines.
- *SchedulingPolicy*. This is a set of policies that define the rules for assigning execution engine time to a set of scheduling jobs. This can be any one of a set of standard policies, including
  - RateMonotonic. Rate-monotonic assignment is a static policy where priorities are assigned on the basis of job periods: the shorter the period is, the higher is the job's priority [49, 65]. This is to be further discussed in Chapter 16.
  - DeadlineMonotonic. Deadline-monotonic assignment is a policy where tasks are assigned priorities according to their deadlines: the highest priority is assigned to the task with the shortest deadline.
  - HKL. This policy was developed by Harbour, Klein, and Lehoczky [38], applicable to tasks with a complex priority structure, where a task may be composed of distinct subtasks, each executing at a different priority level with its own timing requirement.
  - FixedPriority. This is a policy where all tasks have fixed priority. If preemptive is allowed, lower-priority jobs are interrupted by incoming higher-priority jobs.
  - MaximizeAccruedUtility. This is a policy that seeks to maximize the sum of all the tasks' utilities. If the execution engine cannot meet the timing constraints (e.g., deadlines) of all the tasks, it will favor tasks that are more important (from which greater utility can be accrued) than those which are more urgent.
  - MinimumLaxityFirst. This is a policy that selects the task with the minimum laxity to execute next. Laxity is a measure of the flexibility available for scheduling a task. The laxity of a task is defined as the difference between the amount of time left (by its deadline) and the amount of time still needed to complete the task. A laxity of t means that even if the task is delayed by t time units, it will still meet its deadline. A laxity of zero means that the task must begin to execute now or it will risk failing to meet its deadline.
  - MinimumSlackTime. This is a policy that assigns the highest priority to a task with the minimum slack time. Since slack time and laxity typically refer to the same thing, this policy is simply another name for the minimum laxity first policy.

There are other scheduling policies. For instance, earliest deadline first is an approach where the task closest to its deadline is scheduled next for execution.

11.2.1.2 Actions

In execution, a scheduling job unfolds into one or more control paths. Every action execution in a control path consumes a portion of the overall time budget of the job. In schedulability analysis, especially for systems with hard timing constraints, a primary concern is to check that the sum of the time demands for any path must be no more than the overall time budget. Obviously, actions become the basis for specifying and analyzing the timing behavior of a real-time system.

The concept *SAaction* defines actions in terms of time-related properties pertinent to schedulability analysis. These properties, together with those defined for *RTaction* and *CRaction*, are summarized in Table 11.2.

Figure 11.4 illustrates some time-related properties of nonatomic actions, where action 1 has the highest priority and action 3 has the lowest priority. Action 3 initially locked a resource but its execution was preempted by action 1. While action 1 was executing, action 2 was released, and could not execute until action 1 had released the processor. Also, action 2 was preempted

Table 11.2 Time-related properties of actions [5, 34]

Property	Description
Priority	The priority of the action from a scheduling perspective. It can be a static or dynamic value. Defined in «SAaction».
Release time	The time instant at which the action becomes eligible for execution. Defined in «SAaction».
Ready time	The delay between the time when the action is eligible for execution (i.e., the release time) and the actual beginning of execution. Defined in «SAaction».
Worst-case completion time	The overall time taken to execute the action, including all overheads. Defined in «SAaction».
Blocking time	The length of time that the action is blocked waiting for resources. Defined in «SAaction».
Delay time	The length of time that an action eligible for execution waits while acquiring and releasing resources. Defined in «SAaction».
Preempted time	The length of time that the action is preempted to make way for higher-priority actions. Defined in «SAaction».
Laxity	Specifies the type of deadline, which can be hard or soft. Defined in «SAaction».
Relative deadline	Specifies the desired time (relative to the release time) by which the action should be complete. Defined in «SAaction».
Absolute deadline	Specifies the final time instant by which the action must be complete, defined as the relative deadline plus the release time. Defined in «SAaction».
Start	The start time instant of the action execution. Defined in «RTaction».
End	The completion time instant of the action execution. Defined in «RTaction».
Duration	The total duration of the action. Defined in «RTaction».
isAtomic	A Boolean value indicating whether the action can be preempted or not. Defined in «CRaction».

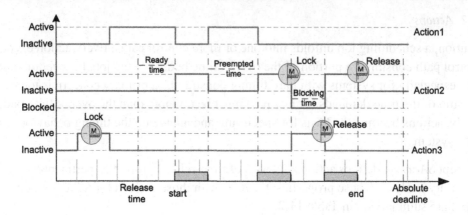

Figure 11.4
Time-related properties of actions.

by action 1 in its execution. After that, action 2 continued its execution until it came to a point where it required the resource being locked by action 3. Action 2 could not resume its execution until action 3 had released the resource.

11.2.1.3 Scheduling job

A scheduling job is the combination of a trigger action (*SAtrigger* object) and a response action (*SAresponse* object). The trigger of a job defines how often that job needs to be executed. The response of a job defines the amount of work in the job.

An *SAtrigger* object is a special action. In addition to the attributes inherited from *SAaction*, *SAtrigger* also defines:

- *isSchedulable*: a model analysis result that indicates whether the job (initiated by this trigger) can be scheduled.
- *endToEndTime*: the worst-case completion time for the complete chain of dependent responses measured from the arrival of the trigger.
- *occurrencePattern*: the pattern of interarrival times between consecutive occurrences of the job represented by this trigger. For instance, it can be characterized by a period or a probability distribution function.

An *SAresponse* object is a special action which models a sequence of action steps that is separately schedulable on an execution engine. Each *SAresponse* has an *SAtrigger* as its cause. The *SAtrigger* bounded to an *SAresponse* can also be taken as the trigger of its first action step. In addition, the completion of an action can be the precedent of other trigger actions.

In addition to the attributes inherited from *SAaction*, *SAresponse* also adds the following specific attributes:

- Utilization: the percentage of time the execution engine (microprocessor) spends doing actions of this response (job).
- *Slack time*: the difference between the amount of time left (by its deadline) and the amount of time still needed to complete the work.
- Spare capacity: the amount of execution time that can be added to this scheduling job without affecting the schedulability of lower-priority jobs in the system.
- *Overlaps*: in the case of soft deadlines, this indicates how many job instances may overlap their execution because of missed deadlines.

Given a scheduling job, its end-to-end execution time can be accumulated by stepping through the sequence of actions defined in the *SAresponse* object.

In general, the action steps of a scheduling job can be executed in more than one schedulable resource (physical thread). On the other hand, a schedulable resource can host a single job or several jobs (sequentially in a single thread or concurrently in a multithreaded process).

11.2.2 Schedulability Stereotypes

The abstract concepts we encountered in the last section are expressed in a series of UML stereotypes, some of which are listed in Table 11.3. Notice that the tag definitions of a stereotype reflect the attributes of the corresponding meta-level concept. For example, the tag definitions of «SAaction» reflect the (reference) attributes defined in Figure 11.3. The stereotype «SAaction», together with its superclass concepts «RTaction» and «CRaction», captures all the time-related properties of the actions given in Table 11.2.

The stereotypes in Table 11.3 offer a general basis for supporting a wide variety of schedulability analysis methods. Users can apply these stereotypes to their modeling elements, such as objects, operations, and messages.

Most of the tag definitions in Table 11.3 have a type *RTtimeValue*; its well-formed values are given in Table 10.3. Another type *RTarrivalPattern* is used to specify the arrival patterns of real-time events; its well-formed values are given in Table 11.4.

Here are some examples arrival patterns of real-time tasks or jobs:

- Bounded arrival pattern: the keyword "bounded" followed by a lower bound (the minimal interval between successive arrivals) and an upper bound (the maximum interval between successive arrivals); for example, {'bounded', (0, 'ms'), (10, 'ms')};
- Bursty arrival pattern: the keyword "bursty" followed by a burst interval and an integer indicating the maximum number of events that can occur during that interval; for example, {'bursty', (10, 'ms'), 30};

Table 11.3 Some stereotypes defined in the RT-UML schedulability subprofile

Stereotype	Base Metaclass	Tags (Optional)	Types
«SAaction» (subclass of «RTaction» and «CRaction»)	Action, ActionExecution, Message, Method, Transition	SApriority: SAblocking: SAready: SAdelay: SArelease: SApreempted: SAworstCase: SAlaxity: SAabsDeadline: SArelDeadline: SAusedResource: SAhost:	Integer RTtimeValue RTtimeValue RTtimeValue RTtimeValue RTtimeValue RTtimeValue Enumeration of {"Hard", "Soft"} RTtimeValue RTtimeValue Refer to an «SAresource» element Refer to an «SAschedulable» element
«SAengine»	Class, Instance, Node	SAschedulingPolicy: SAaccessPolicy: SAaccessPolParam: SArate: SAcontextSwitch: SApriorityRange: SApreemptible: SAutilization: SAschedulable: SAresources:	Enumeration of set S Enumeration of set A Real /*e.g. priority ceiling value*/ Float Time function Integer range Boolean Real (percentage) Boolean Refer to an «SAresource» element
«SAresource»	Class, Instance, Node	SAaccessControl: SAaccessCtrlParam: SAconsumable: SAcapacity: SAacquisition: SAdeacquisition: SApriorityCeiling: SApreemptible:	Enumeration of set A Real /*e.g. priority ceiling value*/ Boolean Integer RTtimeValue RTtimeValue Integer Boolean
«SAresponse» (subclass of «SAaction»)	Action, ActionExecution, Message, Method	SAutilization: SAspare: SAslack: SAoverlaps:	Real (percentage) RTtimeValue RTtimeValue Integer
«SAschedulable» (subclass of «SAresource»)	Class, Instance, Node		
«SAscheduler»	Class, Instance	SAschedulingPolicy: SAexecutionEngine	Enumeration of set S Refer to an «SAengine» element
«SAtrigger» (subclass of «SAaction»)	Message	SAschedulable: SAendToEndTime: SAprecedents: SAoccurrence:	Boolean RTtimeValue Refer to an «SAaction» element RTarrivalPattern
«SAsituation»	Class Diagram, Structured Class and its kind, Sequence Diagram, Activity Diagram		

Set S = {'FIFO,' 'RateMonotonic,' 'DeadlineMonotonic,' 'HKL,' 'FixedPriority,' 'MinimumLaxityFirst,' 'MaximizeAccrued Utility,' 'MinimumSlackTime'}.

Set A = {'FIFO,' 'PriorityInheritance,' 'NoPreemption,' 'HighestLockers,' 'PriorityCeiling'}.

Table 11.4 *RTarrivalPattern*: arrival patterns of real-time events (used as the tag type for SAoccurrence and PAoccurrence). Refer to Table 10.3 for <timeVal> and <PDFstring>)

<RTarrivalPattern>	::=	<bounded-string> | <bursty-string> | <periodic-string> | <irregular-string> | <unbounded-string>
<bounded-string>	::=	" 'bounded' ," <timeVal> "," <timeVal>
<bursty-string>	::=	" 'bursty' ," <timeVal> "," <Integer>
<periodic-string>	::=	" 'periodic' ," <timeVal> ["," <timeVal>]
<irregular-string>	::=	" 'irregular' ," <timeVal> ["," <timeVal>]*
<unbounded-string>	::=	" 'unbounded' ," <PDFstring>

- Irregular arrival pattern: the keyword "irregular" followed by an ordered list of time values representing the successive interarrival times; for example, {'irregular', (2, 'ms'), (5, 'ms'), (9, 'ms'), (11, 'ms')};
- Periodic arrival pattern: the keyword "periodic" followed by a time value indicating the period, which is optionally followed by a second time value representing the maximal deviation (aka jitter) from the period; for example, {'periodic', (50, 'ms'), (100, 'us')};
- Unbounded arrival pattern: the keyword "unbounded" followed by a probability distribution function as specified in Table 10.3; for example, {'unbounded', ('normal', 35, 1.8, 'ms')}, which specifies an unbounded arrival pattern characterized by a normal (Gaussian) distribution with a mean of 35 ms and a standard deviation of 1.8 ms.

Figure 11.5 gives a simple class diagram with the user classes annotated by schedulability stereotypes.

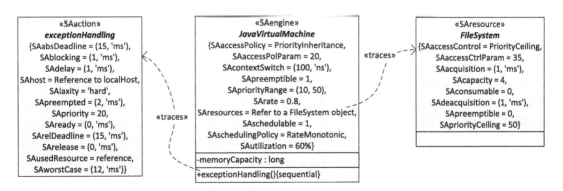

Figure 11.5
An example using schedulability stereotypes.

11.2.3 Using the Schedulability Subprofile

The schedulability stereotypes provide a set of common annotations for modelers to perform very basic schedulability analysis.

The principle objective in schedulability analysis is to check whether a system can meet all of the deadlines defined for the individual scheduling jobs. In so doing, a system designer, assisted by analysis tools, typically needs to analyze the system under several scenarios with different system architectures. For each scenario, the overall system architecture is fixed, while parameter values (say, task priorities) are tuned. The analysis result can be assigned to the *SAschedulable* tag of either a trigger action (representing the schedulability of a job) or an execution engine (representing the schedulability of a set of jobs). The analysis results may also suggest how the system could be improved to make an entity schedulable.

As an example, in Figure 11.6 we sketch a resource-client graph with four tasks, where three tasks—T_1, T_2, and T_3—need to access a shared resource called *PlotCluster*, which is guarded by a semaphore. For instance, the *PlotManager* task needs to update (write to) the plot cluster with newly identified plots; the *Tracker* task needs to process the plots to update tracks; and the *Monitor* task needs to process the plots to produce a graphical presentation of the plots. We assume that the maximum duration of time that T_1 can be blocked by T_2's critical section is 80 ms, and the maximum duration of time that T_2 can be blocked by T_3's critical section is 60 ms. T_4 is an independent task, which is responsible for logging any significant events into a log database.

Suppose that, according to the domain analysis, the characteristics of the four tasks are as follows:

Task T1 (period 200 ms; deadline 200 ms; worst-case execution time 50 ms).

Task T2 (period 400 ms; deadline 400 ms; worst-case execution time 100 ms).

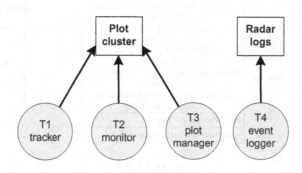

Figure 11.6
A resource-client graph of a radar processing subsystem.

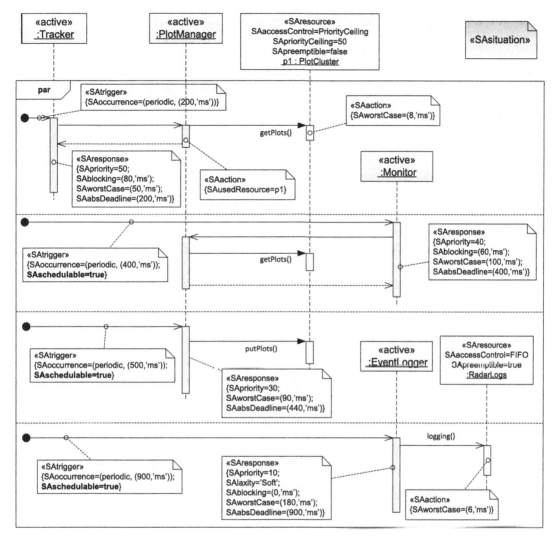

Figure 11.7
A sequence diagram annotated by schedulability stereotypes.

Task T3 (period 500 ms; deadline 440 ms; worst-case execution time 90 ms).

Task T4 (period 900 ms; deadline 900 ms; worst-case execution time 180 ms).

In Figure 11.7, we show a sequence diagram with four parallel partitions, and the task characteristics are captured by appropriate stereotypes and their tag values.

The information captured in Figure 11.7 is sufficient for schedulability analysis. However, it demands knowledge of the principle of rate-monotonic scheduling. We leave this problem open for now, and will come back to it again in Chapter 16.

11.3 RT-UML Profile Performance Modeling Subprofile

While the RT-UML schedulability subprofile is supposed to be used within the design context, the RT-UML performance subprofile can be used both within the design context and for the final system implementation. The performance subprofile [5] defines stereotypes for

- capturing performance *requirements* (performance-related QoS characteristics, such as resource utilizations, waiting times, execution demands and response time) within the design context;
- specifying execution parameters that can be used by performance analysis tools to compute *predicted* performance characteristics;
- documenting performance results computed by analysis tools or *measured* in testing.

11.3.1 RT-UML Profile Metaconcepts for Performance Analysis

Like the schedulability subprofile, the performance subprofile also deals with actions, the execution environment, and resources. Figure 11.8 gives the metaconcepts related to performance analysis.

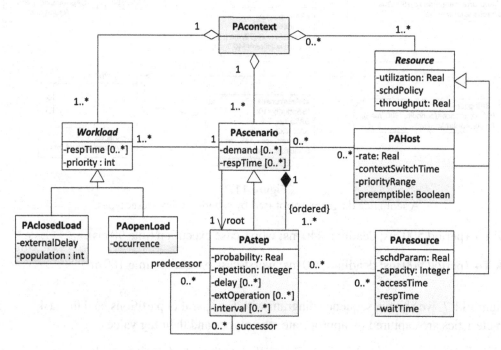

Figure 11.8
Metaconcepts in the performance subprofile as defined in [5].

PAcontext specifies a performance context, which involves a set of resources (both passive and processing resources), a set of workloads, and one or more scenarios that are used to explore the system performance.

For each real-time system, it may operate in different contexts with a wide range of load intensities. For instance, you may specify a "heavily loaded" scenario where the processing demands are very high to explore the worst-case system performance, as well as a few scenarios to evaluate the system performance in normal situations.

Performance measures for a system include CPU utilizations, waiting/delay times, execution time demands on processors (cycles or seconds) and response time. Each measure may be defined in different *versions* at different stages:

- A required value, which comes from the system requirements (e.g. a preset execution demand or budget for a scenario);
- An assumed value, which is derived from experience (e.g., an external delay from an average human user);
- A predicted value, which is typically calculated by a prediction tool (e.g. an estimated execution demand); and
- A measured value, which represents the actual performance or the simulation/testing result.

In addition, in order to produce meaningful statistical results, a scenario is typically evaluated multiple times, producing multiple measured values for each performance measure. A measured value can also be specified in statistical terms, such as means and variances. This explains why some attributes in Figure 11.8 take an array form [0..*].

The two concepts *PAScenario* and *PAstep* are related by three relations:

(1) Generalization. A step is modeled as a special kind of scenario in order to accommodate different levels of abstraction. A designer may choose an appropriate level of abstraction to model action steps. If finer granularity is required, a step can be resolved into a scenario comprising finer-grained steps at the next level of abstraction. The simplest scenario consists of a single atomic step.
(2) Composition. A scenario is composed of an ordered sequence of steps, each of which represents an execution of some action. Those steps may be organized by control structures such as forks, joins, and loops.
(3) Association. A scenario has a reference to its root—the first step.

A scenario has two attributes, where *demand* denotes the total execution demand of the scenario, and *respTime* denotes the total time required to complete the scenario, including all resource waiting, synchronization delays, and the execution time. A scenario can have multiple values for both *demand* and *respTime*. For instance, we may need to specify a 'required' version and a 'measured' version for the response time of a scenario, each version comprising a record for the mean and a record for the variance.

A step inherits all the attributes and associations defined in *Scenario*. In particular, a step executes on a host (*PAhost*, processing resource) and may need to access protected resources. The steps in a scenario conform to a general precedence/successor relationship. This relation has a multiplicity of 0..* on each end, indicating the possibility of multiple concurrent successors and predecessors (stemming from concurrent thread forks and joins respectively). For example, a step that follows a join will have multiple predecessors.

A step has the following attributes:

- *demand*: inherited from *Scenario*, the total execution demand of the step on its host resource;
- *respTime*: inherited from *Scenario*, the total time to execute the step, including all resource waiting, delays, and the execution time;
- *delay*: the value of an *inserted delay* within this step, say, wait for a resource or pause for a user interaction;
- *probability*: in situations where this step is merely one of several successors emanating from the same predecessor step, this is the probability that this step will be executed (the sum of probabilities of all the peer steps has to be 1);
- *repetition*: the number of times the step is repeated;
- *interval*: the time interval between successive repetitions of this step, if repeated;
- *extOperation*: a sequence of external operations, each is of the form "(<opName>, <integer> | <PAperfValue>)," where *opName* is the name of the external operation, followed by an integer indicating the number of repetitions of that operation, or a performance time value (see Table 11.6).

A scenario can be evaluated under different workload situations. A workload specifies the load intensity for the execution of a scenario as well as the required, estimated, or measured response times for that scenario. There are two types of workloads. An *open workload* has a stream of requests that arrive at a given rate in some predetermined pattern (such as a Poisson distribution), while a *closed workload* has a *fixed number* of potential users of the system that cycle between executing the scenario, with an external delay period (sometimes called "think time") between the end of one scenario execution and the start of the next.

A *PAhost* is a processing resource (i.e. processor) with the following attributes:

- *schdPolicy*: scheduling policy, a set of rules for assigning this resource to scenario steps. A number of policies are predefined:
  - FIFO: first in, first out;
  - HeadOfLine: head of the line (nonpreemptive with support for priorities);

- PreemptResume: preempt resume (representing preemptive);
- ProcSharing: processor sharing (representing round robin);
- PrioProcSharing: priority processor sharing (round robin at the same priority level, preemptive otherwise);
   - LIFO: last in, first out.
- *rate*: a relative speed factor for the processor expressed as a percentage of some reference processor.
- *contextSwitchTime*: the length of time (overhead) required by the processor to switch from the execution of one scenario to a different one.
- *priorityRange*: the set of valid priorities (often dependent on the chosen operating system).
- *preemptable*: indicates whether or not the processor is preemptible once it begins to execute an action.
- *utilization*: the result of model analysis, representing the computed utilization of the processing resource expressed as a percentage.
- *throughput*: the rate at which the resource delivers its service.

A protected resource in the performance analysis context has the following attributes:

- *schdPolicy*: the access control policy for handling requests from scenario steps. It is the same as the attribute *accessControlPolicy* of *SAresource* (see Section 11.2.1.1 for details).
- *capacity*: the number of permissible concurrent users (e.g., a counting semaphore).
- *accessTime*: the time delay suffered by a scenario or scenario step in acquiring and releasing a resource.
- *utilization*: the mean number of concurrent users of the resource.
- *waitingTime*: the time from the instant a request for accessing the resource is issued to the time it is granted.
- *responseTime*: the total time expired from the instant the resource is requested till it is released by the user, including the waiting time and using time.

11.3.2 Performance Stereotypes

The shaded metaconcepts in Figure 11.8 are defined as stereotypes, and, together with their tag definitions, are listed in Table 11.5.

Most of the types for the performance tag definitions are primitive types. The type *RTarrivalPattern* was given in Table 11.4. Table 11.6 gives the definition for the type *PAperfValue*, which is used to form value expressions for performance-related QoS characteristics.

A performance value is formed by a source modifier, a type modifier, followed by a real-time value. A source modifier defines how the value was obtained and can be one of *'req,' 'assm,' 'pred,'* or *'msr,'* meaning respectively "required," "assumed," "predicted," and "measured."

Table 11.5 Some stereotypes defined in the RT-UML performance subprofile

Stereotype	Base Metaclass	Tags (Optional)	Types
«PAclosedLoad»	Message, Action, ActionExecution, Method, Operation, Reception	PArespTime [0..*]: PAexternalDelay: PApriority: PApopulation:	PAperfValue PAperfValue Integer Integer
«PAopenLoad»	Message, Action, ActionExecution, Method, Operation, Reception	PArespTime [0..*]: PApriority: PAoccurrence:	PAperfValue Integer RTarrivalPattern
«PAhost»	Class, Instance, Node, Partition	PAutilization [0..*]: PAschdPolicy: PArate: PAcontextSwitchTime: PApriorityRange: PApreemptable: PAthroughput:	Real Enumeration of set H Real PAperfValue Integer range Boolean Real
«PAresource»	Class, Instance, Node, Partition	PAutilization [0..*]: PAschdPolicy: PAschdParam: PAcapacity: PAaccessTime: PArespTime: PAwaitTime: PAthroughput:	Real Enumeration of set R Real /*priority ceiling*/ Integer PAperfValue PAperfValue PAperfValue Real
«PAstep»	Action, ActionExecution, Message, Transition	PAdemand [0..*]: PArespTime [0..*]: PAprobability: PArepetition: PAdelay [0..*]: PAextOperation [0..*]: PAinterval [0..*]:	PAperfValue PAperfValue Real Integer PAperfValue PAextOpValue PAperfValue
«PAcontext»	Class Diagram, Structured Class and its kind, Sequence Diagram, Activity Diagram, Deployment Diagram		

Set H = {'FIFO,' '1HeadOfLine,' 'PreemptResume,' 'ProcSharing,' 'PrioProcSharing,' 'LIFO'}.
Set R = {'FIFO,' 'PriorityInheritance,' 'NoPreemption,' 'HighestLockers,' 'PriorityCeiling'}.

Table 11.6 Performance value types (see Table 10.3 for <timeVal>)

<PAperfValue>	::=	"(" <source-modifier> "," <type-modifier> "," <timeVal> ")"		
<source-modifier>	::=	'req' | 'assm' | 'pred' | 'msr'		
<type-modifier>	::=	'mean' | 'sigma' | 'kth-mom', <Integer> | 'percentile', <Real> | 'max' | 'dist'		
<PAextOpValue>	::=	"(" <String> "," <integer> | <PAperfValue> ")"		

A type modifier specifies the meaning of a value, and can be one of *'mean,' 'sigma,' 'kth-mom'* followed by an integer, *'max,' 'percentile'* followed by a real number, or *'dist,'* which mean, respectively, average, variance, kth moment (k is an integer),[1] maximum, percentile range (a real value), and probability distribution. Multiple statistical properties may be reported for the same measure.

Here are a few examples:

- ('msr', 'mean', (50, 'ms')): a measured mean value of 50 ms;
- ('req', 'sigma', (3, 'ms')): a required variance of 3 ms;
- ('assm', 'max', (100, 'ms')): an assumed maximum value of 100 ms;
- ('pred', 'dist', ('normal', 2.4, 0.8, 'ms')): a predicted normal distribution with a mean of 2.4 ms and a standard deviation of 0.8 ms;
- ('msr', 'percentile', 95, (100, 'ms')): a measured value that 95% of the observations are below 100 ms.

11.3.3 Using the Performance Subprofile

The stereotypes defined in the performance subprofile can be used in UML diagrams to specify *performance requirements* within a design context or performance results for an actual implementation.

Figure 11.9 shows a UML deployment diagram where objects of a radar data processing subsystem are deployed in several computing nodes. The multiplicities on the associations indicate that a *Tracker Computer* node may communicate with one or more *Monitor Computers* and *Smoother Computers*.

Each of the nodes is specified by a «PAhost» stereotype, containing several objects deployed on that node. For example, a monitor computer node contains a *HotAreaDisplay* object and a *WholeRangeDisplay* object. The tagged values of the monitor computer specify the attributes of its processing resource (processor):

- it is preemptible;
- it is regulated by the *PreemptResume* scheduling policy;
- its priority range is between 10 and 300;
- its processing rate is 1 GHz (here for clarity we use the actual CPU frequency instead of a percentage relative to a reference processor); and
- its utilization is 70%.

---

[1] Most performance analysis techniques—for example, simulation tools built upon queuing theory—are statistical, yielding statistical results.

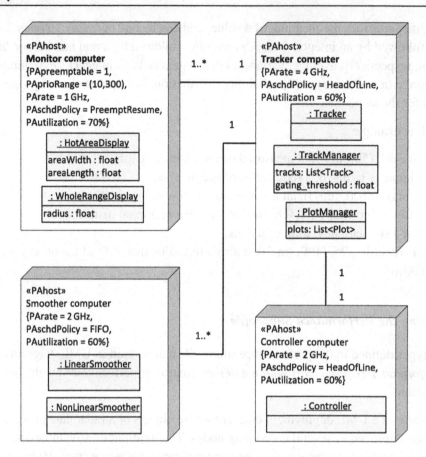

Figure 11.9
UML deployment diagram of the radar data processing subsystem.

In figure 11.9, we see example annotations specified in the <<PAhost>> stereotype, most of which are *execution parameters*. System timing requirements and performance analysis results are typically captured in UML diagrams that can model system scenarios, such as sequence diagrams and activity diagrams. As an example, Figure 11.10 gives a scenario of a radar data processing subsystem.

In UML diagrams, a scenario is represented by its root step—the first step operation or action in the topmost performance context. Any performance attributes of a scenario, including its workload specifications, are attached to the root step. Thus, for the root step action in a performance context, in addition to being stereotyped as a scenario step, it can be stereotyped as a «PAopenLoad» or a «PAclosedLoad» stereotype. For example, in Figure 11.10, the *updateTracks()* operation triggered by *TrackTimer* on the *Tracker* object is annotated by a «PAopenLoad» stereotype, saying that (a) the operation happens periodically with a period of

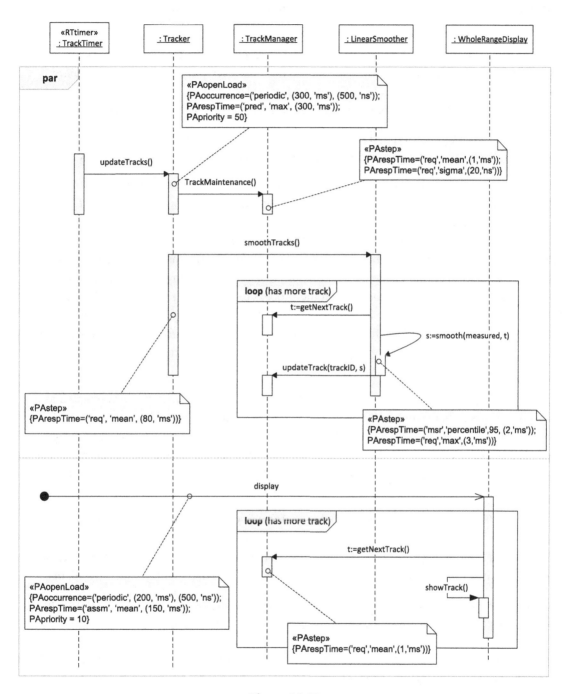

Figure 11.10
Real-time UML: performance annotations.

300 ms and a jitter of 500 ns, (b) it is predicted that the response time is at most 300 ms, and (c) the priority value is 50.

For the other parallel region in Figure 11.10, the «PAopenLoad» stereotype is associated with the found message; it applies to the target operation indirectly, saying that (a) the *display* operation is executed periodically with a period of 200 ms and a jitter of 500 ns, (b) it is assumed that the average response time is 150 ms, and (c) the task priority is 10.

In Figure 11.10, the stereotype «PAstep» is also used for several other operations as well, specifying their respective performance (requirements or measurements) on response time.

It is worth noting that, according to Figure 11.9, the *TrackManager* object and the *WholeRangeDisplay* object are deployed on different computing nodes. The *getNextTrack()* call from the *WholeRangeDisplay* object (also the *LinearSmoother* object) to the *TrackManager* object crosses computing nodes. As messages are sent among objects residing on different computing nodes, the time necessary to construct the message, fragment it (if there is a limitation on the packet size), transmit it, reassemble it on the target node, conduct the required computation, and then repeat the process to return the result is all subsumed within the execution time of the invoked operation. Thus, the timing requirement on the response time of *getNextTrack()* implies that all the above-mentioned steps have to be done within 1 ms on average (this virtually places a requirement on the transmission rate of the media connecting the computing nodes).

The example given in Figure 11.11 is adapted from [5], where the first step is annotated as a closed workload: there can be 100 users viewing videos through Web browsers, and it is assumed that the average external delay is 30 s. This external delay means that on average a user has a delay of 30 seconds between ending one session and starting the next. The first step is also stereotyped by «PAstep», saying that it is required that 95% of the confirmation delays are below 800 ms.

The *receiveFrame(f)* operation on the *aPlayer* object is stereotyped by a «PAstep»: it is required that the execution time demand of this operation have a mean of 15 ms and a deviation of 3 ms. The *showFrame(f)* operation on the *VideoViewer* object also has a «PAstep» stereotype: it is required that 99% of the intervals between frame display are below 30 ms.

The performance stereotypes can also be used in activity diagrams. Figure 11.12 shows an activity diagram, describing the same scenario as that captured by the sequence diagram in Figure 11.11. A difference is that, instead of specifying a performance requirement for the operation 'Receive frame', in Figure 11.12, a performance analysis result is documented: the measured execution demand of this operation on CPU time is 12 ms on average with a standard deviation of 2 ms. It indicates that the implemented system performed better than required.

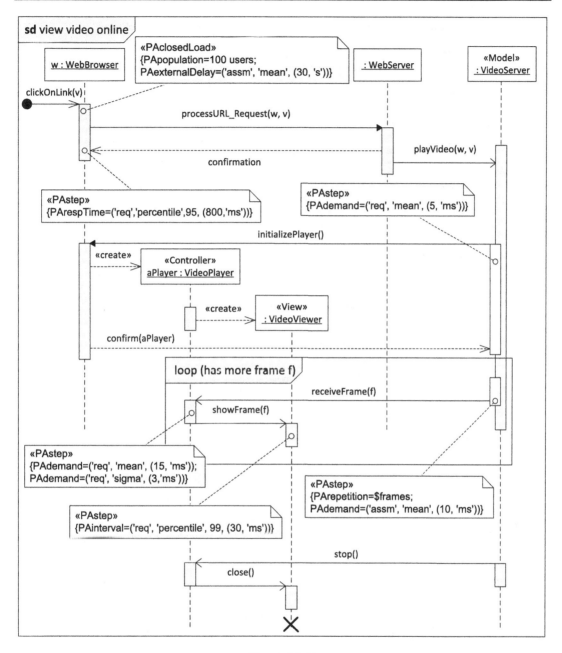

Figure 11.11
Real-time UML: performance annotations.

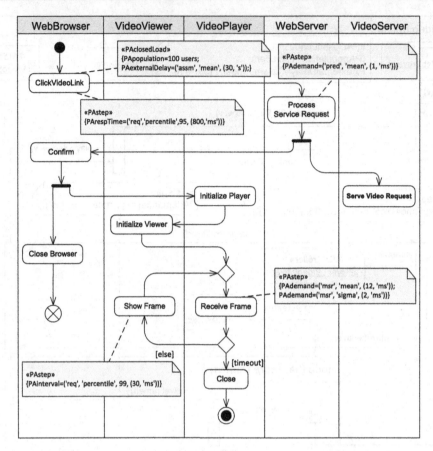

Figure 11.12
Real-time UML: performance annotations in activity diagrams.

Problems

11.1 Explain the meanings of the schedulability stereotypes and tag values as used in the model given in Figure 11.13.

11.2 Explain the meanings of the stereotypes and tag values as used in the model given in Figure 11.14. Refer to Figure 6.33 for the corresponding class definitions.

11.3 In Section 6.2.4.2 we discussed the Liskov substitution principle. The key point is that a client of a superclass object should not be surprised if a subclass object is used instead. When a timing constraint declared in the superclass represents a postcondition (e.g., timing requirements such as the worst-case execution time), it should not be weakened in the subclass [34].

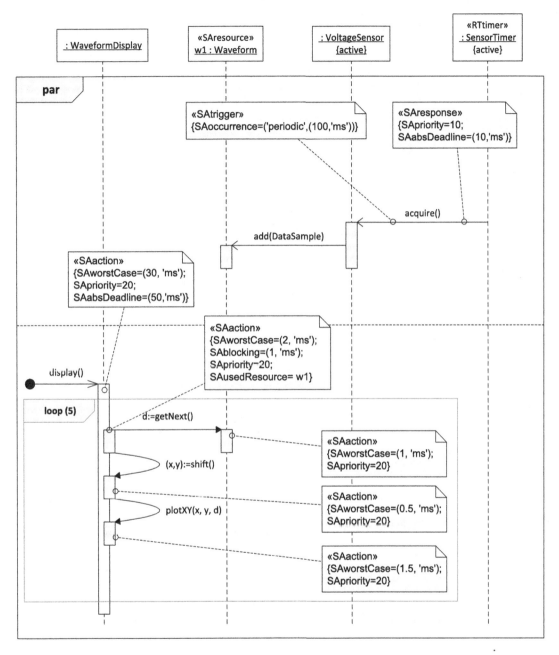

Figure 11.13
A sequence diagram annotated by schedulability stereotypes and tag values.

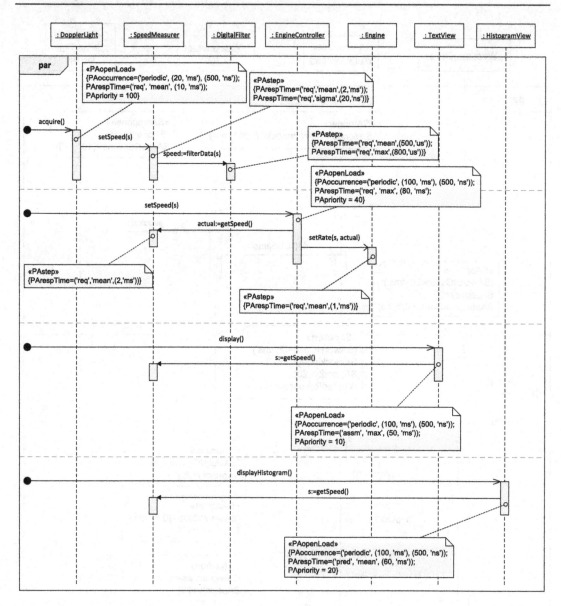

Figure 11.14
A sequence diagram annotated by performance stereotypes and tag values.

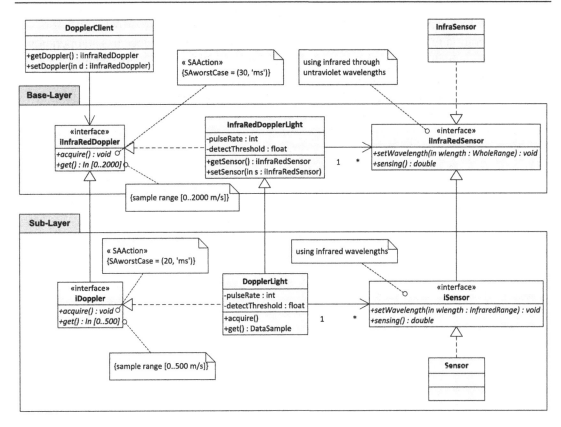

Figure 11.15
A class diagram annotated with schedulability stereotypes.

(a) In Figure 11.15, the class *DopplerLight* implements the interface *iDoppler*; thus, a *DopplerLight* object in the worst case can finish the *acquire()* operation in 20 ms. Similarly, an *InfraRedDopplerLight* object in the worst case can finish the operation *acquire()* in 30 ms. Explain whether the tag values violate the Liskov substitution principle or not. (Hint: Would a client of an *InfraRedDopplerLight* object still meet its timing requirement if the *InfraRedDopplerLight* object is upgraded by a *DopplerLight* object?)

(b) The Liskov substitution principle requires that a subclass use wider types for operation parameters. In Figure 11.15, the *iInfraRedSensor* interface and its subtype *iSensor* define a *setWavelength()* operation that accepts parameters in different ranges. Explain whether the two definitions of the setWavelength() operation violate the Liskov substitution principle or not. If they do, then the client of an *iInfraRedSensor* object could be surprised if an *iSensor* object is used instead. How can you correct the problem?

(c) The Liskov substitution principle requires that a subclass use narrower types for operation returns. In Figure 11.15, the *iInfraRedDoppler* interface has an operation *get()*, which returns a data sample in the range 0-2000m/s, while the subinterface returns a data sample in the range 0-500m/s. Explain whether the two definitions of the *get()* operation violate the Liskov substitution principle or not. If they do, how can you correct the problem?

Real-Time System Design

CHAPTER 12

Software Architectures for Real-Time Embedded Systems

Contents

To create architecture is to put in order. Put what in order? Function and objects.

Le Corbusier

Real-Time Embedded Systems. http://dx.doi.org/10.1016/B978-0-12-801507-0.00012-2

12.1 Real-Time Tasks

The term "task" is a design-time concept. A real-time system typically has multiple tasks, each of which represents a unit of concurrency.

In general, a task is simply a block of instructions (or actions) to be executed by a processor for a specific purpose. A task with real-time constraints is called a real-time task.

A task can be executed multiple times. Each individual run to completion of a task is called a *job* of that task. In such a sense, a task can also be treated as a stream of jobs of the same kind.

Following the classification in [50], we distinguish three types of real-time tasks.

(1) Periodic tasks: A periodic task is a stream of jobs, where the interarrival times between consecutive jobs are almost the same, and are called its period.

(2) Sporadic tasks: A sporadic task is a stream of jobs, where the interarrival times between consecutive jobs may differ widely, and can be arbitrarily small (in short, uneven spurts). A sporadic task is executed in response to events which occur at random instants of time, and the randomness is hard to be characterized by simple probability distribution functions. For example, via the touchpad of an embedded system, one day a person may issue service requests by following the piano notes of Beethoven's Fifth Symphony, but the other day the same person may issue service requests quite differently, say, by following the piano notes of Beethoven's Moonlight Sonata. Both pieces have turbulent yet invigorating parts with rapid progressions from note to note, leading to jobs released in bursts.

(3) Aperiodic tasks: An aperiodic task is a stream of jobs, where the interarrival times between consecutive jobs may follow a known probability distribution function. For example, the serial communication port of an embedded system may receive 3 data packets per second following the Poisson distribution; the interarrival times of the packets can follow the Gaussion distribution with a mean of 100 ms and a standard deviation of 8 ms.

Periodic tasks and sporadic tasks may have hard or soft timing constraints (deadlines). Aperiodic tasks typically have either soft or no deadlines. In the order of importance, periodic tasks with hard deadlines come first, followed by sporadic tasks with hard deadlines, periodic tasks with software deadlines, sporadic tasks with soft deadlines, and lastly aperiodic tasks with soft or no deadlines. For ease of schedulability analysis, if not otherwise specified, the deadlines specified for periodic tasks and sporadic tasks in this book are treated as hard constraints.

12.1.1 Worst-Case Task Execution Time

The execution time of a task (job) is the amount of time required to fully complete its execution, assuming that there is no other task (job) competing for resources. This value highly depends on the speed of the processor on which the task is running, as well as the time-complexity of the task's instructions.

The actual amount of time to complete a job can vary for many reasons. Hardware features, such as cache and pipelining, can affect the actual execution time of consecutive jobs. Software structures, such as conditional and iterative code blocks, can also contribute to the deviation of the actual execution time. For instance, the instruction block of a task may contain conditional branches. Which branch to take depends on the run-time evaluation of the Boolean expression associated with the conditional instruction. Since different branches may take different amounts of time to finish, the actual execution time of one job can be different from that of another job.

For example, the task as shown on the left in Figure 12.1 has four execution paths, which take 40 ms, 35 ms, 30 ms, and 25 ms, respectively, to finish.[1] The execution time of the task on the

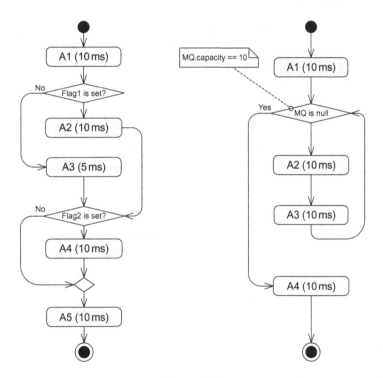

Figure 12.1

The task shown on the left has a worst-case execution time of 40 ms, and the task shown on the right has a worst-case execution time of 220 ms.

---

[1] These values again are estimated on the basis of the expected number of lines of code, the processor speed, the context switching time, the execution time of system calls, etc.

right can be 20 ms, 40 ms, 60 ms, and up to 220 ms, depending on how many messages are in the message queue.

Obviously, the maximum (or worst-case) execution time of the task shown on the left in Figure 12.1 is 40 ms, and the task on the right has a worst-case execution time of 220 ms (when the message queue is full).

In practice, the worst-case execution time of a task is often estimated beforehand, and is used in schedulability analysis at design time. This, undoubtedly, will lead to conservative design. However, such a design makes it possible to conduct thorough schedulability analysis prior to implementation, which is highly important for safety-critical systems.

12.1.2 Task Specification

A periodic task T_i is specified by a tuple (p_i, r_i, e_i, d_i), where

- p_i is the period, which is the length of inter-release times between two consecutive jobs;
- r_i is the release time (aka. phase), which is the instant of time at which the first job becomes available for execution;
- e_i is the worst-case execution time, indicating the demand for processing time;
- d_i is the deadline, which is the instant of time by which the execution of a job has to be finished.

Given a periodic task $T_i = (p_i, r_i, e_i, d_i)$, its jobs, by the order of occurrence, are denoted by J_{i1}, J_{i2}, etc. Normally, every job of a task T_i is released and becomes ready at the beginning of a period. In such a situation, the task is specified by $T_i = (p_i, e_i, d_i)$. When every job of a task T_i has to be completed by the end of its period, the task is specified by $T_i = (p_i, e_i)$.

For an aperiodic task, only the release time and the execution time are needed in schedulability analysis. Hence, an aperiodic task A_i is specified by a tuple (r_i, e_i).

For a sporadic task, its (hard) deadline also needs to be considered in schedulability analysis. A sporadic task S_i is specified by a tuple (r_i, e_i, d_i).

12.1.3 Task Timing Diagrams

A task timing diagram shows the state changes of one or more tasks over time.

A task timing diagram is given in Figure 12.2. The meanings of the timing attributes marked in Figure 12.2 are listed in Table 12.1.

The timing requirements given as an interval, such as *execution time, slack time, jitter time, rise time, initiation time, dwell time, fall time*, and *period*, are often called *duration*

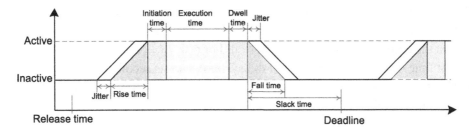

Figure 12.2
A task timing diagram.

constraints. The timing requirements given as an instant, such as *release time*, *deadline*, and state transition instant, are often called *instant constraints*. In Figure 12.2, instant constraints are marked as points on the time axis, and duration constraints are shown by bidirectional arrows with a delimiter line at each end. You may have noticed some similarities between **Table 11.2 and Table 12.1**. Actually a real-time task (or job) comprises a sequence of real-time actions, which may be at multiple levels of granularity. A timing property of a task, say, the execution time, can be the integration of one or more timing priorities of the subordinate actions.

Timing diagrams are very useful to depict the schedule design of a group of tasks. When timing diagrams are used in scheduling analysis, the rise time and fall time of a task are typically suppressed because they are at a much lower magnitude than the other measurements. The attributes such as initiation time and dwell time are also ignored or are treated as part of the execution time of a task.

Table 12.1 Timing information shown in timing diagrams

Attribute	Description
Period	The time interval indicating how often a task is to be performed
Release time	The time instant at which a task becomes eligible for execution
Deadline	The time instant by which a task should be completed
Rise time	The time needed to switch into a new task context
Fall time	The time needed to switch out of the last task context
Jitter time	The time variation for a transition or an event
Initiation time	The time needed to prepare for task execution, i.e., perform initialization actions
Execution time	The time for task execution
Dwell time	The time needed to round up the task execution, e.g., perform management actions (garbage collection). A deadline can be reached during dwell time
Slack time	The time between the end of dwell time and the task deadline

The timing diagram in Figure 12.3 shows the schedule of two periodic tasks and their timing constraints. Every job of the task $T_1 = (20, 6)$ is released at the beginning of a period, and its deadline is at the end of the corresponding period. For task $T_2 = (30, 4, 12, 24)$, its first job J_{21} is released at time 4 (relative to the start of the system), and its second job J_{22} is released at time 34. The inter-release time between jobs J_{21} and J_{22} is equal to the task period. Also note that in general the deadline of a job J_{ij} is calculated by $p_i(j - 1) + d_i$.

Another example is given in Figure 12.4. The timing attributes of the three periodic tasks are shown at the top. Note that they are not part of the timing diagram; they are displayed simply for the convenience of the reader. The portion enclosed within the dashed box is the final task schedule design; again the dashed box is merely for clarity. Above the dashed box there are

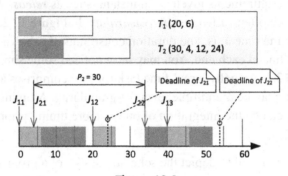

Figure 12.3
Example periodic tasks.

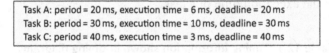

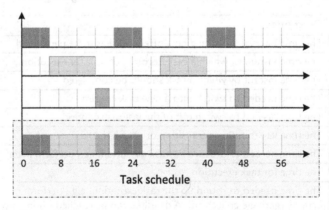

Figure 12.4
A timing diagram showing the design of a task schedule.

three time axes; the schedule of each task is plotted along the corresponding axis. For a system with only one processing unit, special care must be taken to make sure that all the tasks are scheduled interleavingly without breaking any of the timing constraints (task periods and deadlines). We will see many task scheduling diagrams as we learn real-time scheduling later.

12.1.4 Worst-Case Response Time

The performance of a real-time system is typically expressed in terms of the response time. The *response time* of a job is defined as the length of time from the release time of the job to the instant when it finishes.

The response time of a job depends heavily on its execution time, as well as on how jobs are scheduled by the system. For example, consider the job schedule given in Figure 12.5. All jobs of task T_1 are scheduled immediately after they are released; they have the same response time: 10 units of time. Each of the three jobs of task T_2 is scheduled immediately after the completion of a job of task T_1; they have the same response time: 25 units of time. For task T_3,

- its first job J_{31} is released at time 0, scheduled to run at time 25, preempted by job J_{12} at time 30. It resumes its execution at time 40, and finishes at time 55. The response time is thus 55 units of time.
- its second job J_{32} is released at time 90, scheduled to run at time 100, and finishes at time 120. The response time is thus $120 - 90 = 30$ units of time.

The *worst-case response time* of a task is the maximum value among the response times of all its jobs. For the example above, the worst-case response time of task T_3 is 55 units of time.

The concept of response time can be applied to asynchronous events as well. The response time of an event is defined as the length of time from the time instant when it is raised (released) to the instant when it is completely handled. In the rest of this chapter, we will use "worst-case response time" to evaluate several software architectures prevalently used for real-time embedded systems.

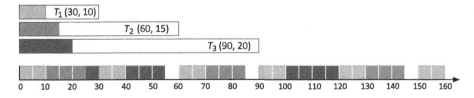

Figure 12.5

Worst-case response time.

12.1.5 Task Implementation

Each job of a task has a single flow of execution. Notice that we here use the word "flow" instead of "thread," because "thread" is an operating system (OS) concept and most real-time embedded systems are implemented without even using an OS.

In the rest of this chapter, we will study three software architectures for real-time embedded systems: round robin, round robin with interrupts, and queue based. In these architectures there is no "thread" concept; the execution flows of many tasks are sequentially organized such that the end of one job is followed by the start of another. In other words, if we overuse the word "thread" here, the whole system has only a single active thread all the time. The system executes those available jobs in a certain order, and repeats this pattern until the system is down.

A real-time OS (RTOS) is the most powerful, and maybe the most expensive, software architecture for real-time embedded systems. We will introduce the thread concept and study RTOS in Chapter 13. For systems built upon an OS, each task can be neatly implemented as a single thread.[2] The run-time behavior of an OS-based system with multiple threads would be in the hands of the OS scheduler, which typically offers acceptable performance to many applications.

12.2 Round-Robin Architecture

Round robin is a well-known principle that is widely adopted in many disciplines. For instance, round robin is one of the simplest scheduling algorithms implemented in many OSs. The rule is quite simple: everything (person, team, task, network node) can take its turn to equally compete for shared resources. As far as real-time embedded systems are concerned, the round-robin principle can be exploited to build software architectures for many simple systems [66].

Below we use a case study to examine its characteristics.

12.2.1 Case Study: Body Thermometer

A digital body thermometer is one of the simplest real-time embedded systems that you will find in almost every house. From a user's perspective, a body thermometer may have the following desired features:

---

[2] A thread must be rooted in a single *active* object that can not only receive events and dispatch them to the participating object(s) of the thread, but also generate its own stimulus events independently from the rest of the system. Thus, the active object of a thread is also called the executor of the embedded sequence of actions [34].

- a button that allows a user to turn on/off the thermometer;
- a display to show the current measure of body temperature;
- a buzzer that starts buzzing to indicate its readiness for user reading;
- a button that allows a user to stop the buzzing;
- a button that allows a user to switch between degree Fahrenheit (°F) mode and degree Celsius (°C) mode;
- a button that allows a user to start a new measure.

For better user experience, it is also a requirement that the body thermometer should respond to each button press within 1 s.

12.2.1.1 Hardware design

In order to meet the user requirements mentioned above, we consider a simple hardware design as specified below:

- The system has a microprocessor with an analog input interface.
- The system has a temperature sensor.
- The system has a status register SSR with a flag bit (s_flag) that indicates whether the sensor is on or off. It also has a data register SDR for storing the latest measure digitized from the analog input from the sensor.
- The system has an LCD for showing the current body temperature.
- The system has a buzzer circuit. It starts buzzing if the temperature measure has not changed for a while (say, 7 s).
- The system has a group of three pushbuttons. The left button allows a user to toggle between degree Fahrenheit mode and degree Celsius mode. The middle button allows a user to stop the buzzer, if it is buzzing, or to start a new measure, if the buzzer is not buzzing. The right button allows the user to turn on/off the thermometer.
- The three buttons share one status register BSR, containing flags that indicate the status of the buttons and the buzzer. The details are given in Figure 12.6. We assume that the flags l_flag, m_flag, and r_flag can be asserted by hardware when the corresponding button is pressed. The other two flags are handled by software.

12.2.1.2 Design of software architecture

As shown in Figure 12.7, a simple software architecture design for the body thermometer is specified in a UML activity diagram.

The system enters into a loop after initialization. The round-robin principle applies to the tasks (activities)[3] inside the loop: they are organized sequentially, taking turns to execute. Task descriptions are given in Table 12.2.

---

[3] In this example most of the activity steps are rather simple, and it is more appropriate to call them actions. However, we stick with the term 'task' within the context of architecture design.

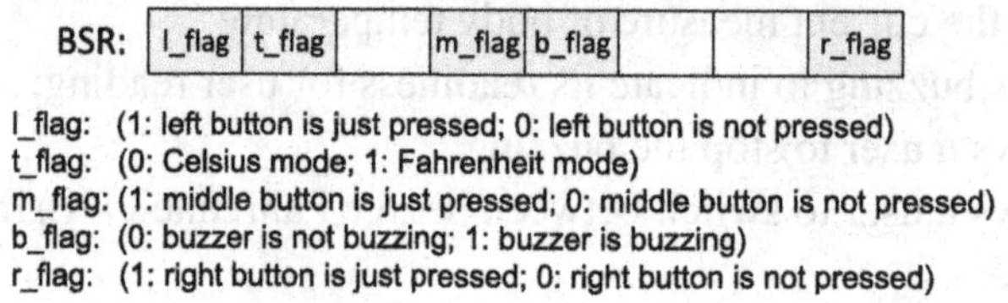

BSR:
l_flag
t_flag
m_flag
b_flag
r_flag
l_flag: (1: left button is just pressed; 0: left button is not pressed)
t_flag: (0: Celsius mode; 1: Fahrenheit mode)
m_flag: (1: middle button is just pressed; 0: middle button is not pressed)
b_flag: (0: buzzer is not buzzing; 1: buzzer is buzzing)
r_flag: (1: right button is just pressed; 0: right button is not pressed)

Figure 12.6
Button status register.

Initialization
Point A
T1: s_flag = Read_SSR ()
s_flag==1
No
Yes
T2: d = Read_SDR ()
T3: t = ConvertTemp(d)
Has changed in last 10 rounds
Yes
No
T4: Turn off sensor Set_SSR(0)
T5: Turn on buzzer Set_BSR(b_flag, 1)
Point B
T6: F = Read_BSR ()
l_flag t_flag m_flag b_flag r_flag
(F & 10000000) != 0
No
Yes
T7: Set_BSR(l_flag,0)
Point C
T8: Set_BSR(t_flag,~F.t_flag)
(F & 01000000) != 0
Yes
No
T9: t = Fahrenheit (t)
Point D
T10: Display (t)
T11: Delay (500 msec)
Point E
(F & 00010000) != 0
Yes
No
T12: Set_BSR(m_flag, 0)
(F & 00001000) != 0
Yes
No
T14: Turn_on sensor Set_SSR(1)
T13: Turn_off buzzer Set_BSR(b_flag, 0)
Point F
(F & 0000001) != 0
No
Yes
Point G

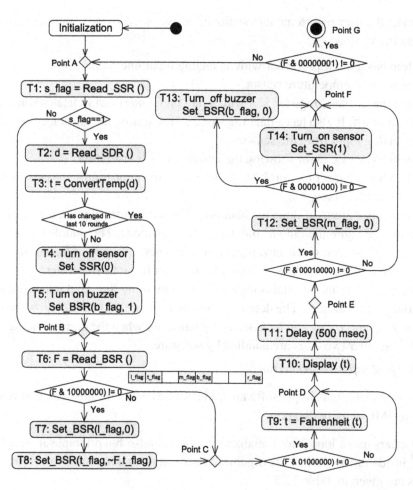

Figure 12.7
A design of a body thermometer.

Table 12.2 Task specification of a body thermometer

Task ID	Task Description	Execution Time	Release Condition
T_1	Read sensor flag	10 ms	Always
T_2	Read sensed temperature	10 ms	Sensor on
T_3	Convert to degree Celsius	50 ms	Sensor on
T_4	Turn off sensor; reset sensor flag	10 ms	No change for 10 rounds
T_5	Turn on buzzer; assert buzzer flag	10 ms	No change for 10 rounds
T_6	Read button status register	10 ms	Always
T_7	Reset l_flag to ack left-button press	10 ms	l_flag asserted
T_8	Toggle t_flag to switch mode	10 ms	l_flag asserted
T_9	Convert temperature to degree Fahrenheit	10 ms	t_flag asserted
T_{10}	Update LCD	100 ms	Always
T_{11}	Delay 500 ms	500 ms	Always
T_{12}	Reset m_flag to ack middle-button press	10 ms	m_flag asserted
T_{13}	Turn off buzzer; reset buzzer flag	10 ms	b_flag asserted
T_{14}	Turn on sensor; assert sensor flag	10 ms	b_flag not asserted

Depending on the run-time evaluation of the branching conditions, some tasks may not execute in every iteration of the loop. We can roughly classify the tasks as follows:

- Tasks T_1, T_6, T_{10}, and T_{11} execute in every iteration. Tasks T_2 and T_3 execute in every iteration when the sensor is on. Task T_9 executes in every iteration when the mode is set to degree Fahrenheit. They can be treated as periodic tasks.
- The execution of T_4 and T_5 is unexpected (no change in the last 10 rounds). They can be taken as aperiodic tasks.
- Tasks T_7, T_8, T_{12}, T_{13}, and T_{14} execute only when a button event happens. For example, T_7 and T_8 execute when a user presses the left button; T_{12} and T_{13} or T_{14} execute when a user presses the middle button. They can be treated as sporadic tasks.

In Table 12.2, we also list the hypothetical worst-case execution time and release conditions for each task. Note that the 'initialization' activity is not listed, because it is outside the round-robin loop and is irrelevant to our analysis.

12.2.1.3 Worst-case task response time

We know from Section 9.1.2 that there are typically three types of events: time events, signal events (internal or external), and change events. All three types of events occur in the design of the body thermometer as given in Figure 12.7. For instance, task T_{11} (delay 500 ms) needs to respond to a time event, and task T_4 raises a change event (turn off sensor) that will affect the system behavior at the decision point before task T_2.

For signal events, there are two subtypes: internal and external. Function calls are internal signals.[4] For example, "Set_SSR(0)" can be taken as a utility function called by task T_4. Although the sensor and the buzzer are hardware devices, they are not supposed to raise events by themselves. Instead, they respond only to software invocations *passively* in our design. Hence, "turn on sensor," "turn off sensor," "turn on buzzer," and "turn off buzzer" are also internal events.

External signal events are those triggered by the external environment. As we know from Chapter 4, external events are asynchronous in the sense that they can occur at any time and typically occur at unanticipated spots of the running program. For example, button presses from a user are external signal events. Devices such as buttons *actively* raise events to the system in response to external triggers (e.g., from the user). External events need to be captured and handled in a timely manner by the system.

Response-time analysis is relatively straightforward for time events, change events, and internal signal events, because they are typically raised and completely processed at fixed locations of the system execution. For example, the sensor can be turned off only in task T_4, and this change is handled immediately in task T_4. It is likely that a timer of 500 ms is launched at the beginning of task T_{11} and the time-out signal from the timer is captured and handled at the end of task T_{11}. Internal signal events are also handled almost immediately.

Thus, in response-time analysis we will focus only on external signal events. For the body thermometer, there is only one type of external event triggered by users: button presses. The task specification given in Table 12.2, together with the UML activity diagram, allows us to examine worst-case scenarios, a few of which are given in Table 12.3.

Take the first entry of Table 12.3 as an example. Suppose a user presses the right button while the system execution is at point A, at which time the context is as follows:

- *s_flag* is set (i.e., sensor is on);
- the temperature measure has not changed for the last 10 rounds; and
- the user also pressed the middle button and the left button while the system execution was in between point C and point A.

For this scenario, starting from point A, the system is turned off after the execution of tasks T_1, T_2, T_3, T_4 (sensor is turned off), T_5 (buzzer is turned on), T_6, T_7, T_8 (switched to degree Fahrenheit mode), $T_9, T_{10}, T_{11}, T_{12}$, and T_{13}. Thus, the path execution time is $10 + 10 + 50 + 10 + 10 + 10 + 10 + 10 + 10 + 100 + 500 + 10 + 10 = 750\,\text{ms}$.

Since the timing requirement for a user button press is 1 s, it seems that all the timing requirements (from a user's perspective) can be satisfied by the round-robin-based design.

---

[4] In systems with interrupts, software interrupts and internal interrupts can also be taken as internal signal events. This category can be further expanded to include OS signals for systems built upon an OS.

Table 12.3 Worst-case event response time

Event	Occurs at	Worst-Case Context	Complete at	Response Time (ms)
Right button pressed (turn off)	Point A	s_flag is set, no change in the last 10 rounds BSR: $\boxed{1}\,\boxed{0}$ $\boxed{1}\,\boxed{0}$ $\boxed{1}$	Point G	$10 + 10 + 50 + 10 + 10 +$ $10 + 10 + 10 + 10 + 100 +$ $500 + 10 + 10 = 750$
Middle button pressed (stop buzzing)	Point F	Buzzing, s_flag is not set, BSR: $\boxed{1}\,\boxed{0}$ $\boxed{1}\,\boxed{1}$ $\boxed{0}$	Point F	$10 + 10 + 10 + 10 + 10 +$ $100 + 500 + 10 + 10 = 670$
Middle button pressed (restart)	Point F	Not buzzing, s_flag is set, no change in the last 10 rounds BSR: $\boxed{1}\,\boxed{0}$ $\boxed{1}\,\boxed{0}$ $\boxed{0}$	Point F	$10 + 10 + 50 + 10 + 10 +$ $10 + 10 + 10 + 10 + 100 +$ $500 + 10 + 10 = 750$
Left button pressed (to degree Fahrenheit)	Point D	s_flag is set, no change in the last 10 rounds BSR: $\boxed{1}\,\boxed{0}$ $\boxed{0}\,\boxed{0}$ $\boxed{0}$	Point D	$100 + 500 + 10 + 10 +$ $10 + 10 + 50 + 10 + 10 +$ $10 + 10 + 10 + 10 = 750$
Left button pressed (to degree Celsius)	Point D	s_flag is set, no change in the last 10 rounds BSR: $\boxed{1}\,\boxed{1}$ $\boxed{1}\,\boxed{0}$ $\boxed{0}$	Point D	$100 + 500 + 10 + 10 +$ $10 + 10 + 50 + 10 + 10 +$ $10 + 10 + 10 = 740$

12.2.1.4 Hardware concurrency

A real-time embedded system is typically composed of many devices, each of which may need the system's attention (processing) whenever something important has just happened—say, a button has just been pressed by a user or a serial communication port has received new data. When something such as this happens, a device will notify the system by raising an external event, which is also called a *service request*.

There are four stages in the processing of an external event: occurrence, detection, acknowledgment, and service to completion.

(1) Occurrence. This stage is completely handled at the hardware level. When a device raises an external event (service request), a "footprint" is typically left in hardware—say, special flags and/or some data are set at a certain memory location (register). A service request (event) is called an *outstanding request* from the time instant it is raised until the time instant it is acknowledged by the system. Depending on its nature, a device may become temporarily irresponsive (i.e., operations disabled) while it has an outstanding service request. There is a good reason for this. While having an outstanding request, if the device were still responsive, it might be forced (say, nothing can prevent a user from pressing a button) to raise another service request, and this new footprint would certainly overwrite the footprint of the outstanding request.

(2) Detection. At this stage, the outstanding service request (event) is detected by the system. This can be done by hardware only (say, interrupts detected by a programmable interrupt controller), or by software (say, evaluating control flags, or reading hardware flags from memory or registers).

(3) Acknowledgment. At this stage,

    (a) the footprint of the service request is preprocessed. The footprint information is either recorded by variables or backed up to a safe location. For instance, a variable can be used to capture the hardware flags in a register (such a variable is also called a software flag). Afterward, the software flags and/or the backup copy will be used in any further processing of the current request. In this way, the handling of the current request is isolated from any potential changes to the hardware device.

    (b) the device is reset and re-enabled, if it was disabled for some reason. In particular, hardware flags in the status register are reset.

    (c) at this time point, we say that the request has been *acknowledged*. This means (i) the system will know (from the software flags) that the task for processing the current request is ready for execution, and (ii) the device can resume its normal operations and is ready to capture the next event. It is clear that any new outstanding request from the device will not affect the handling of the current request.

(4) Service to completion. At this stage, the system starts to schedule the corresponding service task to process the request until it has been serviced completely. Once the service task has started, it can clear, if applicable, the software flags set at the acknowledgment stage, so that they can be reused to indicate the readiness of a new service request.

The four stages discussed above are illustrated in Figure 12.8.

The understanding of the four event processing stages allows us to refactor the design in Figure 12.7 into a new design as shown in Figure 12.9.

This new design organizes tasks by their roles played in the event processing stages. Here, for a device i, we use T_i^d, T_i^a, and T_i^s to denote the tasks corresponding to the detection stage, acknowledgment stage, and service stage, respectively, of device i's event processing. Table 12.4 gives the mapping between the tasks of the original design and those of the new design.

Notice the following facts regarding the new design:

- For some tasks such as T_3^a and T_4^d, there is no counterpart in the original design. Each of these tasks has an execution time of 0. They appear in the new design as placeholders, which are simply for the convenience of our analysis.

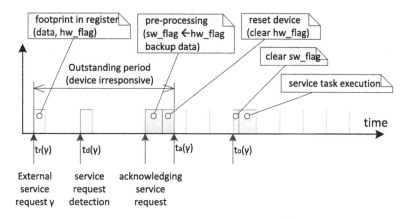

Figure 12.8
Four stages in processing external events.

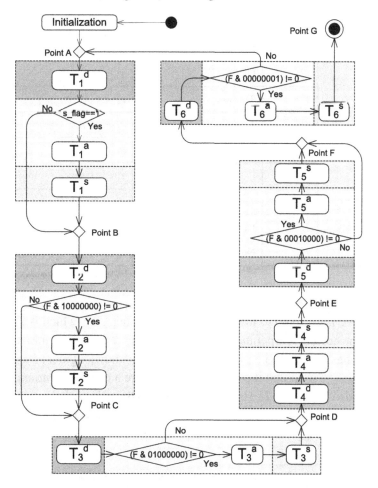

Figure 12.9
Refactored design of the body thermometer.

Table 12.4 Task mapping in structure refactoring

Device	Task ID	Mapped Back to Tasks	Worst-Case Execution Time
Sensor	T_1^d	T_1	$e_1^d = 10$ ms
	T_1^a	T_2	$e_1^a = 10$ ms
	T_1^s	T_3; T_4; T_5	$e_1^s = 70$ ms
Left button	T_2^d	T_6	$e_2^d = 10$ ms
	T_2^a	T_7	$e_2^a = 10$ ms
	T_2^s	T_8	$e_2^s = 10$ ms
Mode filter	T_3^d	T_6 (added)	$e_3^d = 10$ ms
	T_3^a	Null	$e_3^a = 0$
	T_3^s	T_9	$e_3^s = 10$ ms
LCD	T_4^d	Null	$e_4^d = 0$
	T_4^a	Null	$e_4^a = 0$
	T_4^s	T_{10}; T_{11}	$e_4^s = 600$ ms
Middle button	T_5^d	T_6 (added)	$e_5^d = 10$ ms
	T_5^a	T_{12}	$e_5^a = 10$ ms
	T_5^s	(T_{13} | T_{14})	$e_5^s = 10$ ms
Right button	T_6^d	T_6 (added)	$e_6^d = 10$ ms
	T_6^a	Null	$e_6^a = 0$
	T_6^s	Null	$e_6^s = 0$

- Sensor has a task T_1^a devoted to the acknowledgment stage. T_1^a, which is T_2 in the original design, reads the current temperature measure from the register SDR into a variable d.[5] The left button has a task T_2^a (i.e., T_7) devoted to the acknowledgment stage. This task simply resets the *l_flag* so that it can be used to capture a new left-button press event. The middle button also has a task T_5^a (i.e., T_{12}) devoted to the acknowledgment stage. This task simply resets the *m_flag* so that it can be used to capture a new middle-button press event. There is no acknowledgment task for the right button, because the system is simply turned

---

[5] To prevent dirty read, it may disable SDR for writing beforehand and enable writing again afterward.

off when it is pressed. The mode filter and the LCD do not have an acknowledgment task either; they do not need to be acknowledged.

- Some of the original tasks are merged. For example, T_1^s corresponds to the original block composed of T_3, T_4, and T_5. T_1^s's execution time is the block's worst-case execution time, which is at most $50 + 10 + 10 = 70$ ms. T_5^s's execution time is derived similarly.
- There is no counterpart for tasks T_3^d, T_5^d, and T_6^d in the original design. They are added for a reason that will become clear shortly.

We need to use some notation to refer to the important time instants regarding a service request. For a service request γ, let

- $t_r(\gamma)$ denote the time instant at which γ is raised,
- $t_d(\gamma)$ denote the time instant at which γ is detected,
- $t_a(\gamma)$ denote the time instant at which γ is acknowledged, and
- $t_o(\gamma)$ denote the time instant at which the system starts to execute the service task for γ.

Given that a request γ is raised by a device i, while the release time of γ is $t_r(\gamma)$, the release time of the corresponding service task T_i^s is actually $t_a(\gamma)$, the time instant at which T_i^s is ready for execution.

We call the time interval $[t_r(\gamma), t_a(\gamma)]$ the *outstanding period* of γ. When the service requests of a device have a short outstanding period on average, we say that the device has a *high hardware concurrency*, because it can respond to new external triggers very quickly. Let us examine the three buttons to see their worst-case outstanding periods and best-case outstanding periods.

Now, let us assume that T_3^d was not added. Suppose when the system's execution is at point B, a user presses the left button to switch the mode from degree Celsius to degree Fahrenheit. Since the current mode is degree Celsius, the *t_flag* value read to F by T_2^d is 0. In task T_2^s, *t_flag* is changed to 1. However, this flag change would not be noticed by the mode filter block, because the *t_flag* in F is still 0. The mode change would happen the next time when BSR is read again in T_2^d. By addition of T_3^d, the mode change will take effect immediately in the current execution round.

We have added T_5^d and T_6^d for a similar reason. Suppose a user presses the middle button when the system's execution is at point E. From Table 12.5 we see that this is the best-case scenario, where the middle button has its shortest outstanding period. However, if T_5^d were null, then this button event would not be acknowledged immediately by T_5^a because it is not BSR but the variable F that is used at the decision point (F has an obsolete snapshot of BSR); instead it would be captured in the next round at T_2^d and then acknowledged by T_5^a. Consequently, this best-case scenario would have an outstanding period of 760 ms! As an exercise, you may want to check the best-case outstanding period of the right button when T_6^d is null.

Table 12.5 Responsiveness of external button events

Device Event		Raised at	Acknowledged by Task	Length of Outstanding Period	Event Response Time
Left button pressed	Best case	Point B	T_2^a	$e_2^d + e_2^a = 20$	$e_2^d + e_2^a + e_2^s = 30$
	Worst case	End of T_2^a	T_2^a	$\sum_{i=1}^{6}(e_i^d + e_i^a + e_i^s) = 780$	790
Middle button pressed	Best case	Point E	T_5^a	$e_5^d + e_5^a = 20$	$e_5^d + e_5^a + e_5^s = 30$
	Worst case	End of T_5^a	T_5^a	$\sum_{i=1}^{6}(e_i^d + e_i^a + e_i^s) = 780$	790
Right button pressed	Best case	Point F	T_6^d	$e_6^d + e_6^a = 10$	$e_6^d + e_6^a + e_6^s = 10$
	Worst case	Point A	T_6^d	$\sum_{i=1}^{6}(e_i^d + e_i^a + e_i^s) = 780$	780

In sum, tasks T_3^d, T_5^d, and T_6^d are added in the new design simply to shorten the outstanding periods of the corresponding devices. However, this will not do anything good to improve the worst-case outstanding period, which is, for all three buttons, the length of a whole execution round! This means, in worst-case scenarios, users may experience irresponsiveness when a button is pressed, which is an indicator of low hardware concurrency. We will see how to improve hardware concurrency by interrupts in Section 12.3.

12.2.2 General Round-Robin Architecture

You may have already noticed that in Figure 12.9 there is a "detect-acknowledge-service" (DAS) task pattern for each device. In general, we assume that each hardware device of a system can raise its service requests by setting a predefined hardware flag at a specific memory location (register), and the system applies the DAS task pattern to each device to repeatedly detect, acknowledge, and service all the external requests.

Figure 12.10 gives the general round-robin architecture. We have the following observations:

(1) In the circular structure as shown in Figure 12.10, the round-robin principle is applied at the device level. In other words, the architecture is a series of DAS task patterns, and each task runs to completion as an integral unit. In Chapter 13 we will learn that the round-robin principle can be applied at a much finer level of granularity to implement a so-called round-robin task scheduler, where time slices (or time slots or quanta) are assigned to each task in equal portions and in circular order. Since the quantum is generally smaller than the task execution time, each task may take many rounds for it to be completed.

(2) In each round of execution, each device i is handled only once, because there is exactly one DAS task pattern dedicated to it. That is, for a device i, the tasks T_i^d, T_i^a, and T_i^s each

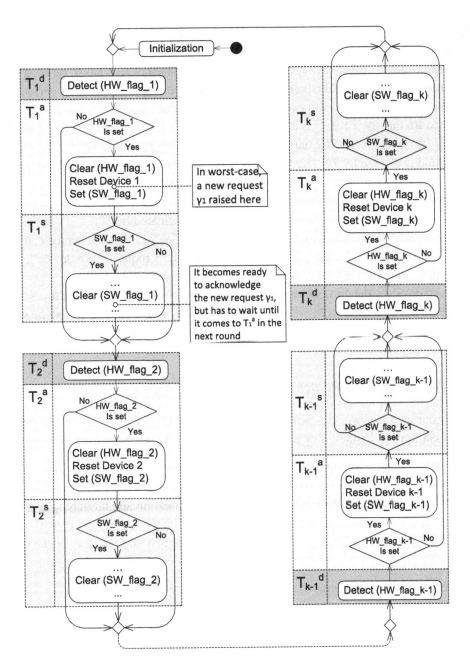

Figure 12.10
Round-robin architecture with a tightly coupled DAS pattern.

appear only once. It is practically possible to have some devices (say, with more urgent deadlines) handled multiple times in each round. As an example, consider a round-robin design with the DAS task pattern appearing twice for a device j. It is easy to show that the worst-case outstanding period and worst-case event response time of device j can be shortened approximately by half. However, such a biased design has to be used very carefully because it would adversely increase the worst-case outstanding periods and worst-case event response times of all the other devices.

(3)   A hardware flag `HW_flag_i` and a software flag `SW_flag_i` are reserved for each device i: `HW_flag_i` is asserted by hardware and is cleared by T_i^a, while `SW_flag_i` is asserted by T_i^a and is cleared by T_i^s. The two flags serve as relay batons so that the external events from device i can be processed in an appropriate order. In particular, device i is irresponsive to new triggers until `HW_flag_i` is cleared by T_i^a; the system is not ready to handle a new request of device i until `SW_flag_i` is cleared by T_i^s.

(4)   Owing to the round-robin nature, immediately after `SW_flag_i` has been cleared,[6] even though the system becomes ready to acknowledge a new request from device i, this will not happen until the control comes to T_i^a in the next execution round.

(5)   For each device i, the corresponding tasks T_i^d, T_i^a, and T_i^s are placed next to each other. However, because of flags `HW_flag_i` and `SW_flag_i`, T_i^d, T_i^a, and T_i^s can be separately arranged as illustrated in Figure 12.11. This will not affect device i's worst-case outstanding period.

12.2.3 Worst-Case Event Response Time

To simplify our analysis, we stick with the architecture given in Figure 12.10, and make the following assumptions explicit. For each device i $(1 \leq i \leq k)$, we have the following:

(1)   In the circular structure, there is exactly one DAS pattern for i.
(2)   The software flag `SW_flag_i` is asserted only in T_i^a, and is cleared only in T_i^s.
(3)   The execution time of a decision point (condition evaluation and branching) is negligible.
(4)   It takes no time to execute a null task. That is, $e_i^d = 0$ if T_i^d is null, and $e_i^a = 0$ if T_i^a is null. Note that we must have $e_i^s > 0$, otherwise device i should not have been considered in the design.

We consider the worst-case scenario as illustrated in Figure 12.12:

•   A new request from a device j, denoted by γ_j', is raised immediately after device j is reset in T_j^a. This new request can be correctly recorded by the hardware flag `HW_flag_j`, and it will not override the current request, denoted by γ_j, because it is recorded by the software flag `SW_flag_j`.

---

[6] The service to the current request of device i may be still in progress.

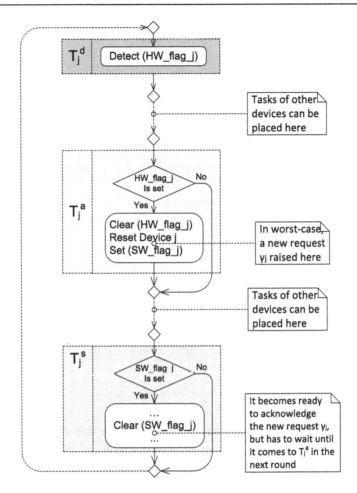

Figure 12.11
Round-robin architecture with loosely coupled DAS patterns.

- T_j^s starts to execute to service the current request γ_j, which takes e_j^s.
- In the worst case, before it comes to T_j^d again, along the execution path the system has to service one request from each of the other devices! For each device i ($1 \leq i \leq k, i \neq j$), the execution time is $(e_i^d + e_i^a + e_i^s)$.
- As it comes to T_j^d again in the next round, T_j^d and T_j^a need to be executed to detect and acknowledge γ_1', and T_j^s needs to be executed to actually service γ_1'. This takes $(e_j^d + e_j^a + e_j^s)$.

Thus, according to the above worst-case analysis, the worst-case outstanding period for a device j is given by

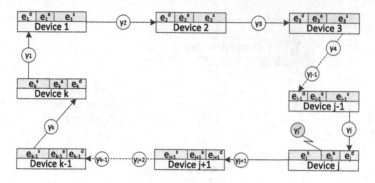

Figure 12.12

Round-robin worst-case response time: a new request γ_j' arrives just after the current request γ_j is acknowledged.

$$\sum_{i=1}^{k}(e_i^{\mathrm{d}} + e_i^{\mathrm{a}} + e_i^{\mathrm{s}}),$$

and the worst-case event response time for a device j is given by

$$e_j^{\mathrm{s}} + \sum_{i=1}^{k}(e_i^{\mathrm{d}} + e_i^{\mathrm{a}} + e_i^{\mathrm{s}}).$$

Obviously, systems based on the round-robin architecture can suffer from low hardware concurrency. This is because the system cannot handle a request from a device immediately; it has to wait until the next time it comes to the DAS task pattern that is dedicated to that device.

12.3 Round Robin with Interrupts

Hardware concurrency is one of the most desirable features in high-performing real-time embedded systems. In order to improve the hardware concurrency of a device, we have to shorten its outstanding period.

One solution is to bring the interrupt concept into play, by which a system could acknowledge external service requests almost immediately. The simplest architecture featuring interrupts is called *round robin with interrupts*. Let us first look at an example.

12.3.1 Case Study: The Simon Game

The Simon game is a memory retention game invented by Ralph H. Baer and Howard J. Morrison to measure and challenge a player's memory retention capacity. The game can generate a growing sequence of events (colors and sounds) that the player has to repeat. There are several versions of the Simon game on the market: some allow a participant to play against another player, and some may feature adjustable skill levels.

We consider a "simplified" version, where the game system has four colored buttons, each producing a particular tone when it is pressed by a player or activated by the system. When the game starts, the system lights up one or more buttons in a random order, after which the player must accurately reproduce that order by pressing the buttons. For each successive round, the system repeats the latest sequence and adds another button to the sequence. This process is repeated until the participant makes an error, or until the sequence reaches a maximum length.

12.3.1.1 Hardware design

Our "simplified" game system has the following major components:

(1) A PIC microcontroller (say, PIC16F84A or PIC16F628).[7]
(2) A power on/off button.
(3) Four colored pushbuttons (red, yellow, green, and orange), each is lit when pushed, to produce visual cues.
(4) A mini piezo, which can produce a particular tone when it is turned on and off at a definite rate (an audial cue associated with each colored button).
(5) A certain number of byte-width registers.

Suppose the colored buttons are coded such that 0, 1, 2, and 3 represent red, yellow, green, and orange, respectively. Then, it requires two bits to record each step, and 12 registers (byte-width registers) can accommodate a series of up to 48 button presses (steps). One register can be used to keep track of the length (number of steps) of the current series.

12.3.1.2 Software architecture design

In Figure 12.13 we give a simple round-robin design with two interrupts, one for handling button press events and one for handling time-out events. Notice that the two interrupt service routines (ISRs) have the same structure: acknowledging the hardware device (e.g., recording the pressed button) then setting a software flag. The ISRs actually play the same role as played by T_i^a in the round-robin architecture. The big difference is that the execution of T_i^a is performed once in each round and in a fixed order, while the ISRs can be triggered many times and always preempt the execution of the round-robin loop.

The architecture design is quite simple. The system plays welcome music when it is powered on. Then, it clears the current series, randomly selects a colored button, and adds it to the

---

[7] The original prototype system, built by Baer, included the Texas Instruments TMS 1000 microprocessor chip, which was of low cost and was used by many games in the 1970s.

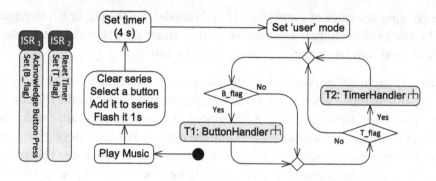

Figure 12.13
An example round-robin architecture with interrupts.

series. The selected button is flashed for 1 s to produce a visual cue (and the corresponding tone is played to produce an audial cue). Then, a timer with a time-out value of 4 s is launched and the "user mode" is set to be the current operation mode. After these initialization steps, the system enters into a round-robin execution loop.

Inside the round-robin execution loop, the system repeatedly checks two flags B_flag and T_flag, which are asserted by ISR_1 and ISR_2, respectively. In other words, when B_flag is true, it indicates that a button has just been pressed; when T_flag is true, it indicates that a time-out event has occurred. Depending on the truth values of B_flag and T_flag, the system executes the two tasks T_1 and T_2 repeatedly in a round-robin manner.

Task T_1 handles button-press events. As shown in Figure 12.14, it clears flag B_flag first (so that the next button press can be correctly recorded by ISR_1). T_1 operates only in user mode, so it terminates immediately if the system is currently not in user mode. In user mode, T_1 first disables the timer (it is canceled because the user has responded within the 4 s "thinking" deadline), then checks whether the pressed button is correct. If the player has pressed a wrong button (failed to recall the correct series), the game is restarted after the system has played "failing" music for 2 s. If the player has pressed a correct button, it is flashed for 1 s, then the system checks whether the series is finished. If it has not finished, a 4 s "thinking" period is started for the user to recall the next correct button in the series; if the series has finished, the system enters into a "replay" mode.

Task T_2 handles time-out events. As shown in Figure 12.15, it clears flag T_flag first (so that the next time-out can be correctly recorded by ISR_2). If the system is in user mode (the 4 s "thinking" deadline is broken), the game is restarted. If the system is in replay mode (it is time to replay the next button), the next button in the series is flashed for 1 s. Next, the system checks whether the series is finished. If it has not finished, a 1 s timer is launched (when it expires, T_flag is set and T_2 is executed again); if it has finished, a new button is randomly selected and is added to the series, and the system enters into user mode.

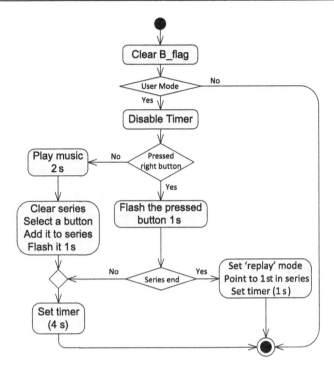

Figure 12.14
Task T_1: handling button press.

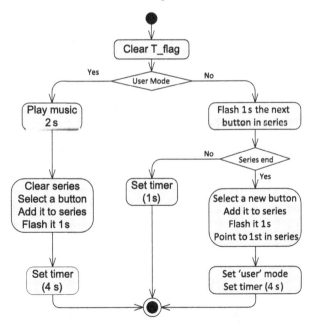

Figure 12.15
Task T_2: handling time-out events.

Obviously, T_1 is an aperiodic task, which is released whenever a user presses a button. It is associated with a soft deadline: a user has to press a button within a 4 s "thinking" time, otherwise the game is restarted. Task T_2 can be viewed as consisting of two subtasks. The user-mode branch is executed whenever a user fails to recall a correct button in 4 s. The replay-mode branch is repeatedly executed every 2 s while the system is replaying a long series.

The worst-case execution time of both tasks is a bit longer than 3 s, which happens when a user has pressed a wrong button.

The worst-case event response time is slightly longer than 6 s; this happens when both a user-mode time-out and a wrong-button press event happen almost at the same time.

12.3.2 General Architecture

In general, the round-robin architecture with interrupts is given in Figure 12.16.

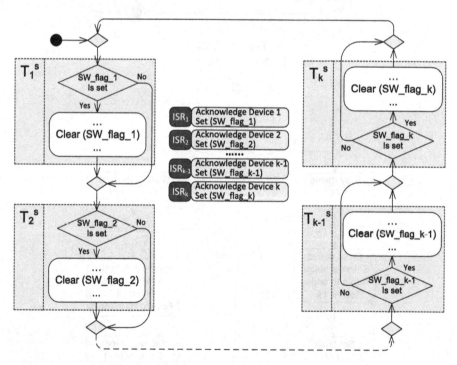

Figure 12.16
Round-robin architecture with interrupts.

This architecture has the following features:

(1)   Similarly to the round-robin architecture, the round-robin principle is applied at the device level.

(2)   Different from the round-robin architecture, T_i^d ($1 \leq i \leq k$) is no longer needed, because the detection of hardware flags is performed by hardware (i.e., PIC microcontroller). Whenever a request from a device i is raised, it is acknowledged by the corresponding interrupt service routine ISR_i.

(3) ISR_i plays exactly the same role as T_i^a. These acknowledgment tasks are no longer placed on the execution path of the round-robin loop. Instead, they are treated separately. Whenever an interrupt request occurs, the execution of the round-robin loop will be preempted by the corresponding ISR.

(4)   The worst case happens if device i has the lowest interrupt priority, and each of the other devices has raised a request that preempts the execution of ISR_i. In such a worst case, the outstanding period of the service request from device i would be

$$\sum_{i=1}^{k} e_i^a.$$

Owing to the nature of the interrupt mechanism, an ISR can be triggered and executed multiple times during one round of the system execution. As illustrated in Figure 12.17(a), an execution of ISR_j ($1 \leq j \leq k$) can appear just after software flag SW_flag_j has been cleared by T_j^s. This execution of ISR_j will acknowledge the service request by setting SW_flag_j again, and this request will not be serviced until it comes to T_j^s in the next round. However, along the execution path, ISR_j can be triggered to execute many times before the control once again comes to T_j^s. Since each trigger of ISR_j indicates a new service request from device j, the question is, how does the system handle multiple requests when the control comes to T_j^s?

There are two solutions. One is to adopt the FIFO policy in the sense that only the first service request is to be serviced by T_j^s while all the subsequent requests are ignored by the system. Another solution is to record all the requests by a counter or an array-like structure and for T_j^s to process them in a batch mode. A side effect of this solution is that the execution time of each round becomes nondeterministic, which may cause some tasks to miss their deadlines.

To simplify our analysis, we assume that the FIFO policy is used. We can actually apply a design pattern to the ISRs in order to enforce that only the first request is honored. As shown on the left in Figure 12.17(b), ISR_j (i.e., T_j^a) simply returns when SW_flag_j is asserted. This implies that another request has already been acknowledged prior to this new request.

Then, when would be the earliest opportunity that a request from a device j can be successfully acknowledged? The answer is that the request has to trigger ISR_j immediately after SW_flag_j has just been cleared! This is illustrated in Figure 12.17(b) and 12.17(c); they show the same situation except that in Figure 12.17(b) the clearance of the software flag

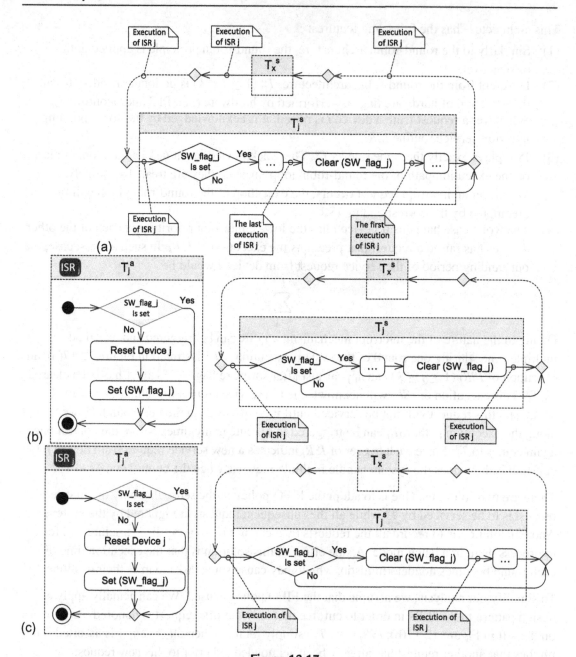

Figure 12.17

Many requests being acknowledged by *ISR$_j$*, the system may choose to honor only the first one or all requests in T_j^s (a); flag clearance placed at the end of T_j^s (b); and flag clearance placed at the beginning of T_j^s (c).

is at the end of task T_j^s, while in Figure 12.17(c) the clearance of the software flag is the first thing to do in T_j^s.

12.3.3 Worst-Case Event Response Time

Let us use Figure 12.18 to analyze the worst-time event response time. Our analysis is based on the following assumptions. For each device i $(1 \le i \le k)$,

(1) task T_i^s appears only once in the circular structure.
(2) software flag SW_flag_i is asserted only in *ISR$_i$*. In addition, the design pattern as shown in the center of Figure 12.18 is used for *ISR$_i$*: SW_flag_i is asserted (to acknowledge the triggering request from device i) only when SW_flag_i is not currently asserted.
(3) software flag SW_flag_i is cleared only in the service task T_i^s. In addition, the clearance of SW_flag_i is placed at the beginning of T_i^s (i.e., the case as illustrated in Figure 12.17(c)).
(4) the execution time of a decision point (condition evaluation and branching) is negligible. In particular, e_i^a (the execution time of *ISR$_i$*) is 0 when SW_flag_i is asserted.

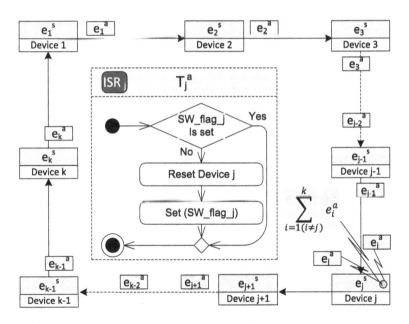

Figure 12.18

Worst-case response time: *ISR$_j$* executes as soon as flag SW_flag_j has been cleared at the beginning of T_j^s.

One worst-case scenario is illustrated in Figure 12.18:

- T_j^s starts to execute to service the current service request from device j. The first thing to do in T_j^s is to clear SW_flag_j.
- As soon as SW_flag_j is cleared, a new service request from device j, denoted by γ_j, triggers the execution of *ISR$_j$*. In the worst case, device j has the lowest interrupt priority, and each of the other devices has raised a request that preempts the execution of *ISR$_j$*. In such a worst case, the total execution time of the ISRs, including the one acknowledging γ_j, is given by $\sum_{i=1}^{k} e_i^a$. After this, the new request γ_j is successfully acknowledged.
- T_j^s completes its service to the "old" request in e_j^s. Now, γ_j is the only request from device j to be serviced.
- Along the execution path, in the worst case, there is at most one request from each device i $(1 \leq i \leq k, i \neq j)$ that can be successfully acknowledged (after servicing the "old" request). Thus, for each i $(1 \leq i \leq k, i \neq j)$, the execution time is $e_i^s + e_i^a$;
- As the control comes to T_j^s again, it starts to service γ_j.
- As soon as SW_flag_j is cleared, another new service request from device j triggers the execution of *ISR$_j$*. The amount of execution time is e_j^a, and this new request is successfully acknowledged.
- T_j^s completes its service to γ_j in e_j^s.

Thus, according to the above worst-case analysis, the worst-case event response time for a device j is given by

$$e_j^s + \sum_{i=1}^{k}(2 \times e_i^a + e_i^s).$$

As we have explained before, owing to the introduction of interrupts, the worst-case outstanding period for a device j $(1 \leq j \leq k)$ is $\sum_{j=1}^{k} e_j^a$—the execution time of all ISRs. As compared with the round-robin architecture, this is significantly less and leads to much better hardware concurrency.

Table 12.6 summarizes the four processing stages for external events.

Table 12.6 Processing stages for external events

Event from Device i	Raised	Detection			Acknowledgment			Service		
	At	By	At	Amount	By	At	Amount	By	At	Amount
Round robin	Any time	Software	T_i^d	$e_i^d > 0$	Software	T_i^a	$e_i^a \geq 0$	Software	T_i^s	$e_i^s > 0$
Round robin with interrupts	Any time	Hardware (in no time)		$e_i^d = 0$	Software	*ISR$_i$*	$e_i^a \geq 0$	Software	T_i^s	$e_i^s > 0$

12.4 Queue-Based Architecture

By introducing a queue data structure, we can further refine the round-robin architecture with interrupts as given in Figure 12.16.

A refactored architecture is given in Figure 12.19, where a queue of length k is used by the system to determine when to execute which task.

The queue is static in the sense that exactly one fixed slot in the queue is assigned to each device. Each slot in the queue stores a function pointer—a pointer to the start memory location of the corresponding task code.

The queue is populated by ISRs. Whenever a request from a device i occurs, ISR_i is executed to acknowledge the device and put a function pointer T_i into the ith slot of the queue.

The queue is examined by the main program, shown on the left in Figure 12.19, to determine whether it needs to execute a service task. The main program contains a loop; it uses an index to keep track of the current slot. For each iteration, it retrieves and clears the current slot j. If it contains a function pointer, the system starts to run the corresponding task to its completion. If the current slot contains a null pointer (there is no service request from device j), the index is increased by 1 and the next iteration starts. The index is wrapped to the first position when it reaches the end of the queue.

This queue-based architecture is equivalent to the architecture given in Figure 12.16. The software flags are replaced by the manipulation of function pointers. The round-robin

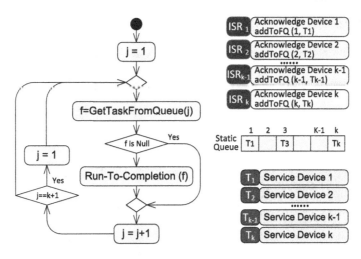

Figure 12.19
Architecture with a static task queue.

structure in Figure 12.16 is implemented here as a static queue which is repeatedly checked by the main program in a fixed order. Consequently, similarly to the round-robin architecture with interrupts, a system built upon this queue-based architecture needs to cycle around the queue to service a new request, and the worst-case event response time for a device j is still $e_j^s + \sum_{i=1}^{k} (2 \times e_i^a + e_i^s)$ (on the basis of assumptions similar to those given in Section 12.3.3).

It is worth noting that the static queue used in this architecture is an exclusive resource shared by the main program and the ISRs. Special care needs to be taken (say, disabling interrupts while the queue is accessed) in the implementation of GetTaskFromQueue() and ISRs.

12.4.1 Nonpreemptive FIFO Queue

The queue used in the architecture given in Figure 12.19 is a static queue, which exhibits at least two limitations. First, the order of service processing is fixed at design time; this is not flexible because at run time it may be necessary to process tasks in an order that is dynamically determined—say, according to how important or urgent they are. Second, a slot j of the static queue can hold at most one service request from device j at the same time. As long as slot j is occupied, any subsequent requests from device j are ignored.

To address these limitations, a dynamic queue can be used instead, so that different requests from the same device can take different positions in the queue. In general, there are two dynamic queue-based architectures: one is based on a FIFO queue and the other is based on a priority queue.

Figure 12.20 gives an architecture based on a FIFO queue. Compared with Figure 12.19, this architecture differs in two main aspects:

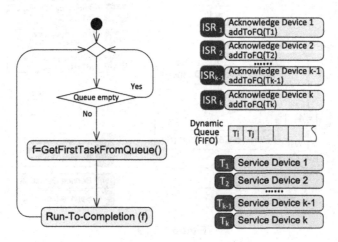

Figure 12.20
An architecture with a FIFO queue.

(1) First, each device no longer has a fixed slot in the queue. Instead, among all the devices, whichever raises the first request, its ISR will add the service task (function pointer) to the first position in the queue, whichever raises the second request, its ISR will add the service task to the second position, and so on. Consequently, all the requests are buffered by the queue in a FIFO manner, and multiple requests from the same device can coexist in the queue.

(2) The main program (shown on the left in Figure 12.20) in each iteration simply removes the first task from the queue and executes the task to its completion. After the execution of GetFirstTaskFromQueue(), the other tasks in the queue are shifted forward by one position (so that the second task becomes the first task and will be executed in the next iteration). In this way, the system executes the tasks from the queue one after the other, serving the external requests in a FIFO order.

12.4.2 Nonpreemptive Priority Queue

The earliest request (task) may not be the most important request (task). A higher priority is typically assigned to a more important task.

Figure 12.21 shows an architecture with a priority queue, which differs from the architecture in Figure 12.20 in two aspects:

(1) First, each ISR needs to add the corresponding task (function pointer) at an appropriate position in the queue such that all the tasks in the queue are ordered decreasingly by their priorities (FIFO for tasks with the same priority).

(2) Since the first position of the queue always contains the task with the highest priority, the main program (shown on the left in Figure 12.21) in each iteration simply removes the

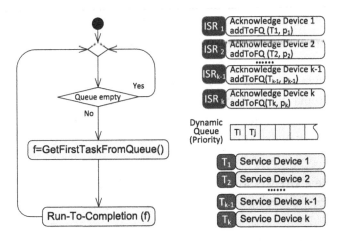

Figure 12.21
Architecture with a nonpreemptive priority queue.

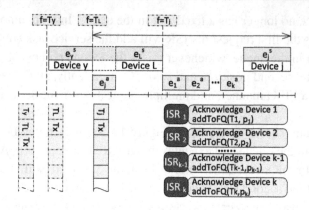

Figure 12.22
Response time for a nonpreemptive priority queue.

first task from the queue and executes the task to its completion. After the execution of GetFirstTaskFromQueue(), the other tasks in the queue are shifted forward by one position. However, the task at the head of the queue might not be executed in the next iteration, because a higher-priority task could be added to the queue during the execution of the current task.

What is the worst-case event response time for the highest-priority task? Figure 12.22 illustrates a worst-case scenario, where device j (i.e., task T_j) has the highest priority:

(1)   At the end of T_y, just after T_L has been retrieved from the queue, a request from device j is raised and the main program is preempted by the execution of ISR_j, which takes e_j^a.

(2)   After ISR_j finishes, the current task T_L is executed, which takes e_L^s.

(3)   Before T_j is scheduled to execute, in the worst case, each device (including device j itself) could raise many requests, and consequently the corresponding ISR could be executed many times. To simplify the analysis, let us assume that a device will not raise another request when it has one request still to be serviced. This can be enforced by applying the design pattern introduced in Figure 12.17(b), where the checking of the software flag is replaced by checking the queue to see the existence of a task from the same device. Then, each ISR_i ($1 \le i \le k, i \ne j$) can be executed once.

(4)   Since task T_j has the highest priority, it is the next task to be executed. Its execution time is e_j^s.

Thus, according to the above worst-case analysis, the event response time for the highest-priority device j is given by

$$e_L^s + e_j^s + \sum_{i=1}^{k} e_i^a.$$

The worst case happens when the task T_L happens to be the longest task, that is, $e_L^s = \max_{i=1}^{k}(e_i^s)$. Note that T_L and T_j could be the service task for the same device.

To conclude, queue-based architectures have three limitations. First, regardless of the type of queue (static, FIFO, or priority), once it is adopted by a system, the *queuing policy* is fixed; at run time the whole system has no other choice, and sticks with it. Second, no task preemption is allowed. Although in priority-queue based architecture, the system can immediately acknowledge an urgent service request, the new task cannot be completely serviced until the system finishes the current task, which in the worst case can be the longest and it may cause the urgent task to break its deadline! Third, in priority-queue-based architecture, starvation could happen to lower-priority tasks.

We will cover real-time operating systems (RTOS) in Chapter 13. The kernel of an RTOS typically offers several scheduling policies, and multiple policies can be applied simultaneously to different tasks of a system. By default, an RTOS also allows task preemption for performance reasons.

Problems

12.1 What is the length of the best-case outstanding period of the right button when T_6^d is null?

12.2 Refer to the architecture given in Figure 12.11, and explain why it will not affect the worst-case outstanding period but can increase the worst-case event response time.

12.3 If we keep all the assumptions given in Section 12.3.3 except that the clearance of a software flag is placed at the end of the corresponding service task (the case as illustrated in Figure 12.17(b)), give a general expression for the worst-case response time for a device *j*.

12.4 Study the UML activity diagram given in Figure 12.23. What is the worst-case response time for a signal SigA? As far as the worst-case event response time is concerned, is it more like an architecture with a static task queue, a FIFO queue, or priority queue? If it is priority-based, explain the priority order used by the design.

12.5 Linked list is a dynamic data structure that can be used to implement queues. Implement a task scheduler based on a FIFO queue. You do not need to use interrupts. Instead, you can program a few use cases to test the task scheduler. Each use case can simply add some function pointers to the FIFO queue.

12.6 Repeat the last problem, but use a priority queue instead.

12.7 Conduct a case study, and explain how the system can be implemented using the round-robin architecture.

12.8 Conduct a case study, and explain how the system can be implemented using the round-robin architecture with interrupts.

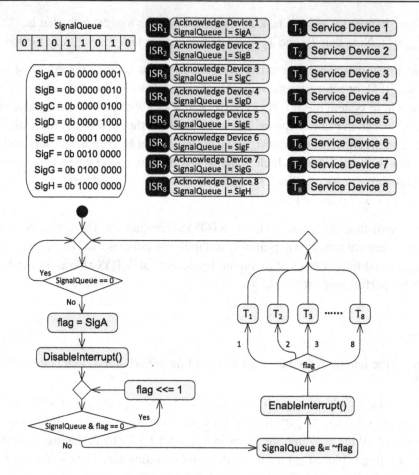

Figure 12.23
Architecture with a signal queue.

12.9   For the round-robin architecture, explain how the DAS pattern is used to handle requests from hardware devices.

12.10 For the round-robin architecture with interrupts, explain how a design pattern is used to enforce that only the first request from a device can be honored.

POSIX and RTOS

Contents

All stable processes we shall predict. All unstable processes we shall control.

John von Neumann

13.1 Introduction to POSIX

POSIX, an acronym for "portable operating system interface," is an open operating system (OS) interface standard that has been developed by the Institute of Electrical and Electronics Engineers (IEEE) and is recognized by the International Organization for Standardization and

Real-Time Embedded Systems. http://dx.doi.org/10.1016/B978-0-12-801507-0.00013-4

339

the American National Standards Institute (ANSI). As a standard software interface, the main objective of POSIX is to promote interoperability and portability of application programs across variants of Unix OSs. Ideally, a program written and compiled for execution on one OS may also be compiled, without changing the source code, for execution on another OS, as long as POSIX is supported by both.

In 1988, the IEEE released its first POSIX standard—IEEE Std 1003.1-1988. Subsequent editions were published in 1990, 1996, and 2001.With the releases of new standards, the term "POSIX" is broadened to refer to a family of related standards, and POSIX.1 is reserved to refer to IEEE Std 1003.1-1988 and its descendants. As of this writing, the latest version of the POSIX.1 standard is IEEE Std 1003.1-2008, also known as POSIX.1-2008 [17].

POSIX.1-2008 consists mainly of POSIX base definitions (such as concurrent execution, file access permissions, memory synchronization, pathname resolution, and headers), system interfaces (such as sockets, threads, signals, streams, and real-time services), and commands and utilities (such as exec, cat, and echo,). Table 13.1 lists some POSIX.1 core services and real-time extensions.[1]

13.1.1 POSIX Processes and Threads

In POSIX, each executing instance of a program is called a process. Each process has its own protected address space.[2] Processes are isolated so that they cannot inadvertently infringe on

Table 13.1 POSIX.1 services

Core Services	Real-Time Services
Process creation and control	Priority scheduling
Thread creation and control	Clocks and timers
Signals and signal handling	Real-time signals
Segmentation violations	Semaphores and mutex
Illegal instructions	Message passing
Bus errors	Shared memory
Floating point exceptions	Synchronized input and output
File and directory operations	Asynchronous input and output
Timers	Process memory locking
Pipes	Interprocess communication
C library (standard ANSI C)	
I/O port interface and control	

---

[1] For the treatment of the POSIX standard [17], in this book I try my best to honor the original term definitions and function descriptions.

[2] At the lowest level, this is handled by the memory management unit (MMU)—a hardware component responsible for translation of virtual addresses to physical addresses, memory protection, cache control, bus arbitration, and bank switching.

each other. Because of this enforced separation, communications between processes can take place only by using OS kernel services.

A POSIX thread is a single flow of execution that runs within a process. When a process starts to run, it is actually running a thread, which is often called the "main" thread because the entry point of a C program is the main() function. The main thread can create additional threads if it is so programmed. Thus, a process is virtually a collection of one or more threads.

As a single flow of execution, each thread within a process has its own stack and values for the various registers. However, the threads running within a process share at least the following:

- the virtual address space of the process (thus switching between threads does not involve changing the memory context);
- variables that are not on the stack (thus the threads can share data and communicate with each other by global variables);
- signal handlers and process-level signal control structures.

When a real-time embedded system is built upon an OS, each task is implemented as a process or simply as a single thread.

13.1.1.1 Process creation

For a large (embedded) system consisting of multiple programs, it is common practice to have one master program, also known as a *starter program*, to start all the other programs. The starter itself is typically the last command of a shell script file (a batch of commands), which is automatically executed when the system is up and running.

A process (actually a thread within the process) can invoke system calls to create new processes. The newly created process is called a child process of the calling process, and the calling process is referred to as the parent of the new process. A child process and its parent process run independently, but a child inherits many attributes from its parent.

POSIX provides the following system calls for starting up new processes:

- fork(). This call creates a new process which *duplicates* the calling process: (a) both processes execute the same application code; (b) their executions are synchronized at this call—both are about to return from the fork() call. The return value of the fork() call can be used to distinguish the calling process from the newly created child process: the return value is 0 in the newly created child process, while the return value is the child process's ID in the parent process. When the new process has just been created, it contains a single thread. If fork() is called from within a multithreaded process, the new process contains a replica of the calling thread only.
- exec() family of functions. The calling process image is *overlaid* with the new process image, which is constructed from an executable file. There shall be no return from a

successful exec. A call to any exec function from a process with more than one thread shall result in all threads being terminated.

- posix_spawn() family of functions. The fork() implementation normally relies on MMU services such as memory swapping or dynamic address translation, which is generally too slow for real-time applications. The posix_spawn() function (and its variants) is a simple, fast implementation without address translation or other MMU services. Upon successful completion, posix_spawn() returns the child process's ID to the parent process.
- system(). This call takes a command, starting up a shell (command language interpreter) to execute the command. This call does not return until the child process has terminated. It can be implemented by a posix_spawn() call, or alternatively by a fork() followed by an exec().

13.1.1.2 Thread creation

Upon being created, a process has a single main or initial thread. An active thread is instructed to execute program code, which may contain system calls to create additional threads. The POSIX function pthread_create() is used for creating a new thread within a process.

Threads within a process do not have a parent-child relationship with each other. A thread can terminate itself. If any thread of a process calls exit(), the process and all threads within it terminate.

13.1.2 POSIX Real-Time Extensions

Many real-time embedded systems have only limited resources and they do not need luxury features such as file systems or independent address spaces for tasks. For such a system, it is desirable to implement it upon an OS that supports only a particular subset of the POSIX functions. The *real-time profile standard* IEEE Std 1003.13-1998 was first published in 1998 to meet this need. It consists of four real-time application environment profiles:

(1) Minimum profile: for building small embedded systems without an MMU, or a file system, or an I/O terminal. Only one process but multiple threads are allowed.
(2) Minimum profile for real-time systems: for building real-time controllers with a file system and I/O terminals. Only one process but multiple threads are allowed.
(3) Limited profile for embedded systems: for building large embedded systems with no file system. Multiple processes and threads are allowed.
(4) Maximum profile: for building large real-time systems and embedded systems with all the features supported.

IEEE Std 1003.1-2008, the latest version of the POSIX.1 standard, also contains a few *amendments* or extensions related to real-time systems:

- IEEE Std 1003.1b-1993 Realtime Extension;
- IEEE Std 1003.1c-1995 Threads;
- IEEE Std 1003.1d-1999 Additional Realtime Extensions; and
- IEEE Std 1003.1j-2000 Advanced Realtime Extensions.

Below, we introduce a few POSIX.1 services specified in the real-time amendments.

13.1.2.1 Priority-based scheduling

Threads are schedulable entities (i.e. resources that host jobs). An OS scheduler selects at most one thread to execute on each processor at any point in time.

Each POSIX thread, upon its creation, is associated with a numeric value called its scheduling priority, which is used by an OS scheduler to determine when and how to schedule the thread. We consider three principles in priority-based scheduling.

(1) *Precedence principle.* When no thread is running, and multiple threads are ready to be scheduled for execution, a thread with a higher priority is always scheduled prior to threads with a lower priority.
(2) *Preemption principle.* When a thread is currently running, and another thread with a higher priority becomes ready for execution, the higher-priority thread preempts the execution of the current thread.
(3) *Fairness principle.* When multiple threads, with the same or different priorities, compete for use of the processing resources, the system's scheduling behavior is regulated by the scheduling policies associated with the threads.

The *precedence principle* is the basis of priority-based scheduling. But what happens if a thread becomes ready while a lower-priority thread is currently running? Let us first examine an approach called *nonpreemptive scheduling* as illustrated in Figure 13.1. Here each task is implemented as a thread. Task T_1 (i.e., the corresponding thread) has the lowest priority, task T_3 has the highest priority, and the priority of task T_2 is in the middle. The first job J_{11} of task T_1 is released at time 0. Since J_{11} is the only job, it is scheduled to execute. Jobs J_{21} and J_{31} are released at time 15 and time 20, respectively. However these two jobs, although each has a higher priority than J_{11}, will not preempt the execution of J_{11}. As soon as J_{11} finishes at time 25, J_{31} is scheduled to execute according to the *precedence principle*. J_{21} is scheduled at time 40 when J_{31} is finished. Notice that J_{21}'s deadline is broken by this schedule.

Figure 13.2 shows another scenario, where the *preemption principle* applies. The three tasks still have the same parameters as those given in Figure 13.1. As job J_{21} is released at time 15, it preempts job J_{11} according to the preemption principle. Similarly, at time 20, job J_{31} preempts job J_{21}, which resumes its execution at time 35 when J_{31} is finished. Similarly, J_{11} resumes its execution at time 40 when J_{21} is finished. According to this schedule, all three tasks can meet their deadlines. This schedule is called *preemptive scheduling*.

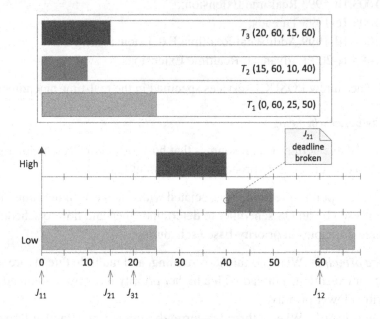

Figure 13.1
Nonpreemptive scheduling.

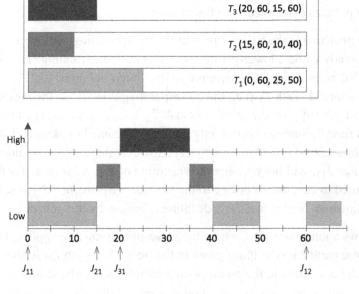

Figure 13.2
Preemptive scheduling.

The *fairness principle* applies when tasks (or threads) are also associated with scheduling policies. The POSIX.1-2008 standard specifies four scheduling policies: SCHED_FIFO, SCHED_RR, SCHED_SPORADIC, and SCHED_OTHER. These policies and the relevant structures are described in Table 13.2.

Each POSIX process/thread can be associated with a scheduling policy and attributes pertinent to the chosen scheduling policy. In particular:

- Upon its creation, a process can be specified with a scheduling policy and attributes (via a posix_spawn call), or can inherit the policy and priority settings of the parent process (via a fork call).
- Via an explicit call to the function sched_setscheduler() or sched_setparam(), a process can set/change the scheduling policy and/or attributes of its own or another process.
- Upon its creation (via a pthread_create call), a thread can be specified with a scheduling policy and scheduling attributes.
- Via an explicit call to the function pthread_setschedparam() or pthread_setschedprio(), a calling thread can set/change the scheduling policy and/or attributes of a thread within the same process.

The scheduling policy and attributes of a process have only an indirect effect, if any, on the scheduling behavior of individual threads. When a process is created, its single thread inherits the process's scheduling policy and the associated scheduling attributes. A newly created thread can also choose to inherit, if they are desirable, the scheduling attributes of the creating thread.

Table 13.2 POSIX.1 scheduling policies and relevant structures/functions

POSIX Definition	Description
SCHED_FIFO SCHED_RR SCHED_SPORADIC SCHED_OTHER	An integer constant indicating the FIFO policy An integer constant indicating the round-robin policy An integer constant indicating the sporadic server policy An integer constant indicating any other policy
struct sched_param	sched_priority: a numeric value for scheduling priority (a higher value represents higher priority) sched_ss_low_priority: low priority for SCHED_SPORADIC sched_ss_repl_period: replenishment period for SCHED_SPORADIC sched_ss_init_budget: initial budget for SCHED_SPORADIC sched_ss_max_repl: maximum replenishments for SCHED_SPORADIC
sched_setparam sched_setscheduler pthread_setschedparam pthread_setschedprio	Set scheduling parameters for the specified process Set scheduling policy and parameters for the specified process Set scheduling policy and parameters for the specified thread Set scheduling priority dynamically for the specified thread

The policies SCHED_FIFO, SCHED_RR, and SCHED_SPORADIC are discussed further in Section 13.4. The POSIX scheduling policy SCHED_OTHER is implementation specific; it may or may not be a real-time scheduling policy.

13.1.2.2 Task synchronization

The POSIX.1-2008 standard defines functions to manage *process synchronization* with semaphores. A semaphore is an OS kernel object that allows mutually exclusive access to shared resources. A POSIX semaphore does not prevent an undesired issue called "unbounded priority inversion," which occurs when a high-priority task has to wait for a lower-priority task to complete its actions protected by the semaphore. See Section 18.3 for the use of semaphores.

The POSIX.1-2008 standard also defines *mutexes* for thread synchronization within a process. A mutex object, if configured appropriately, can also be used for inter-process synchronization. A POSIX mutex has several important attributes:

- Ownership. When a call to pthread_mutex_lock() on a mutex object is successful, the mutex is locked by the calling thread, which becomes the owner of the mutex. A locked mutex can be unlocked only by its owner, via a call to pthread_mutex_unlock().
- Priority ceiling. This defines the minimum priority level at which the critical section guarded by the mutex is executed. In order to avoid priority inversion, at design time the priority ceiling of a mutex should be set to a priority higher than or equal to the highest priority of all the threads that may lock that mutex.
- Protocol. This can be set to one of three protocols by a call to the function pthread_mutexattr_setprotocol(): PTHREAD_PRIO_NONE, PTHREAD_PRIO _INHERIT, and PTHREAD_PRIO_PROTECT. The default protocol value is PTHREAD_PRIO_NONE.

In brief, the priority and scheduling of a thread owning a mutex can be affected as follows:

- The priority and scheduling of a thread is not affected by owning a mutex with the PTHREAD_PRIO_NONE protocol attribute.
- When a thread is blocking higher-priority threads because of owning one or more mutexes with the PTHREAD_PRIO_INHERIT protocol attribute, it shall execute at the higher of its priority and the highest priority of the threads waiting on any of the mutexes owned by this thread and initialized with this protocol.
- When a thread owns one or more mutexes initialized with the PTHREAD_PRIO_ PROTECT protocol, it shall execute at the higher of its priority and the highest of the priority ceilings of all the mutexes owned by this thread and initialized with this protocol, regardless of whether other threads are blocked on any of these mutexes or not.

- If a thread simultaneously owns several mutexes initialized with different protocols, it shall execute at the highest of the priorities that it would have obtained by each of these protocols.

Unbounded priority inversion may be avoided by using mutexes with a PTHREAD_PRIO_INHERIT or PTHREAD_PRIO_PROTECT protocol. See Section 14.5 for details of these protocols. See Section 18.3 for the use of mutexes.

13.1.2.3 Other services

Figure 13.3 gives some objects commonly implemented in the kernel of an OS.

POSIX defines a signaling mechanism for asynchronous interprocess communications: while a task can specify the particular actions to execute when a signal arrives, the task cannot predict when a signal might arrive. POSIX signals can be queued at the receiving process, so that events or requests will not get lost. Each signal is associated with a priority; real-time signals are dequeued in priority order, which allows urgent events to be processed earlier and responded faster. The details of signal processing are covered in Chapter 21.

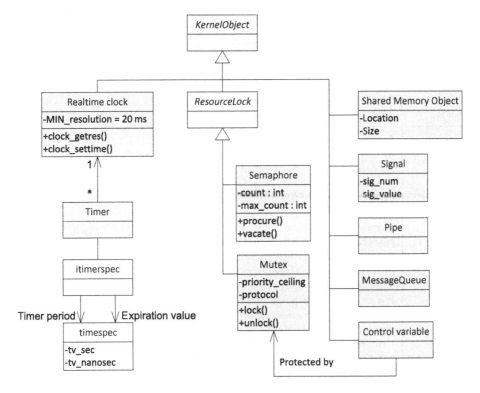

Figure 13.3
Some OS kernel objects.

The POSIX.1-2008 standard requires that an OS implementation should support at least a real-time clock identified by the constant CLOCK_REALTIME, which, by default, represents the amount of time (in seconds or nanoseconds) since the epoch. Real-time clocks can tick at a finite rate, say 1000 Hz. A clock's resolution, also called its granularity, is the inverse of its tick rate. For example, a clock with a tick rate of 1000 Hz has a resolution of 1 ms. See Chapter 22 for more information on time resolution.

Table 13.3 gives some POSIX functions for the use of real-time clocks. For a specific real-time clock, a user can get its resolution, get the current time, or set a new time for the clock. A user can also suspend the execution of a thread by calling sleep functions, which can take relative or absolute time values, and a time value can be in seconds (low resolution) or nanoseconds (high resolution). The remaining time is returned if a relative sleep function is interrupted by a signal. In such a case, if necessary, the remaining time can be used to reissue

Table 13.3 POSIX.1 structures and functions for timers and clocks

POSIX Definition	Description
struct timespec	long tv_sec: seconds long tv_nsec: nanoseconds
struct itimerspec	timespec it_interval: timer period timespec it_value: timer first expiration
timer_create timer_gettime timer_settime	Create a timer for the calling process Get a timer's current time intervals Set/reset an existing timer's time intervals
clock_getres clock_gettime clock_settime sleep nanosleep clock_nanosleep	Return the resolution of a real-time clock Return the current value (struct timespec) for the specified clock Set a new value for the specified clock A relative low-resolution sleep function, which causes the calling thread to be suspended from execution until either the requested time interval (in seconds) has elapsed, or the thread is interrupted by a signal[a] A relative high-resolution sleep function, which causes the calling thread to be suspended from execution until either the requested time interval (in nanoseconds) has elapsed, or the thread is interrupted by a signal. In the latter case, the remaining time is also returned (the requested time minus the time actually slept) This high-resolution sleep function can be in two modes It is an absolute sleep function if the "absolute" flag is set. It causes the current thread to be suspended from execution until either the referenced clock reaches the requested absolute time, or the thread is interrupted by a signal It is a relative sleep function if the "absolute" flag is *not* set. It causes the current thread to be suspended from execution until either the requested time interval has elapsed, or the thread is interrupted by a signal. In the latter case, the remaining time is also returned (the requested time minus the time actually slept)

[a] It eventually makes a kernel call that starts a relative timer.

another function call to meet the original time requirement. If an absolute sleep function (i.e., when clock_nanosleep() is invoked in absolute mode) is interrupted by a signal, it can be simply invoked again with the same absolute time as used in the first call.

POSIX real-time clocks can be used to control the execution of real-time tasks to meet their timing requirements. In particular, there are tasks (threads) in real-time systems that need to be suspended and then activated multiple times in a periodic way. For instance, a display device may need to be refreshed periodically or the status of a noninterrupting device may need to be polled periodically. Such a periodic behavior can be implemented by using real-time clocks. For example, suppose a periodic task $T_1 = (p_1, e_1)$ starts to execute at time t_0 (this is an absolute time instant) and runs to completion at time $t_1 = t_0 + e_1$. It needs to suspend itself until the start of the next period, which is at time $t_0 + p_1$. This can be achieved "precisely" by a call to the clock_nanosleep() function with the absolute time instant $t_0 + p_1$ as the requested time parameter.

It is worth noting that "precise periodic activation" may not be achieved via calls to a relative sleep function, such as sleep(), nanosleep(), or the relative version of clock_nanosleep(). For example, if task T_1 uses nanosleep() instead, prior to calling the nanosleep() function, it must first call clock_gettime() to get the current time, then calculate the difference between the current time and $t_0 + p_1$, and finally call nanosleep() using the computed interval. However, the task could be preempted by other tasks between the two function calls. In such a case, the computed interval would be wrong, and task T_1 would wake up later than desired. This problem would not occur with the absolute clock_nanosleep() function, since only one function call would be necessary to suspend the task until the desired time.

POSIX *timers* can also be used to control the execution of real-time tasks to meet their timing requirements. A process can create many timers[3] to count certain time intervals. The timers created by one process cannot be inherited by its child process.

A timer relies on the use of signals,[4] and it has to reference a real-time clock as its timing basis. As shown in Figure 13.3, a timer can have two time intervals specified: timer period and expiration value, the former is used to define a periodic timer, and the latter specifies a time-out interval for a one-shot timer or the first expiration value for a periodic timer. When the specified time-out interval has elapsed, a timer can generate a "time-out" signal directed to the process that created the timer. For a given timer, at any point in time only a single time-out signal can be queued to the process.

A periodic timer can generate an unspecified number of time-out signals. More information on the use of timers can be found in Chapters 15 and 22.

---

[3] A timer object can be simply an integer that acts as an index in the OS kernel's timer tables.

[4] By default, when a timer expires, a signal of type SIGALRM is raised to the installing process.

The POSIX.1-2008 standard defines *message queues* as an asynchronous interprocess communication mechanism. Processes connecting to a message queue can send structured messages to or read messages from it. Transmission and reception are not synchronized in the sense that the sender does not wait until the receiver has actually retrieved the message from the queue. Transmission and reception of messages can be blocking or nonblocking. POSIX functions for message queues and their uses are described in Chapter 19.

The POSIX.1 standard also specifies memory-mapped files and *shared memory objects* as means for processes to share portions of physical memory. This is very useful when real-time applications need to share large amounts of data with very little overhead. Memory-mapped files provide a mechanism that allows a process to directly map the contents of a file into its address space; file data access is very fast owing to direct memory manipulation. If more than one process maps a file, its contents are shared among them. In particular, data written into the memory object through the address space of one process appear in the address spaces of all processes that map the same portion of the memory object. See Chapter 18 for the use of shared memory objects.

13.1.3 POSIX Compliance and Conformance

POSIX *implementation compliance* means that an OS implementation partially supports the POSIX.1 standard, and documentation is available that shows which POSIX features are supported and which are not.

POSIX *implementation conformance* means that an OS implementation fully supports all the mandatory system interfaces (functions and headers), utilities, and facilities defined within the POSIX.1 standard, as well as some standard extensions.

Depending on the degree of compliance with the POSIX.1 standard, one can classify OSs as partially POSIX.1 compliant or fully (100%) POSIX.1 compliant (conformance). Some OSs, such as Solaris and QNX [8], are fully compliant, and others, such as GNU/Linux and VxWorks, are not fully compliant.

POSIX *application conformance* means that an application strictly conforms to the POSIX.1 standard in the sense that it relies only on the features described in the standard. Application conformance is critical to support software interoperability and portability. Code written for one POSIX-compliant OS will generally port to another POSIX-compliant OS, and can even port to a partially compliant OS with reduced costs. In addition, programmers experienced with one POSIX-compliant OS can directly apply their skill sets to projects involving other POSIX-compliant OSs.

13.2 Task Statics and Dynamics

Many OSs support multitasking, where multiple tasks (threads) compete for the limited computing resources. Tasks within an OS can generally be classified into two types: *system tasks* and *application tasks*. System tasks, such as the OS kernel scheduler and device drivers, offer critical services to application tasks.

In this section, we examine the static structure and dynamic behavior of a task in general. Note that the term "task" is an abstract concept that simply provides a coherent context for us to understand the relevant POSIX concepts; it manifests itself as a process or thread in an OS implementation.

13.2.1 General Task Structure

Figure 13.4 illustrates a UML class diagram showing the static structure of the "task" concept:

*   A task is a schedulable entity. To implement the SchedulableEntity interface, a task has attributes such as a scheduling policy and an assigned priority. A task may temporarily operate with a different priority—the actual priority. A task may also be specified with other parameters pertinent to the scheduling policy chosen.
*   A task can have task parameters such as period, release time, execution time, and deadline. A task may be a periodic task, an aperiodic task, or a sporadic task. At any point in time, a task can be in one of four states: ready, running, blocked, or sleeping.
*   Task is an abstract concept. A thread is (implements) a task. The TID of a thread is an integer number that uniquely identifies the thread.
*   A thread is contained within a process. Each thread is associated with a task routine, where the program counter refers to the current instruction of the routine.
*   A thread has an execution stack, which represents the thread's current execution state. The stack is composed of many stack frames formed during function calls. Each stack frame is composed of many objects, each of which holds the value of a certain register. The execution stack is critical for *context switching*, where the currently running task is suspended and swapped with another task (this replacement task becomes the new running task). During context switching, the values of all the registers used by the processor are saved into the execution stack of the currently running task, and the execution stack of the replacement task, if not empty, is used to restore its latest execution state.
*   A thread also has a control block object which holds thread-level information involved with scheduling, resource access, and performance monitoring, such as the accumulated run time, blocked time, masks for blocked signals, a queue of pending signals, and a list of owning mutexes. A thread control block may also have pointers to the control blocks of other threads, for the sake of implementing dynamic features such as scheduling and blocking.

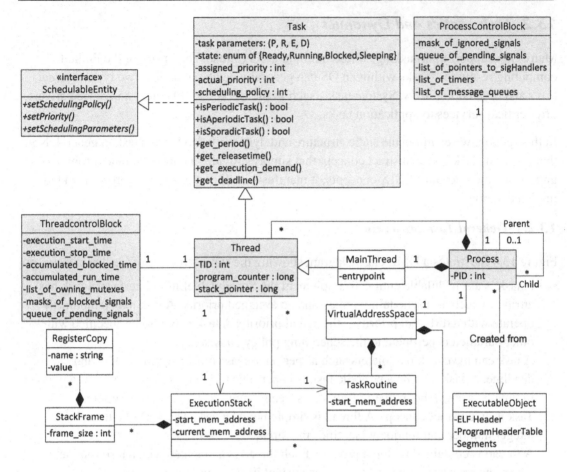

Figure 13.4
Task static structure (class diagram).

- A process is a container for many threads. In particular, each process has a main thread, which is a special thread with an execution entry point.
- A process has an attribute PID—an integer number that uniquely identifies the process. A process can have many child processes, and it may have one parent process.
- The image of a process is created from an executable file object. Once a process image has been formed, it has its own virtual address space, independent of other processes.
- The virtual address space of a process contains an execution stack for each thread running within it, as well as the task routine of each thread.
- A process also has a control block object which holds process-level information used for efficient process management, such as masks of ignored signals, a queue of pending signals, pointers to signal handlers, and process-level resources such as I/O devices, timers, and message queues used by the process.

13.2.2 Task State Transition

At any point in time, a task (thread) can be in one of four states: ready, running, sleeping, or blocked. Before we discuss state transitions, we first introduce some data structures used by an OS to facilitate thread (task) scheduling:

- An *ordered list* is a data structure that contains no, one, or many nodes, which are ordered in a specific way. The node at one end of an ordered list is called the head node, and the node at the other end is the tail node. Existing nodes can be removed from a list and new nodes can be added to a list.
- A *time-ordered list* is a list structure where nodes are ordered by the time each has been on the list. Generally, the head of the list is the node that has been on the list the longest time, and the tail is the node that has been on the list the shortest time.
- A *blocked-thread list* is an ordered list where each node represents a thread that is in the blocked state.
- A *sleeping-thread list* is an ordered list where each node represents a thread that is in the sleeping state.
- A *ready-thread list* is a time-ordered list where each node represents a thread that is in the ready state (i.e., ready for execution).
  - Since multiple tasks might operate with the same priority, there is conceptually one ready-thread list for each priority level. A *level i* ready-thread list contains only threads with priority i.
  - A level i ready-thread list is said to be the highest-priority nonempty ready-thread list (HNERT) if it contains at least one node, and for any $j > i$, the level j ready-thread list is empty.

Figure 13.5 illustrates a UML state diagram showing the dynamic state transitions of a task (thread). It has the following transition rules:

(1) A thread, upon its creation, is ready for execution. It is placed at the tail of the ready-thread list corresponding to its priority.

(2) At any point in time, there is only one running thread on a single-processor system. When there is currently no thread running, for an OS conforming to the POSIX.1 standard, it will select the thread that is at the head of HNERT, removing it from HNERT and then voluntarily transferring the execution control to it.

(3) When the running thread's priority i is no longer the highest—say, a task with a priority higher than i becomes ready—it is *preempted* and becomes the *head* of the level i ready-thread list (this rule conforms to the SCHED_FIFO policy). Another thread is selected as the new running thread.

(4) When the running thread, with priority i, has used up its time quantum enforced by its scheduling policy, it becomes the *tail* of the level i ready-thread list (this rule conforms to the SCHED_RR policy). Another thread is selected as the new running thread.

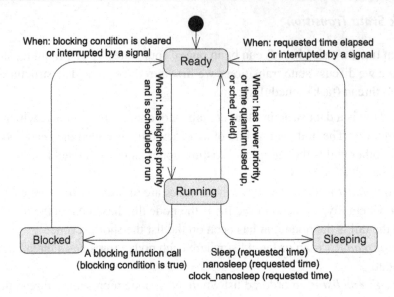

Figure 13.5
Task state diagram.

(5) The running thread is blocked when it calls a blocking function and the associated blocking condition is true. Examples include procuring a semaphore or mutex that is not available, reading from an empty pipe, writing to a pipe with available space less than requested, sending messages to a full message queue, and receiving from an empty message queue. A blocked thread is placed into the blocked-thread list, and another ready thread is selected as the new running thread.

(6) When a blocked thread is unblocked, either the blocking condition is cleared or it is interrupted by a signal,[5] the thread is removed from the blocked-thread list and becomes the *tail* of the ready-thread list corresponding to its priority.

(7) When the running thread voluntarily suspends its execution by calling a sleep function, it starts to sleep and is placed into the sleeping-thread list. Another thread is selected as the new running thread.

(8) When a sleeping thread wakes up, either the requested sleep time has elapsed or it is interrupted by a signal, the thread is removed from the sleeping-thread list and becomes the *tail* of the ready-thread list corresponding to its priority.

(9) When the running thread calls the sched_yield() function, it becomes the tail of the ready-thread list corresponding to its priority, and another thread is selected as the new running thread.

---

[5] Since a blocking function call may be interrupted by a signal, it is good practice to put such a call inside a loop and repeat the call until it returns successfully (i.e., the blocking condition is truly cleared).

(10)   When a running or ready thread's priority has been modified by a call to pthread_setschedprio(),

    (a)   if its priority is raised, the thread becomes the tail of the ready-thread list corresponding to its new priority;

    (b)   if its priority is lowered, the thread becomes the head of the ready-thread list corresponding to its new priority.

(11)   When a running or ready thread's priority or policy has been modified by a function other than pthread_setschedprio(), it becomes the tail of the ready-thread list corresponding to its new priority.

13.3 Real-Time OSs

A general purpose OS is system software that manages user applications and the hardware resources of a computer, defining rules and programming interfaces that allow a program to request OS services and interact with the rest of the system [47, 69]. A general purpose OS that implements some real-time features can be used for soft real-time applications.

For hard real-time systems, however, a Real-Time OS (RTOS), would be necessary in order to respond with *bounded response time* to external events which are typically associated with hard timing constraints (deadlines). For example, a computer numerical control (CNC) machine tool, at a low level, operates stepwisely: the step operation is a periodic task (say, with a period of 1 ms). Within each step, a CNC machine tool has to monitor the current position of the cutting head, calculate the trajectory on the fly, and then drive the head to the next position. In order to follow the programmed cutting patterns precisely, a CNC machine tool certainly requires an RTOS to produce accurately timed sequences of pulses to control the head movement.

What, then, is an RTOS? Undoubtedly, an RTOS is an OS in the first place, composed of multiple software subsystems with the kernel at its core. Depending on the implementation, an RTOS may offer many services commonly found in general-purpose OSs, such as multitasking, priority, task preemption, resource sharing, and intertask communication.

Of course, an RTOS is more than a general-purpose OS. An RTOS, with specialized scheduling algorithms, is intended to run real-time applications with soft and/or hard deadlines. Thus, the key characteristic of an RTOS is that its response time to an event should be *deterministic* (or predictable). In other words, an RTOS should provide reliable mechanisms, such as real-time signals, preemptive scheduling, and nonblocking interprocess communications, for enabling a deterministic response [43, 57].

Figure 13.6 illustrates a potential execution path in response to an event. Let us use Figure 13.6 to examine those factors that may affect the response time of a system to an external event:

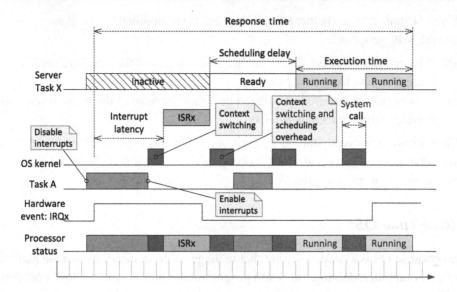

Figure 13.6
A potential execution path in response to an external event.

- The response time to an event from a hardware device X refers to the time interval between the time instant when an interrupt request is raised and the time instant when the request is completely serviced by the server task of device X. It consists of the interrupt latency, the execution time of the interrupt service routine (ISR) for device X, a scheduling delay, and the execution time of the server task for device X.

- Interrupt latency refers to the time interval between the time instant when an interrupt request is raised and the time instant when the first instruction of the corresponding ISR starts to execute. Part of the interrupt latency might be caused by interrupts being disabled by software, as shown in Figure 13.6, or the execution of some ISR for a higher-priority device. This part is completely in the hands of the system designer. No RTOS can compensate for bad designs. A good practice is to keep the code section that disables interrupts as short as possible. Also, priorities for hardware devices should be designed appropriately such that the interrupt latency of critical devices should be reasonably optimized.

- The last portion of the interrupt latency is caused by context switching, where the current execution context should be saved and swapped with another task (which is an ISR in this case). The context switching time is an important performance indicator of an RTOS.

- An RTOS provides default implementation for ISRs (and signal handlers). A user may patch existing ISRs or install new ISRs (or signal handlers). An RTOS is responsible for the execution time of a default ISR (signal handler); a user is responsible for his/her portion.

- The scheduling delay is the time interval where the server task is ready but has not started to execute. It might be caused by the execution of a high-priority task, as well as context switching and scheduling overhead. The scheduling overhead refers to the execution time of the kernel scheduler. While task preemption is mostly determined at design time (in task priority assignment), an RTOS has the responsibility to reduce the scheduling overhead.
- The execution time of the server task depends on the length of the task routine. The code of a task (including the ISR) may invoke system calls to request various OS services. The response time to system calls is another important performance indicator of an RTOS. In addition, a task may communicate with other tasks; each intertask communication introduces a communication delay.

In sum, an RTOS typically achieves a *deterministic* response by

- supporting preemptive priority scheduling: a higher-priority task always preempts a lower-priority task, regardless of user tasks or system tasks.
- offering *bounded latency* on context switching time, scheduling overhead, and system calls. Bounded latency means having a measurable guarantee on latency, which implies that the response time should be
  - short and predictable: context switching time, scheduling overhead, and the completion of system calls should be short and within known latencies.
  - with low jitter: the variance of the response times (for context switching, for scheduling overhead, and for each type of system call) under all possible circumstances should be very small. For instance, a system call may react on average in 10 ms but can have spikes up to 20 ms. The worst-case values are often used in schedulability analysis.
- offering efficient ISRs: the execution time of a default ISR (signal handler) should be short and bounded.
- supporting deterministic synchronization: multiple tasks can communicate within a predictable time.

The most widely adopted RTOS include RTLinux, Windows CE, LynxOS, VxWorks, and QNX, to mention only a few [24]. For example, VxWorks is a commercial RTOS that supports preemptive priority scheduling with 256 priority levels, as well as semaphores, message queues, and high-speed intertask communications. VxWorks is available for many processor platforms, including x86, PowerPC, ARM, MIPS, Pentium, and SPARC. QNX Neutrino [8] is widely used as the basis for many embedded real-time applications such as automotive electromechanical components, medical instruments, defense systems, and nuclear power plants. QNX is built upon a microkernel architecture, as shown in Figure 13.7, which allows a system to be scaled to very small sizes and still provide multitasking, preemptive scheduling, and fast context-switching. QNX Neutrino runs on many modern processor platforms, including PowerPC, x86, MIPS, ARM, and XScale.

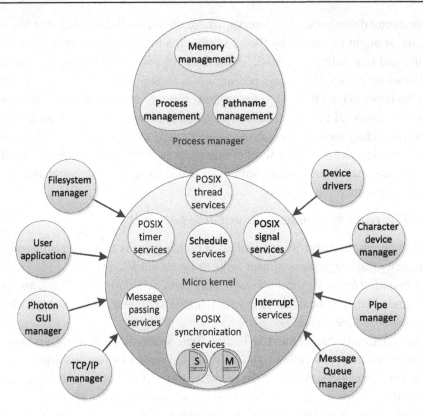

Figure 13.7
The QNX microkernel architecture.

As we already know, an RTOS represents only one type of architecture for embedded software. You do not have to use an RTOS to implement a real-time embedded system that is small in size and functionality or with only soft timing constraints. If using an RTOS is the best option for a mission-critical system, in order to make a "wise" selection among the RTOSs available on the market, you may need to consider factors such as the real-time characteristics (bounded latency), portability (processor support), scalability (OS memory usage), and support (development tools, device drivers, and board support packages), as well as your budget constraints.

13.4 POSIX Real-Time Scheduling Policies

In Section 13.1.2.1, we discussed the precedence principle and the preemption principle, which apply when the tasks ready for execution have distinct priorities. The purpose of a scheduling policy is to define fairness rules for tasks with the same priority (say,

SCHED_FIFO and SCHED_RR) or for asks with a bounded execution budget (say, SCHED_SPORADIC).

13.4.1 FIFO Scheduling Policy

Since the SCHED_FIFO (for "First In First Out") scheduling policy is only applicable for scheduling tasks with the same priority, we focus here on a level k ready-thread list—tasks with a priority k.

Suppose the level k ready-thread list has the following threads: $[T_1, T_2, \ldots, T_n]$. We use T_0 to refer to the running thread if it has the same priority k. According to the definition of ready-thread list, if $1 \leq i < j \leq n$, T_i must have been on the list for a time longer than T_j.

For tasks on a level k ready-thread list, the SCHED_FIFO policy states that a task T_i can never be preempted by any task T_j ($j > i$). In other words, a task T_{i+1} can only be scheduled to execute until after T_i has run to completion.

Figure 13.8 gives an example, where the system consists of three periodic tasks with the same priority.

Job J_{11} is ready (released) at time 0. Since T_1 is the only task on the ready list, it becomes the running thread. At time 1, job J_{21} is released and is added to the ready list. At time 2, J_{11} finishes, and J_{21} becomes the new running job. Also, job J_{31} is released at time 2 and is added to the ready list. At time 4, J_{21} finishes, and J_{31} becomes the new running job. At time 6, job J_{12} is released and is added to the ready list. As an exercise, you can continue this analysis. One thing is clear: no job released later has ever preempted a job released earlier.

Note that job J_{14} has its deadline broken at time 24.

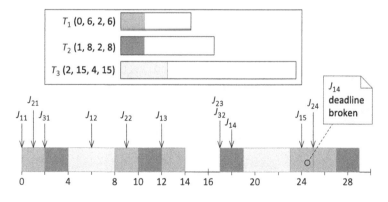

Figure 13.8
An example illustrating the SCHED_FIFO policy.

13.4.2 Round-Robin Scheduling Policy

Similarly to SCHED_FIFO, the SCHED_RR (round-robin) scheduling policy is only applicable for scheduling tasks with the same priority. For an OS implementing the SCHED_RR policy, the length of a time slot, also called a *time quantum*, is returned by the POSIX function sched_rr_get_interval().

The SCHED_RR policy states that (1) all the tasks on each ready-thread list are scheduled in turn according to their orders; (2) when a new running thread needs to be selected, if the level k ready-thread list is HNERT, the head thread is removed from the list and becomes the running thread; (3) a running thread, as soon as it has used up its time quantum, becomes the tail of the ready-thread list for its priority.

Figure 13.9 gives an example, where the system consists of three periodic tasks with the same priority.

Job J_{11} is released at time 0 and becomes the running job. At time 1, job J_{21} is released and is added to the ready list. Starting from this time instant, the two jobs, J_{11} and J_{21}, will take turns to execute. At time 2, job J_{31} is released and also joins the round-robin game. A job is out of the round-robin game as soon as it runs to completion.

The SCHED_RR policy virtually offers a time-sharing mechanism, ensuring that a task will not monopolize the processor when there are other threads with the same priority. Such a mechanism can be implemented by using the clock (timer) interrupts.

Also note that if a system has tasks with different priorities, the round-robin scheduling of tasks on a ready-thread list could be preempted ("interrupted") by the execution of tasks with a higher priority.

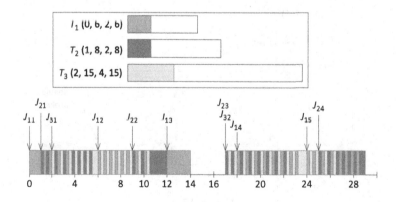

Figure 13.9
An example illustrating the SCHED_RR policy.

13.4.3 Sporadic Server Scheduling Policy

The scheduling policy SCHED_SPORADIC (sporadic server) allows a task to change its priority at run time.

We focus on a specific thread T_i that uses the SCHED_SPORADIC scheduling policy. T_i is specified with the following scheduling parameters:

- a normal priority, denoted by μ;
- a low priority, denoted by ω, where $\omega < \mu$;
- a replenishment period, denoted by p; and
- an initial execution budget, denoted by e.

The thread T_i obtains its initial execution budget e upon its creation. The budget value e will be consumed as T_i is running. The *effective priority* of the thread T_i is μ when $e > 0$, and ω otherwise. Depending on its effective priority, when T_i is ready, it shall be on either the level μ ready list or the level ω ready list.

The execution budget of T_i can be replenished by replenishment actions. Each replenishment action is defined by $\langle t, v \rangle$, where t is the time to perform this replenishment action, and v is the amount to be replenished.

The scheduling behavior of T_i is regulated by the following rules:

(1) Each time T_i is placed at the *tail* of the level μ ready list, that time instant is recorded by τ. τ is called T_i's *latest activation time*. Upon its creation, T_i has $\tau = 0$. τ changes over time.

(2) When T_i is running with priority μ, its budget is consumed at the rate of one per unit of execution time.

(3) When the budget of T_i, running with priority μ, becomes 0, it becomes the tail of the level ω ready list. The OS schedules a replenishment action $\langle t, v \rangle$, where $t = \tau + p$, and v equals the budget consumed since the latest activation time τ. If t is before the current time, the replenishment action is carried out immediately.

(4) When T_i, running with priority μ, becomes a blocked thread, the OS schedules a replenishment action $\langle t, v \rangle$, where $t = \tau + p$, and v equals the budget consumed since the latest activation time τ. If t is before the current time, the replenishment action is carried out immediately.

(5) When T_i, running with priority μ, is preempted, it becomes the *head* of the level μ ready list. In this case τ is not changed.

(6) When T_i is at the head of the level ω ready list and this list is HNERT, it becomes a running thread. Its budget remains unchanged when T_i is running with priority ω.

(7) When it is time to perform a replenishment action $\langle t, v \rangle$ (i.e., the current time is not earlier than t), T_i's budget is changed: $e = e + v$. If T_i's effective priority is ω, it is changed to μ, and T_i becomes the *tail* of the level μ ready list.

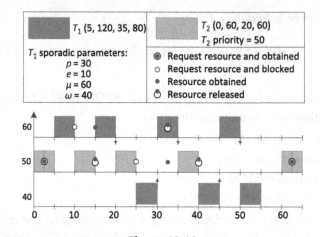

Figure 13.10
An example illustrating the SCHED_SPORADIC scheduling policy.

As an example, consider a system consisting of two tasks (threads) T_1 and T_2 (assume that the execution overheads of system tasks can be ignored if an RTOS is used). The two tasks share a mutually exclusive resource protected by a semaphore. As shown in Figure 13.10, task T_1's release time, period, execution-time demand, and deadline are 5, 120, 35, and 80, respectively; task T_2's task parameters are 0, 60, 20, and 60, respectively. Task T_2 has a fixed priority of 50; task T_1 uses the SCHED_SPORADIC policy with parameters as follows:

- $\mu = 60$;
- $\omega = 40$;
- $p = 30$; and
- $e = 10$.

The scheduling scenario given in Figure 13.10 is explained below:

(1) At time 0 when the system starts, thread T_2 is ready and becomes the running thread. At time 2.5, T_2 requests and has successfully obtained the resource. We also have $\tau = 0$.
(2) At time 5, thread T_1 is ready. It preempts the execution of T_2 and becomes the new running thread.
(3) From time 5 to 10, T_1 executes with priority 60. Its execution budget reduces to 5 at time 10.
(4) At time 10, T_1 requests the resource which is currently locked by T_2. T_1 is blocked and placed on the blocked-thread list. The OS schedules a replenishment action $\langle t, v \rangle$, where $t = \tau + p = 0 + 30 = 30$, and $v = 5$, which is the budget consumed since time 0. At this time point, T_2 is the only thread ready for execution, so it becomes the running thread.
(5) From time 10 to 15, T_2 executes.

(6) At time 15, T_2 releases the resource, which causes T_1 to be removed from the blocked list and become the tail of the level 60 ready list. Now, we have $\tau = 15$. In addition, T_1 immediately preempts T_2 and becomes the running thread.

(7) From time 15 to 20, T_1 executes with priority 60. Its execution budget becomes 0 at time 20.

(8) At time 20, since it has consumed all its budget, T_1 becomes the tail of the level 40 ready list. The OS schedules a replenishment action $\langle t, v \rangle$, where $t = \tau + p = 15 + 30 = 45$, and $v = 5$, which is the budget consumed since time 15. At this time point, T_2 is the head of HNERT, and becomes the running thread.

(9) From time 20 to 25, T_2 executes.

(10) At time 25, T_2 requests the resource which is currently locked by T_1. T_2 is blocked and placed on the blocked-thread list. At this time point, T_1 is the only thread ready for execution, so it becomes the running thread.

(11) From time 25 to 30, T_1 executes with priority 40. Its execution budget remains 0.

(12) At time 30, the OS performs the replenishment action $\langle 30, 5 \rangle$, T_1's budget becomes 5 and its priority is restored to 60. Also, we have $\tau = 30$.

(13) From time 30 to 35, T_1 executes with priority 60. Its execution budget becomes 0 at time 35. At time 32.5, T_1 releases the resource, which causes T_2 to be removed from the blocked list and become the tail of the level 50 ready list.

(14) At time 35, since it has consumed all its budget, T_1 becomes the tail of the level 40 ready list. The OS schedules a replenishment action $\langle t, v \rangle$, where $t = \tau + p = 30 + 30 = 60$, and $v = 5$, which is the budget consumed since time 30. At this time point, T_2 is the head of HNERT, and becomes the running thread.

(15) From time 35 to 40, T_2 executes.

(16) At time 40, T_2 releases the resource and runs to completion. At this time point, T_1 is the only thread ready for execution, so it becomes the running thread.

(17) From time 40 to 45, T_1 executes with priority 40. Its execution budget remains 0.

(18) At time 45, the OS performs the replenishment action $\langle 45, 5 \rangle$, T_1's budget becomes 5 and its priority is restored to 60. Also, we have $\tau = 45$.

(19) From time 45 to 50, T_1 executes with priority 60. Its execution budget becomes 0 at time 50.

(20) At time 50, since it has consumed all its budget, T_1 becomes the tail of the level 40 ready list. The OS schedules a replenishment action $\langle t, v \rangle$, where $t = \tau + p = 45 + 30 = 75$, and $v = 5$, which is the budget consumed since time 45. At this time point, T_1 is the only ready thread, and becomes the running thread.

(21) From time 50 to 55, T_1 executes with priority 40. Its execution budget remains 0.

(22) At time 55, T_1 runs to completion.

(23) At time 60, the second job of T_2 is released and starts to run.

13.5 Other Real-Time Scheduling Policies

13.5.1 Minimum Laxity First

The *laxity* of a task is defined as the difference between the amount of time left (by its deadline) and the amount of time still needed to complete the task. More formally, given a task $T_i = (p_i, r_i, e_i, d_i)$, its slack time is defined as

$$(d_i - t) - e'_i,$$

where t is the time elapsed since the start of the current cycle, and $e'_i \leq e_i$ is its remaining demand for execution time.

Laxity is a measure of the flexibility available for scheduling a task. A laxity of t means that even if the task is delayed by t time units, it can still meet its deadline. A laxity of zero means that the task must begin to execute now or it will risk failing to meet its deadline.

The *minimum laxity first* scheduling policy, also known as least laxity first or least slack time scheduling, is a dynamic-priority approach, where a task with a smaller laxity has a higher priority. Obviously, the task with the minimum laxity has the highest priority.

To implement this policy, the OS scheduler needs to continuously monitor the laxities of all ready jobs and compare them with the laxity of the running job. It reassigns priorities to jobs whenever the relative order of their laxities changes, allowing jobs with equal laxities to execute in a round-robin manner. Obviously, the run-time overhead due to continuous computation is significant, and too much round-robin scheduling incurs unnecessary context-switching cost. In practice, an OS may implement a nonstrict approach, where the laxities of ready jobs are checked only at the time when a job is released or completed.

Figure 13.11 gives an example scheduling that conforms to the nonstrict minimum laxity first policy. The table at the bottom of Figure 13.11 gives the job laxities at the corresponding "critical" time instants.

At the beginning when the system starts, the first jobs of the three tasks are released, and their respective laxities are 2.5, 7, and 9. Since J_{11} has the minimum laxity, it gets the highest priority and runs to completion at time 1.5. The laxities of ready jobs are checked at time 1.5, and J_{21} has the minimum laxity. Thus, J_{21} gets the highest priority and runs until time 4, at which time J_{12} is released. This renders an opportunity to recheck the laxities of ready jobs. Newly released job J_{12} has the minimum laxity; it thus gets the highest priority and runs to completion at time 5.5. As an exercise, trace the scenario to see how the job laxities change and how the changes affect the scheduling of the three tasks.

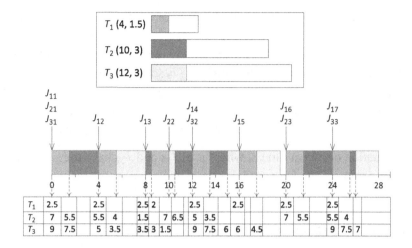

Figure 13.11
Minimum laxity first.

13.5.2 Earliest Deadline First

Another commonly implemented approach is called *earliest deadline first* (EDF), which is also a dynamic-priority scheduling policy. The EDF policy requires that a task (job) with an earlier deadline has a higher priority. Obviously, the task with the earliest deadline has the highest priority.

Figure 13.12 gives an example EDF-based scheduling. The table at the bottom of Figure 13.12 gives the job deadlines at the corresponding "critical" time instants.

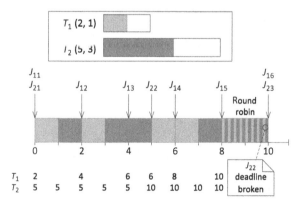

Figure 13.12
Earliest deadline first.

At the beginning when the system starts, the first jobs of the two tasks are released, and their respective deadlines are 2 and 5. Since J_{11} has an earlier deadline, it gets a higher priority and runs to completion at time 1. At time 1, J_{21} is the only ready job; it runs until time 2, at which time job J_{12} is released. With an earlier deadline, J_{12} is scheduled to run, and it finishes at time 3. J_{21} resumes its execution at time 3 and continues its execution until time 5, at which time job J_{21} finishes and job J_{22} is released. Meanwhile, job J_{13} is released at time 4 with a deadline of 6. At time 5, J_{13} starts to run because it has an earlier deadline than job J_{22}.

An interesting situation occurs at time 8, where jobs J_{15} and J_{22} have the same deadline. Since they have the same priority, the OS schedules them in a round-robin manner. Job J_{15} finishes at time 10; job J_{22}, however, breaks its deadline.

13.5.3 Scheduling with Deadline-Monotonic Assignment

The *deadline-monotonic assignment* approach is a static-priority scheduling policy, where priorities are assigned to tasks according to their deadlines known at design time: a higher priority is assigned to a task with an earlier deadline. Unlike the EDF approach, task priorities in deadline-monotonic assignment are fixed.

Figure 13.13 gives an example of deadline-monotonic-assignment-based scheduling. According to their deadlines, task T_2 has the highest priority, task T_3 has the lowest priority, and the priority of task T_1 in the middle.

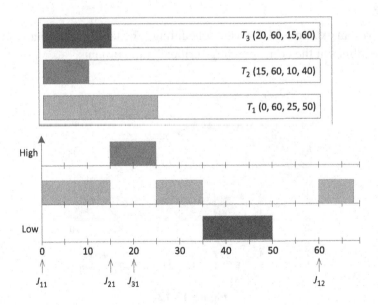

Figure 13.13
Deadline-monotonic scheduling.

At the beginning when the system starts, job J_{11} is released and starts to run. At time 15, J_{21} is released, and preempts J_{11}. Job J_{31} is released at time 20 and becomes the tail of the ready list for its priority. As soon as J_{21} finishes at time 25, J_{11} becomes the running job because it has a higher priority than J_{31}. At time 35, J_{11} finishes and J_{31} starts to run.

13.5.4 Scheduling with Rate-Monotonic Assignment

The rate-monotonic assignment approach is also a static-priority scheduling policy [49, 65], where priorities are assigned to tasks according to their periods known at design time: a higher priority is assigned to a task with a shorter period (i.e., higher rate). Like deadline-monotonic assignment, task priorities in rate-monotonic assignment are fixed.

Figure 13.14 gives an example of rate-monotonic-assignment-based scheduling. According to their periods, task T_1 has the highest priority, task T_3 has the lowest priority, and the priority of task T_2 in the middle.

At the beginning when the system starts, all three jobs—J_{11}, J_{21}, and J_{31}—are released. J_{11} starts to run because it has the highest priority. Upon the completion of J_{11} at time 7, J_{21} starts to run because it has a higher priority than J_{31}. At time 19, J_{21} finishes and J_{31} starts to run. At time 24, J_{12}, the second job of T_1, becomes ready and starts to run.

More schedulability analysis principles based on rate-monotonic assignment are covered in Chapter 16.

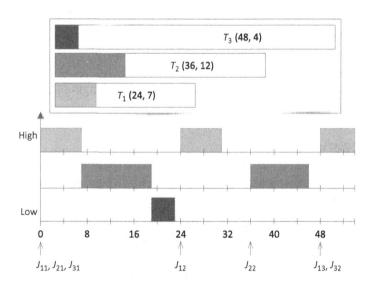

Figure 13.14
Rate-monotonic scheduling.

Problems

13.1 What is POSIX? Why is it important for an application to conform to the POSIX standard?

13.2 Explain ways to create a new POSIX process and a new POSIX thread.

13.3 What is priority-based scheduling? What is the precedence principle? What is the preemption principle? What is the fairness principle?

13.4 When a thread owns a mutex, how might its priority and scheduling be affected?

13.5 How may one use the POSIX sleep functions to implement periodic tasks (threads)?

13.6 What is POSIX compliance? What is POSIX conformance?

13.7 What is POSIX application conformance?

13.8 What is context switching? Which components of the task structure are used to implement context switching?

13.9 When can a thread be transitioned from a ready state to a running state? When can a thread be transitioned from a running state to a ready state?

13.10 When can a thread be transitioned from a running state to a blocked state? When can a thread be transitioned from a blocked state to a ready state?

13.11 When can a thread be transitioned from a running state to a sleeping state? When can a thread be transitioned from a sleeping state to a ready state?

13.12 Why does a running thread, if preempted, need to be placed at the head of the ready list for its priority? Why does a running thread, if it has used up its time quantum, need to be placed at the tail of the ready list for its priority?

13.13 What is the definition of an OS? What is the definition of an RTOS?

13.14 What is the key characteristic of an RTOS? What factors need to be considered in order to implement an RTOS with good real-time performance?

13.15 What is the SCHED_FIFO policy?

13.16 What is the SCHED_RR policy? How can one tell the length of the time quantum used in round-robin scheduling?

13.17 What is the SCHED_SPORADIC policy? In order to use this policy, what parameters need to be specified for the thread?

13.18 What is minimum-laxity-first scheduling policy? Refer to Figure 13.11, and continue the analysis in the text to understand how job priorities are dynamically changed according to their current laxities.

13.19 What is EDF scheduling? Why is it classified as a dynamic-priority approach?

13.20 What is deadline-monotonic assignment scheduling? Why is it classified as a static-priority approach?

13.21 What is rate-monotonic assignment scheduling? What is the difference between rate-monotonic assignment and deadline-monotonic assignment?

Multitasking

Contents

> *Coming together is a beginning. Keeping together is progress. Working together is success.*
> *Henry Ford*

14.1 Introduction to Multitasking

Multitasking refers to a design or implementation strategy by which a system is decomposed into multiple tasks that work together to offer system services.

We already know that there exist context-switching overheads when an operating system (OS) kernel needs to swap between multiple tasks. Why bother to consider multitasking, then? This is because a system can generally benefit from multitasking in the following ways:

Real-Time Embedded Systems. http://dx.doi.org/10.1016/B978-0-12-801507-0.00014-6

(1) Multitasking is one way to reduce complexity by *separation of concerns*:
  (a) Different types of external events can be handled separately. When a system needs to respond to multiple I/O devices (e.g., keyboard, display, pushbuttons, serial ports), each can be considered separately and implemented as a single task or a group of related tasks.
  (b) Isolation of the ripple effect becomes possible with a multitasking solution. The impact of a failed task can be isolated from independent tasks; this may not cause the whole system to crash.
  (c) Multitasking advocates parallel development. Once intertask interfaces have been determined, individual tasks can be separately designed/coded/tested/debugged on the host platform.
(2) Multitasking brings *service concurrency*. Real-time embedded systems are inherently *event driven*, in the sense that they need to respond to events or service requests concurrently. To support service concurrency, a system typically employs an advanced architecture, running tasks to handle requests according to their relative importance. In particular, a single-processor system can achieve service concurrency by allocating time quanta to tasks such that those tasks at the same priority level can be executed interleavingly. More and more systems are built with a multiprocessor architecture (say, SMP), which brings the dual benefits of faster performance and failure resilience. In such a case, a multitasking implementation allows an OS to achieve a better concurrency by allocating the available processors to individual threads.
(3) Multitasking brings *scheduling flexibility*. Real-time embedded systems are also *time driven*. The round-robin architecture could be used to form a *monolithic task* by serializing all the activities in a single execution loop. This, however, becomes cumbersome and even impossible as systems grow in size and hard timing constraints come into play. In contrast, multitasking offers greater flexibility, allowing a system to employ advanced priority-based scheduling policies to meet the critical timing constraints of individual tasks.

14.2 Multitask Design

Multitask design is a critical activity in the real-time embedded system development process (see Figure 2.1). In general, it might be necessary to conduct multitask design multiple times until the set of tasks becomes schedulable.

Below, we examine the multitask design activity in detail. As shown in Figure 14.1, multitask design involves three activities: task identification, task transformation, and task parameter estimation.

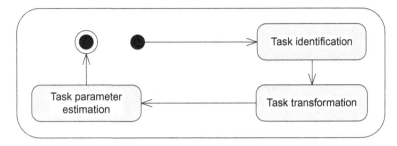

Figure 14.1
Multitask design subprocess.

14.2.1 Task Identification

The objective of task identification is to elicit functional and nonfunctional requirements from the given problem statement. Toward this goal, designers typically start with a system context diagram. This activity produces an initial system design documented by a preliminary task diagram.

14.2.1.1 System context diagram

In software engineering and systems engineering, a system context diagram is a diagram that shows the anticipated interactions between the system to be developed and its external environment. The context diagram of a system can serve as a starting point for identifying potential external events, which are the driving factors for deriving a complete set of system requirements and constraints.

An embedded system interacts with its environment only through its peripheral devices. Hence, to draw a context diagram for an embedded system, we need to consider only the peripheral devices available on the target board.

Peripheral I/O devices can be active or passive:

- *Passive I/O devices* do not generate interrupts. For example, joysticks may be implemented as passive input devices, while LEDs may be implemented as passive output devices. A passive input device typically raises "change" events (see Section 9.1.2). More often than not, embedded software needs to periodically poll a specific register associated with a passive device in order to detect a change event. It is worth noting that polling a passive device can be tricky. On the one hand, a change event might be missed if the polling frequency is too low, but on the other hand, the system may suffer from the polling overhead if the frequency is too high.

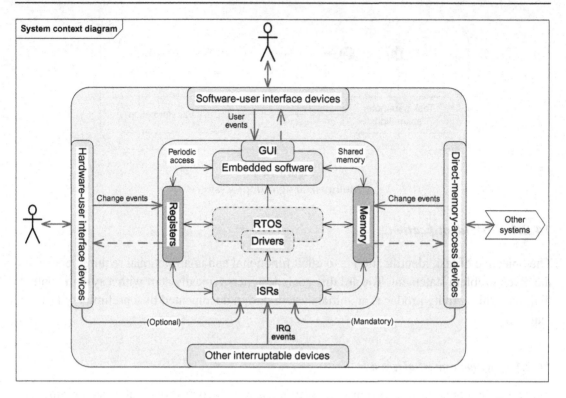

Figure 14.2
A generic system context diagram for embedded systems.

- *Active I/O devices* can raise interrupt request (IRQ) events, periodically or aperiodically, to inform the system that new input is ready or it is ready for the next output. For example, pushbuttons and a keypad can be implemented as active input devices.

Figure 14.2 gives a generic system context diagram for embedded systems, where the peripheral devices are classified into four groups:

(1) Hardware-user interface devices. These devices allow a human user (or a robot) to input requests to or get feedbacks from the system. Examples include pushbuttons, joysticks, keypads, and LEDs. A passive input device in this category typically raises "change" events, and an active input device can raise IRQ events.

(2) Software-user interface devices. These devices facilitate a human user (or a robot) to interact with the system via its graphical user interface (GUI). Examples include mice, on-board touchscreens, and LCDs. An active input device in this category (e.g., mouse, touchscreen) raises IRQ events, relaying user requests to the application.

(3) Direct memory access (DMA) devices. These active I/O devices internally operate in DMA mode, and externally allow the system to exchange massive data-streams with

other computing systems. Examples include network interface cards, AV ports, USB ports, serial ports, and sensors. An input device in this category typically raises IRQ events—say, notifying the completion of a packet sending or receiving operation. Some may also raise change events, which can be detected by the embedded software via shared memory objects.

(4) Other interruptable devices. This category includes all the other active devices that are able to generate IRQ events for the system, such as real-time clocks.

In Figure 14.2, we also purposely disclose the internal structure of the embedded system, showing that (a) the embedded software and the interrupt service routines (ISRs) can independently access device registers or system memory, and (b) the embedded software may be optionally built upon a real-time OS (RTOS), which offers device drivers and other services to the user application. When it comes to a specific embedded system, the context diagram should not expose the system's internal structure.

The radar data processing subsystem, as shown in Figure 1.3, is a complex real-time system on its own. A real-time system is not necessarily a real-time embedded system. For example, the deployment diagram shown in Figure 11.9 gives one solution where the data processing subsystem is implemented as a network of computers. We pursue an embedded approach in this chapter, using the data processing subsystem as an example to explain the process of multitask design.

A simplified context diagram is given in Figure 14.3, which involves the following peripheral devices:

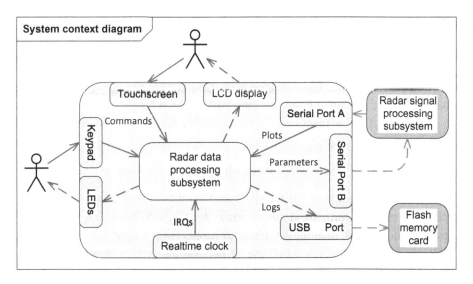

Figure 14.3
An example system context diagram.

- a touchscreen for a user to issue commands;
- an LCD for a user to view statistics and radar maps;
- a keypad for a user to adjust parameter values;
- some LEDs for showing critical system status;
- a real-time clock for generating clock interrupts;
- a USB port for outputting event logs to a flash memory;
- a serial port for sending parameters to the signal processing subsystem; and
- a serial port for accepting plots from the signal processing subsystem.

If the target development board for a system is prebuilt rather than designed from scratch, it may feature more peripheral devices than needed. For such a system, its context diagram should cover only those peripheral devices that are pertinent to the functionality of the system.

14.2.1.2 Task diagram

A *task diagram* is a UML class/object diagram that shows task structures and intertask resource dependency. In general, a task diagram may include the following types of entities:

- Active task. Such a task, annotated with «active», represents an active concept (object) to be implemented as a process or thread. Active tasks include application-specific tasks and system tasks such as a task scheduler and device drivers.
- ISR. Such a task, annotated with «ISR», is triggered by external events from a specific interruptable device.
- Server task. Such a task, annotated with «driver», is tied with an ISR (say, by a signal event). Upon being signaled, it continues the processing activity left over by the ISR to complete its service to an external event.
- Service object. Such an object is used for task synchronization or communication. Examples include signals, message queues, semaphores, mutexes, conditional variables, pipes, and shared memory objects.

With a system context diagram at hand, we can work on a preliminary task diagram by identifying the potential active tasks. A common strategy is to *introduce an active task for each event source or event target*.

For the context diagram in Figure 14.3, we can identify the following potential active objects:

- a UIHandler object, handling GUI events issued via the touchscreen;
- a Monitor object, refreshing the statistics and radar maps on the LCD;
- a KeypadHandler object, managing parameter inputs from a user;
- an LEDReporter object, manipulating the LEDs to show system status;
- a RealtimeClockISR object for the real-time clock;
- an EventLogger object, programming logs to the flash memory device plugged into the USB port;

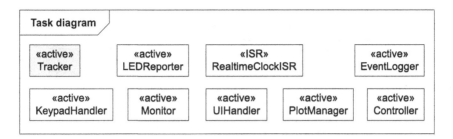

Figure 14.4
A preliminary task diagram.

- a Controller object, adjusting and sending parameters to the signal processing subsystem;
- a PlotManager object, receiving new plots from the signal processing subsystem; and
- a Tracker object, processing new plots to update tracks.

These identified objects are included in the preliminary task diagram shown in Figure 14.4. We next study some task transformation techniques to evolve this task diagram further.

14.2.2 Task Transformation

14.2.2.1 Serial port design pattern

Embedded software often needs to interact with serial devices. A design pattern called a *serial port design pattern* defines a generic solution to refine tasks that interface with a serial port device.

This design pattern is normally used to implement a serial device driver program, involving a SerialPortDriver object, which manages several SerialPort objects. Figure 14.5 illustrates the concepts involved in the serial port design pattern. The SerialPortDriver object is responsible for installing the ISR for serial ports. As a "server task," it also keeps waiting for signal events originating from the ISR. Each SerialPort object is responsible for a specific serial port, sending data to or receiving data from the serial device. Each SerialPort object also has two queue structures: a ReceiveQueue for storing messages received from the serial port, and a TransmitQueue for buffering the messages to be transmitted.

Let us follow the UML sequence diagram given in Figure 14.6 to understand how serial data transmission works.[1] Upon initialization, the SerialPortDriver object installs the serial ISR code for handling IRQs from serial ports. The diagram is then partitioned into three parallel regions, corresponding to three execution flows at run time:

---

[1] This is for non-DMA serial devices. DMA serial devices should work similarly.

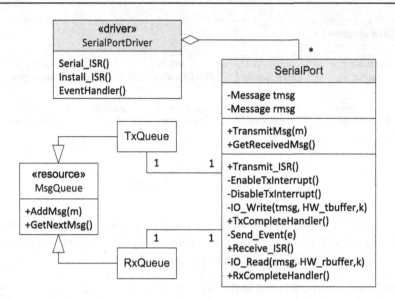

Figure 14.5
Serial port design pattern.

- *User task thread.* At any time, a user task may call the function TransmitMsg(m) to send out a message through a specific serial port. The corresponding SerialPort object, upon being requested, first adds the received message *m* to TxQueue. If this is the first message in TxQueue (i.e., the queue was empty before this request), the message is moved to *tmsg*, which holds the current message being transmitted, and the serial port's *transmit-empty interrupt* is enabled.

- *Hardware level.* Each serial port has a one-byte "transmit shift register" and a *k*-byte FIFO *hardware transmit buffer* (say, 16 bytes for high-speed ports). The shift register holds the byte being transmitted: bits are taken one-by-one and sent out bit-by-bit on the serial line until the last bit has been sent, at which time the shift register tries to grab another byte from the hardware buffer. Whenever the last byte is taken (i.e., the hardware buffer becomes empty), a *transmit-empty interrupt* is raised to the processor.

- *Hardware interrupt thread.* Upon being interrupted, the processor will stop what it was doing and start running Serial_ISR(), which checks the status registers of the serial ports to find out what has happened. In this case, it finds out that the serial port's transmit buffer is empty, and the function Transmit_ISR() of the corresponding SerialPort object is invoked. Transmit_ISR() will first write up to *k* bytes of new data to the hardware buffer (*k*-byte chunks are taken in order from *tmsg*). New bytes should arrive when the serial port is still transmitting the byte in the shift register bit-by-bit. Hence, when the shift register needs another byte to transmit, it will always find one from the transmit buffer (unless there are no more bytes to send). After the I/O write, *tmsg* is checked, and a signal event,

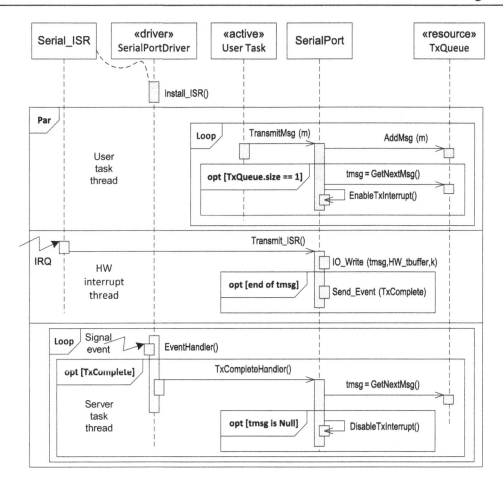

Figure 14.6
A scenario of the serial port design pattern.

"TxComplete," is sent out if *tmsg* becomes empty (the whole message has been transmitted).

• *Server task thread.* Upon receiving a signal event "TxComplete," the SerialPortDriver object will invoke the function TxCompleteHandler() of the corresponding SerialPort object, which will then grab another message from TxQueue to *tmsg*. The serial port's *transmit-empty interrupt* is disabled if there is no more message to transmit.

Receiving data from a serial port works similarly. Each serial port also has a *k*-byte FIFO *hardware receive buffer* (say, 16 bytes for high-speed ports). Whenever a byte is fully received from the external cable it goes into the FIFO receive buffer. If the buffer becomes almost full (say 14 bytes in a 16-byte buffer), the port will raise a *receive-ready interrupt* to the processor. Upon being interrupted, the processor will start running the serial ISR. The interrupt thread, in this case, will find that the serial port's receive buffer is ready, and the function Receive_ISR() is invoked. All data are then picked up from the hardware receive buffer via an I/O read operation, which will empty the receive buffer, making it ready for storing more incoming bytes. If the serial device driver fails to remove data promptly, data loss will happen when the receiver buffer overflows (overruns). If a message is fully received, Receive_ISR() will also send a signal event, "RxComplete." Upon receiving an "RxComplete" event, the server task thread will invoke the function RxCompleteHandler() of the corresponding SerialPort object, which will then save the received message *rmsg* to RxQueue. A user task can call GetReceivedMsg() to retrieve a message received from a specific serial port.

By applying the serial port design pattern, we can evolve the example task diagram further to include a new ISR and a server task for managing the serial ports. The other devices can be treated similarly. Figure 14.7 gives the second version of our task diagram, where board support package (BSP) is used to organize the device drivers and related objects together. The BSP of an evaluation board is typically available from the vendor of the board or the RTOS to be adopted. When you work on a task diagram, the device drivers can be disregarded if you have planned to use the off-the-shelf solutions from an RTOS or a BSP. However, when it comes to schedulability analysis, all driver tasks should be considered if the system under development has hard timing constraints.

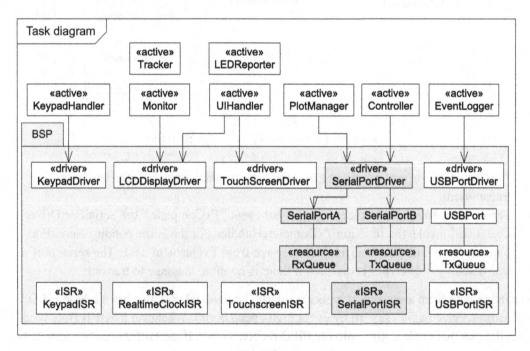

Figure 14.7
A task diagram after applying the serial port design pattern.

It is also worth noting that a single server task (SerialPortDriver) is introduced in the serial port design pattern to handle all the serial ports. This observation leads us to a strategy; let us name it the *joint manager strategy*: create a single thread to manage multiple resources of the same type. This strategy is useful for applications that need to process inputs from disparate sources or to dispatch outputs to multiple targets.

14.2.2.2 Task splitting strategies

Task splitting strategies are techniques for transforming lengthy tasks into multiple shorter tasks.

In the serial port design pattern given in Figure 14.5, code for serial ports is split into two tasks (threads): the hardware interrupt thread is executed every time a hardware interrupt is raised by a serial port (at multibyte level), and the server task thread is executed every time a message is completely received or transmitted (at message level). Jobs of the server task arrive at a much lower frequency (especially for long messages) than jobs of the interrupt thread.

In general, we have the *single-frequency strategy*: if a task routine contains sections that are executed at different frequencies, then it is reasonable to split it into multiple routines, each with a defined frequency of execution. This strategy is very useful for eliciting new periodic tasks. For example, a Tracker object has several responsibilities: plot gating, track smoothing, track prediction, and maintenance. The code for track smoothing may be executed every 300 ms, the code for track management may be executed every 400 ms, and the codes for all the other tasks are executed every 200 ms. If this is the case, we should split the Tracker task, which allows us to add two new active tasks: TrackSmoother and TrackManager. Figure 14.8 shows the third version of our task diagram for the radar data processing subsystem.

The *sequence strategy* can be used to split a lengthy task T_i into k sequentially dependent tasks $(T_i^1, T_i^2, \ldots, T_i^j, \ldots, T_i^k)$ such that T_i^j, where $1 < j \le k$, is not ready for execution until T_i^{j-1} runs to completion. If T_i is a periodic task, T_i^1 cannot be executed until T_i^k of the last period runs to completion.

There is yet another strategy called the *importance-level strategy*: if a task routine contains sections that represent activities at different levels of importance, those sections should be treated as different tasks. This strategy is closely tied with the assignment of task priorities, allowing a system to handle more important activities with a higher priority. This would not be possible if activities with dissimilar importance levels were twisted inside one task.

The opposite of task splitting is task merging. Sometimes, it might be useful to merge multiple tasks into one task for good reasons. Here are a few possible situations:

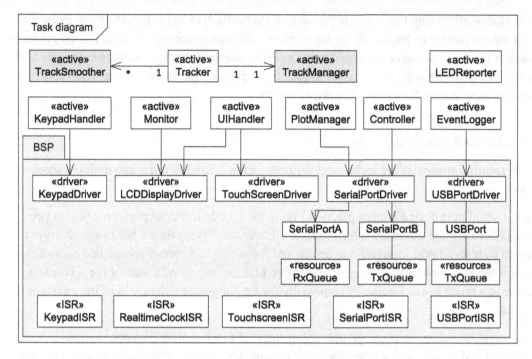

Figure 14.8
A task diagram after task splitting.

- Multiple periodic tasks with the same period can be merged in order to reduce task switching overhead.
- Multiple tasks driven by the same external event (e.g., a timer) can be merged in order to reduce task switching overhead.
- Multiple closely coupled tasks that need to exchange messages can be merged in order to reduce intertask synchronization or communication overhead.

14.2.3 Task Parameter Estimation

From Section 12.1, we know that a periodic task has four parameters: period, release time, (worst-case) execution time, and deadline. Aperiodic or sporadic tasks may have no periods or deadlines. These task parameters, together with priority assignment, are critical information to be used in schedulability analysis.

The release time of a periodic task by default is the same as the start of its period. The release time of an aperiodic task may follow a certain probability distribution function. The release time of a sporadic task can be rather unexpected. Indeed, task release time is not a value to be estimated; it is more like a hypothetical value to be considered in schedulability analysis. The deadline of a task specifies a timing constraint that has to be satisfied or the system fails and the system users inevitably suffer from it. Likewise, it is not a value to be estimated; it is rather a nonfunctional requirement imposed by the external environment of the system. Hence, task parameter estimation reduces to the estimation of task execution time and period, if applicable.

As explained in Section 12.1.1, the (worst-case) task execution time depends on both hardware features and software structures. Given a task, UML activity diagrams can be used to calculate the approximate execution time of the longest execution path. If an RTOS is to be adopted, the estimation can also benefit from the known real-time characteristics of the RTOS, say, context switching time, the execution times of system calls. See Section 12.1.1 for examples.

Estimating task periods is application specific. It is easy for some tasks that are inherently periodic—say, an LCD is refreshed at 60 Hz. In some tricky situations we might have to make some assumptions or fix certain design parameters. For instance, in the serial port design pattern, if the incoming or outgoing messages over a serial port are structured with a fixed length, then the server task can be treated as a periodic task with a known period—the time it takes to completely transmit/receive a message. The *single-frequency strategy* also simplifies the period estimation of a complex task.

The estimated task parameters can be documented by using real-time UML notations in sequence diagrams. An example is given in Figure 14.9.

As a side note, when the embedded software becomes (partially) available as time goes on, task parameter estimation can be further improved by using hardware or software probes to audit the execution time of certain tasks.

Note that priority is not a task parameter; instead, it is assigned by a designer to tune the schedulability of a set of tasks. Here are a few criteria for priority assignment:

- Priority assignment is not necessary if a clock-driven approach is to be used (see Chapter 15).
- Priority assignment is enforced by a certain well-defined principle if a deadline-monotonic or a rate-monotonic approach is to be used (see Chapter 16).
- Critical tasks, with hard deadlines that must be met, should have relatively higher priorities.
- Urgent tasks, with relatively earlier deadlines, should have relatively higher priorities.
- Lengthy tasks, with a long execution time, usually have relatively lower priorities.

14.3 Multitask Resource Sharing

It is common for multiple tasks (threads) to share resources, such as the storage medium (disk, a memory page), a device (card reader), a global variable, or a file descriptor. A resource is exclusive if it can be used by only one task at a time.

When multiple tasks simultaneously use an exclusive resource, the system may behave abnormally if the integrity of the resource is compromised (say, the resource itself or any data

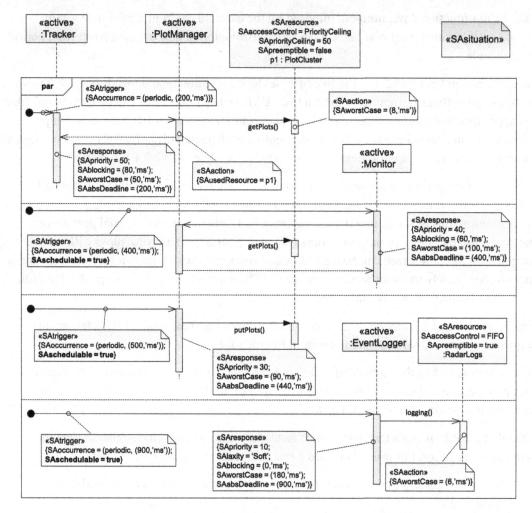

Figure 14.9
A sequence diagram annotated with schedulability stereotypes.

contained within the resource are corrupted). For example, suppose that a system has several threads, each of which makes a call to the same library routine to access a global variable. The value of the global variable might be corrupted if the routine is not *thread-safe* in the sense that it does not employ some sort of synchronization mechanism to prevent the threads from modifying the shared data at the same time. Another example is given in Figure 14.6, where the TxQueue object is an exclusive resource used by at least two tasks: the user task thread and the server task thread.

Within a task routine, a block of code that accesses a certain shared resource is called a *critical section* (or critical region). *Resource locks*, such as semaphores and mutexes, are

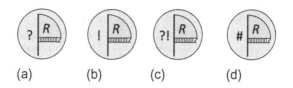

Figure 14.10
Notation for the use of resource locks.

normally used to protect the integrity of resources shared among multiple tasks. Figure 14.10 gives our notation for using resource locks, where the notation in Figure 14.10(a) is used for the action of requesting for a resource lock R, the notation in Figure 14.10(b) is used for the action of securing a resource lock R; the notation in Figure 14.10(c) is used for the combined action of requesting a resource lock R and securing it immediately, and the notation in Figure 14.10(d) is used for the action of releasing a resource lock R.

However, protecting resources by locks comes with two critical issues: resource deadlocks and priority inversion. In this section, we examine these two issues, and in the next two sections we introduce solutions for them.

14.3.1 Resource Deadlocks

A *resource deadlock* refers to a situation where no task can run to completion owing to each task holding some resources that are currently requested by others. A resource-client graph as introduced in Section 10.2.5 can be used to analyze potential resource deadlock situations.

In general, a resource deadlock can arise if four conditions hold simultaneously:

(1) Mutual exclusion: the resources under consideration can be used by only one task at a time.
(2) No resource preemption: a resource can be released only by the task owning it (typically after that task has finished its use of the resource).
(3) Hold and wait: a task, while holding at least one resource, is waiting to acquire additional resources.
(4) Circular wait: there exists a chain $(T_0, T_1, \ldots, T_k, T_0)$ of waiting tasks such that T_0 is waiting for a resource held by T_1, T_1 is waiting for a resource held by T_2, and so on, and T_k is waiting for a resource held by T_0.

Figure 14.11 shows a deadlock situation, where a blank block on the lifeline of a task indicates that the task is executing, and a shaded block indicates the execution of a critical section. At time t_1, task T_2 takes the lock R_2 and enters its critical section. At t_2, T_2 is preempted by task T_1, which at t_3 takes the lock R_1 and enters its own critical section. While T_1 is still within its critical section, at t_5 it attempts to enter a nested critical section protected

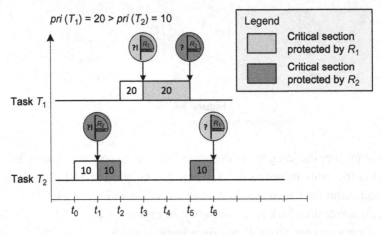

Figure 14.11
A deadlock situation.

by lock R_2, which is currently held by T_2. Hence, T_1 is blocked at time t_5 and T_2 starts to run. At time t_6, T_2 also attempts to enter a nested critical section protected by lock R_1. Since R_1 has not been released by T_1, neither can run to completion, and a deadlock is formed.

14.3.2 Priority Inversion

POSIX-compliant RTOSs implement a *priority-based preemptive scheduler*, which ensures that among all the threads that are ready to run, the one with the highest priority is always the task that is actually running. When a higher-priority thread is released, the scheduler will preempt the currently running thread in the middle of its execution.

However, priority-based preemptive scheduling could be compromised when tasks share exclusive resources that are protected by locks: requesting a resource locked by a lower-priority task would prevent a higher-priority ready task from running when it should. This issue is called *priority inversion*, referring to the phenomenon where a higher-priority task must *wait* for the lower-priority task [64].

Consider a situation in a single-processor environment where global data are protected by a resource lock R and shared by a higher-priority thread T_1 and a lower-priority thread T_2. As illustrated in Figure 14.12(a), if T_2 gains access first and then T_1 requests access to the shared data, T_1 would be blocked until T_2 completes its use of the shared data.

Figure 14.12(b) illustrates an even worse situation where *unbounded priority inversion* arises. Suppose a low-priority task T_2 and a high-priority task T_1 use a shared resource. T_2 runs first and locks the resource at time t_1. T_2 is preempted by T_1 at time t_2. However, T_1 is blocked at time t_3 when it attempts to gain access to the resource (T_1 has to wait until T_2 completes its

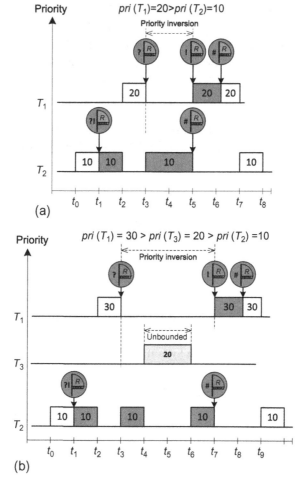

Figure 14.12
Priority inversion (a) and unbounded priority inversion (b).

use of the resource). Before T_2 finishes with the resource, another task T_3 with a priority in between becomes ready at time t_4 and preempts T_2.[2] While T_3 (and perhaps additional intermediate-priority tasks) runs, the highest-priority task T_1 remains in a blocking state.

Priority inversion is a critical issue because it may lead to the missing of task deadlines. One notorious priority inversion example occurred on the Mars Pathfinder mission in July 1997 [56]. The Pathfinder mission is best known for the little rover that took high-resolution color

---

[2] Some OSs (e.g., VxWorks) may offer a special lock called a "preemption lock," which freezes the scheduler to prevent preemption of the current task by other tasks. If task T_2 also uses a preemption lock, it will not be preempted by T_3. However, the preemption lock itself would cause T_3's priority to be inverted.

pictures of the Martian surface and relayed them back to Earth. In the spacecraft, various devices communicated over an MIL-STD-1533 data bus. Activity on this bus is managed by two high-priority tasks. The first one is the *bus scheduler task*, which controls the setting up of transactions on the MIL-STD-1553 bus. The second one is the *bus distribution manager task*, which handles the collection of the transaction results. Every cycle, these two high-priority tasks check to ensure that the other has completed its execution.

The bus distribution manager task communicated through a pipe with a low-priority meteorological science task known as the ASI/MET. The OS kernel used a semaphore to protect access to (the file descriptors of) the pipe. One day, the system was reset and all the computing activities scheduled on that day were terminated owing to a priority inversion issue. Briefly, the scenario started with the low-priority science task which had acquired the semaphore in its computation. While the science task was in the process of giving the semaphore but had not finished, it was preempted by a couple of medium-priority tasks. In the meantime, the bus distribution manager task attempted to send the newest ASI/MET data to the same pipe and was blocked upon taking the semaphore, because it was still held by the science task. The medium-priority tasks were heavily loaded[3] and continued to run, causing unbounded priority inversion. Fortunately, when the other high-priority task, the bus scheduler task, was awakened, it found that the distribution manager task had not completed its cycle (a hard deadline in the system) and forced a system reset.

14.4 Addressing Resource Deadlocks

14.4.1 Deadlock Prevention

The simplest way of addressing resource deadlocks is *deadlock prevention*, which is performed at design time.

The four conditions given in Section 14.3.1 are necessary conditions for a resource deadlock to occur. Thus, to prevent deadlocks from happening, we have to ensure that at any point in time at least one of the four conditions is not satisfied.

The first condition can be denied only for nonexclusive resources. General approaches have been suggested to deny the remaining three conditions:

- To attack the "no preemption" condition, a task has to voluntarily release the resources it is holding if its request for additional resources is denied.
- To attack the "hold and wait" condition, each task must request all its required resources at once and cannot proceed until all have been made available.

---

[3] This was an unexpected worst-case scenario where the data rates from the Martian surface were higher than anticipated. The situation was aggravated also by the proportionally increasing number of science activities.

- To attack the "circular wait" condition, one solution is to predefine a total order on all the resources, and all tasks are enforced to follow the predefined order to request resources.

14.4.2 Deadlock Detection

The second approach to addressing resource deadlocks is to detect deadlocks at run time. There is a simple *deadlock detection* approach which is based on the following assumptions:

- A system may have multiple types of resources. If there are multiple resource copies (or instances) of the same type, they are deemed identical and indistinguishable.
- The maximum resource use of each task is known.
- A task has to wait if the resource it requested is not available.
- After a task gets all its resources, it must return them in a finite amount of time.

In general, let $\{T_i | 1 \leq i \leq n\}$ denote a set of tasks, and let m be the number of resource types in the system under consideration. We define the following data structures:

(1) System resource vector: $R = (R_1, \ldots, R_j \ldots, R_m)$, where R_j ($1 \leq j \leq m$) denotes the total number of resource instances of type j.

(2) Available resource vector: $A = (A_1, \ldots, A_j \ldots, A_m)$, where A_j ($1 \leq j \leq m$) denotes the number of resource instances of type j that are currently available.

(3) Current allocation matrix:

$$C = \begin{bmatrix} C_{11} & C_{12} & \cdots & C_{1m} \\ C_{21} & C_{22} & \cdots & C_{2m} \\ \vdots & \vdots & \cdots & \vdots \\ C_{n1} & C_{n2} & \cdots & C_{nm} \end{bmatrix}$$

where the ith row of C, denoted by C_i, indicates the number of resource instances of each type that are possessed by task T_i. Particularly, C_{ij} is the number of resource instances of type j that are currently allocated to task T_i.

(4) Additional demand matrix:

$$D = \begin{bmatrix} D_{11} & D_{12} & \cdots & D_{1m} \\ D_{21} & D_{22} & \cdots & D_{2m} \\ \vdots & \vdots & \cdots & \vdots \\ D_{n1} & D_{n2} & \cdots & D_{nm} \end{bmatrix}$$

where the ith row of D, denoted by D_i, indicates the number of resource instances of each type that are still needed by task T_i. Particularly, D_{ij} is the number of resource instances of type j that are needed by task T_i.

We define a relation $A \leq B$ on two vectors A and B: $A \leq B$ holds if and only if $A_j \leq B_j$ for $1 \leq j \leq m$. Since each resource instance is either allocated or available, for all $1 \leq j \leq m$, we have an invariant that holds all the time:

$$A_j + \sum_{i=1}^{n} C_{ij} = R_j.$$

The procedure DEADLOCK-DETECTION given below can be used for detecting resource deadlocks:

DEADLOCK-DETECTION (R, A, C, D)

```
1    τ = Ø // τ is a list, initially empty.
2    for i from 1 to n: Finish[i] = false
3    while(|τ| < n)
4        r = Search(D) // find a task such that Finish[r] == false and D_r ≤ A
5        if (r == 0) // no such task exists
6            break;
7        else // after task r completes, return allocated resources to the pool
8            for j from 1 to m: A[j] = A[j] + C[r,j]
9            Finish[r] = true, τ.tail = r
10   if (|τ| < n)
11       All tasks not in τ are deadlocked
```

For each iteration, it looks for a task whose resource needs can be satisfied by the currently available resources. If no such task exists and there are unfinished tasks, then a resource deadlock is detected (there is a deadlock among tasks not in τ). If such a task is found, it can run until completion, at which time it is marked as finished, and returns the resources it is holding to the resource pool. There exists no deadlock if all the tasks are able to run to completion. The list τ actually gives a deadlock-free schedule for the tasks.

14.4.3 Deadlock Avoidance

Resources are typically requested one at a time. *Deadlock avoidance* is a conservative approach where granting a resource to a task is denied if this may cause the system to run into a deadlock situation.

14.4.3.1 Safe state

A task sequence $\langle T_{i1}, T_{i2}, \ldots, T_{in} \rangle$ is *safe* if for each $T_{ik}(1 \leq k \leq n)$ the resources that T_{ik} still needs can be satisfied by the currently available resources together with those resources held by all the tasks prior to T_{ik} in the sequence.[4]

---

[4] If the resource needs of T_{ik} are not immediately available, then T_{ik} can wait until all the tasks prior to it have finished.

A system is in a *safe state* if there exists a safe sequence of all tasks. If a system is in a safe state, no deadlocks can occur. If a system is in an unsafe state, the tasks may run into deadlock situations.

14.4.3.2 The banker's algorithm

In order to avoid deadlocks, when a task requests a resource or resources, the request is granted only if immediate resource allocation would lead the system into a new safe state. This is known as the banker's algorithm [28, 29]:

Is-Safe (R, A, C, D)

```
1    τ = Ø // τ is a list, initially empty.
2    for i from 1 to n: Finish[i] = false
3    while(|τ| < n)
4         r = Search(D) // find a task such that Finish[r] == false and D_r ≤ A
5         if (r == 0) // no such task exists
6              break;
7         else // return allocated resources to the pool
8              for j from 1 to m: A[j] = A[j] + C[r,j]
9              Finish[r] = true, τ.tail = r
10   if (|τ| < n)
11        return false // A is not safe
12   else return true // A is safe
```

Banker (R, A, C, D, α_{ij})

```
1    A' = A, A'[j] = A'[j] − α_ij
2    D' = D, D'[ij] = D'[ij] − α_ij
3    C' = C, C'[ij] = C'[ij] + α_ij
4    // Check to see whether to allow transition from C to C'
5    s = Is-Safe (R, A', C', D')
6    if (s == true)
7         grant task T_i with the request for resource R_j;
8    else reject the request α_ij from task T_i
```

Is-Safe is the same as Deadlock-Detection, except that it returns a Boolean value to indicate whether the system is in a safe state or not. For each $\alpha_{ij} \geq 0$, a request from task T_i for resources of type j, the Banker procedure simply invokes Is-Safe to check whether granting this request will lead the system into a safe state or not.

Consider the simple example given below.

$$R = \begin{bmatrix} 7 & 3 & 4 & 2 \end{bmatrix} \qquad R = \begin{bmatrix} 7 & 3 & 4 & 2 \end{bmatrix} \qquad R = \begin{bmatrix} 7 & 3 & 4 & 2 \end{bmatrix}$$

$$A = \begin{bmatrix} 2 & 0 & 2 & 0 \end{bmatrix} \qquad A = \begin{bmatrix} 2 & 0 & \boxed{1} & 0 \end{bmatrix} \qquad A = \begin{bmatrix} 2 & 0 & \boxed{0} & 0 \end{bmatrix}$$

$$C = \begin{bmatrix} 3 & 0 & 1 & 1 \\ 0 & 1 & 0 & 0 \\ 1 & 1 & 1 & 0 \\ 1 & 1 & 0 & 1 \\ 0 & 0 & 0 & 0 \end{bmatrix} \qquad C = \begin{bmatrix} 3 & 0 & 1 & 1 \\ 0 & 1 & \boxed{1} & 0 \\ 1 & 1 & 1 & 0 \\ 1 & 1 & 0 & 1 \\ 0 & 0 & 0 & 0 \end{bmatrix} \qquad C = \begin{bmatrix} 3 & 0 & 1 & 1 \\ 0 & 1 & 1 & 0 \\ 1 & 1 & 1 & 0 \\ 1 & 1 & 0 & 1 \\ 0 & 0 & \boxed{1} & 0 \end{bmatrix}$$

$$D = \begin{bmatrix} 1 & 1 & 0 & 0 \\ 0 & 1 & 1 & 2 \\ 3 & 1 & 0 & 0 \\ 1 & 0 & 1 & 0 \\ 2 & 1 & 1 & 0 \end{bmatrix} \qquad D = \begin{bmatrix} 1 & 1 & 0 & 0 \\ 0 & 1 & \boxed{0} & 2 \\ 3 & 1 & 0 & 0 \\ 1 & 0 & 1 & 0 \\ 2 & 1 & 1 & 0 \end{bmatrix} \qquad D = \begin{bmatrix} 1 & 1 & 0 & 0 \\ 0 & 1 & 0 & 2 \\ 3 & 1 & 0 & 0 \\ 1 & 0 & 1 & 0 \\ 2 & 1 & \boxed{0} & 0 \end{bmatrix}$$

From the left safe state, suppose task T_2 requests one instance of resource type R_3, i.e., $\alpha_{23} = 1$. This would lead the system into the new state given in the middle, which is a safe state (with the sequence $\langle T_4, T_1, T_5, T_2, T_3 \rangle$). Thus, this request can be granted.

Suppose within the middle state task T_5 also requests one instance of resource type R_3, i.e., $\alpha_{53} = 1$. This would lead the system into the state shown on the right, which is not a safe state. Because of this, the request α_{53} would be rejected by the system.

14.5 Addressing Priority Inversion

In order to prevent unbounded priority inversion, tasks need to follow certain rules when accessing shared resources protected by locks. Such rules are referred to as *resource access protocols*, also known as *resource patterns*. Below we will introduce a few resource access protocols typically implemented by mutexes, and examine how they address the issue of unbounded priority inversion.

In general, a task T_i may contain several critical sections, where the jth critical section is denoted by $z_{i,j}$, which is protected by a resource lock (mutex) denoted by $R_{i,j}$.

Definition 14.1. Given a critical section $z_{i,j}$ of task T_i protected by a lock $R_{i,j}$, a task T is said to be blocked by T_i, or specifically blocked by $z_{i,j}$ that is locked by $R_{i,j}$, if T_i has a lower priority than T but T has to wait for T_i to complete $z_{i,j}$.

The following two rules apply to all the protocols to be discussed next:

(1) A higher-priority task T is able to preempt lower-priority tasks immediately when T becomes ready.

(2) Each critical section is protected by a lock. Upon entering its critical section, a task must first obtain the lock guarding the critical section. Upon exiting a critical section, the task must release the lock.

14.5.1 Priority Inheritance Protocol

The POSIX.1-2008 standard supports a protocol called the *priority inheritance protocol* (PIP), which works as follows:

(1) *Grant clause.* When a task T is attempting to enter a critical section, if the lock is available, T will obtain the lock and enter its critical section.

(2) *Block clause.* When a task T_j is attempting to enter a critical section, if the lock has already been taken by another task T_i, T_j is blocked by T_i.

(3) *Inheritance clause.* If task T_i blocks a set of higher-priority tasks, it inherits (uses) the highest priority of the tasks in this set. When T_i exits a critical section, it resumes the priority it had immediately before it entered the critical section.

(4) Priority inheritance is transitive. In particular, suppose there are three tasks T_1, T_2, and T_3, where T_1 has the highest priority, T_3 has the lowest priority, and T_2's priority level is somewhere in between. If task T_3 blocks task T_2, which in turn blocks task T_1, T_3 would inherit the priority of T_1 via T_2.

Here are a few properties of the PIP:

Lemma 14.1. *A task T can be blocked by a lower-priority task T_L, only if, when T becomes ready, T_L is executing within a critical section that can block T.*

Lemma 14.2. *Under the PIP, a high-priority task T can be blocked by a lower-priority task T_L for at most the duration of one critical section, regardless of the number of locks they may share.*

Theorem 14.1. *(bounded priority inversion). Given a task T and n lower-priority tasks $\{T_1, \ldots, T_n\}$, under the PIP, each of the n lower-priority tasks can block task T for at most the duration of one critical section.*

Theorem 14.2. *Under the PIP, if there are m locks which can block a task T, then T can be blocked at most once by each of the m locks.*

Note that the grant clause is a "greedy" rule. For this reason, the PIP by itself does not prevent deadlocks. Let us consider Figure 14.11 again. The scenario conforms to the PIP. In particular, at time t_4, according to the block clause, T_1 is blocked by T_2, which inherits T_1's priority and continues to execute the rest of its critical section. At time t_5, T_2 also attempts to enter a nested critical section protected by lock R_1. Since neither can run to completion, a deadlock is formed.

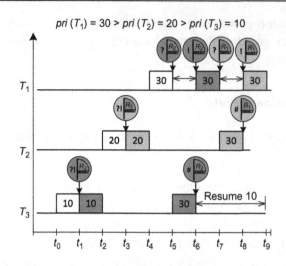

Figure 14.13
A chain of blocking situation.

The PIP can also cause chains of blocking (many preemptions). As shown in Figure 14.13, task T_1 needs to access resources protected by R_2 and R_1 sequentially. Suppose task T_2 preempts task T_3 while it is within its critical section protected by R_2. While T_2 is running within its own critical section protected by R_1, it is preempted by a new task T_1. When T_1 attempts to enter its critical section, it will be sequentially blocked by tasks T_3 and T_2 in a chain.

In general, a task that needs to access n resources may be blocked n times, and the maximal blocking is equal to the summed execution time of all the critical sections within lower-priority tasks. The run-time overhead due to the number of preemptions associated with chains of blocking may cause a system to break task deadlines. Better protocols are needed in order to minimize the amount of blocking time.

14.5.2 Highest Locker Protocol

The *highest locker protocol* (HLP) is also called "immediate priority inheritance," the "immediate ceiling priority protocol," or the "priority protect protocol" [64].

Definition 14.2. The *priority ceiling* of a resource lock R_j, denoted by ceil(R_j), is defined to be the highest priority of all those tasks that may use R_j.[5]

---

[5] Sometimes the priority ceiling is defined to be one level above the highest priority of all those tasks that may use R_j. This is not necessary as long as no task voluntarily yields the CPU while holding a lock. It may be used to ensure a unique priority for all tasks at all times, which simplifies the analysis. Otherwise, it can be complicated, given that tasks of equal priority might be handled differently by different schedulers.

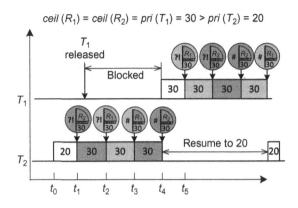

Figure 14.14
HLP prevents deadlocks.

The priority ceiling value of a lock is determined at design time when the set of tasks that may use the lock have been identified.

The HLP has the following rules:

(1) *Grant clause.* When a task T is attempting to enter a critical section protected by a lock R, if R is available, T will obtain the lock and start to execute the critical section.

(2) *Promote clause.* As soon as a task T obtains a lock R, its priority is changed dynamically to the maximum of its current priority and ceil(R). When T exits a critical section, it resumes the priority it had at the point of entry into the critical section.

Figure 14.14 shows how the HLP prevents deadlocks. At time t_1, task T_2 takes the lock R_2, and its priority becomes ceil(R_2) according to the promote clause. Shortly after this, task T_1 is released. T_1 cannot preempt T_2 because they have the same priority. Task T_1 has to wait until T_2 resumes its original priority when it releases the lock R_2.

Figure 14.15 illustrates that a task can be blocked at most once. Tasks T_2 and T_1 are released after T_3 has taken the lock R_2. The taking of R_2 promotes T_3's priority to the same level as that of task T_1. Note that task T_2 is *not* blocked during the inactive period after t_3. According to the definition of "blocking," a task can be blocked only by a lower-priority task.

In conclusion, the HLP prevents unbounded priority inversion and deadlocks, and it ensures that a task can be blocked by lower-priority tasks at most once. The drawback is that the HLP may cause unnecessary blocking. For example, suppose we have
$\text{pri}(T_E) = \text{ceil}(R) > \text{pri}(T_H) > \text{pri}(T_M) > \text{pri}(T_L)$, where the resource lock R is shared among tasks T_E, T_H, and T_L, but task T_M is an independent task (with no access to R). Suppose R is now taken by the low-priority task T_L. If the PIP were used, T_L would not change its priority until either T_E or T_H attempts to take R. In the HLP, T_L's priority is promoted to ceil(R) as

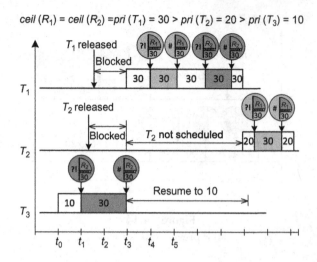

Figure 14.15
Tasks are blocked at most once.

soon as it successfully acquires the lock R. This "early promotion" will unnecessarily prevent task T_M from running even though task T_M has nothing to do with the lock R.

We next introduce the priority ceiling protocol (PCP), which combines the benefits of the HLP and the PIP.

14.5.3 Priority Ceiling Protocol

Given a task T_j, let $\Phi(T_j)$ denote the set of all the resource locks currently owned by tasks *other than* T_j, and let R_j^* be the lock with the highest priority ceiling among the locks in the set $\Phi(T_j)$. Obviously, $\forall R \in \Phi(T_j)$, $\text{ceil}(R_j^*) > \text{ceil}(R)$ if $R_j^* \neq R$.

While $\text{ceil}(R_j)$, the priority ceiling of a resource lock R_j, is a design-time concept, $\text{ceil}(R_j^*)$ is a run-time concept. For a specific task T_j that requests a resource lock at time t, $\text{ceil}(R_j^*)$ is the highest ceiling value of those resource locks currently held by tasks other than T_j.

The PCP [49, 64] has the following rules:

(1) *Grant clause.* When a task T_j is attempting to enter a critical section protected by a lock R, it obtains the lock if its priority $\text{pri}(T_j)$ is strictly higher than the priority ceilings of all the locks currently held by tasks other than T_j, i.e., $\text{pri}(T_j) > \text{ceil}(R_j^*)$.

(2) *Block clause.* When a task T_j is attempting to enter a critical section protected by a lock R, it will be blocked if $\text{pri}(T_j) \leq \text{ceil}(R_j^*)$. In this case, the task T_j is said to be blocked by the task that is holding the lock R_j^*.

(3) *Inheritance clause.* If task T_i blocks a set of higher-priority tasks, it inherits the highest priority of the tasks in this set. When T_i exits a critical section, it reacquires the priority it had before the critical section was entered.

(4) Priority inheritance is transitive.

Notice that the grant clause and the block clause are different from those of the PIP. The PCP is more cautious in granting access to shared resources.

Definition 14.3. *Transitive blocking* is said to occur if a task T_i is blocked by T_j, which, in turn, is blocked by another task T_k.

For the PCP, we have the following properties:

Lemma 14.3. *The PCP prevents transitive blocking.*

Theorem 14.3. *The PCP prevents deadlocks.*

Theorem 14.4. *Under the PCP, a task can be blocked by lower-priority tasks at most once.*

Figure 14.16 shows a scenario where tasks T_1 and T_2 both have critical sections (say, accessing some shared resources) protected by locks R_1 and R_2. Task T_1 has a higher priority and needs to manipulate the resource locks in the order "lock R_1; lock R_2; release R_2; release R_1." Task T_2 has a lower priority and needs to use the locks in the reverse order. Since both tasks need to use the two locks, the priority ceilings of both locks are equal to the priority of task T_1.

Task T_2 takes lock R_2 at time t_1. At t_2, task T_2 is preempted by task T_1. At time t_3, T_1 is attempting to take lock R_1. At this point in time, we have $R_1^* = R_2$. Because the priority of T_1 is not higher than the priority ceiling of R_2 ($\text{pri}(T_1) = \text{ceil}(R_2)$), T_1 is blocked by T_2 according

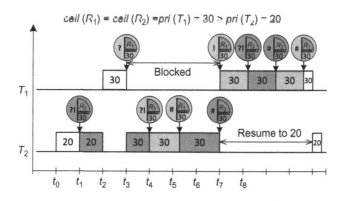

Figure 14.16
PCP prevents deadlocks.

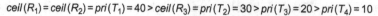

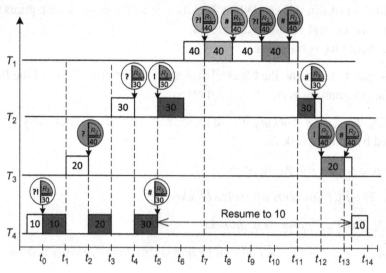

Figure 14.17
A more complicated PCP scenario.

to the block clause. Because T_1 is not granted lock R_1 but is blocked instead, the potential deadlock as shown in Figure 14.11 is avoided. Because T_2 blocks T_1, T_2 inherits T_1's priority according to the inheritance clause, and resumes execution at t_3. Task T_1 is blocked until t_7, at which time T_2 exits its critical section and reacquires its original priority. Then immediately, T_2 is preempted by T_1. From this point on, T_1 will run to completion, and then T_2 will resume its execution until its completion.

Figure 14.17 shows a more complicated scenario, where the priority of task T_4 changes from 10 to 20, to 30, and then returns to 10.

Task T_4 takes lock R_3 at time t_0. Task T_4 runs with priority 10 until it comes to t_1, at which time it is preempted by task T_3. Task T_3 at time t_2 attempts to take lock R_2. According to the block clause, T_3 is blocked by T_4, which interits the priority of T_3 and resumes its execution. At time t_3, task T_4 is preempted by task T_2. Task T_2 attempts to take lock R_3 at time t_4. According to the block clause, T_2 is blocked by T_4, which interits the priority of T_2 and resumes its execution. Task T_4 releases lock R_3 at time t_5. Task T_2 takes lock R_3 and executes until it is preempted by task T_1 at time t_6. Task T_1 attempts to take lock R_1 at time t_7. According to the grant clause rule, the lock is procured by T_1. At time t_{11}, task T_1 runs to completion. Task T_2 then resumes its execution and finishes at time t_{12}. Task T_3 takes lock R_2 at time t_{12} and runs to completion at some time after t_{13}. Task T_4 resumes its execution once T_3 completes.

As an exercise, you may check how the chain of blocking in Figure 14.13 can be prevented by using the PCP.

The POSIX.1-2008 standard specifies the PIP and the PCP as two options for mutexes. The PCP can be used for situations where the highest reliability is required. However, the PCP may not be supported by an RTOS because of its complexity. In such a case, developers may have to implement it by themselves.

Problems

14.1 What is multitasking? Why is multitasking important for real-time embedded systems?

14.2 How does multitasking bring service concurrency?

14.3 What is a system context diagram? How can it help in identifying tasks?

14.4 What is a task diagram? How does one derive a task diagram from a context diagram?

14.5 How many threads are involved in the serial port design pattern? Explain each of them.

14.6 What is the benefit of using a multibyte hardware transmit buffer in a serial port?

14.7 Explain the single-frequency strategy and the importance-level strategy.

14.8 What are the necessary conditions for a resource deadlock to occur? How may one attack each in order to prevent resource deadlocks?

14.9 What is priority inversion? What is unbounded priority inversion?

14.10 Explain how the DEADLOCK-DETECTION algorithm works. Explain how the BANKER algorithm works.

14.11 How does the PIP work? Can it prevent unbounded priority inversions? Can it prevent deadlocks? What is the problem with this protocol?

14.12 How does the HLP work? Can it prevent unbounded priority inversions? Can it prevent deadlocks? What is the problem with this protocol?

14.13 How does the PCP work? Can it prevent unbounded priority inversions? Can it prevent deadlocks?

14.14 Figure 14.18 shows a resource-client graph. Suppose each of the resources R_1, R_2, and R_3 is protected by a distinct resource lock. We also use R_1, R_2, and R_3 to refer to the resource locks. What are the respective priority ceilings for the three resource locks?

14.15 Assume the PCP is used for the task design in Figure 14.18. The timing diagram in Figure 14.19 shows a possible schedule. Explain what happens at time t_3. What is the priority of task T_3 before t_3? What is the priority of task T_3 between t_3 and t_{10}? What is the priority of task T_3 after t_{10}?

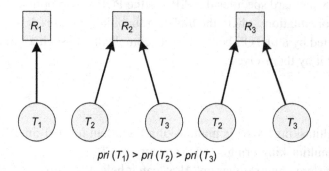

Figure 14.18
A task design in a resource-client graph.

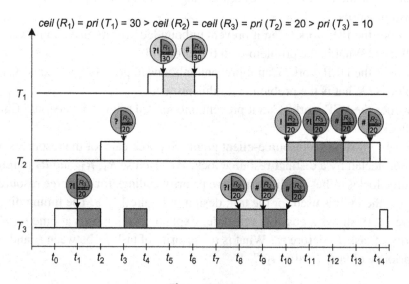

Figure 14.19
A scenario of the PCP.

CHAPTER 15

Real-Time Scheduling: Clock-Driven Approach

Contents

Know the true value of time; snatch, seize, and enjoy every moment of it. No idleness, no
laziness, no procrastination; Never put off till tomorrow what you can do today.

Lord Chesterfield

15.1 Introduction to Cyclic Scheduling

Cyclic scheduling is an approach to scheduling periodic tasks according to a *static schedule table* precomputed off-line. The schedule table specifies exactly when each job of a periodic task executes. During run time, the schedule table is used by the scheduler to dispatch the released periodic jobs. When there is no periodic job ready, the processor can take its time to execute aperiodic jobs.

Real-Time Embedded Systems. http://dx.doi.org/10.1016/B978-0-12-801507-0.00015-8

15.1.1 Assumptions

Throughout this chapter, we make the following assumptions:

(1) The system $\mathbb{T}$ contains k independent periodic tasks, where the number k is fixed at design time: $\mathbb{T} = \{T_i = (p_i, r_i, e_i, d_i) | 1 \leq i \leq k\}$.

(2) The parameters of all periodic tasks are known prior to schedulability analysis. Each task T_i releases its first job J_{i1} at the time instant r_i, and then releases jobs one after another every p_i units of time. Recall that T_i is denoted by (p_i, e_i) $(1 \leq i \leq k)$ when $r_i = 0$ and d_i is equal to the end of its period p_i.

(3) At run time, aperiodic jobs may be released at unexpected time instants.

(4) Each job becomes ready for execution immediately at its release time.

(5) Jobs are scheduled to execute on one processor. Aperiodic jobs can be scheduled only when no periodic job is ready, and their execution can be preempted by periodic jobs.

In cyclic scheduling, the schedule table for a system has to be repeatable, so that the scheduler can come back to the first entry when it has finished the last entry of the table. Given a task set, the minimum length of time that is sufficient to create an infinitely repeatable schedule is called its *hyperperiod*. Formally, the hyperperiod of a task set $\mathbb{T}$ is defined by

$$H(\mathbb{T}) = \operatorname*{LCM}_{T_i \in \mathbb{T}}(p_i),$$

where LCM is an operator for the "least common multiple."

Within a hyperperiod, jobs from each periodic task can be released exactly an integer number of times. For example, consider the four periodic tasks given in Figure 15.1, where

$$\mathbb{T}_1 = \begin{cases} T_1 = (20, 5), \\ T_2 = (25, 9), \\ T_3 = (100, 5), \\ T_4 = (100, 10). \end{cases}$$

The hyperperiod $H(\mathbb{T}_1) = \text{LCM}(20, 25, 100, 100) = 100$. Within 100 units of time, task T_1 has five jobs released, task T_2 has four jobs released, and tasks T_3 and T_4 each have exactly one job released. This pattern repeats itself one hyperperiod after another.

Figure 15.1
Hyperperiod of a set of periodic tasks.

15.1.2 Preemptable Aperiodic Jobs

Before we come to the schedule of periodic tasks, we first look at aperiodic jobs, which are released randomly and have no hard deadlines. Aperiodic jobs, once released, are typically placed in a queue-like data structure in the order (say, FIFO) that is appropriate for the application under consideration. Whenever the processor is available for aperiodic jobs, the job at the head of this queue executes.

A task server, called an *aperiodic server*, can be introduced to regulate the execution of aperiodic jobs. The procedure APERIODIC-SERVER given below implements a simple task server, where the operating system (OS) signaling mechanism is exploited to implement the behavior of preemption/resumption (the signaling mechanism is covered in Chapter 21). Briefly, signals are, just like hardware interrupts, events delivered to a task asynchronously. When a task receives a signal, its normal flow of execution is interrupted by the OS, and the control is transferred to the corresponding signal handler.

APERIODIC-SERVER ()

```
 1    s = init_Semaphore(0)
 2    // AperiodicQueue is populated by other tasks or ISRs
 3    AperiodicQueue = init_JobQueue()
 4    install_signal_handler(SIG_Preemption, PREEMPTION-HANDLER)
 5    install_signal_handler(SIG_Resumption, RESUMPTION-HANDLER)
 6    sem_wait(s); // this server is controlled by signals issued by task scheduler
 7    for (; ; )
 8        job = AperiodicQueue.get() // blocking when null
 9        RUN-JOB-TO-COMPLETION (job) // preemptible by signals
10    end
```

PREEMPTION-HANDLER (*Semaphore s*)
```
    sem_wait(s); // take the semaphore
```

RESUMPTION-HANDLER (*Semaphore s*)
```
    sem_post(s); // release the semaphore
```

The procedure APERIODIC-SERVER makes use of a semaphore, which is initialized to 0; this ensures that the task server is blocked at line 6 upon system start. Whenever another task (say, the clock-driven scheduler to be discussed later) sends to this aperiodic server a signal SIG_Resumption, in response, the signal-handler RESUMPTION-HANDLER is invoked and the semaphore is released (unlocked), which is then consumed by the code at line 6. This allows the aperiodic server to enter into the loop structure and execute the next aperiodic job, if available.

While the aperiodic server is executing an aperiodic job, the clock-driven scheduler may send out a signal SIG_Preemption. In response, the aperiodic server is interrupted by the OS, and the control is transferred to the signal-handler PREEMPTION-HANDLER, where the handler is blocked, waiting for the semaphore. The aperiodic server will not resume its execution until it receives a signal SIG_Resumption.

In sum, through the two signals SIG_Preemption and SIG_Resumption, a task scheduler can control when to allow aperiodic jobs to execute. Let us first consider an ad-hoc clock-driven task scheduler.

15.2 Ad-hoc Clock-Driven Scheduling

Given a task set $\mathbb{T} = \{T_i = (p_i, e_i) | 1 \leq i \leq k\}$, we can use the following procedure to derive a schedule table:

(1)   Determine the hyperperiod: $H(\mathbb{T}) = \underset{T_i \in \mathbb{T}}{\mathrm{LCM}}(p_i)$.

(2)   Work out a timing diagram such that
    (a)   The length of the timing diagram is equal to a hyperperiod.
    (b)   The amount of time allocated to each job J_{ix} is equal to e_i.
    (c)   No periodic job can be preempted (no run-time context switching).
    (d)   Every job finishes by its deadline.
    (e)   When there are many feasible schedules, it is preferable to choose one where the processor idles nearly periodically. This can help to minimize the response time of aperiodic jobs.

(3)   If there is no feasible schedule, the task set cannot be scheduled by the clock-driven approach. Otherwise, a schedule table can be derived from the timing diagram.

As an example, consider the task set

$$\mathbb{T}_2 = \begin{cases} T_1 = (6, 2), \\ T_2 = (12, 3), \\ T_3 = (18, 4). \end{cases}$$

First, the hyperperiod of task set $\mathbb{T}_2$ is $H(\mathbb{T}_2) = \mathrm{LCM}(6, 12, 18) = 36$. Within a hyperperiod, task T_1 has six jobs released, task T_2 has three jobs released, and task T_3 has two jobs released.

Then, we need to work out a schedule such that all the jobs can meet their deadlines. Figure 15.2 gives one feasible schedule. There also exist other schedules. For example, the idle time between jobs J_{21} and J_{12} could be between jobs J_{11} and J_{21}.

The schedule given in Figure 15.2 can then be converted into a schedule table as shown in Table 15.1. Note that the entries of the "Job list" column are either periodic jobs or "I," which

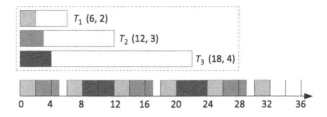

Figure 15.2
A timing diagram showing the schedule of periodic tasks.

**Table 15.1 Schedule table
for task set $\mathbb{T}_2$**

k	Timer List	Job List
0	0	J_{11}
1	2	J_{21}
2	5	/
3	6	J_{12}
4	8	J_{31}
5	12	J_{13}
6	14	J_{22}
7	17	/
8	18	J_{14}
9	20	J_{32}
10	24	J_{15}
11	26	J_{23}
12	29	/
13	30	J_{16}
14	32	/

indicates time intervals where the processor idles. As mentioned before, aperiodic jobs can be scheduled to execute when it comes to an "idle" entry.

15.2.1 Ad-hoc Clock-Driven Scheduler

The implementation of a clock-driven scheduler is quite straightforward.

The procedure CLOCK-SCHEDULER given below starts with initialization, where the job list and timer list are populated from the schedule table.

After initialization, the scheduler enters into a forever loop. For each iteration of the loop, the scheduler first reads the current entry from the job list. It then launches the next timer, so that it can proceed to the next entry once it had received a time-out signal from the timer. If the current entry of the job list is idle and the aperiodic server is currently not running, it is time to send a signal to the aperiodic server so that it can start or resume the execution of an aperiodic job. If the current entry of the job list is a periodic job, the aperiodic server is preempted if it is running, and then the periodic job runs to completion. At the end of the loop, the scheduler waits for a time-out signal to start the next iteration.

CLOCK-SCHEDULER (*ScheduleTable*: *st*)

```
1   N = st.getLength()
2   JobList = st.getJobList()
3   TimerList = st.getTimerList()
4   AperiodicServerIsRunning = false
5   k = 0                // the current index position of the table
6   for (; ; )
7       curJob = JobList(k)
8       k = k + 1  (mod N)   // setup the new timer
9       LAUNCH-NEXT-TIMER (TimerList(k))
10      if (curJob.isNull()) AND (AperiodicServerIsRunning == false)
11          AperiodicServerIsRunning == true
12          Signal(APERIODIC-SERVER, SIG_Resumption)
13      if curJob.isNotNull()
14          if (AperiodicServerIsRunning == true)
15              AperiodicServerIsRunning == false
16              Signal(APERIODIC-SERVER, SIG_Preemption)
17          RUN-JOB-TO-COMPLETION (curJob)
18      WAIT-TIMEOUT-SIGNAL ()                // blocking
19  end
```

15.2.2 Execution Overhead

It is worth noting that by RUN-JOB-TO-COMPLETION (*job*) we mean that the job has to be executed until it is done without any preemption from other periodic jobs. This can be implemented differently, depending on whether the system is built upon an OS or not.

If an OS is not used (say, the system implements a round-robin architecture with interrupts as discussed in Chapter 12), the timer used in the algorithm given in Section 15.2.1 could be replaced by chaining a user-level timer interrupt, and the designer should ensure that the implemented interrupt service routine (ISR) can signal the scheduler with an appropriate time resolution (say, every 20 ms) as required by the job schedules. Since multithreading is not possible, a technique termed *in situ execution* is used, where the scheduler runs the jobs in its schedule table in a round-robin manner. In particular, every time a notification is received from the ISR, the scheduler checks whether it is time to run the next job. If it is, the next job starts to run, and upon its completion, the scheduler moves to the next entry.

If an OS is used, as explained in Chapter 13, each job can be executed in a dedicated thread. In such a case, each thread should signal the scheduler upon job completion. The system can create a new thread for each job, or be more efficient, merely create a thread for each periodic task.

However, even though an OS is used, the in situ execution approach is still appealing, especially when the jobs do not have hard deadlines. In situ execution is also preferable when the overhead associated with thread creation and context switching is intolerably high.

One drawback of the ad-hoc clock-driven scheduler is that it needs to make scheduling decisions at specific time instants according to the precomputed schedule table. The timers launched by the procedure Clock-Scheduler can vary widely, depending on the actual task execution times. Often we want our system to be able to make scheduling decisions periodically so that (a) enforcement actions can be taken periodically to handle occasional job overruns, and (b) the response time of aperiodic jobs can be fairly reduced. This leads to the frame-based scheduling approach.

15.3 Frame-Based Scheduling

In *frame-based scheduling*, the schedule table is composed of *frames* of equal size, and multiple jobs can be scheduled sequentially within each frame (there is no preemption within a frame).

Below, we assume that scheduling decisions are made only at the beginning of every frame. In particular, at the beginning of a frame j,

- the scheduler needs to check whether every job scheduled in the previous frame $j - 1$ has finished by its deadline. When there is a *frame overrun*,[1] at the frame boundary, the scheduler may take some appropriate actions to recover from it.
- the scheduler needs to check whether every job scheduled in the current frame j has indeed been released and is ready for execution.

15.3.1 Constraints on Frame Size

Frame-based scheduling would reduce to a round-robin-like job scheduler if each frame could hold only one job. At the other extreme, it is no longer a cyclic scheduling if there is only one frame which is equal to the hyperperiod. The frame size ought to be somewhere in between. Given a task set $\mathbb{T} = \{T_i = (p_i, e_i, d_i) | 1 \leq i \leq k\}$, it is suggested to pick a frame size F such that it can satisfy the following constraints:

---

[1] Some job scheduled in the previous frame has executed for longer than the time allocated to it by the precomputed schedule.

$$H \pmod{F} = 0, \tag{15.1}$$

$$F \geq \max_{i=1}^{k}(e_i), \tag{15.2}$$

$$2F - \gcd(p_i, F) \leq d_i, \tag{15.3}$$

where the operator gcd is the greatest common divisor.

Constraint (15.1) says that the hyperperiod ought to have an integer number of frames. This ensures that the repeatable schedule pattern is within one hyperperiod. Constraint (15.2) says that the frame size ought to be sufficiently long so that every job can start and finish within a single frame. This ensures that no job will cross a frame boundary. Constraint (15.3) implies that the frame size ought to be sufficiently small so that there is at least one frame between the release time and the deadline of every job. This ensures that there is a chance for the scheduler to determine whether the job is ready for execution, or has finished by its deadline.

As an example, let us work out a frame size for the task set $\mathbb{T}_1$:

(1) Constraint (15.1): Since $H = 100$, from $H \pmod{F} = 0$ we have $F \in \{2, 4, 5, 10, 20, 25, 50, 100\}$.

(2) Constraint (15.2): From $F \geq \max_{i=1}^{k}(e_i)$, we have $F \geq \max(5, 9, 10) = 10$. The candidate frame sizes reduce to $F \in \{10, 20, 25, 50, 100\}$.

(3) Constraint (15.3): $2F - \gcd(p_i, F) \leq d_i$. Since there are three different periods, we have the following three inequalities:

$$2F - \gcd(20, F) \leq 20,$$
$$2F - \gcd(25, F) \leq 25,$$
$$2F - \gcd(100, F) \leq 100.$$

(a) Try $F = 10$. It satisfies all three inequalities.

(b) Try $F = 20$. It does not satisfy the second inequality.

Hence, $F = 10$ can be used as the frame size for task set $\mathbb{T}_1$.

Figure 15.3 shows a frame-based schedule ($F = 10$) for this task set. Note the following:

- There are 10 frames within a hyperperiod.
- Each frame contains one or two jobs.
- Within each frame, jobs are scheduled sequentially. No job is preempted.
- Every job starts and finishes within a single frame. No job ever crosses a frame boundary.
- The execution of every job starts in a frame which begins after the release time of the job.
- Every job is scheduled to finish in a frame which ends before the deadline of the job.

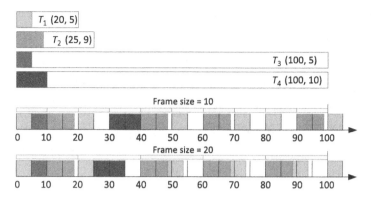

Figure 15.3
A frame-based schedule for $\mathbb{T}_1$.

It is worth noting that the three constraints are only heuristics for finding a *desirable frame size*. The constraints are *not* necessary conditions for *schedulability*. For the task set $\mathbb{T}_1$, even though F cannot be 20, Figure 15.3 actually gives a schedule when $F = 20$.

The constraints on frame sizes are not sufficient conditions for *schedulability* either. Consider the following task set:

$$\mathbb{T}_3 = \begin{cases} T_1 = (4, 1, 4), \\ T_2 = (5, 2, 7), \\ T_3 = (20, 3, 20), \\ T_4 = (20, 2, 20). \end{cases}$$

Let us first work out a frame size for task set $\mathbb{T}_3$:

(1)   Constraint (15.1): Since $H = 20$, from $H \pmod F = 0$ we have $F \in \{2, 4, 5, 10, 20\}$.
(2)   Constraint (15.2): From $F \geq \max_{i=1}^{k}(e_i)$, we have $F > \max(1, 2, 3) = 3$. The candidate frame sizes reduce to $F \in \{4, 5, 10, 20\}$.
(3)   Constraint (15.3): $2F - \gcd(p_i, F) \leq d_i$. Since there are three different periods, we have the following three inequalities:

$$2F - \gcd(4, F) \leq 4,$$
$$2F - \gcd(5, F) \leq 7,$$
$$2F - \gcd(20, F) \leq 20.$$

(a)   Try $F = 4$. It satisfies all three inequalities.
(b)   Try $F = 5$. It does not satisfy the first two inequalities.

Hence, $F = 4$ can be used as the frame size for task set $\mathbb{T}_3$.

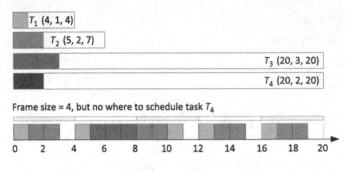

Figure 15.4
No workable schedule when $F = 4$.

However, as indicated in Figure 15.4, no schedule can be found for task T_4.

There may exist multiple frame sizes satisfying the constraints. Consider the following task set:

$$\mathbb{T}_4 = \begin{cases} T_1 = (30, 10), \\ T_2 = (60, 15), \\ T_3 = (90, 20). \end{cases}$$

Let us first examine the constraints on the frame size for task set $\mathbb{T}_4$:

(1) Constraint (15.1): Since $H = 180$, from $H \pmod{F} = 0$ we have
$F \in \{2, 3, 4, 5, 6, 9, 10, 12, 15, 18, 20, 30, 36, 45, 60, 90, 180\}$.

(2) Constraint (15.2): From $F \geq \max_{i=1}^{k}(e_i)$, we have $F \geq \max(10, 15, 20) = 20$. The candidate frame sizes reduce to $F \in \{20, 30, 36, 45, 60, 90, 180\}$.

(3) Constraint (15.3): $2F - \gcd(p_i, F) \leq d_i$. Since there are three different periods, we have the following three inequalities:

$$2F - \gcd(30, F) \leq 30,$$
$$2F - \gcd(60, F) \leq 60,$$
$$2F - \gcd(90, F) \leq 90.$$

    (a) Try $F = 20$. It satisfies all three inequalities.
    (b) Try $F = 30$. It satisfies all three inequalities.
    (c) Try $F = 36$. It does not satisfy the first two inequalities.

The frame size can be either 20 or 30.

As shown in Figure 15.5, a schedule can be found for task set $\mathbb{T}_4$ when its frame size is either 20 or 30.

Table 15.2 gives the schedule table derived from the frame-based schedule given in Figure 15.5 when $F = 30$.

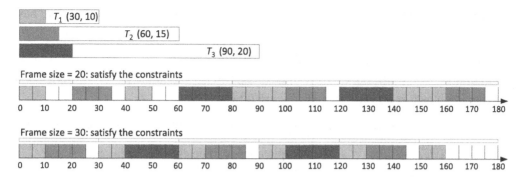

Figure 15.5
A task set with multiple frame sizes.

Table 15.2 Frame-based
schedule table for $\mathbb{T}_4$ ($F = 30$)

k	Timer List	Job List
0	0	J_{11}, J_{21}
1	30	J_{12}, J_{31}
2	60	J_{13}, J_{22}
3	90	J_{14}, J_{32}
4	120	J_{15}, J_{23}
5	150	J_{16}

15.3.2 Robust Frame-Based Schedule

As discussed in the last section, a schedule can be found even though the frame size chosen does not satisfy the constraints, and there may exist multiple frame-based schedules with each frame size satisfying the constraints.

We have seen several good properties in the frame-based schedule for task set $\mathbb{T}_1$ as shown in Figure 15.3. We now give some criteria for evaluating the quality of a frame-based schedule.

Given a frame-based schedule of a task set, for any job J_x, let

- t_x^r denote the time it is released;
- t_x^s denote the time it is scheduled to execute;
- t_x^c denote the time when it runs to completion;
- t_x^d denote its deadline.

Also, given a frame-based schedule of a task set, for any time instant t,

- let $f_p(t)$ denote the time instant of the previous frame boundary immediately before t;
- let $f_n(t)$ denote the time instant of the next frame boundary immediately after t.

Given a frame-based schedule of a task set, it is a *robust schedule* if, for any job J_x, we have

(1) criterion R.1: $t_x^c \leq f_n(t_x^c) \leq t_x^d$. Its completeness can be checked at the next frame boundary, which is no later than its deadline.

(2) criterion R.2: $t_x^r \leq f_p(t_x^s) \leq t_x^s$. Its readiness can be checked at the previous frame boundary, which is no early than its release time.

Given a frame-based schedule of a task set, it is a *feasible schedule* if, for any job J_x, we have

(1) criterion Fl.1: $t_x^c \leq f_n(t_x^c) \leq t_x^d$. Its completeness can be checked at the next frame boundary, which is no later than its deadline.

(2) criterion Fl.2: $f_p(t_x^s) < t_x^r \leq t_x^s$, and the system is contiguously busy with some other job(s) during $[f_p(t_x^s), t_x^s]$. In this case, although its readiness cannot be checked, it is guaranteed that J_x is not scheduled to run until after it is released.

or we have

(1) criterion F2.1: $f_p(t_x^c) = t_x^s < t_x^c \leq t_x^d \leq f_n(t_x^c)$. It is the first job in the frame and its completeness can be checked at the next frame boundary, which, however, is no early than its deadline.

(2) criterion F2.2: $t_x^r \leq f_p(t_x^s) \leq t_x^s$. Its readiness can be checked at the previous frame boundary, which is no early than its release time.

Consider the two frame-based schedules given in Figure 15.3. The schedule when $F = 10$ is a robust schedule, because all the jobs satisfy criteria R.1 and R.2. The schedule when $F = 20$ is only a feasible schedule, because

- the job of task T_3 and all the jobs of task T_1 satisfy criteria R.1 and R.2;
- jobs J_{21} and J_{24} satisfy criteria R.1 and R.2. However, jobs J_{22} and J_{23} both satisfy criteria F2.1 and F2.2.

Hence, the frame size $F = 10$ is more desirable; a smaller frame size actually allows the scheduler to have more chances of finding erroneous conditions and take corrective actions, if necessary.

Consider the two frame-based schedules given in Figure 15.5. The schedule when $F = 20$ is only a feasible schedule, because

- all the jobs of tasks T_2 and T_3 satisfy criteria R.1 and R.2;
- jobs J_{11} and J_{12} satisfy criteria R.1 and R.2. However, jobs J_{13} and J_{15} both satisfy criteria F2.1 and F2.2, while jobs J_{14} and J_{16} both satisfy criteria F1.1 and F1.2.

The schedule when $F = 30$ is a robust schedule and is more desirable, because all the jobs satisfy criteria R.1 and R.2.

In sum, a robust schedule is better than a feasible schedule. If more than one frame size can be used and each can lead to a robust schedule, it is typically more desirable to select as large a frame size as possible. This way, the context-switching overhead can be kept low, because as many jobs as possible can execute from start to finish without interruption.

15.3.3 Frame-Based Scheduler

Since frames are of equal size, the schedule table can be implemented as a circular structure with nodes evenly spread over time. As an example, Figure 15.6 shows a circular structure for the schedule in Table 15.2.

In general, the scheduler can make use of a periodic timer, which sends a time-out signal every time a frame expires. Upon receiving a time-out signal, the scheduler can move ahead to schedule those jobs associated with the next frame.

FRAME-SCHEDULER0 (*ScheduleTable*: *st*)

```
1   N = st.getTableLength()
2   θ = st.getFrameSize()
3   FrameList = st.getFrameList()
4   LAUNCH-PERIODIC-TIMER (θ)
5   k = 0          // the current index position of the table
6   for (; ; )
7       curFrame = FrameList(k)
8       while curFrame ≠ ∅
9           job = GET-NEXT-JOB (curFrame)
10          RUN-JOB-TO-COMPLETION (job)
11      end
12      k = k + 1 (mod N)        // next entry
13      WAIT-TIMEOUT-SIGNAL ()                    // blocking
14  end
```

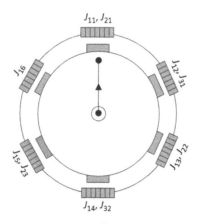

Figure 15.6
Using a circular structure to implement frame-based scheduling.

The procedure FRAME-SCHEDULER0 sketches a simple implementation, where the scheduler enters into a forever loop after the installation of a periodic timer (the time-out value is equal to the frame size). For each iteration of the loop, the scheduler simply iterates over all the jobs associated with the current frame, and runs each job to completion. Whenever it receives a time-out signal, the scheduler will move forward to the next frame. As a laboratory exercise, implement FRAME-SCHEDULER0 in a real-time OS.

15.4 Scheduling Aperiodic Jobs

Since aperiodic jobs have no hard deadlines, they are normally scheduled after all the periodic jobs scheduled in a frame have finished.

Figure 15.7 shows a robust schedule of task set $\mathbb{T}_4$. Assume that one aperiodic job A_1 is released at time 55 and demands five units of execution time, and another aperiodic job A_2 is released at time 80 and demands 17.5 units of execution time.

The FIFO-based schedule of the aperiodic jobs is given first in Figure 15.7:

(1) The processor is not idle until J_{22} finishes at time 85. At this time both A_1 and A_2 are released and are ready for execution. Since aperiodic jobs are scheduled in a FIFO manner, A_1 is scheduled immediately after J_{22} and finishes at the end of the third frame.

(2) The processor becomes idle again at time 145. A_2 is scheduled at this time and is preempted at the end of the frame.

(3) The execution of job A_2 is resumed at time 160 and runs to completion at 172.5.

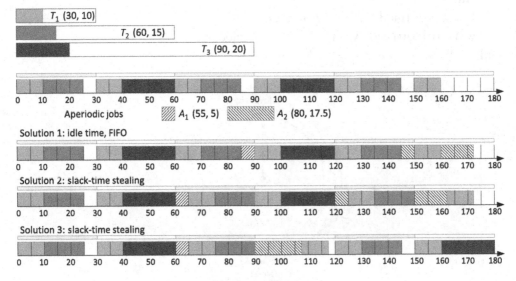

Figure 15.7

Scheduling aperiodic jobs.

However, there is no advantage to completing periodic jobs early. As long as the deadlines of periodic jobs are honored, it is preferable to complete aperiodic jobs as soon as possible. One technique is called *slack stealing*, where the aperiodic jobs are scheduled ahead of periodic jobs whenever possible. The guideline of this approach is that every periodic job should be scheduled in a frame that ends no later than its deadline.

Let the total amount of time allocated to the periodic jobs scheduled in frame j be $\kappa(j)$. We have the following rules:

- At the beginning of a frame j, the available slack time, denoted by $\eta(j)$, is equal to $F - \kappa(j)$. If there is an aperiodic job ready for execution at this time, the aperiodic job can be executed for an amount of time up to $\eta(j)$; this will not cause any periodic job to miss its deadline.
- The available slack time $\eta(j)$ is reduced while it is consumed by the execution of an aperiodic job.
- Upon the completion of a job, as long as $\eta(j) > 0$, a released aperiodic job can be executed.
- When there are no more aperiodic jobs or $\eta(j)$ becomes 0, a periodic job scheduled in the current frame starts to execute. $\eta(j)$ remains the same during the execution of periodic jobs.

Now, let us examine the second schedule of the aperiodic jobs given in Figure 15.7:

(1) A_1 is released at time 55. At time 60, the scheduler enters into the third frame, where we have $\eta(3) = 30 - e_1 - e_2 = 5$. Since $\eta(3) > 0$ and A_1 is ready for execution, A_1 is scheduled to execute and runs to completion at time 65.

(2) $\eta(3) = 0$ at time 65, so periodic job J_{13} starts to execute, followed by job J_{22}.

(3) The aperiodic job A_2 is released at time 80. A_2 cannot be scheduled at time 90, because $\eta(4) = 0$.

(4) At time 120, since $\eta(5) = 5$, A_2 is scheduled for five units of execution time, and is preempted by J_{15} at time 125.

(5) At time 150, since $\eta(6) = 20$, A_2 is scheduled to execute until it finishes at 162.5.

According to the slack-stealing schedule, both aperiodic jobs have a better response time. Actually, there is an even better schedule, as shown at the bottom of Figure 15.7, where both aperiodic jobs are scheduled at the beginning of the frame immediately after their respective release time. Of course, this solution is no longer a static schedule; it requires the schedule table to be dynamically adjusted in favor of the randomly released aperiodic jobs.

Similarly to ad-hoc CLOCK-SCHEDULER, the algorithm FRAME-SCHEDULER given below makes use of the signaling mechanism available in a real-time OS to manipulate the execution of AperiodicServer. Also, aperiodic jobs are always scheduled at the end of a frame, if there is slack time available. FRAME-SCHEDULER can be easily changed to implement the slack stealing technique such that the execution of aperiodic jobs is always scheduled at the beginning of a frame.

FRAME-SCHEDULER (*ScheduleTable*: *st*)

```
 1   N = st.getTableLength()
 2   θ = st.getFrameSize()
 3   FrameList = st.getFrameList()
 4   LAUNCH-PERIODIC-TIMER (θ)
 5   AperiodicServerIsRunning = false
 6   k = 0                    // the current index position of the table
 7   for (; ; )
 8       curFrame = FrameList(k)
 9       if (curFrame = ∅) AND (AperiodicServerIsRunning == false)
10           AperiodicServerIsRunning == true
11           Signal(APERIODIC-SERVER,SIG_Resumption)
12       if curFrame ≠ ∅
13           if (AperiodicServerIsRunning == true)
14               AperiodicServerIsRunning == false
15               Signal(APERIODIC-SERVER,SIG_Preemption)
16           while curFrame ≠ ∅
17               job = GET-NEXT-JOB (curFrame)
18               RUN-JOB-TO-COMPLETION (job)
19           end
20           // use the slack time of this frame for aperiodic jobs
21           if (curFrame.lefttime > δ) // δ is threshold for switching cost
22               AperiodicServerIsRunning == true
23               Signal(APERIODIC-SERVER,SIG_Resumption)
24       end
25       k = k + 1  (mod N)   // next entry
26       WAIT-TIMEOUT-SIGNAL ()              // blocking
27   end
```

15.5 Task Splitting

Consider the following task set:

$$
\mathbb{T}_5 = \begin{cases} T_1 = (4, 1), \\ T_2 = (6, 2), \\ T_3 = (24, 5). \end{cases}
$$

Let us first examine the constraints on the frame size for task set $\mathbb{T}_5$:

(1) Constraint (15.1): Since $H = 24$, from $H \pmod F = 0$ we have $F \in \{2, 3, 4, 6, 8, 12, 24\}$.
(2) Constraint (15.2): From $F \geq \max_{i=1}^{k}(e_i)$, we have $F \geq \max(1, 2, 5) = 5$. The candidate frame sizes reduce to $F \in \{6, 8, 12, 24\}$.
(3) Constraint (15.3): $2F - \gcd(p_i, F) \leq d_i$. Since there are three different periods, we have the following three inequalities:

$$2F - \gcd(4, F) \leq 4,$$
$$2F - \gcd(6, F) \leq 6,$$
$$2F - \gcd(24, F) \leq 24.$$

Try $F = 6$. It does not satisfy any of the three inequalities.

No feasible frame size can be derived from the constraints.

For situations like this, in order to use frame-based scheduling, we have to transform some tasks of the system to produce a new task set. For the original task set $\mathbb{T}_5$, constraint (15.2) requires that the frame size be no less than 5. This lower bound can be lowered if we can change the execution time of task T_3. This leads to a technique called task splitting.

In general, a task $T_i = (p_i, e_i, d_i)$ can be split into m pieces, where each piece is a periodic task $T_{ij} = (p_i, e_{ij}, d_i)$, and

$$e_i = \sum_{1 \leq j \leq m} e_{ij}.$$

Obviously, each job execution of the original task T_i is equivalent to the combined execution of one job from each T_{ij}.

As an example, Figure 15.8 illustrates the task splitting technique. Solution 1 transforms $\mathbb{T}_5$ to a new task set

$$\mathbb{T}_5' = \begin{cases} T_1 = (4, 1), \\ T_2 = (6, 2), \\ T_{31} = (24, 1), \\ T_{32} = (24, 3), \\ T_{33} = (24, 1). \end{cases}$$

Let us examine the constraints on the frame sizes for task set $\mathbb{T}_5'$:

(1) Constraint (15.1): Since $H = 24$, from $H \pmod F = 0$ we have $F \in \{2, 3, 4, 6, 8, 12, 24\}$.
(2) Constraint (15.2): From $F \geq \max_{i=1}^{k}(e_i)$, we have $F \geq \max(1, 2, 3) = 3$. The candidate frame sizes reduce to $F \in \{3, 4, 6, 8, 12, 24\}$.

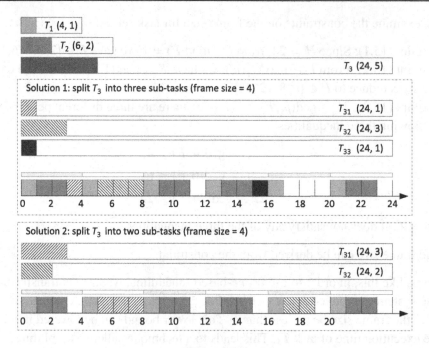

Figure 15.8
Multiple ways of task splitting.

(3)   Constraint (15.3): $2F - \gcd(p_i, F) \leq d_i$. Since there are three different periods, we have the following three inequalities:

$$2F - \gcd(4, F) \leq 4,$$
$$2F - \gcd(6, F) \leq 6,$$
$$2F - \gcd(24, F) \leq 24.$$

(a)   Try $F = 3$. It does not satisfy any of the three inequalities.
(b)   Try $F = 4$. It satisfies all of three inequalities.
(c)   Try $F = 6$. It does not satisfy any of the three inequalities. Stop trying.

Hence, the frame size for $\mathbb{T}'_5$ can be 4, and a schedule is given in the middle of Figure 15.8. It can be verified that the schedule is a robust schedule.

Given a task set, there may exist multiple ways to split a task. Solution 2 given in Figure 15.8 shows another transformation:

$$\mathbb{T}''_5 = \begin{cases} T_1 = (4, 1), \\ T_2 = (6, 2), \\ T_{31} = (24, 3), \\ T_{32} = (24, 2). \end{cases}$$

For this splitting, the frame size is still 4, and a schedule is given at the bottom of Figure 15.8. It can be verified that it is also a robust schedule.

Problems

15.1 Work out a feasible frame-based schedule for the task set

$$\mathbb{T}_6 = \begin{cases} T_1 = (60, 12), \\ T_2 = (150, 30), \\ T_3 = (300, 30). \end{cases}$$

15.2 Implement a frame-based scheduler as sketched in FRAME-SCHEDULER0. Your program should use in situ execution for periodic jobs. Consider using task set $\mathbb{T}_6$ for testing.

15.3 Implement a frame-based scheduler as sketched in FRAME-SCHEDULER0. Your program should use a dedicated thread for each periodic job. Consider using task set $\mathbb{T}_6$ for testing.

15.4 Work out a feasible frame-based schedule for task set $\mathbb{T}_3$. (Hint: try frame size 5).

15.5 Refer to Figure 15.9.
   (a) Is the schedule a robust schedule? Explain why or why not.
   (b) Is the schedule a feasible schedule? Explain why or why not. If it is not, work out a feasible schedule.

15.6 Work out a feasible frame-based schedule for the task set

$$\mathbb{T}_7 = \begin{cases} T_1 = (6, 1), \\ T_2 = (12, 2), \\ T_3 = (18, 2). \end{cases}$$

15.7 Work out a feasible frame-based schedule for the task set

$$\mathbb{T}_8 = \begin{cases} T_1 = (8, 1), \\ T_2 = (16, 4), \\ T_3 = (24, 9). \end{cases}$$

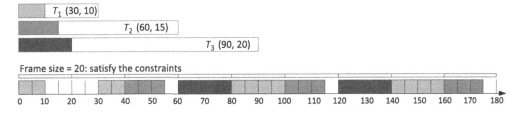

Figure 15.9
Schedule of a set of periodic tasks.

15.8 Work out a feasible frame-based schedule for the task set

$$\mathbb{T}_9 = \begin{cases} T_1 = (5, 1), \\ T_2 = (10, 2), \\ T_3 = (20, 6). \end{cases}$$

CHAPTER 16

Real-Time Scheduling: Rate-Monotonic Approach

Contents

The key is not to prioritize what's on your schedule, but to schedule your priorities.

Stephen R. Covey

16.1 Priority Assignment

In Chapter 15, we discussed clock-driven scheduling (ad-hoc or frame based), where a repeatable schedule table is used to lay out the execution of a set of periodic tasks. One important feature of clock-driven scheduling is that there is no task preemption among the periodic tasks. In other words, the notion of priority plays no role in clock-driven scheduling.

In this chapter, we consider systems where periodic tasks can be associated with different priority levels. We focus on fixed-priority scheduling, where the priority of a task is statically determined at design time.

Let us first look at the issue of priority assignment. Consider a system consisting of two periodic tasks:

$$\mathbb{T}_1 = \begin{cases} T_1 = (40, 20), \\ T_2 = (80, 30). \end{cases}$$

According to our notation, tasks T_1 and T_2 are released at the first instant of their respective periods, and their job deadlines are the same as the end of their periods.

Since the system contains only two tasks, there are only two different priority assignments:

(1) $\mathrm{pri}(T_1) < \mathrm{pri}(T_2)$: task T_2 has a higher priority than task T_1. A schedule of $\mathbb{T}_1$ is given in Figure 16.1. Both tasks (jobs J_{11} and J_{21}) are ready at time 0. J_{21} is scheduled first because of its higher priority. J_{21} runs to completion at 30, at which time J_{11} starts to execute. However, J_{11} misses its deadline at time 40. Also note that the second job, J_{12}, of task T_1 starts at 50 and ends at 70.

(2) $\mathrm{pri}(T_1) > \mathrm{pri}(T_2)$: task T_1 has a higher priority than task T_2. A schedule of $\mathbb{T}_1$ is given in Figure 16.2. Both tasks (jobs J_{11} and J_{21}) are ready at time 0. J_{11} is scheduled first because of its higher priority. J_{11} runs to completion at 20, at which time J_{21} starts to execute. At time 40, the second job, J_{12}, of task T_1 is released, which preempts the execution of J_{21}. After J_{12} runs to completion at time 60, J_{21} resumes its execution and finishes at time 70.

Obviously, the priority assignment $\mathrm{pri}(T_1) > \mathrm{pri}(T_2)$ is preferable because the jobs of both tasks can finish before their respective deadlines. This example also implies that the issue of priority assignment is critical in the design of real-time systems: an inappropriate priority assignment might cause a system to break its timing constraints. Fortunately, there are some

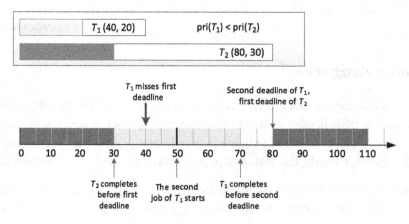

Figure 16.1

Priority assignment: $\mathrm{pri}(T_1) < \mathrm{pri}(T_2)$.

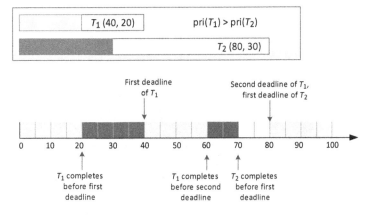

Figure 16.2
Priority assignment: $\text{pri}(T_1) > \text{pri}(T_2)$.

principles that can guide us to assign priorities at design time. For instance, the deadline-monotonic assignment (DMA) principle states that a task with the earliest deadline should have the highest priority. Another principle is called the rate-monotonic assignment (RMA) principle, which is our focus below.

16.2 RMA Principle

Given a system consisting of k periodic tasks $\mathbb{T} = \{T_i = (p_1, e_i)|1 \leq i \leq k\}$, the RMA principle states that a task with the highest request rate (job release frequency) should have the highest priority. Since the job release frequency is inversely related to the task period, the RMA principle implies that a task with a shorter period has a higher priority.

A system with a task set $\mathbb{T}$ is *schedulable* if all the tasks in $\mathbb{T}$ can be scheduled without breaking any deadlines. At design time, it is very important to conduct schedulability analysis to examine whether a priority assignment is feasible or not. If a task set $\mathbb{T}$ is not schedulable, it is better to reconsider the domain problem to see whether the task design itself is appropriate. If necessary, some tasks in $\mathbb{T}$ should be adjusted or transformed in order to make the task set $\mathbb{T}$ schedulable.

Let us take a look at some examples.

Consider the task set

$$\mathbb{T}_2 = \begin{cases} T_1 = (10, 3), \\ T_2 = (15, 4), \\ T_3 = (20, 2). \end{cases}$$

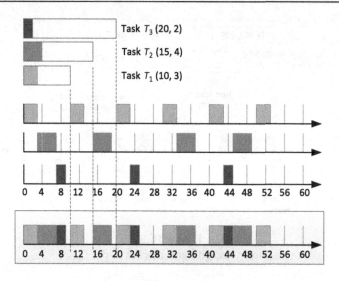

Figure 16.3
RMA-based task schedule: a schedulable task set.

Is task set $\mathbb{T}_2$ schedulable by the RMA principle?

First, according to the RMA principle, we know that $\text{pri}(T_1) > \text{pri}(T_2) > \text{pri}(T_3)$ holds. This priority assignment allows us to work out a timing diagram as shown in Figure 16.3:

(1)   Since task T_1 has the highest priority, it is considered first. Its timing diagram can be drawn such that jobs from T_1 are always scheduled to execute as soon as they become ready (released at the beginning of each period).

(2)   Task T_2 is considered next, within the context of the timing diagram of task T_1. In other words, the execution of T_2 cannot interfere with the execution of T_1. As a job J_{2j} from T_2 is ready, (a) if T_1 is not executing, J_{2j} can be immediately scheduled to execute, and (b) if a job J_{1i} from T_1 is executing, J_{2j} starts immediately after J_{1i} runs to completion. In addition, if a job J_{2j} is not finished when it comes to a time interval that is already occupied by a job J_{1i}, then job J_{2j} is preempted, and the rest of J_{2j} is scheduled immediately after J_{1i}.

(3)   Since task T_3 has the lowest priority, it is considered last, within the context of the timing diagram of tasks T_1 and T_2. In other words, a released job from T_3 can be scheduled only within interval(s) where no job from T_1 or T_2 is executing.

The timing diagrams of T_1, T_2, and T_3 can be combined to produce a timing diagram for the whole task set $\mathbb{T}_2$, which is given at the bottom of Figure 16.3. According to the schedule, all the job deadlines are honored. Thus, task set $\mathbb{T}_2$ is schedulable by the RMA principle.

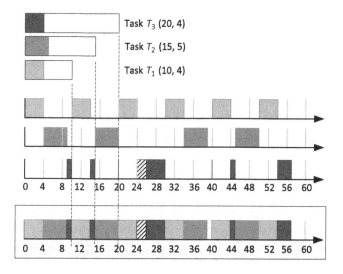

Figure 16.4
RMA-based task schedule: a nonschedulable task set.

Consider another task set,

$$\mathbb{T}_3 = \begin{cases} T_1 = (10,4), \\ T_2 = (15,5), \\ T_3 = (20,4). \end{cases}$$

According to the RMA principle, we know that $\mathrm{pri}(T_1) > \mathrm{pri}(T_2) > \mathrm{pri}(T_3)$ holds. This priority assignment allows us to work out a timing diagram as shown in Figure 16.4:

(1) Since task T_1 has the highest priority, it is considered first. Its timing diagram can be drawn such that jobs from T_1 are always scheduled to execute as soon as they are ready (released at the beginning of each period).

(2) Task T_2 is considered next, within the context of the timing diagram of task T_1. In other words, a released job from T_2 can be scheduled only within interval(s) where no job from T_1 is executing.

(3) Since task T_3 has the lowest priority, it is considered last, within the context of the timing diagram of tasks T_1 and T_2. Job J_{31} is released at time 0; however, it is not scheduled until J_{21} runs to completion at time 9. The execution of job J_{31} lasts for only one time unit and is preempted at time 10 by job J_{12}. J_{31} resumes its execution after J_{12} finishes at time 14. The execution of job J_{31} is again preempted by job J_{22} at time 15. The execution of J_{31} is not resumed until time 24. This, however, breaks the deadline of J_{31}!

The timing diagrams of T_1, T_2, and T_3 can be combined to produce a timing diagram for the whole task set $\mathbb{T}_3$, which is given at the bottom of Figure 16.4. According to the schedule, the deadline of job J_{31} is broken. Thus, we cannot conclude that task set $\mathbb{T}_3$ is schedulable by the RMA principle.

Up to this point, we know that some task sets (e.g., $\mathbb{T}_2$) are schedulable according to the RMA principle, while some are not (e.g., $\mathbb{T}_3$). The question is, are there any generic guidelines to help us determine the schedulability of a task set? We give an answer next.

16.3 Rate-Monotonic Analysis

Given a periodic task $T_i = (p_i, e_i)$, the processing demand (or *CPU utilization*) incurred by T_i is defined as e_i/p_i.

Given a set of k independent, preemptable periodic tasks $\mathbb{T} = \{T_i = (p_i, e_i)|1 \leq i \leq k\}$, we have the following theoretical result [49].

Theorem 16.1 (Bound test). *A task set* $\mathbb{T}$, *when scheduled by the RMA principle, can always meet the task deadlines if*

$$\sum_{i=1}^{k}\left(\frac{e_i}{p_i}\right) = \frac{e_1}{p_1} + \cdots + \frac{e_k}{p_k} \leq k(2^{1/k} - 1). \tag{16.1}$$

Theorem 16.1 actually says that a task set is schedulable according to the RMA principle if the accumulative CPU utilization is no more than a bound given by $k(2^{1/k} - 1)$. As shown in Table 16.1, the bound on the accumulative utilization rapidly converges to 0.69 as k, the size of the task set, becomes large. Here we can actually see a big limitation of fixed-priority scheduling: it is not always possible to fully utilize the processing resource when the system contains only periodic tasks.

Let us now look at some examples. Consider the task set

$$\mathbb{T}_4 = \begin{cases} T_1 = (12, 3), \\ T_2 = (15, 2), \\ T_3 = (30, 10). \end{cases}$$

Table 16.1 The bound of accumulative CPU utilization as k changes

k	1	2	3	4	$\cdots$	∞
$k(2^{1/k} - 1)$	1.00	0.83	0.78	0.76	$\cdots$	$\log_e(2) = 0.69$

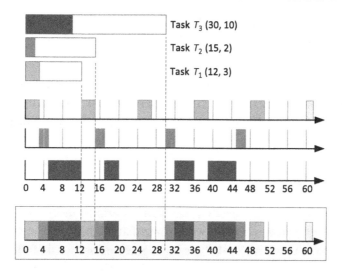

Figure 16.5
A RMA-based schedule of $\mathbb{T}_4$.

We calculate the accumulative utilization of $\mathbb{T}_4$, which is $\sum_{i=1}^{3}\left(\frac{e_i}{p_i}\right) = 3/12 + 2/15$ $+10/30 = 0.72$. It is less than the bound 0.78. Thus, according to Theorem 16.1, task set $\mathbb{T}_4$ is schedulable by the RMA principle. Figure 16.5 gives an RMA-based schedule of $\mathbb{T}_4$.

RMA is an *optimal* static-priority assignment approach in the sense that a task set cannot be scheduled by any other static-priority assignment if it cannot be scheduled by the RMA principle. The DMA mentioned before is another optimal static-priority assignment principle. DMA-based scheduling is optimal when deadlines are less than periods.

What if the accumulative utilization of a task set exceeds the corresponding bound? For example, we consider the following task set:

$$\mathbb{T}_5 = \begin{cases} T_1 = (10, 2.5), \\ T_2 = (20, 5), \\ T_3 = (30, 9). \end{cases}$$

The accumulative utilization of $\mathbb{T}_5$ is 0.8, which is larger than the bound 0.78. Does this mean that task set $\mathbb{T}_5$ is not schedulable by the RMA principle? The answer is no. The bound given in Theorem 16.1 is rather pessimistic and it is only a *sufficient* condition. The accumulative utilization in practice is about 88% on average. This implies that a task set might be still schedulable by the RMA principle even though its accumulative utilization is above the bound.

In the next section, we discuss the *completion-time test*, which gives a less-restrictive condition for determining schedulability. Given a periodic task set, Theorem 16.1 can be used first to check its schedulability. If the accumulative utilization is greater than the bound, the completion-time test can be used to conduct further schedulability analysis.

16.4 Completion-Time Test

Given a system with a set of independent, preemptable periodic tasks $\mathbb{T} = \{T_i = (p_i, e_i, d_i) | d_i \leq p_i, 1 \leq i \leq k\}$, we assume that all tasks in $\mathbb{T}$ are indexed in order of decreasing priority. In other words, task T_1 is the highest-priority task, followed by T_2, and so on. When the RMA principle applies, we have $\forall T_i, T_j \in \mathbb{T}, p_i < p_j$ holds if $i < j$.

First consider the system given in Figure 16.6. Recall that the response time of a job (task) refers to the time interval between the instant when the job is released and the instant when the job runs to completion. As indicated in Figure 16.6, the response times of jobs from task T_1 are always 6, because it is the task with the highest priority. The response times of jobs from task T_2 are as follows:

- J_{21}: 8 (is released at time 0 and finishes at time 8).
- J_{22}: 3 (is released at time 25 and finishes at time 28).
- J_{23}: 2 (is released at time 50 and finishes at time 52).
- J_{24}: 2 (is released at time 75 and finishes at time 77).

Similarly, the response times of jobs from task T_3 are as follows:

- J_{31}: 20 (is released at time 0 and finishes at time 20).
- J_{32}: 18 (is released at time 30 and finishes at time 48).

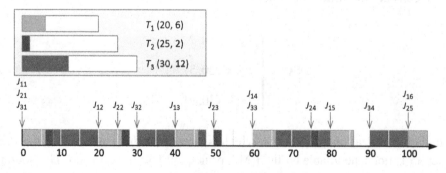

Figure 16.6
Critical instants.

- J_{33}: 20 (is released at time 60 and finishes at time 80).
- J_{34}: 20 (is released at time 90 and finishes at time 110).

It seems that the first jobs of tasks T_2 and T_3 have the worst (longest) response time among all their respective sequence of jobs. This is not a coincidence: all three tasks have their first job released at time 0, and they all start to compete for the processing time simultaneously.

Theoretically, there are many possible combinations of job release times for a task set $\mathbb{T}$. The *worst-case combination* happens when all tasks in $\mathbb{T}$ have their job released at the same time. In particular, since the response time of a job is not affected by any jobs with a lower priority, the worst-case response time happens to a job J_{ij} when all jobs in $\mathbb{T}$ that have higher priorities than J_{ij} are released simultaneously with J_{ij}. This co-release time is called a *critical instant* in the literature [49].

A critical instant of a task exists in general. Formally, we have

Theorem 16.2 (Existence of critical instant). *Given a set $\mathbb{T}$ of independent, preemptable periodic tasks where every job finishes before the next job of the same task is released, a critical instant of any task T_i occurs when one of its jobs J_{ix} is released at the same time as a job in every higher-priority task in $\mathbb{T}$.*

Given a task $T_i \in \mathbb{T}$, suppose the release time $t_\perp$ of a job J_{ix} is a critical instant of T_i, we use $w_i(t)$, where $t_\perp < t \le t_\perp + p_i$, to denote the total demand for processor time by J_{ix} and all the higher-priority jobs during time interval $[t_\perp, t]$:

$$w_i(t) = e_i + \sum_{j=1}^{i-1} \left\lceil \frac{t - t_\perp}{p_j} \right\rceil e_j, \quad \text{for } t_\perp < t \le t_\perp + p_i.$$

Here, the term $\left\lceil \frac{t-t_\perp}{p_j} \right\rceil$ represents the number of times task T_j ($j < i$) arrives from $t_\perp$ till time t; therefore, $\left\lceil \frac{t-t_\perp}{p_j} \right\rceil e_j$ represents the total demand of processor time by task T_j from $t_\perp$ till time t.

For ease of analysis, the critical instant $t_\perp$ is typically set to time 0 (i.e., all the tasks of T_1 up to T_i have their first jobs simultaneously released at time 0). Then we have

$$w_i(t) = e_i + \sum_{j=1}^{i-1} \left\lceil \frac{t}{p_j} \right\rceil e_j, \quad \text{for } 0 < t \le p_i. \tag{16.2}$$

The function $w_i(t)$ is called the *time-demand function* of task T_i.

Because the job J_{i1} released at a critical instant has the worst-case response time among all jobs in T_i, it is easy to see that all jobs in T_i can meet their deadlines if J_{i1} can meet its deadline. Formally, we have

Theorem 16.3 (Completion-time test). *For a set* $\mathbb{T}$ *of independent, preemptable periodic tasks, a task* T_i *can meet all its deadlines (i.e., it is schedulable), if a job of* T_i *released at a critical instant can meet its deadline—that is, if* $w_i(t) \leq t$ *holds for some* $t \leq d_i$.

The condition "$w_i(t) \leq t$ holds for some $t \leq d_i$" implies that at some time point before the deadline, the time supply, which is t, becomes equal to or greater than the demand $w_i(t)$ for processor time. That is, the first job can finish by its deadline.

What then is the schedulability of T_i if the condition given in Theorem 16.3 is not met? In other words, is T_i still schedulable if $w_i(t) > t$ holds for all $0 < t \leq d_i$? Because $w_i(t)$ indicates the worst-case time demand, task T_i may still be schedulable if a critical instant of T_i (the worst-case scenario) never happens. Theorem 16.2 states that a critical instant of a task occurs as long as the variations in the inter-release times of tasks are not negligibly small (see the proof in [50]). Hence, if we assume that the variations in the inter-release times of tasks are not negligibly small, we can conclude that, if $w_i(t) > t$ holds for all $0 < t \leq d_i$, T_i (and the whole system $\mathbb{T}$) cannot be feasibly scheduled by the RMA principle.

Theorem 16.3 is a more generic result, which can be applied to any static-priority assignment, including the RMA-based priority assignment. Theorem 16.3 directly leads to the following procedure for determining the schedulability of a task set $\mathbb{T}$:

(1) Test one task at a time starting from the highest-priority task T_1 in order of decreasing priority.

(2) Assume that task T_{i-1} is schedulable, and do the following to determine the schedulability of task T_i:

    (a) Solve the equation $w_i(t) = t$ iteratively by

        (i) $t_0 = \sum_{j=1}^{i} e_j$;

        (ii) $t_1 = w_i(t_0)$;

        (iii) $t_2 = w_i(t_1)$;

        (iv) $t_3 = w_i(t_2)$;

        (v) $\ldots$;

        (vi) $t_k = w_i(t_{k-1})$;

        (vii) stop when $t_k > d_i$ or $t_k = w_i(t_{k-1}) = t_{k-1}$.

    (b) If $t_k > d_i$, then T_i is not schedulable.

    (c) Otherwise, we have a solution, denoted by t^*, to the equation $w_i(t^*) = t^*$. This means that at time t^* the total demand for processor time by tasks with a priority equal to or higher than that of T_i is exactly equal to t^*. The time point t^* is often called the fix point of the function $w_i(t)$. Since $t^* \leq d_i$ holds, task T_i is schedulable.

(3) $\mathbb{T}$ is schedulable if all tasks in it are schedulable.

As an example, let us now consider task set $\mathbb{T}_3$.

First, the accumulative utilization of $\mathbb{T}_3$ is 0.93, which is larger than the bound 0.78. We next need to conduct the completion-time test:

(1) Start from the highest priority task T_1. T_1 is obviously schedulable.
(2) Consider task T_2 next.
   (a) Solve the equation $w_2(t) = t$. Start with $t_0 = \sum_{j=1}^{2} e_j = 9$. Then, $t_1 = w_2(t_0) = 9 = t_0$. So, t^* is 9.
   (b) Task T_2 is schedulable because $t^* \leq d_2 = 15$.
(3) Consider T_3 next.
   (a) Solve the equation $w_3(t) = t$. Start with $t_0 = \sum_{j=1}^{3} e_j = 13$. Then, $t_1 = w_3(t_0) = w_3(13) = 17$ and $t_2 = w_3(t_1) = w_3(17) = 22$.
   (b) The iteration stops since we already have $t_2 = 22 > d_3 = 20$. Task T_3 is thus not schedulable. Indeed, the first fix point of the equation is 30, which is far beyond T_3's deadline.
(4) Thus, task set $\mathbb{T}_3$ is not schedulable.

As another example, let us consider task set $\mathbb{T}_5$.

First, the accumulative utilization of $\mathbb{T}_5$ is 0.8, which is larger than the bound 0.78. We next need to conduct the completion-time test:

(1) Start from the highest-priority task T_1. T_1 itself is obviously schedulable.
(2) Consider T_2 next.
   (a) Solve the equation $w_2(t) = t$. Start with $t_0 = \sum_{j=1}^{2} e_j = 7.5$. Then, $t_1 = w_2(t_0) = w_2(7.5) = 7.5 = t_0$. So, t^* is 7.5.
   (b) Task T_2 is schedulable because $t^* \leq d_2 = 20$.
(3) Consider T_3 next.
   (a) Solve the equation $w_3(t) = t$. Start with $t_0 = \sum_{j=1}^{3} e_j = 16.5$. Then, $t_1 = w_3(t_0) = w_3(16.3) = 19$ and $t_2 = w_3(t_1) = w_3(19) = 19 = t_1$. So, t^* is 19.
   (b) Task T_3 is schedulable because $t^* \leq d_3 = 30$.
(4) Thus, task set $\mathbb{T}_5$ is schedulable.

An RMA-based schedule of $\mathbb{T}_5$ is given in Figure 16.7.

16.5 Period Transformation

What if a task set $\mathbb{T}$ cannot pass the completion-time test? In such a case, most likely, $\mathbb{T}$ is not schedulable, and we have to reconsider the domain problem to see whether the task design itself is appropriate. If applicable, some tasks in $\mathbb{T}$ can be adjusted to form a new task set.

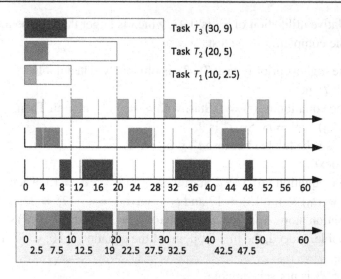

Figure 16.7
An RMA-based schedule of $\mathbb{T}_5$.

Consider the task set

$$\mathbb{T}_7 = \begin{cases} T_1 = (10, 3.5, 10), \\ T_2 = (14, 7, 13). \end{cases}$$

Task T_2 has a lower priority according to the RMA principle. We can use the completion-time test to check whether task T_2 is schedulable or not:

$$t_0 = e_1 + e_2 = 10.5,$$

$$t_1 = w_2(t_0) = w_2(10.5) = 2 \times e_1 + e_2 = 14,$$

$$t_2 = w_2(t_1) = 2 \times e_1 + e_2 = 14 = t_1.$$

However, the fix-point solution 14 is larger than the deadline (13) of T_2. Thus, T_2 (and $\mathbb{T}_7$) is not schedulable. Indeed, as shown in Figure 16.8, the first job deadline of T_2 is broken.

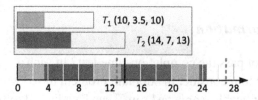

Figure 16.8
Task T_2 breaks its first deadline.

If, from the domain problem, we further know that task T_2 is more important or critical than task T_1, then we have to seek a priority assignment that is different from the RMA-based assignment: $\mathrm{pri}(T_2) > \mathrm{pri}(T_1)$. We can still use the completion-time test to check whether $\mathbb{T}_7$ is schedulable or not. Everything is still the same except that this time we need to compare the fix point 14 with T_1's deadline. It turns out that $\mathbb{T}_7$ is still not schedulable under this new priority assignment.

A technique called *period transformation* [63] can be applied to situations where the task with a longer period is more important than the tasks with a shorter period. The idea is to transform the high-importance task by shortening its period.

In general, given a system with a set of independent, preemptable periodic tasks $\mathbb{T} = \{T_i = (p_i, e_i, d_i) | 1 \le i \le k\}$, let $T_m = (p_m, e_m, d_m) \in \mathbb{T}$ denote the high-importance task to be transformed. The procedure is as follows:

(1) Split the task T_m equally into x pieces, each with an execution length of e_m/x.
(2) Create a new task $T'_m = (p_m/x, e_m/x, d_m - (x - 1) \times p_m/x)$. Note that the new deadline of T'_m is $d_m - (x - 1) \times p_m/x$; this guarantees that the deadline of the xth release of T'_m equals $(x - 1) \times p_m/x + [d_m - (x - 1) \times p_m/x]$, which equals exactly d_m, the deadline of the original task T_m.
(3) Replace task T_m in $\mathbb{T}$ by T'_m. Note that in order to finish the original task T_m, the new task T'_m should be released/scheduled x times such that the first piece of the original task is executed in the first period of T'_m, the second piece of the original task is executed in the second period of T'_m, and so on. The original task finishes at the end of the xth period of T'_m.

By period transformation, the task set becomes $\mathbb{T}'_7$, where

$$\mathbb{T}'_7 = \begin{cases} T_1 = (10, 3.5, 10), \\ T'_2 = (7, 3.5, 6). \end{cases}$$

Here, the original task T_2 is split into two pieces, each of which is characterized by the new task T'_2. Note that the completion of two releases of T'_2 is counted as one completion of the original task.

After the period transformation, task T'_2 has a higher priority according to the RMA principle. We can use the completion-time test again to check whether task set $\mathbb{T}'_7$ is schedulable or not. We start with the highest-priority task T'_2, which, by itself, is schedulable. We consider T_1 next:

$$t_0 = e_1 + e_2 = 7,$$

$$t_1 = w_2(t_0) = w_2(7) = 7 = t_0.$$

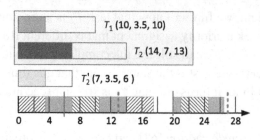

Figure 16.9
Period transformation.

Obviously, the fix-point solution 7 is earlier than T_1's deadline 10. Thus, T_1 (and $\mathbb{T}'_7$) is schedulable. The timing diagram of its schedule is given in Figure 16.9.

As another example, consider the task set

$$\mathbb{T}_8 = \begin{cases} T_1 = (50, 25, 50), \\ T_2 = (70, 30, 70). \end{cases}$$

As shown in Figure 16.10, when the RMA-based priority assignment is used, task T_2 misses its first deadline. In order to honor the timing constraints, we employ the period transformation approach and change the original task set into

$$\mathbb{T}'_8 = \begin{cases} T_1 = (50, 25, 50), \\ T'_2 = (35, 15, 35). \end{cases}$$

After the period transformation, task T'_2 has a higher priority according to the RMA principle. We can use the completion-time test to check whether task set $\mathbb{T}'_8$ is schedulable or not. We start with the highest-priority task T'_2, which, by itself, is schedulable. We consider T_1 next:

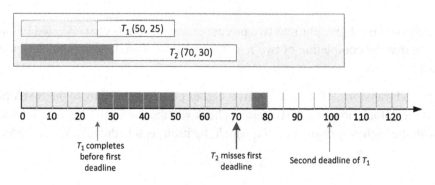

Figure 16.10
Task T_2 breaks its first deadline.

$$t_0 = e_1 + e_2 = 40,$$

$$t_1 = w_2(t_0) = w_2(40) = 55,$$

$$t_2 = w_2(t_1) = w_2(55) = 80,$$

$$t_3 = w_2(t_2) = w_2(80) = 95,$$

$$t_4 = w_2(t_3) = w_2(95) = 95 = t_3.$$

Obviously, the fix-point solution 95 is after T_1's deadline 50. Thus, T_1 (and $\mathbb{T}'_8$) is not schedulable.

Let us change the original task set into

$$\mathbb{T}''_8 = \begin{cases} T_1 = (50, 25, 50), \\ T''_2 = (14, 6, 14). \end{cases}$$

To apply the completion-time test, we start with the highest-priority task T''_2, which, by itself, is schedulable. We consider T_1 next:

$$t_0 = e_1 + e_2 = 31,$$

$$t_1 = w_2(t_0) = w_2(31) = 43,$$

$$t_2 = w_2(t_1) = w_2(43) = 49,$$

$$t_3 = w_2(t_2) = w_2(49) = 49 = t_3.$$

The fix-point solution 49 is before T_1's deadline 50. Thus, T_1 (and $\mathbb{T}''_8$) is schedulable. The timing diagram of its schedule is given in Figure 16.11.

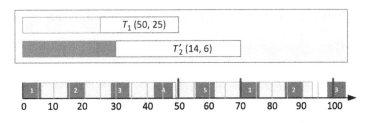

Figure 16.11
Split T_2 into five pieces.

16.6 Generalized Schedulability Analysis

Up to this point, the task sets we have considered are composed of periodic tasks that are independent and have distinct levels of priority. In this section, we take a step further to consider systems where some of the periodic tasks might be blocked by some other tasks, or equal priority may be assigned to some tasks.

16.6.1 Tasks with Blocking Time

Given a system with a set of k periodic tasks $\mathbb{T} = \{T_i = (p_i, e_i) | 1 \leq i \leq k\}$, we next examine extensions to Theorems 16.1 and 16.3 to cover situations with tasks being blocked due to nonpreemptive lower-priority tasks. As before, the tasks in $\mathbb{T}$ are scheduled by the RMA principle and for all $1 \leq i < j \leq k$, we have $\mathrm{pri}(T_i) > \mathrm{pri}(T_j)$.

In real-time systems, tasks often need to interact with each other—say, accessing shared resources protected by locks (semaphores or mutexes).[1] While a task (job) is using a mutually exclusive resource, it becomes *nonpreemptable* and a higher-priority task requesting the same resource is blocked until the nonpreemptable section of the lower-priority task finishes. This delay due to blocking may cause the higher-priority task to miss its deadlines. Thus, to analyze the schedulability of a task T_j, we must consider not only all those tasks with a higher priority than T_j, but also the blocking time due to the nonpreemptable sections of lower-priority tasks.

First, it is worth noting that a job J_{ix} of task T_i can be blocked by a lower-priority job *only once*. In particular, J_{ix} is blocked if when it is released a lower-priority job is executing its nonpreemptable portion at the time. As soon as this nonpreemptable portion finishes, the job J_{ix} or a higher-priority job immediately preempts the lower-priority job and no other lower-priority job could ever be scheduled to execute until J_{ix} finishes.

Second, for the sake of determining whether task T_i is schedulable, we again focus on the worst-case scenario, where the job J_{ix} happens to be blocked by a job that has the *longest nonpreemptable portion* among all lower-priority jobs.

Formally, let β_m^i denote the longest time that task T_i can be blocked by a task T_m (in its execution of a nonpreemptable portion). We have $0 \leq \beta_m^i \leq e_m$, and β_m^i can be different from β_m^j when $i \neq j$. Let b_i denote the worst-case blocking time of task T_i due to nonpreemptivity, which is given by:

$$b_i = \max_{i < m \leq k} \beta_m^i.$$

---

[1] In such situations, the priority ceiling protocol is typically applied in order to prevent unbounded priority inversion (see Section 14.5 for details).

In particular, for the task set $\mathbb{T} = \{T_i = (p_i, e_i) | 1 \leq i \leq k\}$, we have

- $b_k = 0$;
- $b_{k-1} = \beta_k^{k-1}$;
- $b_{k-2} = \max\left(\beta_{k-1}^{k-2}, \beta_k^{k-2}\right)$;
- $\ldots$;
- $b_1 = \max\left(\beta_2^1, \beta_3^1, \ldots, \beta_k^1\right)$.

16.6.1.1 Generalized bound test

As far as CPU utilization is concerned, the effect of blocking time b_i can be modeled as though task T_i's utilization were increased by an amount b_i/p_i. Hence, Theorem 16.1 can be generalized to derive the following theoretical result [64].

Theorem 16.4 (Bound test with blocking). *A task set $\mathbb{T}$ scheduled by the RMA principle will always meet its deadlines if for all tasks T_i ($1 \leq i \leq k$) we have*

$$\frac{e_1}{p_1} + \frac{e_2}{p_2} + \cdots + \frac{e_i}{p_i} + \frac{b_i}{p_i} \leq i(2^{1/i} - 1). \tag{16.3}$$

The first i items in the above inequality together represent the effect of preemptions from all higher-priority tasks and task T_i's own execution, while the last item represents the worst-case blocking time due to the lower-priority tasks.

As an example, consider the task set

$$\mathbb{T} = \begin{cases} T_1 = (4, 1), \\ T_2 = (8, 2), \\ T_3 = (16, 4), \end{cases}$$

where $b_1 = 3$ and $b_2 = 2$.

We have

$$\frac{e_1}{p_1} + \frac{b_1}{p_1} = 1,$$

$$\frac{e_1}{p_1} + \frac{e_2}{p_2} + \frac{b_2}{p_2} = 0.75 \leq 2(2^{1/2} - 1) = 0.83,$$

$$\frac{e_1}{p_1} + \frac{e_2}{p_2} + \frac{e_3}{p_3} = 0.75 \leq 3(2^{1/3} - 1) = 0.78.$$

According to Theorem 16.4, the task set is schedulable.

Since $\frac{b_i}{p_i} \leq \max \left(\frac{b_1}{p_1}, \ldots, \frac{b_{k-1}}{p_{k-1}} \right)$ and $k(2^{1/k} - 1) \leq i(2^{1/i} - 1)$ for all $1 \leq i \leq k$, from Theorem 16.4 we have

Lemma 16.1 (Rough bound test). *A task set $\mathbb{T}$ scheduled by the RMA principle will always meet its deadlines if*

$$\frac{e_1}{p_1} + \cdots + \frac{e_k}{p_k} + \max \left(\frac{b_1}{p_1}, \ldots, \frac{b_{k-1}}{p_{k-1}} \right) \leq k(2^{1/k} - 1). \tag{16.4}$$

16.6.1.2 Generalized completion-time test

With the blocking time taken into consideration, Theorem 16.3 is still applicable, except that the time-demand function given by Equation (16.2) needs to be modified in order to take into account the effect of blocking:

$$w_i(t) = e_i + b_i + \sum_{j=1}^{i-1} \left\lceil \frac{t}{p_j} \right\rceil e_j, \quad \text{for } 0 < t \leq p_i. \tag{16.5}$$

As an example, consider the task set

$$\mathbb{T} = \begin{cases} T_1 = (10, 3), \\ T_2 = (20, 4), \\ T_3 = (40, 10), \\ T_4 = (50, 8), \end{cases}$$

where $b_1 = 4$, $b_2 = 10$, and $b_3 = 5$.

Task T_1 $(i = 1)$ is schedulable because

$$t_0 = \sum_{j=1}^{1} e_i = e_1 = 3,$$

$$t_1 = w_1(t_0) = e_1 + b_1 + \sum_{j=1}^{0} \left\lceil \frac{t_0}{p_j} \right\rceil e_j$$

$$= 3 + 4 + 0 = 7,$$

$$t_2 = w_1(t_1) = e_1 + b_1 + \sum_{j=1}^{0} \left\lceil \frac{t_1}{p_j} \right\rceil e_j$$

$$= 3 + 4 + 0 = 7 = t_1 < d_1 = 10.$$

Task T_2 ($i = 2$) is schedulable because

$$t_0 = \sum_{j=1}^{2} e_i = e_1 + e_2 = 7,$$

$$t_1 = w_2(t_0) = e_2 + b_2 + \sum_{j=1}^{1} \left\lceil \frac{t_0}{p_j} \right\rceil e_j$$

$$= 4 + 10 + \left\lceil \frac{7}{10} \right\rceil 3 = 4 + 10 + 3 = 17,$$

$$t_2 = w_2(t_1) = e_2 + b_2 + \sum_{j=1}^{1} \left\lceil \frac{t_1}{p_j} \right\rceil e_j$$

$$= 4 + 10 + \left\lceil \frac{17}{10} \right\rceil 3 = 4 + 10 + 6 = 20,$$

$$t_3 = w_2(t_2) = e_2 + b_2 + \sum_{j=1}^{1} \left\lceil \frac{t_2}{p_j} \right\rceil e_j$$

$$= 40 + 10 + \left\lceil \frac{20}{10} \right\rceil 3 = 20 = t_2 \leq d_2 = 20.$$

Task T_3 ($i = 3$) is schedulable because

$$t_0 = \sum_{j=1}^{3} e_i = e_1 + e_2 + e_3 = 17,$$

$$t_1 = w_3(t_0) = e_3 + b_3 + \sum_{j=1}^{2} \left\lceil \frac{t_0}{p_j} \right\rceil e_j$$

$$= 10 + 5 + \left\lceil \frac{17}{10} \right\rceil 3 + \left\lceil \frac{17}{20} \right\rceil 4 = 10 + 5 + 6 + 4 = 25,$$

$$t_2 = w_3(t_1) = e_3 + b_3 + \sum_{j=1}^{2} \left\lceil \frac{t_1}{p_j} \right\rceil e_j$$

$$= 10 + 5 + \left\lceil \frac{25}{10} \right\rceil 3 + \left\lceil \frac{25}{20} \right\rceil 4 = 10 + 5 + 9 + 8 = 32,$$

$$t_3 = w_3(t_2) = e_3 + b_3 + \sum_{j=1}^{2} \left\lceil \frac{t_2}{p_j} \right\rceil e_j$$

$$= 10 + 5 + \left\lceil \frac{32}{10} \right\rceil 3 + \left\lceil \frac{32}{20} \right\rceil 4 = 10 + 5 + 12 + 8 = 35,$$

$$t_4 = w_3(t_3) = e_3 + b_3 + \sum_{j=1}^{2} \left\lceil \frac{t_3}{p_j} \right\rceil e_j$$

$$= 10 + 5 + \left\lceil \frac{35}{10} \right\rceil 3 + \left\lceil \frac{35}{20} \right\rceil 4 = 35 = t_3 < d_3 = 40.$$

Task T_4 ($i = 4$) is schedulable because

$$t_0 = \sum_{j=1}^{4} e_i = e_1 + e_2 + e_3 + e_4 = 25,$$

$$t_1 = w_4(t_0) = e_4 + b_4 + \sum_{j=1}^{3} \left\lceil \frac{t_0}{p_j} \right\rceil e_j$$

$$= 8 + 0 + \left\lceil \frac{25}{10} \right\rceil 3 + \left\lceil \frac{25}{20} \right\rceil 4 + \left\lceil \frac{25}{40} \right\rceil 10 = 8 + 9 + 8 + 10 = 35,$$

$$t_2 = w_4(t_1) = e_4 + b_4 + \sum_{j=1}^{3} \left\lceil \frac{t_1}{p_j} \right\rceil e_j$$

$$= 8 + 0 + \left\lceil \frac{35}{10} \right\rceil 3 + \left\lceil \frac{35}{20} \right\rceil 4 + \left\lceil \frac{35}{40} \right\rceil 10 = 8 + 12 + 8 + 10 = 38,$$

$$t_3 = w_4(t_2) = e_4 + b_4 + \sum_{j=1}^{3} \left\lceil \frac{t_2}{p_j} \right\rceil e_j$$

$$= 8 + 0 + \left\lceil \frac{38}{10} \right\rceil 3 + \left\lceil \frac{38}{20} \right\rceil 4 + \left\lceil \frac{38}{40} \right\rceil 10 = 38 = t_2 < d_4 = 50.$$

16.6.2 Tasks with Earlier Deadlines

We now consider the situation where a task's deadline is before the end of its period.

Given a set of k preemptable periodic tasks $\mathbb{T} = \{T_i = (p_i, e_i, d_i) | 1 \leq i \leq k, d_i \leq p_i\}$, for a task T_i, let $q_i = (p_i - d_i)$ denote the amount of time between its deadline and the start of its next period.

The schedulability of a task T_i with an earlier deadline can be determined by considering it as having a deadline at the end of the period but being blocked by lower-priority tasks for a duration of q_i. Hence, this effect can also be modeled as though task T_i's utilization were increased by an amount q_i/p_i. Therefore, Theorem 16.1 can be generalized to accommodate an earlier deadline [65].

Theorem 16.5 (Bound test with earlier deadlines). *A task set* $\mathbb{T}$ *scheduled by the RMA principle will always meet its deadlines if for all task* T_i *(1 ≤ i ≤ k) we have*

$$\frac{e_1}{p_1} + \frac{e_2}{p_2} + \cdots + \frac{e_i}{p_i} + \frac{q_i}{p_i} \leq i(2^{1/i} - 1). \tag{16.6}$$

If a task T_i has both a blocking time b_i and an earlier deadline, the combined effect can be modeled as increasing task T_i's utilization by $(b_i + q_i)/p_i$. Therefore, Theorem 16.1 can be further generalized.

Theorem 16.6 (Generalized bound test). *A task set* $\mathbb{T}$ *scheduled by the RMA principle will always meet its deadlines if for all task* T_i *(1 ≤ i ≤ k) we have*

$$\frac{e_1}{p_1} + \frac{e_2}{p_2} + \cdots + \frac{e_i}{p_i} + \frac{b_i + q_i}{p_i} \leq i(2^{1/i} - 1). \tag{16.7}$$

To conduct the completion-time test, in the case of a task with an earlier deadline, Theorem 16.3 can be used directly without any modification to the time-demand function given by Equation (16.2); in the case of a task with a blocking time (and an earlier deadline), the function given by Equation (16.5) has to be used instead.

Figure 16.12 shows the relationship of the conditions used in Lemma 16.1, Theorem 16.6, and Theorem 16.3. Basically, a task set that can pass the schedulability test by Theorem 16.3 may

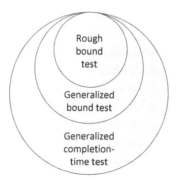

Figure 16.12
Relationship of test conditions.

not satisfy the condition given by Theorem 16.6 or Lemma 16.1. The example given in the next section illustrates this point.

16.6.3 Example

In Section 11.2.2, we captured the timing behaviors of four periodic tasks, the characteristics of which are repeated here:

$$\mathbb{T}_6 = \begin{cases} T_1 = (200\,\text{ms}, 50\,\text{ms}, 200\,\text{ms}), \\[1mm] T_2 = (400\,\text{ms}, 100\,\text{ms}, 400\,\text{ms}), \\[1mm] T_3 = (500\,\text{ms}, 90\,\text{ms}, 440\,\text{ms}), \\[1mm] T_4 = (900\,\text{ms}, 180\,\text{ms}, 900\,\text{ms}). \end{cases}$$

Recall that tasks T_1, T_2, and T_3 each have a critical section accessing shared objects (resources) guarded by locks. The maximum time that T_1 can be blocked is 80 ms, and the maximum time that T_2 can be blocked is 60 ms. Thus, we have $b_1 = 80$ ms and $b_2 = 60$ ms.

According to the RMA principle, we can derive the following priority assignment:

$$\text{pri}(T_1) > \text{pri}(T_2) > \text{pri}(T_3) > \text{pri}(T_4).$$

Assume that the priority ceiling protocol is used. Let us now analyze the schedulability of this set of tasks.

16.6.3.1 Rough bound test

The condition of the rough bound test is

$$\frac{e_1}{p_1} + \cdots + \frac{e_k}{p_k} + \max\left(\frac{b_1}{p_1}, \ldots, \frac{b_{k-1}}{p_{k-1}}\right)$$

$$= \frac{50}{200} + \frac{100}{400} + \frac{90}{500} + \frac{180}{900} + \max\left(\frac{80}{200}, \frac{60}{400}\right)$$

$$= 0.25 + 0.25 + 0.18 + 0.2 + \max(0.4, 0.15)$$

$$= 1.28$$

$$> k(2^{1/k} - 1)$$

$$= 4(2^{1/4} - 1) = 0.76.$$

Obviously, the rough bound test fails for task set $\mathbb{T}_6$.

16.6.3.2 Generalized bound test

To pass the generalized bound test, we need to check whether

$$\frac{e_1}{p_1} + \frac{e_2}{p_2} + \cdots + \frac{e_i}{p_i} + \frac{b_i + q_i}{p_i} \leq i(2^{1/i} - 1)$$

holds for tasks T_1, T_2, T_3, and T_4.

Task T_1 ($i = 1$) is schedulable because

$$\frac{e_1}{p_1} + \frac{e_2}{p_2} + \cdots + \frac{e_i}{p_i} + \frac{b_i + q_i}{p_i}$$

$$= \frac{e_1}{p_1} + \frac{b_1 + q_1}{p_1}$$

$$= \frac{50}{200} + \frac{80 + 0}{200}$$

$$= 0.65$$

$$< (2 - 1) = 1.$$

Task T_2 ($i = 2$) is schedulable because

$$\frac{e_1}{p_1} + \frac{e_2}{p_2} + \cdots + \frac{e_i}{p_i} + \frac{b_i + q_i}{p_i}$$

$$= \frac{e_1}{p_1} + \frac{e_2}{p_2} + \frac{b_2 + q_2}{p_2}$$

$$= \frac{50}{200} + \frac{100}{400} + \frac{60 + 0}{400}$$

$$= 0.75$$

$$< 2(2^{1/2} - 1) = 0.83.$$

Task T_3 ($i = 3$) is *not* schedulable because

$$\frac{e_1}{p_1} + \frac{e_2}{p_2} + \cdots + \frac{e_i}{p_i} + \frac{b_i + q_i}{p_i}$$

$$= \frac{e_1}{p_1} + \frac{e_2}{p_2} + \frac{e_3}{p_3} + \frac{b_3 + q_3}{p_3}$$

$$= \frac{50}{200} + \frac{100}{400} + \frac{90}{500} + \frac{0 + 60}{500}$$

$$= 0.8$$

$$> 3(2^{1/3} - 1) = 0.78.$$

Task T_4 ($i = 4$) is *not* schedulable because

$$\frac{e_1}{p_1} + \frac{e_2}{p_2} + \cdots + \frac{e_i}{p_i} + \frac{b_i + q_i}{p_i}$$

$$= \frac{e_1}{p_1} + \frac{e_2}{p_2} + \frac{e_3}{p_3} + \frac{e_4}{p_4} + \frac{b_4 + q_4}{p_4}$$

$$= \frac{50}{200} + \frac{100}{400} + \frac{90}{500} + \frac{180}{900} + \frac{0 + 0}{500}$$

$$= 0.88$$

$$> 4(2^{1/4} - 1) = 0.76.$$

Thus, the generalized bound test fails for task set $\mathbb{T}_6$.

16.6.3.3 *Generalized completion-time test*

To pass the generalized completion-time test, for each task T_i, we need to check whether $w_i(t) \le t$ holds for some $t \le d_i$, where $w_i(t)$ is given by Equation (16.5).

Task T_1 ($i = 1$) is schedulable because

$$t_0 = \sum_{j=1}^{1} e_i = e_1 = 50,$$

$$t_1 = w_1(t_0) = e_1 + b_1 + \sum_{j=1}^{0} \left\lceil \frac{t_0}{p_j} \right\rceil e_j$$

$$= 50 + 80 + 0 = 130,$$

$$t_2 = w_1(t_1) = e_1 + b_1 + \sum_{j=1}^{0} \left\lceil \frac{t_1}{p_j} \right\rceil e_j$$

$$= 50 + 80 + 0 = 130 = t_1 < d_1 = 200.$$

Task T_2 ($i = 2$) is schedulable because

$$t_0 = \sum_{j=1}^{2} e_i = e_1 + e_2 = 150,$$

$$t_1 = w_2(t_0) = e_2 + b_2 + \sum_{j=1}^{1} \left\lceil \frac{t_0}{p_j} \right\rceil e_j$$

$$= 100 + 60 + \left\lceil \frac{150}{200} \right\rceil 50 = 100 + 60 + 50 = 210,$$

$$t_2 = w_2(t_1) = e_2 + b_2 + \sum_{j=1}^{1} \left\lceil \frac{t_1}{p_j} \right\rceil e_j$$

$$= 100 + 60 + \left\lceil \frac{210}{200} \right\rceil 50 = 100 + 60 + 100 = 260,$$

$$t_3 = w_2(t_2) = e_2 + b_2 + \sum_{j=1}^{1} \left\lceil \frac{t_2}{p_j} \right\rceil e_j$$

$$= 100 + 60 + \left\lceil \frac{260}{200} \right\rceil 50 = 260 = t_2 < d_2 = 400.$$

Task T_3 $(i = 3)$ is schedulable because

$$t_0 = \sum_{j=1}^{3} e_i = e_1 + e_2 + e_3 = 240,$$

$$t_1 = w_3(t_0) = e_3 + b_3 + \sum_{j=1}^{2} \left\lceil \frac{t_0}{p_j} \right\rceil e_j$$

$$= 90 + 0 + \left\lceil \frac{240}{200} \right\rceil 50 + \left\lceil \frac{240}{400} \right\rceil 100 = 90 + 100 + 100 = 290,$$

$$t_2 = w_3(t_1) = e_3 + b_3 + \sum_{j=1}^{2} \left\lceil \frac{t_1}{p_j} \right\rceil e_j$$

$$= 90 + 0 + \left\lceil \frac{290}{200} \right\rceil 50 + \left\lceil \frac{290}{400} \right\rceil 100 = 290 = t_1 < d_3 = 440.$$

Task T_4 $(i = 4)$ is schedulable because

$$t_0 = \sum_{j=1}^{4} e_i = e_1 + e_2 + e_3 + e_4 = 420,$$

$$t_1 = w_4(t_0) = e_4 + b_4 + \sum_{j=1}^{3} \left\lceil \frac{t_0}{p_j} \right\rceil e_j$$

$$= 180 + 0 + \left\lceil \frac{420}{200} \right\rceil 50 + \left\lceil \frac{420}{400} \right\rceil 100 + \left\lceil \frac{420}{500} \right\rceil 90 = 180 + 150 + 200 + 90 = 620,$$

$$t_2 = w_4(t_1) = e_4 + b_4 + \sum_{j=1}^{3} \left\lceil \frac{t_1}{p_j} \right\rceil e_j$$

$$= 180 + 0 + \left\lceil \frac{620}{200} \right\rceil 50 + \left\lceil \frac{620}{400} \right\rceil 100 + \left\lceil \frac{620}{500} \right\rceil 90 = 180 + 200 + 200 + 180 = 760,$$

$$t_3 = w_4(t_2) = e_4 + b_4 + \sum_{j=1}^{3} \left\lceil \frac{t_2}{p_j} \right\rceil e_j$$

$$= 180 + 0 + \left\lceil \frac{760}{200} \right\rceil 50 + \left\lceil \frac{760}{400} \right\rceil 100 + \left\lceil \frac{760}{500} \right\rceil 90 = 760 = t_2 < d_4 = 900.$$

In conclusion, task set $\mathbb{T}_6$ is schedulable according to the priority assignment $\text{pri}(T_1) > \text{pri}(T_2) > \text{pri}(T_3) > \text{pri}(T_4)$.

16.6.4 Tasks with Equal Priorities

A real-time operating system typically implements a hierarchical scheduling scheme, where the application system contains a number of disjoint subsets of periodic tasks:

$$\mathbb{T} = \{\mathbb{T}^m | \mathbb{T}^i \cap \mathbb{T}^j = \emptyset, 1 \le i \ne j \le n\}.$$

Each subset $\mathbb{T}^m$ $(1 \le m \le n)$ is called a subsystem. In a *priority-driven/round-robin* system, the subsystems, each of which is treated as a whole unit, are scheduled in a priority-driven manner, while the tasks within each subsystem are scheduled in a round-robin manner in the intervals assigned to the subsystem.

For the sake of schedulability analysis, the whole system $\mathbb{T}$ can be analyzed first, with each subsystem $\mathbb{T}^m$ treated as a single periodic task. After this, the approach described below can be used to check the schedulability of each subsystem separately.

Given a task set (a subsystem) $\mathbb{T}$ containing a set of k preemptable periodic tasks $\mathbb{T} = \{T_i = (p_i, e_i)|1 \le i \le k\}$, where the tasks in $\mathbb{T}$ have fixed but may be nondistinct priorities, since jobs of equal priority are scheduled either on a FIFO basis or on a round-robin basis, a job J_{ix} of task T_i can be delayed by a job of another task with an equal priority. Obviously, the *worst-case delay* suffered by a job J_{ix} due to equal-priority tasks occurs when a job from each of these tasks is released immediately before J_{ix}.

Let $\mathbb{T}_E(i) \subseteq \mathbb{T}$ denote the subset of tasks, other than T_i, that have the same priority as T_i, and let $\mathbb{T}_H(i) \subseteq \mathbb{T}$ denote the subset of tasks that have a higher priority than T_i.

The *worst-case delay* suffered by a job J_{ix} is equal to the sum of the execution times of all the tasks in $\mathbb{T}_E(i)$. With this being considered, we can extend the time-demand function given by Equation (16.2) to:

$$w_i(t) = e_i + b_i + \sum_{T_j \in \mathbb{T}_E(i)} e_j + \sum_{T_j \in \mathbb{T}_H(i)} \left\lceil \frac{t}{p_j} \right\rceil e_j, \quad \text{for } 0 < t \le p_i, \qquad (16.8)$$

where, as before, b_i still denotes the worst-case blocking time of task T_i due to nonpreemptivity.

Problems

16.1 Consider the following systems of periodic tasks:
  (a) $\mathbb{T}_1 = \{(80, 30), (90, 30), (120, 20)\}$,
  (b) $\mathbb{T}_2 = \{(80, 30), (120, 40), (150, 40)\}$,
  (c) $\mathbb{T}_3 = \{(80, 40), (100, 20), (120, 25)\}$.
  Which systems are schedulable by the RMA principle?

16.2 A system $\mathbb{T}$ has four periodic tasks:

$$\mathbb{T} = \begin{cases} T_1 = (8, 1), \\ T_2 = (12, 3), \\ T_3 = (18, 3), \\ T_4 = (24, 5). \end{cases}$$

  (a) What is the total utilization of the system $\mathbb{T}$?
  (b) Draw a timing diagram to show the RMA-based schedule of $\mathbb{T}$.

16.3 A system $\mathbb{T}$ has three periodic tasks:

$$\mathbb{T} = \begin{cases} T_1 = (10, 5), \\ T_2 = (20, 5), \\ T_3 = (40, 10). \end{cases}$$

  (a) Is the system $\mathbb{T}$ schedulable according to the RMA principle?
  (b) Draw a timing diagram to show the RMA-based schedule of $\mathbb{T}$.

16.4 A system $\mathbb{T}$ has four periodic tasks:

$$\mathbb{T} = \begin{cases} T_1 = (10, 2), \\ T_2 = (12, 4), \\ T_3 = (15, 5), \\ T_4 = (30, 2). \end{cases}$$

  Use the completion-time test to check the schedulability of the tasks.

16.5 A system $\mathbb{T}$ has four periodic tasks:

$$\mathbb{T} = \begin{cases} T_1 = (6,2), \\ T_2 = (10,2), \\ T_3 = (16,4), \\ T_4 = (36,6). \end{cases}$$

Use the completion-time test to check the schedulability of the tasks.

16.6 A system $\mathbb{T}$ has three periodic tasks:

$$\mathbb{T} = \begin{cases} T_1 = (5,2), \\ T_2 = (8,3), \\ T_3 = (20,6), \end{cases}$$

Use the completion-time test to check the schedulability of the tasks.

16.7 A system $\mathbb{T}$ has seven periodic tasks:

$$\mathbb{T} = \begin{cases} T_1 = (5,1), \\ T_2 = (12,0.4), \\ T_3 = (20,3), \\ T_4 = (25,0.5), \\ T_5 = (36,4), \\ T_6 = (50,3), \\ T_7 = (60,8). \end{cases}$$

To put the system into a concrete context, we can assume that tasks T_1, T_3, and T_7 are normal domain-specific tasks. In addition, the system also needs to handle periodic interrupt requests (IRQs) from two source devices A and B. The IRQs from device A happens every 12 units of time, and the corresponding interrupt service routine (ISR) takes 0.4 unit of time to complete. This task is represented by T_2. Also, there is a deferred service task for IRQs from device A: it is triggered every three IRQs from device A, and it takes 4 units of time to finish. This task is represented by T_5. Notice that the period of T_5 is three times of the period of T_2. Similarly, the IRQs from device B happens every 25 units of time, and the corresponding ISR takes 0.5 unit of time to complete. This task is represented by T_4. Also, there is a deferred service task for IRQs from device B: it is triggered every two IRQs from device B, and it takes 3 units of time to finish. This task is represented by T_6.

Given the above context, explain whether the task set $\mathbb{T}$ is schedulable or not. *Hint*: The task set can be analyzed according to the RMA principle. However, when an interrupt arrives, it is serviced immediately (an ISR always preempts user tasks) and in a nonpreemptable fashion. So, task T_1 can be blocked by T_2 for 0.4 unit of time, and by T_4 for 0.5 unit of time. Similarly, task T_3 can be blocked by T_4 for 0.5 unit of time.

16.8 A system $\mathbb{T}$ has two periodic tasks:

$$\mathbb{T} = \begin{cases} T_1 = (10, 5), \\ T_2 = (14, 6). \end{cases}$$

(a) Is the system $\mathbb{T}$ schedulable according to the RMA principle?

(b) Suppose task T_2 is more important than task T_1. Use the period transformation approach to modify the task set so that the periodic tasks are schedulable. Draw a timing diagram to show the RMA-based schedule of the tasks.

CHAPTER 17

Real-Time Scheduling: Sporadic Server

Contents

> *The concept of randomness and coincidence will be obsolete when people can finally define a formulation of patterned interaction between all things within the universe.*
>
> *Toba Beta*

17.1 Sporadic Tasks

As we have already learned, aperiodic tasks are released at random time instants and have no or soft deadlines. Depending on the problem at hand, some nonperiodic tasks may have hard deadlines relative to their release times. We refer to those nonperiodic tasks with hard deadlines as *sporadic tasks*.

A sporadic task S_i is denoted by (r_i, e_i, d_i), where the task parameters are release time, execution time, and deadline, respectively.

Figure 17.1 illustrates a task set with three sporadic tasks:

- $S_1 = (7, 2.5, 20)$: S_1 is released at time instant 7, has an execution demand of 2.5 units of time, and has an absolute deadline of 20.
- $S_2 = (13, 1, 35)$: S_2 is released at 13, has an execution demand of one unit of time, and has an absolute deadline of 35.
- $S_3 = (15, 3, 30)$: S_3 is released at 15, has an execution demand of three units of time, and has an absolute deadline of 30.

Real-Time Embedded Systems. http://dx.doi.org/10.1016/B978-0-12-801507-0.00017-1

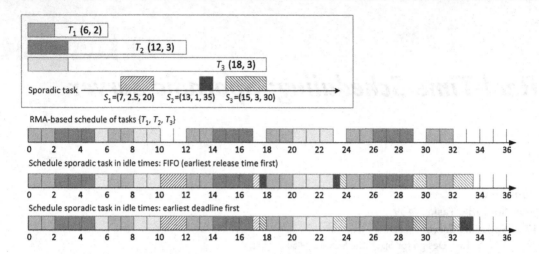

Figure 17.1
Scheduling sporadic tasks.

Sporadic tasks can be scheduled whenever the processor is idle. However, doing so can easily cause a sporadic task to miss its deadline. A rate-monotonic assignment (RMA)-based schedule of the periodic tasks T_1, T_2, and T_3 is shown in the middle of Figure 17.1. Utilizing the idle times, we can schedule the three sporadic tasks in two ways:

(1) FIFO: According to their release times, S_1 is scheduled first at time point 10 and is finished at time point 17.5 (being preempted from 12 to 17). S_2 is scheduled next at 17.5 and is finished at time point 23.5 (being preempted from 18 to 23). S_3 is scheduled at 23.5 and resumes its execution at time point 29. Obviously sporadic task S_3 has missed its hard deadline.

(2) Earliest deadline first: S_1 is scheduled first at time point 10 and is finished at time point 17.5. S_3 is scheduled next. However, S_3 would still miss its hard deadline.

In general, given that we already have a schedule for periodic tasks (say, RMA-based schedule as in Figure 17.1), it is often inevitable to break the deadlines of sporadic tasks if they could be considered only during idle times. We need a better approach where the sporadic tasks can be treated as first-class citizens.

17.2 Sporadic Server

A *sporadic server* is a periodic task created specifically for the purpose of executing sporadic tasks (jobs) [50, 67, 68].

A sporadic server, denoted by Φ_s, is given by (p_s, e_s, θ, ρ), where

- (p_s, e_s) defines a periodic task with a period of p_s and an execution time of e_s, where e_s is also called the server's *execution budget*;
- $\theta = \{S_i = (r_i, e_i, d_i) | 1 \leq i \leq k\}$ is a set to be populated with sporadic jobs released at run time; and
- ρ is a set of rules for regulating the operations of Φ_s.

Since a sporadic server is treated as a periodic task, it is straightforward to employ the RMA principle to schedule the sporadic server together with the other periodic tasks that are pertinent to the domain problem at hand. In addition, from Figure 17.1, it seems that the earliest-deadline-first scheduling policy [20, 50] offers a better chance to meet job deadlines. Thus, for the rest of this chapter, we have two assumptions:

(1) A sporadic server, as a schedulable unit, is scheduled together with the other periodic tasks according to the RMA principle.
(2) As a sporadic server Φ_s is running, it executes sporadic jobs in θ according to the earliest-deadline-first policy.

17.2.1 Task Design for Sporadic Servers

For a real-time system with a sporadic server, a critical step in task design is to determine the sporadic server's task parameters: p_s and e_s.

Since the RMA principle is adopted, the period p_s is directly related to the static priority of the sporadic server. The period p_s has to be designed carefully to ensure that the whole set of periodic tasks, including the sporadic server itself, are schedulable. In Chapter 16, we learned a few approaches to schedulability analysis of periodic task sets. Those approaches can also be used here to determine the feasibility of a sporadic server design.

Let us first consider an extreme situation where a sporadic server, whenever scheduled, will always use up its budget to execute the sporadic jobs from θ.

The task design given in Figure 17.2(a) does not satisfy the condition given in Theorem 16.1 (the total utilization is 0.95, which is above the bound), but it can pass the completion test of Theorem 16.3.[1] Thus, this is a feasible design.

The task design given in Figure 17.2(b) satisfies neither the condition given in Theorem 16.1 nor the condition given in Theorem 16.3. So, this is not a feasible design (actually, the deadline of task T_2 is broken).

---

[1] Start with tasks $\{T_1, \Phi_s\}$ and consider task T_2. We have $t_0 = 60$, $t_1 = 85$, and $t_2 = 95 = t_3 < 100$, and thus $\{T_1, \Phi_s, T_2\}$ is schedulable.

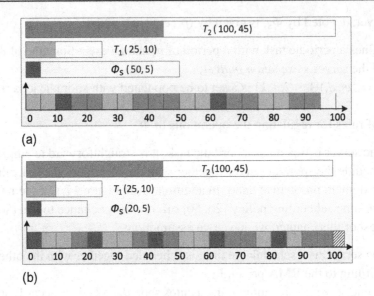

Figure 17.2
Two task designs for a sporadic server.

However, the extreme situation considered above rarely happens in practice. Even though sporadic jobs can be released in a bursty manner, the interval between consecutive bursts can be arbitrarily long. In such a situation, a sporadic server may find no job to execute when it comes to its turn. Normally a sporadic server is implemented such that it immediately gives up its occupation of the processor whenever there are no more jobs to execute. Consequently, the *effective* CPU utilization contributed by the sporadic server can be much less than e_s/p_s.

Let us assume that on average the *effective* CPU utilization contributed by a sporadic server is given by $\delta \cdot e_s/p_s$, where $\delta \leq 1$ represents a discount factor. The factor δ can be viewed as an indicator of the quality of service offered by Φ_s. Note that δ is an overall statistic. When $\delta = 0.5$, it is not that the sporadic server simply utilizes half of its execution budget for each period. Instead, it could be the case that the server is fully in execution in some periods while completely idle in other periods.

The techniques we learned in Chapter 16 are still applicable, except that in the calculation, e_s/p_s (the portion from Φ_s) is replaced by $\delta \cdot e_s/p_s$. Consider again the sporadic server Φ_s given in Figure 17.2(b). If we further know that $\delta = 0.5$, the task design becomes feasible (the task set can pass the completion test of Theorem 16.3).

17.2.2 Acceptance Test

Now, we assume that the task design of a sporadic server is fixed and the set of periodic tasks are schedulable when δ is considered.

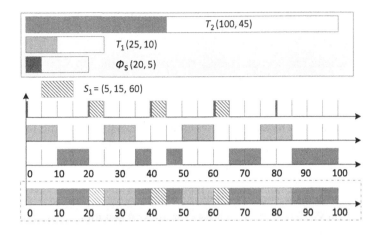

Figure 17.3
An example with a broken deadline.

A feasible design does not guarantee that all the sporadic jobs will meet their respective deadlines. As explained before, the task set in Figure 17.3 is schedulable when $\delta = 0.5$. However, the server cannot meet the deadline of the sporadic job S_1. Note that at time 0, since there is no sporadic job to execute, the execution budget of the first period of Φ_s is not utilized.

Since it might be unaffordable to break the deadlines of sporadic jobs, a sporadic server will reject a sporadic job if it cannot honor the specified deadline. Just like a service provider with its customers, it is better not to sign the contract rather than to break it. Actually, it is not a bad thing at all for a server to reject a sporadic job; the system may have another sporadic server[2] that can handle and meet the deadline of the sporadic job.

In general, a sporadic server will reject a sporadic job if it cannot pass a so-called *acceptance test* as explained below.

Given a sporadic server $\Phi_s = (p_s, e_s, \theta)$, where $\theta = \{S_i = (r_i, e_i, d_i) | 1 \le i \le m\}$ is a set of sporadic jobs already accepted by Φ_s, the slack of a sporadic job $S_k = (r_k, e_k, d_k)$ at time t is defined by (note that S_k may or may not belong to θ)

$$\omega(S_k, t) = \lfloor (d_k - t)/p_s \rfloor e_s - e_k - \sum_{S_i \in \theta \cdot d_i < d_k} (e_i - \xi_i),$$

where ξ_i is the execution time of the completed portion of the existing sporadic job S_i. Note that $\lfloor (d_k - t)/p_s \rfloor e_s$ is the total possible budget of the server before the deadline d_k of the job

---

[2] It may run on another microprocessor in a multiprocessor environment or on another computing node in a grid-computing configuration.

S_k. This budget has to be distributed among the job S_k and all the jobs with a deadline earlier than d_k.

At time t, to decide whether to accept a newly released sporadic job $S_x = (t, e_x, d_x)$, the server first computes the slack time $\omega(S_x, t)$. The new job S_x cannot be accepted if $\omega(S_x, t)$ is less than 0.

If $\omega(S_x, t) \geq 0$, the server next has to check whether any existing sporadic job with a deadline later than d_x may be adversely affected by the acceptance of S_x. Put them together, sporadic task $S_x = (t, e_x, d_x)$ is accepted at time t if we have:

$$\begin{cases} \omega(S_x, t) \geq 0, & and \\ \omega(S_j, t) - e_x \geq 0 & \text{holds for all } S_j \in \theta \text{ with } d_j \geq d_x. \end{cases} \tag{17.1}$$

Note that θ is dynamically maintained at run time:

- A sporadic job S_x is added to θ if it passes the acceptance test.
- The server Φ_s always selects from θ the job with the earliest deadline to execute. This job is also called the server's current job.
- When it comes to the server's turn, the current job is scheduled to execute.
- When the server's current job has run to completion, it is immediately removed from θ.

Now, let us apply the acceptance test to the example in Figure 17.3. The job is released at time instant 5. Since θ is empty at time 5, we have

$$\omega(S_1, 5) = \lfloor (60 - 5)/20 \rfloor 5 - 15 = -5,$$

which is less than 0, so S_1 cannot be accepted by the server $\Phi_s(20, 5)$.

Let us consider the sporadic tasks in Figure 17.4. S_1 is released at time 2, at which time $\theta = \emptyset$, and we have

$$\omega(S_1, 2) = \lfloor (50 - 2)/20 \rfloor 5 - 7.5 = 2.5,$$

which is larger than 0. Since there is no existing sporadic job with a deadline later than 50, S_1 is accepted by the server $\Phi_s(20, 5)$.

The job S_2 is released at time 25, at which time $\theta = \{S_1\}$, where S_1 has an earlier deadline than S_2. We have

$$\omega(S_2, 25) = \lfloor (70 - 25)/20 \rfloor 5 - 2 - (7.5 - 5) = 5.5,$$

which is larger than 0. Since there is no existing sporadic job with a deadline later than 70, S_2 is accepted by the server $\Phi_s(20, 5)$.

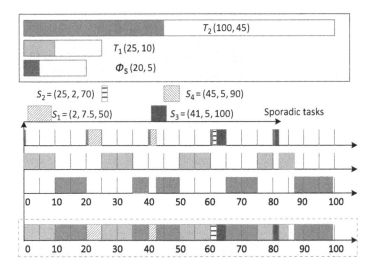

Figure 17.4
Acceptance test of a sporadic server.

S_3 is released at time 41, at which time $\theta = \{S_1, S_2\}$, where both S_1 and S_2 have an earlier deadline than S_3. We have

$$\omega(S_3, 41) = \lfloor (100 - 41)/20 \rfloor 5 - 5 - [(7.5 - 6) + (2 - 0)] = 1.5,$$

which is larger than 0. Since there is no existing sporadic job with a deadline later than 100, S_3 is accepted by the server $\Phi_s(20, 5)$.

S_4 is released at time 45, at which time $\theta = \{S_2, S_3\}$ (note that S_1 has finished at time 42.5 and is removed from θ), where S_2 has an earlier deadline than S_4. We have

$$\omega(S_4, 45) = \lfloor (90 - 45)/20 \rfloor 5 - 5 - (2 - 0) = 3,$$

which is larger than 0. Since S_3 has a deadline later than 90, the deadline of the new job S_4, we need to examine whether S_3 may be adversely affected by the acceptance of S_4. We have

$$\omega(S_3, 45) = \lfloor (100 - 45)/20 \rfloor 5 - 5 - (2 - 0) = 3,$$

so $\omega(S_3, 45) - 5 = -2 < 0$. Thus, S_4 cannot be accepted by the server $\Phi_s(20, 5)$. Indeed, as can be seen from the timing diagram in Figure 17.4, the processing time is almost fully utilized.

The behavior of a sporadic server (p_s, e_s, θ, ρ) is regulated by ρ, a set of predefined consumption and replenishment rules. We next study three sporadic servers regulated by different rule sets.

17.3 A Naive Sporadic Server

We use $\Phi_s^0 = (p_s, e_s, \theta, \rho_0)$ to denote our first sporadic server. The set ρ_0 includes the following rules:

(1) E1: The execution of the sporadic server is suspended as soon as its budget becomes empty.

(2) R1: At the beginning of each period, the sporadic server has its budget e_s fully replenished.

(3) C1: When the server is executing a sporadic job, its budget is consumed at the rate of one per unit of execution time.

(4) C2: When there is no sporadic job to execute (θ is empty), the budget is fully consumed instantaneously.

The sporadic server $\Phi_s^0 = (25, 5, \rho_0)$ shown in Figure 17.5 has a period of 25 time units and a budget of five time units. Now, let us apply the rule set ρ_0 to understand its behavior:

(1) At the beginning, Φ_s^0 is scheduled first. Since there is no sporadic job at time instant 0, rules C2 and E1 apply and Φ_s^0 is suspended instantaneously. As a consequence, job J_{11} is scheduled to execute.

(2) The sporadic job S_1 is released at time point 1 with a deadline of 60 and an execution demand of 10 units of time.

(3) Once job J_{11} has run to completion at time 10, the execution of job J_{21} follows.

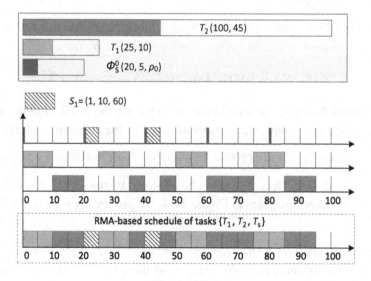

Figure 17.5

A naive sporadic server.

(4) Φ_s^0 is scheduled again at time point 20, where Φ_s^0 is replenished according to rule R1. The execution of J_{21} is preempted, and Φ_s^0 starts to execute the sporadic job S_1. Rule C1 applies as Φ_s^0 is executing, until time point 25, where its budget becomes empty. According to rule E1, Φ_s^0 stops its execution at time 25. There is still a need for five units of time to finish the rest of S_1.

(5) At time point 25, the second job, J_{12}, of task T_1 is scheduled to execute. It finishes at time 35, where job J_{21} resumes its execution.

(6) At time 40, Φ_s^0 is fully replenished according to rule R1 and preempts the execution of J_{21}. Φ_s^0 finishes the rest of sporadic job S_1 at time point 45, where Φ_s^0 is suspended according to rules C1 and E1.

(7) From time 45 on until the end of the hyperperiod, since there are no more sporadic jobs to execute, the execution time is used by tasks T_1 and T_2 only.

Now, let us reconsider the tasks given in Section 17.1.

17.3.1 Task Design

As shown in Figure 17.6, a sporadic server is designed such that it has a period of 4 and execution budget of 2. According to this design, Φ_s^0 will have the highest scheduling priority.

17.3.2 Acceptance Test

(1) $S_1 = (7, 2.5, 20)$ is released at time 7, at which time $\theta = \emptyset$, and we have

$$\omega(S_1, 7) = \lfloor (20 - 7)/4 \rfloor 2 - 2.5 = 3.5,$$

which is larger than 0. Since there is no existing sporadic job with a deadline later than 20, S_1 is accepted by the server $\Phi_s^0(4, 2, \rho_0)$.

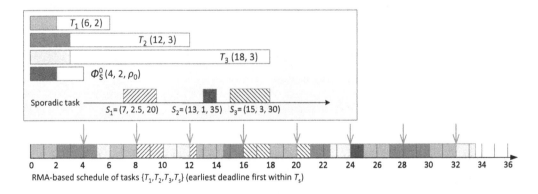

Figure 17.6
A design of a naive sporadic server.

(2) $S_2 = (13, 1, 35)$ is released at time 13, at which time $\theta = \emptyset$ (note that S_1 has finished at time 12.5 and has been removed from θ), and we have

$$\omega(S_2, 13) = \lfloor (35 - 13)/4 \rfloor 2 - 1 = 9,$$

which is larger than 0. Since there is no existing sporadic job with a deadline later than 35, S_2 is accepted by the server $\Phi_s^0(4, 2, \rho_0)$.

(3) $S_3 = (15, 3, 30)$ is released at time 15, at which time $\theta = \{S_2\}$, where there is no job with an earlier deadline than S_3. We have

$$\omega(S_3, 15) = \lfloor (30 - 15)/4 \rfloor 2 - 3 = 3,$$

which is larger than 0. Since S_2 has a deadline later than 30 (the deadline of the new job S_3), we need to examine whether S_2 may be adversely affected by the acceptance of S_3. We have

$$\omega(S_2, 15) = \lfloor (35 - 15)/4 \rfloor 2 - 1 = 9,$$

so $\omega(S_2, 15) - 3 = 6 > 0$. Thus, S_3 is accepted by the server at time 15.

A feasible schedule is given at the bottom of Figure 17.5. As an exercise, you can check which rule(s) is (are) applied at the beginning of each period of the server Φ_s^0.

The naive sporadic server Φ_s^0 could not hold its budget when there is no sporadic job to execute. This might be undesirable in certain applications. We next examine a smarter sporadic server.

17.4 A Fixed-Priority Sporadic Server

Consider a system that consists of a set of independent preemptable periodic tasks $\mathbb{T} = T_H \cup \{\Phi_s^f\} \cup T_L$, where

$$\Phi_s^f = (p_s, e_s, \theta, \rho_f),$$
$$T_H = \{T_i = (p_i, e_i) | 1 \le i \le m\},$$
$$T_L = \{T_j = (p_j, e_j) | m < j \le n\}.$$

The tasks in $\mathbb{T}$ satisfy the following constraints:

$$\forall T_i = (p_i, e_i) \in T_H \cdot p_s > p_i,$$
$$\forall T_j = (p_j, e_j) \in T_L \cdot p_j > p_s.$$

In other words, all tasks in T_H have a shorter period than Φ_s^f, and all tasks in T_L have a longer period than Φ_s^f.

We assume that the task set $\mathbb{T}$ is scheduled according to the RMA principle, and the task Φ_s^f is designated as a sporadic server for sporadic tasks (jobs). Obviously, T_H is the subset of tasks in $\mathbb{T}$ that have a higher priority than Φ_s^f, while T_L is the subset of tasks in $\mathbb{T}$ that have a lower priority than Φ_s^f.

Before we describe the consumption/replenishment rules for the sporadic server Φ_s^f, we first define some notation:

- t always refers to the current time instant.
- β denotes the current budget amount of Φ_s^f.
- t_r denotes the latest replenishment time of Φ_s^f.
- t_a denotes the adjusted latest replenishment time of Φ_s^f.
- t_f denotes the first instant after t_r at which the server Φ_s^f begins to execute. Note that at t_f, Φ_s^f must have a full budget ($\beta = e_s$).
- t_n denotes the next replenishment time of Φ_s^f. t_n always equals $t_a + p_s$.
- A time interval from t_i to t_j is *busy* with respect to a task set F, denoted by $[t_i \xrightarrow{F} t_j]$, if some job from F is executing during that interval.
- $[t_{i_1} \xrightarrow{F} t_{i_2} \xrightarrow{F} t_{i_3}]$ denotes two busy intervals $[t_{i_1} \xrightarrow{F} t_{i_2}]$ and $[t_{i_2} \xrightarrow{F} t_{i_3}]$ that are *contiguous* (the latter one begins immediately after the earlier one ends).
- In general, $[t_{i_1} \xrightarrow{F} t_{i_2} \xrightarrow{F} \cdots \xrightarrow{F} t_{i_k}]$ denotes a contiguous sequence of busy intervals of length k.
- Let $\text{LLCB}(F, t_{i_1}, t_{i_k}, t) = \emptyset[t_{i_1} \xrightarrow{F} t_{i_2} \xrightarrow{F} \cdots \xrightarrow{F} t_{i_k}]\emptyset t$, which denotes the *latest longest contiguous* sequence of busy intervals of F that started before the current time instant t in the sense that

$$\not\exists t_{i_0} \cdot (t_{i_0} < t_{i_1}) \wedge [t_{i_0} \xrightarrow{F} t_{i_1}], \quad \text{and}$$

$$\not\exists t_{j_1}, t_{j_2} \cdot (t_{i_k} \leq t_{j_1} < t_{j_2} \leq t) \wedge [t_{j_1} \xrightarrow{F} t_{j_2}].$$

- Define $t_{run}^H = t_{i_1}$, where $\text{LLCB}(T_H, t_{i_1}, t_{i_k}, t)$ holds (i.e., $\emptyset[t_{i_1} \xrightarrow{T_H} t_{i_2} \xrightarrow{T_H} \cdots \xrightarrow{T_H} t_{i_k}]\emptyset t$ holds). In other words, t_{run}^H refers to the beginning instant of the earliest busy interval among the latest contiguous sequence of busy intervals of T_H that started before t.
- Define t_{end}^H, the end of the latest busy interval of T_H, as

$$t_{end}^H = \begin{cases} t_{i_k} & \text{if } \text{LLCB}(T_H, t_{i_1}, t_{i_k}, t) \text{ and } t_{i_k} < t, \\ \infty & \text{otherwise.} \end{cases}$$

The rule set ρ_f includes the following rules:

1. E1: The server Φ_s^f is able to execute a sporadic job from θ only when its budget β is not empty.
2. C1: When the server Φ_s^f is executing, its execution budget β is consumed at the rate of one per unit time until the budget is exhausted.

3. C2: When the server Φ_s^f is not executing, if it had executed since t_r and $t_{end}^H < t$, its execution budget β is consumed at the rate of one per unit time until the budget is exhausted.

4. C3: For other cases, the server Φ_s^f holds its execution budget.

5. R1: When the system $\mathbb{T}$ begins execution for the first time ($t = 0$), the server's budget is replenished: $\beta = e_s, t_r = t_a = t$.

6. R2: When it comes to t_f (i.e., $t = t_f$), set the adjusted latest replenishment time

$$
t_a = \begin{cases} \max(t_r, t_{run}^H) & \text{if } (t_{end}^H = t_f) \ (T_H \text{ has just stopped}), \\ t_f & \text{if } t_{end}^H < t_f \ (T_H \text{ stopped a while ago}). \end{cases}
$$

Note that since $t_n = t_a + p_s$, the next replenishment time is effectively modified owing to the adjustment of t_a.

7. R3: as time passes,

   (a) when the budget becomes exhausted at t ($\beta = 0$), if $t_n < t_f$, the budget is immediately replenished: $\beta = e_s, t_r = t_a = t$.

   (b) when the system $\mathbb{T}$ starts to run at t, if $\mathbb{T}$ was idle before t, the budget is replenished: $\beta = e_s, t_r = t_a = t$ (a fresh restart).

   (c) when it comes to the replenishment time ($t = t_n$), the budget is replenished: $\beta = e_s, t_r = t_a = t$.

The best way to understand the above convolving rules is to look at an example in operation. In Figure 17.7, we have three periodic tasks and a sporadic server, which are scheduled according to the RMA principle. Note that $T_H = \{T_1, T_2\}$ and $T_L = \{T_3\}$. The value changes of the time-related variables are detailed in Table 17.1.

Now, let us examine the timing diagram step-by-step:

(1) At the beginning, the server Φ_s^f has its full budget of 3 and t_r is set to 0. The first job of T_1 is scheduled to execute because it has the highest priority.

(2) Job J_{21} starts to execute when J_{11} runs to completion at time 1.

(3) Job J_{31} starts to execute when J_{21} runs to completion at time 3.

(4) Job J_{31} is preempted at time 6 by J_{12}. At the same time, the sporadic job S_1 is released. From time 0 to 6, the sporadic job list is empty and Φ_s^f is suspended. According to rule C3 Φ_s^f's budget stays at 3. Although Φ_s^f becomes ready at 6, it has to wait for the higher-priority task T_1.

(5) Φ_s^f starts to execute S_1 at time 7. At time 7^+ (just after time instant 7), because this is the first instant after $t_r = 0$ at which Φ_s^f starts to execute, rule R2 applies, where we have $t_{end}^H = 7 = t_f$ since LLCB($T_H, 6, 7, 7^+$) holds. Thus, $t_a = \max(t_r, t_{run}^H) = \max(0, 6) = 6$, and the next replenishment time t_n is set to 16.

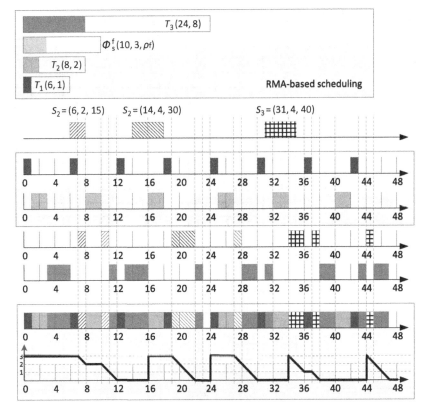

Figure 17.7
A fixed-priority sporadic server in operation.

(6) The execution of Φ_s^f only lasts until time 8, at which time it is preempted by J_{22}. According to rule C1, the server's budget is consumed in the interval $[7, 8]$ and becomes 2 at time 8.

(7) While J_{22} is executing from time 8 to 10, Φ_s^f's budget stays at 2 (rule C3).

(8) Φ_s^f resumes its execution at time 10, and sporadic job S_1 runs to completion at time 11. Rule C1 applies during $[10, 11]$.

(9) At time 11, Φ_s^f is suspended since there are no more sporadic jobs, and job J_{31} gets a chance to run. Rule C2 applies during $[11, 12]$ because Φ_s^f had executed since time 0 and $t_{\text{end}}^H = 10 < t$ (we have LLCB($T_H, 8, 10, 11$)). Consequently, the server's budget becomes empty at time 12.

(10) Job J_{13} runs during $[12, 13]$. Φ_s^f's budget stays at 0 (rule C3).

(11) Job J_{31} is scheduled to execute during $[13, 14]$. Φ_s^f's budget stays at 0 (rule C3).

(12) At time 14, sporadic job S_2 is released. However, J_{31} continues to run from time 14 to 16, because the server Φ_s^f is suspended according to rule E1.

Table 17.1 Application of consumption and replenishment rules in ρ_f

Step	Time Instant or Interval	Rule	t_r	t_f	t_{run}^H	t_{end}^H	t_a	$t_n = t_a + p_s$	β
1	0	R1	0		0	0	0	10	3
2	0-3	C3	0		0	∞	0	10	3
3	3-6	C3	0		0	3	0	10	3
4	6-7	C3	0		6	∞	0	10	3
5	7⁺ (just after 7)	R2	0	7	6	7	$\max(t_r, t_{run}^H) = 6$	16	3
6	7-8	C1	0	7	6	7	6	16	↓
7	8-10	C3	0	7	8	∞	6	16	2
8	10-11	C1	0	7	8	10	6	16	↓
9	11-12	C2	0	7	8	10	6	16	↓
10	12-13	C3	0	7	12	∞	6	16	0
11	13-14	C3	0	7	12	13	6	16	0
12	14-16	E1	0	7	12	13	6	16	0
13	16	R3c	16		12	13	16	26	3
14	16-19	C3	16		16	∞	16	26	3
15	19⁺	R2	16	19	16	19	$\max(t_r, t_{run}^H) = 16$	26	3
16	19-22	C1	16	19	16	19	16	26	↓
17	22-24	E1	16	19	16	19	16	26	0
18	24	R3b	24		16	19	24	34	3
19	24-27	C3	24		24	∞	24	34	3
20	27⁺	R2	24	27	24	27	$\max(t_r, t_{run}^H) = 24$	34	3
21	27-28	C1	24	27	24	27	24	34	↓
22	28-30	C2	24	27	24	27	24	34	↓
23	30-31	C3	24	27	30	∞	24	34	0
24	31-32	E1	24	27	30	31	24	34	0
25	32-34	C3	24	27	32	∞	24	34	0
26	34	R3c	34		32	∞	34	44	3
27	34⁺	R2	34	34	32	34	$\max(t_r, t_{run}^H) = 34$	44	3
28	34-36	C1	34	34	32	34	34	44	↓
29	36-37	C3	34	34	36	∞	34	44	1
30	37-38	C1	34	34	36	37	34	44	↓
31	38-40	E1	34	34	36	37	34	44	0
32	40-43	E1	34	34	40	∞	34	44	0
33	43-44	E1	34	34	40	43	34	44	0
34	44	R3c	44		40	43	44	54	3
35	44⁺	R2	44	44	40	43	$t_f = 44$	54	3
36	44-45	C1	44	44	40	43	44	54	↓
37	45-47	C2	44	44	40	43	44	54	↓
38	47-48	C3	44	44	40	43	44	54	0

(13) At time 16, it is the replenishment time ($t_n = 16$), so the server's budget becomes 3, and t_r is set to 16.

(14) From time 16 to 19, Φ_s^f's budget stays at 3 (rule C3).

(15) Φ_s^f starts to execute S_1 at time 19. At time 19⁺ (just after time instant 19), because this is the first instant after $t_r = 16$ at which Φ_s^f starts to execute, rule R2 applies, where we have $t_{end}^H = 19 = t_f$ since LLCB$(T_H, 16, 19, 19^+)$ holds. Thus, $t_a = \max(t_r, t_{run}^H) = \max(16, 16) = 16$, and the next replenishment time t_n is set to 26.

(16) From time 19 to 22, Φ_s^f is executing, and according to rule C1 its budget becomes empty at time 22.

(17) At time 22, job J_{31} resumes its execution and runs to its completion at time 23. The whole system is idle from 23 to 24.

(18) At time 24, job J_{15} is released and starts to execute. Note that rule R3b applies at time 24 since the system was idle from 23 to 24. So the server's budget becomes 3, t_r is set to 24, and t_n becomes 34.

(19) From time 24 to 27, jobs from T_H execute. Φ_s^f's budget stays at 3 (rule C3).

(20) Φ_s^f starts to execute at time 27. At time 27^+ (just after time instant 27), because this is the first instant after $t_r = 24$ at which Φ_s^f starts to execute, rule R2 applies, where we have $t_{end}^H = 27 = t_f$ since $LLCB(T_H, 24, 27, 27^+)$ holds. Thus, $t_a = \max(t_r, t_{run}^H) = \max(24, 24) = 24$, and the next replenishment time t_n is adjusted to 34.

(21) From time 27 to 28, Φ_s^f is executing, and according to rule C1 its budget becomes 2 at time 28.

(22) Rule C2 applies during $[28, 30]$ because Φ_s^f had executed since time 24 and $t_{end}^H = 27 < t$ (we have $LLCB(T_H, 24, 27, t)$). Consequently, the server's budget becomes empty at time 30.

(23) Job J_{16} runs during $[30, 31]$. Rule C3 applies to Φ_s^f.

(24) S_3 is released at time 31. It, however, cannot be scheduled by Φ_s^f according to rule E1.

(25) Job J_{25} executes during $[32, 34]$. Φ_s^f's budget stays at 0 (rule C3).

(26) At time 34, it is the replenishment time ($t_n = 34$), so the server's budget becomes 3, and t_r is set to 34.

(27) Also, Φ_s^f starts to execute at time 34. At time 34^+ (just after time instant 34), because this is the first instant after $t_r = 34$ at which Φ_s^f starts to execute, rule R2 applies, where we have $t_{end}^H = 34 = t_f$ since $LLCB(T_H, 32, 34, 34^+)$ holds. Thus, $t_a = \max(t_r, t_{run}^{II}) = \max(34, 32) = 34$, and the next replenishment time t_n is adjusted to 44.

(28) From time 34 to 36, Φ_s^f is executing, and according to rule C1 its budget becomes 1 at time 36.

(29) From time 36 to 37, job J_{17} from T_H executes. Φ_s^f's budget stays at 1 (rule C3).

(30) From time 37 to 38, Φ_s^f is executing, and according to rule C1 its budget becomes 0 at time 38.

(31) From time 38 to 40, Φ_s^f is not executing according to rule E1.

(32) From time 40 to 43, Φ_s^f is not executing according to rule E1.

(33) From time 43 to 44, Φ_s^f is not executing according to rule E1.

(34) At time 44, it is the replenishment time ($t_n = 44$), so the server's budget becomes 3, and t_r is set to 44.

(35)   Also, Φ_s^f starts to execute at time 44. At time 44^+ (just after time instant 44), because this is the first instant after $t_t = 44$ at which Φ_s^f starts to execute, rule R2 applies, where we have $t_{end}^H = 43 < t_f$ since $\text{LLCB}(T_H, 40, 43, 44^+)$ holds. Thus, $t_a = t_f = 44$, and the next replenishment time t_n is adjusted to 54.

(36)   From time 44 to 45, Φ_s^f is executing, and according to rule C1 its budget becomes 2 at time 45.

(37)   Rule C2 applies during $[45, 47]$ because Φ_s^f had executed since time 44 and $t_{end}^H = 43 < t$ (we have $\text{LLCB}(T_H, 40, 43, t)$). Consequently, the server's budget becomes empty at time 47.

(38)   The system is idle during $[47, 48]$.

The rules defined in ρ_f are not hard to implement. The algorithm given by SPORADIC-AUDITING is an event-driven implementation.

SPORADIC-AUDITING (T_H, Φ_s^f, T_L)

```
 1   T = T_H ∪ {Φ_s^f} ∪ T_L   // T is the whole system
 2   t_n = p_s                 // t_n is the anticipated next replenishment time
 3   t = 0                     // t is the current time instant
 4   t_r = 0                   // t_r is the latest replenishment time
 5   t_a = 0                   // t_a is the adjusted latest replenishment time
 6   β = e_s                   // rule R1: get the full budget initially
 7   curJob = Null
 8   lastJob = Null
 9   systemIsIdle = True       // the system T is idle initially
10   for (; ; )
11        event = RECEIVE-SIGNAL ()
12        switch (event)
13            case: Unit-Time-Tick event
14                 t++
15                 if ((isRunning(Φ_s^f)) or (β < e_s and t_end^H < t))
16                      β = max(0, β − 1)              // rule C1 and rule C2
17                 if ((β = 0) and (t_n < t_f))        // rule R3a
18                      t_r = t_a = t; β = e_s; t_n = t_a + p_s
19                 if (t = t_n)                        // rule R3c
20                      t_r = t_a = t; β = e_s; t_n = t_a + p_s
21            case: Starts-To-Run-Job(α)               // job α starts to run
22                 lastJob = curJob
23                 curJob = α
24                 if (systemIsIdle is True)           // rule R3b
25                      systemIsIdle = False
```

```
26                    t_r = t_a = t; β = e_s; t_n = t_a + p_s
27              if (curJob ∈ T_H and lastJob ∉ T_H)
28                    t^H_run = t; t^H_end = ∞
29              if (curJob = Φ^f_s)
30                    if (t^H_end = ∞) t^H_end = t              // T_H has stopped
31                    if (β = e_s)
32                          // Rule R2: the first instant when Φ^f_s starts to run after t_r
33                          t_f = t
34                          if (t^H_end < t_f)
35                                t_a = t_f
36                          else
37                                t^H_end = t; t_a = max(t_r, t^H_run)
38                          t_n = t_a + p_s
39              if (curJob ∈ T_L)
40                    if (t^H_end = ∞) t^H_end = t              // T_H has stopped
41        case: Stops-To-Run(T)                                // no task is running
42              systemIsIdle = True
43              t^H_end = t
44    end
45 end
```

17.5 A Dynamic-Priority Sporadic Server

The sporadic server discussed in the last section cannot execute sporadic jobs when its budget becomes empty. Next, we look at a sporadic server that can execute sporadic jobs even though its budget is empty.

Consider a system that consists of a set of independent tasks $\mathbb{T} = T_H \cup \{\Phi^d_s\} \cup T_L$, where $\Phi^d_s = (p_s, e_s, \theta, \rho_d, \mu_s)$, and the tasks in $\mathbb{T}$ satisfy the following constraints:

$$\forall T_i \in T_H \cdot \mathrm{pri}(T_i) > \mu_s,$$
$$\forall T_j \in T_L \cdot \mathrm{pri}(T_j) < \mu_s.$$

Let ω denote a priority that is one level lower than the minimum task priority: $\omega = \min_{T_j \in T_L} \mathrm{pri}(T_j) - 1$. The sporadic server Φ^d_s can execute sporadic jobs with its normal priority μ_s when its budget is larger than 0; it can also execute sporadic jobs with its priority lowered to ω when its budget is 0.

A replenishment action, denoted by γ, is defined by $\langle t_\gamma, v_\gamma \rangle$, where t_γ is the replenishment time, and v_γ is the replenishment amount.

We use τ to denote the *latest activation time* of Φ_s^d. When the system starts, $\tau = 0$. τ is reset to the current time t when (a) Φ_s^d's priority is changed from ω to μ_s owing to a replenishment action applied at t, or (b) Φ_s^d becomes unblocked at t (Φ_s^d can be blocked owing to the request for unavailable resources).

The set ρ_d includes the following rules:

(1)   R1: Φ_s^d has its budget e_s fully replenished when the system starts.
(2)   R2: When it is time to apply the next replenishment action $\langle t_\gamma, v_\gamma \rangle$—that is, $t \geq t_\gamma$—the server's budget is adjusted such that $e_s = e_s + v_\gamma$. If its priority was ω immediately before t, its priority is restored to its normal level μ_s, and the time t is set to be the new value of the latest activation time τ.
(3)   C1: When Φ_s^d is executing a sporadic job with its normal priority μ_s, its budget e_s is consumed at the rate of one per unit of execution time.
(4)   C2: When Φ_s^d is executing a sporadic job with priority ω, it holds its budget.
(5)   C3: The server Φ_s^d holds its budget when it is not executing.
(6)   E1: When e_s becomes 0 at t and there is still a sporadic job to execute, $\text{pri}(\Phi_s^d) = \omega$, and a replenishment action $\gamma = \langle t_\gamma, v_\gamma \rangle$ is appended, where $t_\gamma = \tau + p_s$, and the replenishment amount v_γ is set to the amount consumed between τ and t. If t_γ is before the current time, the replenishment action is applied immediately.
(7)   E2: When $e_s > 0$ and Φ_s^d is *blocked* at t, a replenishment action $\gamma = \langle t_\gamma, v_\gamma \rangle$ is appended, where $t_\gamma = \tau + p_s$, and the replenishment amount v_γ is set to the amount consumed between τ and t. If t_γ is before the current time, the replenishment action is applied immediately.

Again, the best way to understand the rules in ρ_d is to look at an example in operation. In Figure 17.8, we have three periodic tasks and a sporadic server, where we have $T_H = \{T_1, T_2\}$ and $T_L = \{T_3\}$.

Let us examine how the budget changes as time goes on:

(1)   At time 0, according to rule R1, the server Φ_s^d's budget is fully replenished: $e_s = 3$. Also, $\tau = 0$.
(2)   From time 0 to 7, the server Φ_s^d is not executing, so it holds its budget according to rule C3. Meanwhile, sporadic job S_1 is released at time 6.
(3)   From time 7 to 8, the server Φ_s^d executes S_1 with its normal priority μ_s. According to rule C1, its budget becomes 2 at time 8.
(4)   At time 8, a higher-priority job J_{22} preempts Φ_s^d.
(5)   From time 8 to 10, the server Φ_s^d is not executing, so it holds its budget according to rule C3.
(6)   From time 10 to 11, the server Φ_s^d executes S_1 with its normal priority μ_s. According to rule C1, its budget becomes 1 at time 11.

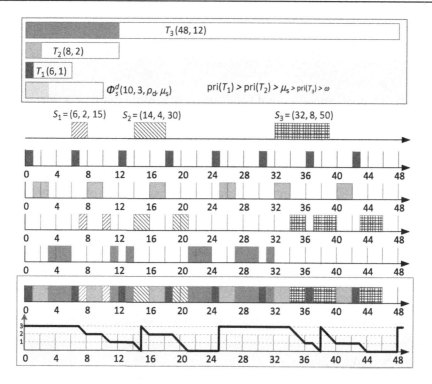

Figure 17.8
A dynamic-priority sporadic server in operation.

(7) At time 11, job S_1 finishes.

(8) From time 11 to 14, the server Φ_s^d is not executing, so it holds its budget according to rule C3.

(9) At time 14, sporadic job S_2 is released and the server Φ_s^d starts to execute S_2 with its normal priority μ_s.

(10) From time 14 to 15, the server Φ_s^d executes S_2. According to rule C1, its budget becomes 0 at time 15.

(11) At time 15, according to rule E1, Φ_s^d's priority becomes ω, and a replenishment action $\langle 10, 3 \rangle$ is scheduled. Since $10 < 15$, this action is applied immediately. Then, according to rule R2, the server Φ_s^d's budget becomes 3, its priority is restored to μ_s, and $\tau = 15$.

(12) From time 15 to 16, the server Φ_s^d executes S_2. According to rule C1, its budget becomes 2 at time 16.

(13) From time 16 to 19, jobs from T_H execute.

(14) From time 19 to 21, the server Φ_s^d executes S_2, which runs to completion. According to rule C1, its budget becomes 0 at time 21.

(15) At time 21, according to rule E1, Φ_s^d's priority becomes ω, and a replenishment action $\langle 25, 3 \rangle$ is scheduled.

(16) From time 21 to 25, the server Φ_s^d is not executing, so it holds its budget according to rule C3.

(17) At time 25, according to rule R2, the server Φ_s^d's budget becomes 3, its priority is restored to μ_s, and $\tau = 25$.

(18) From time 25 to 34, the server holds its budget according to rule C3. Meanwhile, sporadic job S_3 is released at time 32.

(19) From time 34 to 36, the server Φ_s^d executes S_3. According to rule C1, its budget becomes 1 at time 36.

(20) From time 36 to 37, job J_{17} executes.

(21) From time 37 to 38, the server Φ_s^d executes S_3. According to rule C1, its budget becomes 0 at time 38.

(22) At time 38, according to rule E1, Φ_s^d's priority becomes ω, and a replenishment action $\langle 35, 3 \rangle$ is scheduled. Since $35 < 38$, this action is applied immediately. Then, according to rule R2, the server Φ_s^d's budget becomes 3, its priority is restored to μ_s, and $\tau = 38$.

(23) From time 38 to 40, the server Φ_s^d executes S_3. According to rule C1, its budget becomes 1 at time 40.

(24) From time 40 to 43, the server holds its budget according to rule C3.

(25) From time 43 to 44, the server Φ_s^d executes S_3 with priority μ_s. According to rule C1, its budget becomes 0 at time 44.

(26) At time 44, according to rule E1, Φ_s^d's priority becomes ω, and a replenishment action $\langle 48, 3 \rangle$ is scheduled.

(27) From time 44 to 46, the server Φ_s^d executes S_3 with priority ω. According to rule C2, the server holds its budget, which is 0.

(28) At time 48, according to rule R2, the server Φ_s^d's budget becomes 3, its priority is restored to μ_s, and $\tau = 48$.

It is worth noting that in this chapter we assume that there is only one sporadic server task in the system, which may or may not run upon a real-time operating system. Recall that in Chapter 13 we learned that a POSIX-compliant real-time operating system also implements a scheduling policy called a sporadic server scheduling policy, which can be employed by multiple tasks (threads). The rules in ρ_d are actually adapted from the sporadic server scheduling policy specified by the POSIX.1-2008 standard [17].

Problems

17.1 Consider a system containing three periodic tasks: $T_1 = (25, 10)$, $T_2 = (40, 5)$, and $T_3 = (50, 7.5)$. The system also has a sporadic server $\Phi_s^0 = (20, 5, \rho_0)$, which is scheduled with the other periodic tasks according to the RMA principle.

(a) Analyze the schedulability of the system.

(b) Suppose the system has two sporadic jobs $S_1 = (30, 7.5)$ and $S_2 = (75, 6)$. Work out a timing diagram showing the schedule of the system.

(c) Suppose the server changes to $\Phi_s^0 = (10, 2.5, \rho_0)$. Redo (a) and (b).

(d) Suppose the server changes to $\Phi_s^0 = (50, 12.5, \rho_0)$. Redo (a) and (b).

17.2 Consider a system containing three periodic tasks: $T_1 = (30, 10)$, $T_2 = (40, 5)$, and $T_3 = (50, 5)$. The system also has a sporadic server $\Phi_s^0 = (20, 8, \rho_0)$, and $\delta = 0.5$. Is this system schedulable?

17.3 Consider a system containing three periodic tasks: $T_1 = (10, 3)$, $T_2 = (30, 10)$, and $T_3 = (60, 6)$. The system also has a sporadic server $\Phi_s^0 = (20, 4, \rho_0)$. While the system is running, three sporadic jobs arrive: $S_1 = (3, 3, 40)$, $S_2 = (13, 5, 50)$, and $S_3 = (15, 10, 35)$. For each of the sporadic jobs, examine whether it can be accepted by the server or not.

17.4 A system has two periodic tasks, $T_1 = (30, 6)$ and $T_2 = (120, 40)$, and a sporadic server, $\Phi_s^0 = (60, 10, \rho_0)$, which is scheduled with the other tasks according to the RMA principle. What is the time instant at which a sporadic job $S_1 = (8, 12)$ runs to completion?

17.5 Consider a system containing three periodic tasks: $T_1 = (10, 3)$, $T_2 = (30, 5)$, and $T_3 = (60, 6)$. The system also has a sporadic server $\Phi_s^f = (20, 5, \rho_f)$, which is scheduled with the other tasks according to the RMA principle. Suppose there are three sporadic jobs $S_1 = (10, 7, 30)$ and $S_2 = (15, 3, 60)$, and $S_3 = (20, 8, 70)$. Work out a timing diagram showing the schedule of the system. Would any deadline be broken?

17.6 Consider a system containing three periodic tasks: $T_1 = (10, 3)$, $T_2 = (30, 5)$, and $T_3 = (60, 6)$. The system also has a sporadic server $\Phi_s^d = (20, 5, \mu_s, \rho_d)$, where $\mathrm{pri}(T_1) > \mathrm{pri}(T_2) > \mu_s > \mathrm{pri}(T_3)$. Suppose there are three sporadic jobs $S_1 = (10, 7, 30)$, $S_2 = (15, 3, 60)$, and $S_3 = (20, 8, 70)$. Work out a timing diagram showing the schedule of the system. Would any deadline be broken?

Implementation Patterns

Resource Sharing

Contents

When we leverage, we aggregate and organize existing resources to achieve success.

Richie Norton

18.1 Shared Variables

Multiple threads within the same process can share global variables, through which data can flow back and forth among them.

We consider a classic example as illustrated in Figure 18.1, where a producer thread generates data then places the data in a shared buffer object, and a consumer thread reads data from the buffer. An example code is given in Listing 18.1, where the producer writes numbers from 1 to

Real-Time Embedded Systems. http://dx.doi.org/10.1016/B978-0-12-801507-0.00018-3

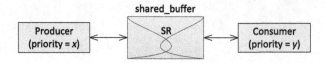

Figure 18.1
Two tasks using a shared variable.

10 to the shared buffer, and the consumer reads the shared buffer and calculates the sum of the values read so far.

```
1  #include <stdio.h>
   #include <pthread.h>
3  #include <time.h>

5  //Data is shared by child processes producer and consumer
   int shared_buffer = -1;
7
   void set_shared_buffer(int value) {
9    fprintf(stdout, "Producer writes\t%2d", value);
     shared_buffer = value;
11 }
   int get_shared_buffer() {
13   fprintf(stdout, "Consumer reads\t%2d\n", shared_buffer);
     return shared_buffer;
15 }
   int get_shared_buffer2() {
17   return shared_buffer;
   }
19
   void * consumer(void *notused) {
21   int sum = 0;
     int firstget, secondget;
23   int r, count;
     for (count = 1; count <= 10; count++) {
25     r = rand() % 4;
       sleep(r);
27     firstget = get_shared_buffer();
       sleep(1); //simulate longer processing
29     secondget = get_shared_buffer2();
       sum += secondget;
31     fprintf(stdout, "Consumer process\t\t\t%2d,%2d,%2d\n",
         firstget, secondget, sum);
33   }
     fprintf(stdout, "\n%s %d\n%s\n", "Consumer read values total", sum, "Terminating
       Consumer");
35 }

37 void * producer(void *notused) {
```

```
     int sum = 0;
39   int r, count;
     for (count = 1; count <= 10; count++) {
41     r = rand() % 4;
       sleep(r);
43     set_shared_buffer(count);
       sum += count;
45     fprintf(stdout, "\t%2d\n", sum);
     }
47   fprintf(stdout, "Producer done\nTerminating Producer\n");
   }
49
   int main() {
51   fprintf(stdout, "Action\t\tValue\tSum of Produced\tfirst, second,Sum\n");
     fprintf(stdout, "------\t\t-----\t---------------\t------------------\n");
53
     pthread_attr_t attr;
55   struct sched_param sp;

57   //initializes the thread attributes to their default values
     pthread_attr_init(&attr);
59   pthread_attr_setinheritsched(&attr, PTHREAD_EXPLICIT_SCHED);

61   //Set the priority to 50 to the producer thread
     sp.sched_priority = 50;
63   pthread_attr_setschedparam(&attr, &sp);
     // create the producer thread
65   pthread_create(NULL, &attr, producer, NULL);

67   //Set the priority to 20 to the consumer thread
     sp.sched_priority = 20;
69   pthread_attr_setschedparam(&attr, &sp);
     // create the consumer thread
71   pthread_create(NULL, &attr, consumer, NULL);

73   // let the threads run for a bit
     sleep(30);
75   return 1;
   }
```

Listing 18.1
Sample code of using shared variable: multiple threads.

Figure 18.2 shows a screenshot of a run of the program, which reveals some critical issues:

(1) *Data integrity*. In each round, the consumer reads and prints the value of the shared variable twice. The two values are the same for some rounds but are different for others.

(2) *Data loss*. Some data written by the producer are lost on the consumer's side. For example, as far as the consumer's first reads are concerned, the numbers 2, 5, 7, and 8 are

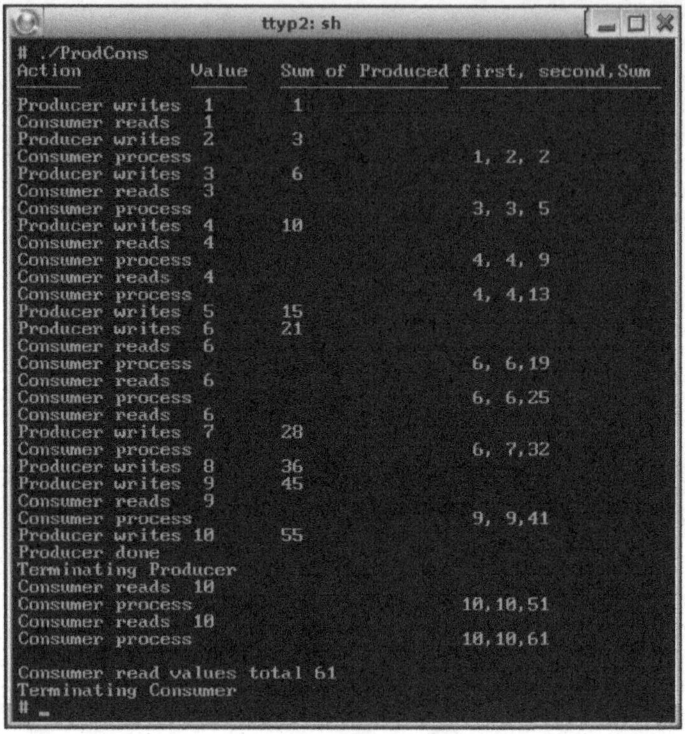

Figure 18.2
A run of the program in Listing 18.1.

lost; as far as the consumer's second reads are concerned, the numbers 1, 5, and 8 are lost. The reason is that the producer has written new data into the shared buffer before the consumer reads the previous data.

(3) *Data duplication.* As far as the consumer's second reads are concerned, the numbers 4, 6, and 10 are each read twice. This is because the buffer is duplicately read by the consumer before the producer produces the next value.

For the above reasons, the total value read by the consumer is different from the total value written by the producer. In order to protect data integrity, the section of code that accesses the shared resource should be protected such that it cannot be accessed by the producer and the consumer at the same time. In order to eliminate data loss or duplication, the behavior of the producer and the consumer should be synchronized such that each value the producer thread writes to the shared buffer is consumed exactly once by the consumer thread.

The producer-consumer problem will be revisited in this chapter and a feasible solution will be gradually developed.

18.2 Shared Memory

An object (a regular file or memory object) can be directly mapped into the address space of one or more processes. Memory mapping makes the mapped region of the object directly addressable by a process. This eliminates unnecessary data read, write, or copy operations, and thus offers an efficient means of passing data between processes running on the same machine.

Table 18.1 lists a few POSIX functions related to memory mapping.

In particular, the mmap() function is for memory mapping, which has the following signature:

```
void * mmap( void * addr, size_t len, int prot, int flags, int fildes, off_t off),
```

where

- addr is a pointer to an address in the calling process's address space where the object is to be mapped. The address actually mapped and returned can be different from addr. It is good practice to let the operating system (OS) choose a suitable address by specifying a NULL value for addr.
- len is the number of bytes to map into the caller's address space.
- prot is the protection bits defining the access types for the memory region being mapped:
  - PROT_EXEC: the region can be executed;
  - PROT_NONE: the region cannot be accessed;
  - PROT_READ: the region can be read;
  - PROT_WRITE: the region can be written.
- flags contains various flags for handling the memory region being mapped:

Table 18.1 POSIX.1 function calls for memory mapping

POSIX Function	Description
open	Open a regular file object
posix_typed_mem_open	Open a typed memory object. The creation/definition of typed memory objects is OS specific
shm_open	Create a new shared memory object or open an existing shared memory object
shm_unlink	Remove a shared memory object by name
mmap	Map a region within an object into the caller's address space
mprotect	Change the access protections on a memory mapping
msync	Write out data in a mapped region to the permanent storage for the mapped file. This ensures data integrity of the file
munmap	Remove any mappings for pages in the specified address range

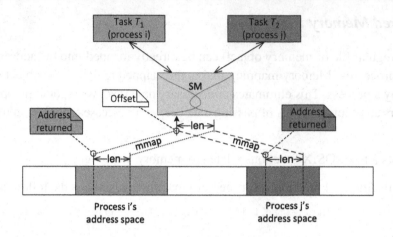

Figure 18.3
Mapping a shared memory object.

- – MAP_PRIVATE: the mapping is private to the calling process;
- – MAP_SHARED: the mapping may be shared by many processes;
- – MAP_FIXED: the object is mapped to the address starting exactly at *addr*, or the function call fails. For portability, this flag is discouraged.
- fildes is the file descriptor for the object being mapped. The object can be a regular file, a shared memory object, or a typed memory object.
- off is the offset into the object from which the mapping starts.

A call of the form

```
pa=mmap(addr, len, prot, flags, fd, off)
```

establishes a mapping between the memory space of the calling process at the address *pa* for *len* bytes to the memory object represented by the file descriptor *fd* at offset *off* for *len* bytes. This is illustrated in Figure 18.3, where an object is mapped, respectively, into the address spaces of two processes. Note that the offset used in the memory mapping operation is not the start of the object being mapped.

The object being mapped can be a regular file, a shared memory object, or a typed memory object. A shared memory object is created by shm_open(), and it can be opened by multiple processes by name. The creation of a typed memory object is OS specific. In QNX, for example, typed memory objects are defined and created in the startup code.

18.2.1 Mapping File Objects

By memory mapping, all or part of a named file can be inserted logically into a process's address space. Reading from and writing to a mapped file is very efficient because of direct memory operations. Pages of memory are written to the file only if their contents have been modified.

When a single file is mapped into more than one process, the mapped file can be used as a means for data communications.

Let us have a look at an example. Listing 18.2 is a *starter* program. It creates a file named log.txt, writes an integer 0 to the file, and then closes the file. Afterward, it spawns three child processes from the code given in Listing 18.3.

```c
1 #include <stdio.h>
  #include <stdlib.h>
3 #include <spawn.h>
  #include <fcntl.h>
5 #include <sys/wait.h>

7 int main(int argc, char **argv) {
    char *args[] = { "user", NULL };
9   int i, status, fd, value = 0;
    struct inheritance inherit;
11  pid_t pid;

13  //open a file and set initial value
    fd = open("log.txt", O_RDWR | O_CREAT, S_IRWXU);
15  write(fd, &value, sizeof(int));
    close(fd);

17
    // create 3 child processes
19  for (i = 0; i < 3; i++) {
      inherit.flags = 0;
21    if ((pid = spawn("user", 0, NULL, &inherit, args, environ)) == -1)
        perror("spawn() failed");
23    else
        printf("spawned child, pid = %d\n", pid);
25  }
    while (1) {
27    if ((pid = wait(&status)) == -1) {
        perror("Starter: ");
29      exit(EXIT_FAILURE);
      }
31    printf("User %d terminated\n", pid);
    }
33 }
```

Listing 18.2
Sample code of using shared variable: multiple processes.

As shown in Listing 18.3, a child process opens the file log.txt and maps it into its own process space. Note that the whole file, which contains only an integer value, is mapped. Inside the loop, the pointer returned from the call mmap() is used to read and increase the integer value.

```
1  // qcc -o user user.c
   #include <stdio.h>
3  #include <stdlib.h>
   #include <fcntl.h>
5  #include <sys/mman.h>

7  int main(int argc, char **argv) {
     int fd, i, nloop = 3, *ptr;
9
     //open a file and map it into memory
11   fd = open("log.txt", O_RDWR, S_IRWXU);
     ptr = mmap(NULL, sizeof(int), PROT_READ | PROT_WRITE, MAP_SHARED, fd, 0);
13   close(fd);

15   for (i = 0; i < nloop; i++) {
       printf("User %d starts processing data: %d\n", getpid(),
17         (*ptr)++);
       sleep(2); //simulate processing
19     printf("User %d stops processing data\n", getpid());
       sleep(1);
21   }
     exit(0);
23 }
```

Listing 18.3
Sample code of using shared variable: multiple processes.

Figure 18.4 shows a screenshot of a run of the program, which indicates that the integer contained in the file can be shared among the three child processes, which, together, increase the integer value accumulatively.

Figure 18.4
A run of the program given in Listings 18.2 and 18.3.

18.2.2 *Shared Memory Objects*

As another example, Listings 18.4 and 18.5 demonstrate the use of a shared memory object.

Listing 18.4 is the server code. It first creates a shared memory object named "MyShm." This object is then mapped into the server's process space, where the length of the mapped memory object is 27 bytes [72]. The pointer *shm* returned from the call mmap() references the first byte of the shared memory object.

The server then writes 26 uppercase characters to the shared memory object. It then sleeps and waits for an asterisk to be written to the first byte of the shared memory object.

```
1  //server.c
   //adapted from Dave Marshall, Programming in C [72]
3  #include <stdio.h>
   #include <stdlib.h>
5  #include <fcntl.h>
   #include <sys/mman.h>
7  #define SHMSZ 27
   char SHM_NAME[] = "MyShm";
9
   int main() {
11   char ch;
     int shmid;
13   char *shm, *s;
15   if ((shmid = shm_open(SHM_NAME, O_RDWR | O_CREAT, S_IRWXU))<0) {
       perror("shm_open");
17     return EXIT_FAILURE;
     }
19
     /* Set the size of the shared memory object */
21   if( ftruncate( shmid, SHMSZ ) == -1 ) {
       perror("out of memory");
23     return EXIT_FAILURE;
     }
25   //map the shared memory object to the process memory
     if ((shm = mmap(NULL, SHMSZ, PROT_READ|PROT_WRITE, MAP_SHARED,
27       shmid, 0)) == MAP_FAILED) {
       perror("memory mapping");
29     return EXIT_FAILURE;
     }
31
     //start writing into memory
33   s = shm;
     int r;
35   for (ch = 'A'; ch <= 'Z'; ch++) {
       *s++ = ch;
37     r = rand() % 3;
       sleep(r);
39   }
```

```
41    while (*shm != '*') sleep(1);
      shm_unlink(SHM_NAME);
43    return EXIT_SUCCESS;
    }
```

Listing 18.4
Sample code of using named shared memory: server.

Listing 18.5 is the client code. It first opens the shared memory object named "MyShm." This object is then mapped into the client's process space, where the length of the mapped memory object is still 27 bytes. The pointer *shm* returned from the call mmap() references the first byte of the shared memory object.

The client then reads and prints out the characters from the shared memory object one by one. In the end, the client writes an asterisk to the first byte of the shared memory object; this will terminate the server process.

```
     //client.c
2    //adapted from Dave Marshall, Programming in C [72]
     #include <stdio.h>
4    #include <fcntl.h>
     #include <sys/mman.h>
6
     #define SHMSZ 27
8    char SHM_NAME[] = "MyShm";

10   int main() {
       int shmid;
12     char *shm, *s;

14     /* open the shared memory object */
       shmid = shm_open(SHM_NAME, O_RDWR, 0);
16     if (shmid == -1) {
         perror("client: error opening the shared memory object\n");
18       exit(1);
       }
20     /* get a pointer to a piece of the shared memory, note that we
        only map in the amount we need to */
22     shm = mmap(0,SHMSZ,PROT_READ|PROT_WRITE, MAP_SHARED, shmid, 0);
       if (shm == MAP_FAILED) {
24       perror("client: mmap failed\n");
         exit(1);
26     }

28     //Dirty data: the 1st read and 2nd read can be different
       int i, r;
30     for (i = 0; i < 20; i++) {
```

```
      printf("1st read:-- %s\n",shm);
32    r = rand() % 3;
      sleep(r);
34    printf("2nd read:-- %s\n",shm);
      }
36
      //once done, inform the server
38    *shm = '*';
      exit(0);
40 }
```

Listing 18.5
Sample code of using named shared memory: client.

18.3 Semaphore

When multiple tasks gain access to a shared resource at the same time, the integrity of the resource is endangered. We have already seen a multithreading example in Section 18.1, where two consecutive reads of the shared variable can be different. The violation of data integrity can happen to multiple processes as well. In the example given in Section 18.2.1, for each iteration, each child process reads the value of the shared integer only once. Should you add another read operation, you will find that the two consecutive reads of the shared integer can be different.

The block of code that accesses a shared resource is often called a critical section. Two critical sections that access the same exclusive resource are called *contending critical sections*. We obviously need some tools to protect shared resources such that no two contending critical sections can be executed at the same time.

One such tool is called a *semaphore*, introduced by Dutch computer scientist Edsger Dijkstra. The semaphore concept is illustrated in Figure 18.5, which also shows a related concept called mutex (this will be discussed in Section 18.3).

The semaphore concept has the following features.

- A semaphore is a system object managed by the OS kernel.
- A semaphore, upon creation, can be specified with an initial value denoted by *count*. At any time, a semaphore is *available* only if $count > 0$; it is unavailable when $count \leq 0$. Depending on the potential context, an application designer can typically impose an external constraint on the use of a semaphore—the maximum value that it may take. We use *max_count* to refer to this capacity.
- As a resource lock, a semaphore has a task waiting list, containing those tasks that are blocked because of the unavailability of the semaphore.

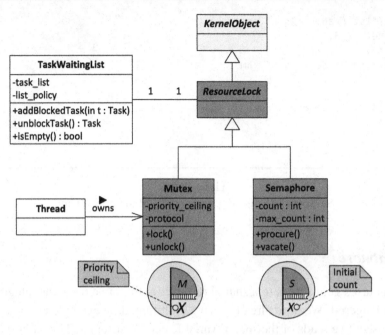

Figure 18.5
Semaphore, mutex, and their notations.

- There are two operations defined on semaphores: procure and vacate. POSIX uses the terms "wait" and "post." Other texts may use terms such as "down" and "up," "acquire" and "release," or "pend" and "post." The semaphore value (count) can be viewed as the number of keys to a lock protecting a shared resource, and each key represents a privilege to access the protected resource. The procure and vacate operations allow tasks to acquire keys from or return keys to a semaphore.
  - A task executes the procure operation on a semaphore to request a privilege. If the semaphore is currently available (i.e., $count > 0$), the semaphore value is decreased by 1, and the task has granted access to the protected resource. If the semaphore is currently unavailable (i.e., $count \leq 0$), the task is added to the task waiting list.
  - A task executes the vacate operation on a semaphore to release a privilege. If there are no tasks in the waiting list, this operation simply increases the semaphore value by 1 (up to the capacity max_count). If there are tasks in the waiting list, one task blocked waiting for the semaphore is unblocked and returns successfully from its procure operation (note that in this case, we have $count = 0$ both before and after the vacate operation).

The UML state diagram in Figure 18.6 shows how the procure and vacate operations may change the state of a semaphore.

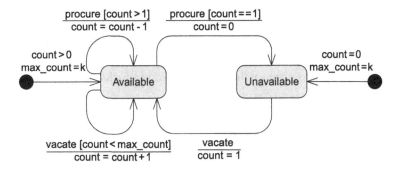

Figure 18.6
UML state diagram for a semaphore.

18.3.1 Task Synchronization

Semaphores can be used as a tool for intertask synchronization.

Figure 18.7 shows an example, where task T_2 waits for a semaphore that only task T_1 can release. A line directed into a semaphore indicates a vacate operation by the task at its tail, and a line directed out of a semaphore indicates a procure operation by the task at its head.

In particular, T_1 has a lower priority and it vacates the semaphore at time 1 relative to the start of its period; T_2 has a higher priority and it procures the semaphore at time 4 relative to the start of its period. The semaphore has an initial value of 0. The associated constraint says that it can only take a value of 0 or 1. A semaphore with such a constraint is often called a *binary semaphore*.

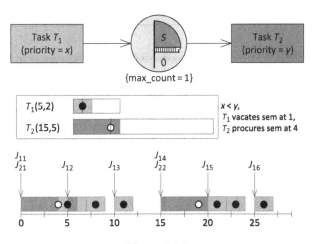

Figure 18.7
Using a semaphore for task synchronization.

The timing diagram shows the scenario as explained below:

(1) At time 0, both tasks release a job. The job J_{21} is scheduled to execute because of its higher priority. The semaphore value is 0 initially.

(2) At time 4, J_{21} attempts to procure the semaphore and blocks because the semaphore is not available. Job J_{11} starts to execute.

(3) At time 5, J_{11} vacates the semaphore. This unblocks J_{21}, which preempts J_{11}. Note that the second job, J_{12}, of task T_1 is released.

(4) J_{21} runs to completion at time 6. Then J_{11} runs to completion at time 7.

(5) The semaphore value becomes 1 because of the execution of J_{12} (from time 7 to 9).

(6) The third job of task T_1 is released at time 10. Its vacate operation has no effect on the semaphore because it has already reached the maximum capacity.

(7) At time 15, jobs J_{14} and J_{22} are released.

(8) At time 19, J_{22} is able to procure the semaphore successfully. It runs to completion at time 20.

Figure 18.8 illustrates a so-called *rendezvous synchronization pattern*.

Tasks T_1 and T_2 both have a synchronization point which is regulated by two binary semaphores sem_1 and sem_2. In particular, the two tasks have the following pseudocode.

```
T1()                    T2()
{ ...                   { ...
  vacate (sem2);          vacate (sem1);
  procure(sem1);          procure(sem2);
  ...                     ...

}                       }
```

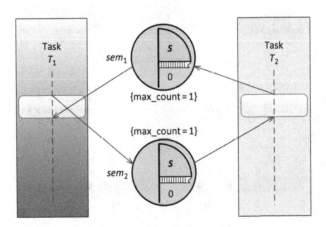

Figure 18.8
Rendezvous synchronization pattern.

The behavior of the two tasks can be analyzed as follows:

(1) Suppose T_1 has a higher priority than T_2, and T_1 comes to its synchronization point first.
   (a) T_1 executes the vacate operation on sem_2, which ends up with a value of 1.
   (b) T_1 executes the procure operation on sem_1. Since sem_1 is not available, T_1 is blocked.
   (c) When T_2 comes to its synchronization point, it first executes the vacate operation on sem_1, which will unblock T_1. Since T_1 has a higher priority, it preempts the execution of T_2.
   (d) After (a job of) T_1 has run to completion and T_2 has been scheduled to execute, the procure operation successfully decreases sem_2 to 0.
(2) Suppose T_2 has a higher priority than T_1, and T_2 comes to its synchronization point first.
   (a) T_2 executes the vacate operation on sem_1, which ends up with a value of 1.
   (b) T_2 executes the procure operation on sem_2. Since sem_2 is not available, T_2 is blocked.
   (c) When T_1 comes to its synchronization point, it first executes the vacate operation on sem_2, which will unblock T_2. Since T_2 has a higher priority, it preempts the execution of T_1.
   (d) After (a job of) T_2 has run to completion and T_1 has been scheduled to execute, the procure operation successfully decreases sem_1 to 0.

In sum, regardless of which task has a higher priority, the one that comes to its synchronization point first has to wait for the other task.

18.3.2 Flow Control

A semaphore can also be used to control the data flow from one task to another. Figure 18.9 shows such a design, where the sender task needs to send *fixed-size* data packets to the receiver task through a shared memory object.

The receiver task uses a shared memory object as its receive buffer, which can hold at most k data packets. Also, a semaphore is used by the receiver to regulate the behavior of the sending task. Initially both the semaphore's value and its maximum capacity can be set to k. The

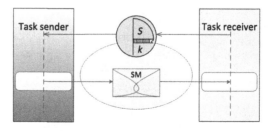

Figure 18.9
Flow control of the use of a shared resource.

sender has to execute a procure operation to get a "token" from the semaphore before sending a data packet. On the other side, the receiver always executes a vacate operation to increase the semaphore value every time it successfully retrieves a data packet from the shared memory object. If the sender has a faster pace of sending data, it will eventually be blocked on the semaphore when the number of tokens becomes zero. In the long run, the sender's pace will be regulated close to the receiver's pace of processing.

18.3.3 Resource Protection

As resource locks, semaphores can also be used to protect the integrity of shared resources. When used as such, the value of a semaphore is initialized to 1 to indicate that the resource being protected is initially available.

Figure 18.10 shows a design where each task first has to procure the binary semaphore prior to accessing the shared resource. When the semaphore is locked by one task, the other task that attempts to procure the semaphore blocks until the first task has finished its use of the resource and vacated the semaphore.

Sometimes, a resource being protected may have many instances that are exactly the same as far as the resource user is concerned. For example, a system may have a memory pool containing a dozen of 100-byte memory blocks. Each of the blocks can equally satisfy a user's memory request that is no more than 100 bytes. In such a case, we say the resource "100-byte memory block" has 12 instances.

For a multi-instance resource, the value of a semaphore is initialized to be the number of resource instances that are available for use. Figure 18.11 shows such a design. A task that wishes to obtain a resource (instance) first has to procure a "permission token" from the semaphore. The task blocks if all the k instances are currently in use.

It is worth noting that the semaphore as used in Figure 18.11 can only indicate how many resource instances are still free. In practice, some other mechanism or data structure should be

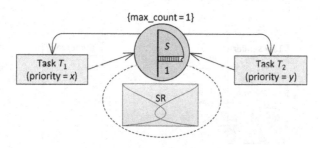

Figure 18.10
Protect a single resource.

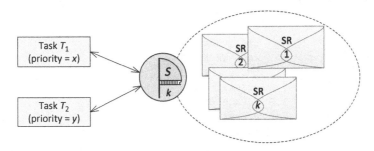

Figure 18.11
Protect a multi-instance resource.

employed to keep track of which instance is free and which is not, so that a task will not be granted access to an instance that is currently in use.

Sometimes, a task needs to use multiple (types of) resources at the same time. The task designer has to be careful to avoid resource deadlocks. In Section 14.3.1, we learned the four necessary conditions causing a resource deadlock. If it is inevitable to have the "hold-and-wait" situation, we have to break the "circular wait" condition. One way is to define an order for the use of the resources and ensure that this order is obeyed by all the tasks that need to use the involved resources in the hold-and-wait fashion.

Figure 18.12 shows such a design, where each resource is protected by a binary semaphore, and both tasks follow the same order to procure the semaphores. In particular, the two tasks may have the following pseudocode.

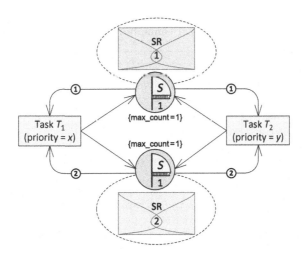

Figure 18.12
Define an order for semaphores to prevent circular wait.

```
T1()                        T2()
{ ...                       { ...
  procure(sem1);              procure(sem1);
  access1(SR1);               access2(SR1);
  procure(sem2);              procure(sem2);
  access1(SR2);               access2(SR2);
  vacate (sem2);              vacate (sem2);
  access1(SR1);               access2(SR1);
  vacate (sem1);              vacate (sem1);
  ...                         ...
}                           }
```

18.3.4 POSIX Functions for Semaphores

POSIX defines two kinds of semaphores: named semaphore and unnamed semaphore. A named semaphore has a unique name and appears in the file system just like a file. It is typically used between processes that are not necessarily related.

An unnamed or anonymous semaphore has no name and does not appear in the file system. It is typically used among threads within the same process, or between processes that are related either by a parent-child relationship (say, by fork or posix_spawn) or by a sibling relationship (i.e. a process creates an unnamed semaphore, which is then used by its child processes).

Table 18.2 lists some POSIX functions for using semaphores. The function sem_open() is used to create or open a named semaphore, and sem_init() is used to create and initialize an unnamed semaphore. The functions sem_wait() and sem_post() correspond to the procure and vacate operations, respectively. Be very careful when an application uses named semaphores.

Table 18.2 POSIX.1 semaphore function calls

POSIX Function	Description
sem_open sem_init	Create a new named semaphore or open an existing named semaphore Create and initialize an unnamed semaphore
sem_getvalue sem_post sem_wait	Get the value of a semaphore Increment (vacate) the semaphore value Decrement (procure) the semaphore value. If the semaphore value is not positive, the calling thread blocks until either it can decrement the semaphore value or the call is interrupted by a signal
sem_timedwait	Decrement (procure) the semaphore value. If the semaphore value is not positive, the calling thread blocks until it can decrement the semaphore value, the call is interrupted by a signal, or the specified time-out expires
sem_trywait	Decrement (procure) the semaphore value if the semaphore's value is positive, otherwise it simply returns
sem_destroy sem_close sem_unlink	Destroy an unnamed semaphore Close a named semaphore, which is no longer usable in the calling process Remove a named semaphore from the system.

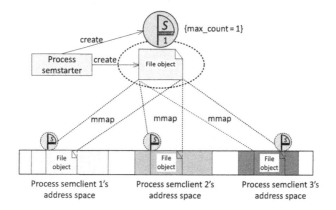

Figure 18.13
Use a semaphore to protect a memory-mapped file object.

If a process fails to call sem_unlink() to remove a named semaphore from the system (say, the process terminated abnormally), another process (which maybe created from the same executable object) that calls sem_open() to create a new semaphore with the same name could end up with accessing the 'leftover' semaphore with an undesired "state" (i.e. the obsolete count value is used instead in the calling process).

18.3.5 Semaphore Examples

In this section, we take a look at some examples of using named and unnamed semaphores.

18.3.5.1 Use a named semaphore to protect a memory-mapped file

Figure 18.13 illustrates a design where the process named *semstarter* creates a file object which is protected by a binary semaphore named *mysem*.

```
#include <stdio.h>
#include <stdlib.h>
#include <spawn.h>
#include <unistd.h>
#include <fcntl.h>
#include <semaphore.h>
#include <sys/wait.h>

int main(int argc, char **argv) {
  char SEM_NAME[] = "mysem";
  char *args[] = { "semclient", NULL };
  int i, status, fd, value = 0;
  struct inheritance inherit;
  pid_t pid;
  sem_t *sem;
```

```
16
    //open a file and set initial value
18  fd = open("log.txt", O_RDWR | O_CREAT, S_IRWXU);
    write(fd, &value, sizeof(int));
20  close(fd);

22  //create & initialize a semaphore
    sem = sem_open(SEM_NAME, O_CREAT, 0644, 1);
24  if (sem == SEM_FAILED) {
      perror("unable to create semaphore");
26    sem_unlink(SEM_NAME);
      exit(1);
28  }

30  // create 3 child processes
    for (i = 0; i < 3; i++) {
32    inherit.flags = 0;
      if ((pid = spawn("semclient", 0, NULL, &inherit, args, environ)) == -1)
34      perror("spawn() failed");
      else
36      printf("spawned child, pid = %d\n", pid);
    }
38  while (1) {
      if ((pid = wait(&status)) == -1) {
40      perror("Starter: ");
        sem_unlink(SEM_NAME);
42      exit(EXIT_FAILURE);
      }
44    printf("Semclient %d terminated\n", pid);
    }
46 }
```

Listing 18.6
Sample code of using a named semaphore: starter.

The code of semstarter is given in Listing 18.6. It spawns three child processes from the code given in Listing 18.7. Let us refer to the child processes by the names semclient1, semclient2, and semclient3, respectively. According to the code, each of the three child processes maps the file log.txt into its own process space, opens the named semaphore created by semstarter, and then uses the semaphore to gain exclusive access to the mapped file object. In particular, for each round of the loop, the manipulation of the file object is enclosed by a procure operation and a vacate operation. This helps to preserve the integrity of the file object.

```
// qcc -o semclient semclient.c
2 #include <stdio.h>
```

```
  #include <stdlib.h>
4 #include <semaphore.h>
  #include <fcntl.h>
6 #include <sys/mman.h>

8 int main(int argc, char **argv) {
    int fd, i, nloop = 3, *ptr;
10  sem_t *sem;
    char SEM_NAME[] = "mysem";

12
    //open a file and map it into memory
14  fd = open("log.txt", O_RDWR, S_IRWXU);
    ptr = mmap(NULL, sizeof(int), PROT_READ | PROT_WRITE, MAP_SHARED, fd, 0);
16  close(fd);

18  //open an existing semaphore
    sem = sem_open(SEM_NAME, 0, 0644, 0);
20  if (sem == SEM_FAILED) {
      perror("reader:unable to open semaphore");
22    sem_close(sem);
      exit(1);
24  }

26  for (i = 0; i < nloop; i++) {
      sem_wait(sem);
28    printf("Semclient %d entered critical section: %d\n", getpid(),
          (*ptr)++);
30    sleep(2); //simulate processing
      printf("Semclient %d leaving critical section\n", getpid());
32    sem_post(sem);
      sleep(1);
34  }
    sem_close(sem);
36  exit(0);
  }
```

Listing 18.7
Sample code of using a named semaphore: semclient.

Figure 18.14 shows a screenshot of a run of the program. As compared with Figure 18.4, the big difference is that here another task cannot start to access the file object until after the current task has completed its use, whereas in Figure 18.4, more than one task can access the file object at the same time.

18.3.5.2 Use a named semaphore to protect a shared memory object

The code given in Listings 18.8 and 18.9 is the same as the code in Listings 18.4 and 18.5 (see Section 18.2.2), except that a named binary semaphore is used to protect the access to the shared memory object.

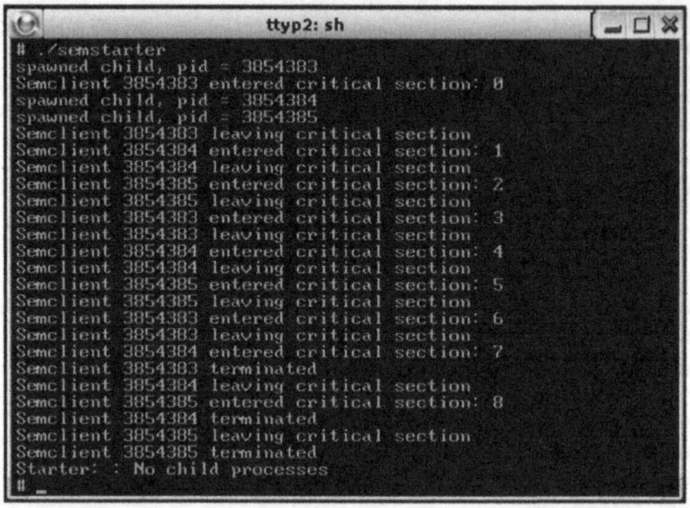

Figure 18.14
A run of the program given in Listings 18.6 and 18.7.

```
1  //semserver.c
   #include <stdio.h>
3  #include <stdlib.h>
   #include <semaphore.h>
5  #include <sys/mman.h>
   #include <fcntl.h>
7
   #define SHMSZ 27
9  char SEM_NAME[] = "MySem";
   char SHM_NAME[] = "MyShm";
11
   int main() {
13   char ch;
     int shmid;
15   char *shm, *s;
     sem_t *sem;
17
     //sem_unlink(SEM_NAME);
19   //create & initialize semaphore
     sem = sem_open(SEM_NAME, O_CREAT, 0644, 1);
21   if (sem == SEM_FAILED) {
       perror("unable to create semaphore");
23     sem_unlink(SEM_NAME);
       exit(EXIT_FAILURE);
25   }
```

```
27   if ((shmid = shm_open(SHM_NAME, O_RDWR | O_CREAT, S_IRWXU)) < 0) {
       perror("shm_open");
29     exit(EXIT_FAILURE);
     }
31   if (ftruncate(shmid, SHMSZ) < 0) {
       perror("ftruncate");
33     exit(EXIT_FAILURE);
     }
35   //map the shared memory object to the process memory and return a pointer to it
     if ((shm = mmap(NULL, SHMSZ, PROT_READ | PROT_WRITE, MAP_SHARED, shmid, 0))
37       == MAP_FAILED) {
       perror("mmap");
39     exit(EXIT_FAILURE);
     }
41
     //start writing into memory
43   s = shm;
     int r;
45   for (ch = 'A'; ch <= 'Z'; ch++) {
       sem_wait(sem);
47     *s++ = ch;
       r = rand() % 3;
49     sleep(r);
       sem_post(sem);
51   }
     while (*shm != '*') {
53     sleep(1);
     }
55   sem_close(sem);
     sem_unlink(SEM_NAME);
57   shm_unlink(SHM_NAME);
     exit(0);
59 }
```

Listing 18.8
Using a named semaphore: writing to a shared memory object.

```
1  // semuser.c
   #include <stdio.h>
3  #include <stdlib.h>
   #include <semaphore.h>
5  #include <sys/mman.h>
   #include <fcntl.h>
7
   #define SHMSZ 27
9  char SEM_NAME[] = "MySem";
   char SHM_NAME[] = "MyShm";
11
```

```
   int main() {
13   int shmid;
     char *shm, *s;
15   sem_t *sem;

17   //create & initialize existing semaphore
     sem = sem_open(SEM_NAME, 0, 0644, 0);
19   if (sem == SEM_FAILED) {
       perror("reader:unable to execute semaphore");
21     sem_close(sem);
       exit(EXIT_FAILURE);
23   }

25   /* open the shared memory object */
     shmid = shm_open(SHM_NAME, O_RDWR, 0);
27   if (shmid == -1) {
       perror("client: error opening the shared memory object\n");
29     exit(EXIT_FAILURE);
     }
31   /* get a pointer to a piece of the shared memory, note that we
     only map in the amount we need to */
33   shm = mmap(0, SHMSZ, PROT_READ | PROT_WRITE, MAP_SHARED, shmid, 0);
     if (shm == MAP_FAILED) {
35     perror("client: mmap failed\n");
       exit(EXIT_FAILURE);
37   }

39   //Protected: the 1st read and the 2nd read are the same
     int i, r;
41   for (i = 0; i < 20; i++) {
       sem_wait(sem);
43     printf("1st read:-- %s\n",shm);
       r = rand() % 3;
45     sleep(r);
       printf("2nd read:-- %s\n",shm);
47     sem_post(sem);
     }
49
     //once done, inform the server
51   *shm = '*';
     sem_close(sem);
53   exit(0);
   }
```

Listing 18.9
Using a named semaphore: reading from a shared memory object.

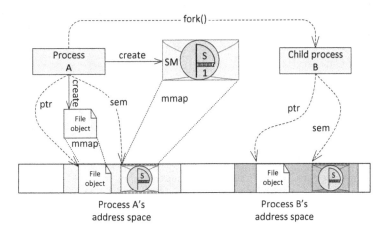

Figure 18.15
Unnamed semaphore embedded inside a shared memory object.

18.3.5.3 Use an unnamed semaphore embedded in a shared memory object

Figure 18.15 shows a design where process A creates a shared memory object which contains an unnamed semaphore. This semaphore is used to protect the content of a mapped file object. When process A creates a child Process B via a call to fork(), both the mapped file object and the shared memory object will have a duplicate in process B's address space.

```
   #include <semaphore.h>
2  #include <stdio.h>
   #include <errno.h>
4  #include <stdlib.h>
   #include <unistd.h>
6  #include <sys/types.h>
   #include <sys/stat.h>
8  #include <fcntl.h>
   #include <sys/mman.h>
10
   int main(int argc, char **argv) {
12   int fd, i, nloop = 3, value = 0, *ptr;
     sem_t *sem;
14   int shm;

16   //open a file and map it into memory
     fd = open("log.txt", O_RDWR | O_CREAT, S_IRWXU);
18   write(fd, &value, sizeof(int));
     ptr = mmap(NULL, sizeof(int), PROT_READ | PROT_WRITE, MAP_SHARED, fd, 0);
20   close(fd);

22   //create a shared memory region of size sem_t
```

```
     // shm_open takes a name for the shared memory region,
24   // Other processes just need to know and use the same name.
     if ((shm = shm_open("myshm", O_RDWR | O_CREAT, S_IRWXU)) < 0) {
26     perror("shm_open");
       exit(1);
28   }
     if (ftruncate(shm, sizeof(sem_t)) < 0) {
30     perror("ftruncate");
       exit(1);
32   }

34   //map the shared memory object to the process memory and return a pointer to it
     if ((sem = mmap(NULL, sizeof(sem_t), PROT_READ | PROT_WRITE, MAP_SHARED,
36     shm, 0)) == MAP_FAILED) {
     perror("mmap");
38   exit(1);
     }
40
     /* create, initialize an unnamed semaphore */
42   if (sem_init(sem, 1, 1) < 0) {
     perror("semaphore initialization");
44   exit(0);
     }
46   if (fork() == 0) { /* child process*/
       for (i = 0; i < nloop; i++) {
48       sem_wait(sem);
         printf("child entered critical section: %d\n", (*ptr)++);
50       sleep(2);
         printf("child leaving critical section\n");
52       sem_post(sem);
         sleep(1);
54     }
       sem_close(sem);
56     exit(0);
     }
58   /* back to parent process */
     for (i = 0; i < nloop; i++) {
60     sem_wait(sem);
       printf("parent entered critical section: %d\n", (*ptr)++);
62     sleep(2);
       printf("parent leaving critical section\n");
64     sem_post(sem);
       sleep(1);
66   }
     sem_destroy(sem);
68   exit(0);
   }
```

Listing 18.10
Unnamed semaphore embedded in a shared memory object.

An example implementation of this design is given in Listing 18.10. Note that both the parent process and the child process use the same unnamed semaphore to synchronize their access to the mapped file object.

Also, the call to sem_init() in line 42 has three arguments, where *sem* is a pointer to the unnamed semaphore to be generated, and the last argument specifies the initial value of the semaphore. The second argument is a flag: if it has value 0, the unnamed semaphore is shared among threads of the creating process; if it has a nonzero value, the unnamed semaphore can be shared by multiple processes. Since there is no name to refer to an unnamed semaphore, to be accessible to other processes, it has to be located in a shared memory object.

18.4 Mutex

Semaphores come with a big limitation: no ownership. This exposes two critical problems. First, a semaphore can be procured and vacated by any tasks. If an unrelated task (maybe a bad citizen) sneaks in and performs procure or vacate operations on a semaphore that is supposed to be used only by a group of well-defined tasks, the whole purpose of using the semaphore will be annihilated.

Second, as shown in Figure 18.16, a task may have several places (say, in a function call) that need to gain access to the same resource protected by a semaphore. The task will be blocked when for the second time it attempts to procure the semaphore. However, as a single thread of execution, as long as the task has already been granted permission once, it should be safe for the task to access the resource. In other words, a task should not be blocked on a resource to

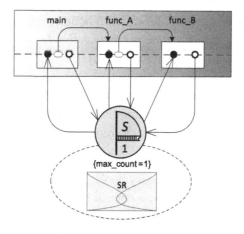

Figure 18.16
Recursive requests for a resource lock.

which it has already gained access. Such a desirable feature is not available in the semaphore because it has no information about its current user or owner.

This leads us to another type of resource lock: *mutex*. As a synchronization or resource protection tool, mutex is like an *unnamed binary semaphore* but is more powerful for the following reasons:

(1)   A thread gains the ownership of a mutex when it first locks the mutex.
(2)   A mutex can be unlocked (released) only by its owner thread.
(3)   If configured so, a mutex can be locked by the owner thread recursively.
(4)   A mutex can be specified to use protocols, such as the priority inheritance protocol or the priority ceiling protocol, to prevent the notorious priority inversion problem.

Mutexes are normally implemented as a low-level primitive from which other thread synchronization utilities can be built. One such example is the *condition variable*, which is covered in Section 18.5.

18.4.1 Mutex Usage Pattern

Similarly to semaphores, mutexes in general can be used for intertask synchronization and resource protection. As an example, let us consider how to use a mutex to protect the heap—an area of dynamically allocatable memory.

The memory management subsystem of an OS generally offers the following services:

(1)   It uses control blocks (each containing an allocation table typically implemented as a bitmap, an array, a heap data structure, or a linked list) to indicate which memory blocks are in use and which blocks are free.
(2)   Whenever a thread calls a memory-allocation function such as malloc() to request x bytes of memory, it employs a specific strategy (such as first fit, worst fit, or best fit) to locate a free block that is no less than x and returns the starting address.
(3)   Whenever a thread calls a memory-deallocation function such as free(), it marks the specified space available for further allocation.
(4)   During memory allocation and deallocation, it may also reorganize the memory blocks and modify its control block accordingly to reduce memory fragmentation.

If the sizes of the required memory blocks are predictable at design time, the heap area of a system can be organized as fixed-size memory pools. The design illustrated in Figure 18.17 has three memory pools, each described by a control block managing a list of fixed-size memory blocks (shaded blocks are in use). The control block of a memory pool contains information such as the block size, the total number of memory blocks, and the number of free memory blocks. The control blocks themselves are linked together and sorted by size. Given a

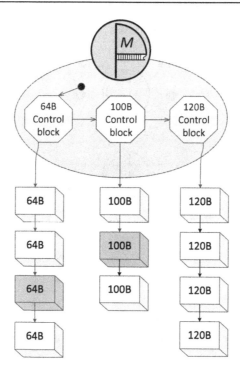

Figure 18.17
Fixed-size memory pools.

memory-allocation request, the list of control blocks is searched to locate a free memory block that is adequate for the request. Because multiple tasks might issue memory-allocation requests simultaneously, their access to the list of control blocks must be regulated. This is where the mutex comes into play: it ensures that a free memory block would be used to satisfy at most one request at any given time.

For a system operating in a dynamic environment, its heap memory area may occasionally become exhausted. In such a situation, the calling task of the malloc() function fails, and it may have to restart in order to recover from the memory allocation failure. However, the memory exhaustion condition may only occur momentarily. It would be more desirable if a memory-allocation function allows a task to block and wait for memory to become available.

Figure 18.18 illustrates a design that employs a semaphore and a mutex to implement a blocking memory-allocation function. It works as follows:

- One semaphore and one mutex are required for each memory pool. The design in Figure 18.18 covers only a single memory pool containing memory blocks of 64 bytes.
- The count value of the semaphore is initialized to k—the total number of free memory blocks at the creation of the memory pool. The semaphore value always equals the

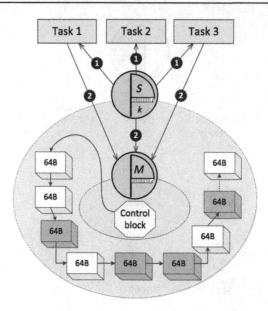

Figure 18.18
Memory management with waiting facility.

number of free memory blocks: it is decreased by 1 when a memory block is allocated and is increased by 1 when a block is freed.

- A task follows the sequence below to obtain a free memory block:
  - It firstly has to procure a "permission token" from the semaphore. A successful procure operation reserves a free memory block from the pool. The task blocks on the semaphore if its value becomes 0 (all the k blocks are currently in use).
  - The task next needs to lock the mutex in order to access the control block. The task blocks on the mutex if it is currently owned by another task. Here, we use a line directed toward a mutex to indicate that the task at the tail attempts to lock the mutex.
  - The task, as the owner of the mutex, modifies the control block, changing the flag of a free memory block from "available" to "allocated." The task then unlocks the mutex. After this point, the task can use the allocated memory block safely.
- A task follows the sequence below to free a memory block:
  - The task needs to lock the mutex in order to access the control block.
  - The task, as the owner of the mutex, modifies the control block, changing the flag of the memory block from "allocated" to "available."
  - The task vacates a "permission token" back to the semaphore.
  - Finally, the task unlocks the mutex.

A memory pool is simply a special multi-instance resource (see Figure 18.11). We can further generalize the design in Figure 18.18 into a design pattern as shown in Figure 18.19.

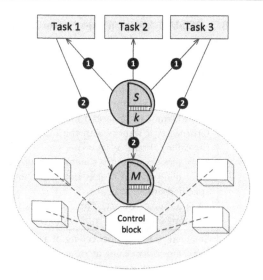

Figure 18.19
Multi-instance resource protection pattern.

The mutex is used for protecting the control structure that keeps track of which resource instance is free and which is not.

Whenever a task needs to obtain a resource instance from the pool, it firstly has to get a "permission token" from the semaphore. With a token in hand, the task is guaranteed to be able to get a resource instance. It then has to lock the mutex to examine the control structure in order to find out which resource instance is free. The control structure itself is a shared resource that has to be protected to ensure that one resource instance could only be granted to at most one user at any given time.

18.4.2 POSIX Functions for Mutexes

Table 18.3 lists a few POSIX functions for mutexes.

In particular, the function pthread_mutex_lock() is used to lock an initialized mutex object, and pthread_mutex_unlock() is used by the owner thread to release a mutex.

Mutexes are typically used for synchronization among threads within the same process. A mutex object has a "process-shared" attribute, which can be configured to allow the mutex to be shared by threads in different processes. The process-shared attribute of a mutex by default is set to PTHREAD_PROCESS_PRIVATE. If a mutex object is contained in a shared memory object (just like the unnamed semaphore example in Figure 18.15), and its process-shared

Table 18.3 POSIX.1 function calls for mutexes

POSIX Function	Description
pthread_mutex_init	Initialize a mutex object
pthread_mutex_unlock	Unlock a mutex object
pthread_mutex_lock	Lock a mutex object. If the mutex is already locked by another thread, the calling thread blocks until the mutex becomes available. Otherwise, it returns with the mutex object in the locked state with the calling thread as its owner
pthread_mutex_timedlock	Lock a mutex object. If the mutex is already locked by another thread, the calling thread blocks until the mutex becomes available or the specified time-out expires
pthread_mutex_trylock	Lock a mutex object. The call returns immediately if the mutex is currently locked
pthread_mutex_setprioceiling	Set/change the priority ceiling of a mutex object
pthread_mutexattr_setprioceiling	Set the priority ceiling attribute of a mutex attributes object
pthread_mutexattr_setprotocol	Set the protocol attribute of the mutex attributes object
pthread_mutexattr_setpshared	Set the process-shared attribute
pthread_mutexattr_settype	Set the mutex type attribute (say, to allow recursive locks)

attribute is set to PTHREAD_ PROCESS_SHARED, then the mutex object can be used by threads from more than one process.

18.4.3 An Example of Using a Mutex

As an example of using a mutex, let us revisit the producer-consumer problem we discussed in Section 18.1.

In Figure 18.20, a mutex object is used to protect the shared variable. Specifically, the producer or the consumer cannot access the shared variable unless it owns the mutex.

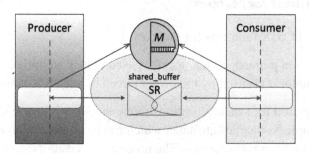

Figure 18.20
Exclusive resource access by mutexes.

```
 1 #include <stdio.h>
   #include <pthread.h>
 3 #include <time.h>

 5 //Data is shared by child processes producer and consumer
   int shared_buffer = -1;
 7 pthread_mutex_t mutex = PTHREAD_MUTEX_INITIALIZER;

 9 void set_shared_buffer(int value) {
     fprintf(stdout, "Producer writes\t%2d", value);
11   shared_buffer = value;
   }
13 int get_shared_buffer() {
     fprintf(stdout, "Consumer reads\t%2d\n", shared_buffer);
15   return shared_buffer;
   }
17 int get_shared_buffer2() {
     return shared_buffer;
19 }

21 void * consumer(void *notused) {
     int sum = 0;
23   int firstget, secondget;
     int r, count;
25   for (count = 1; count <= 10; count++) {
       r = rand() % 4;
27     sleep(r);

29     pthread_mutex_lock(&mutex);
       //start of critical section
31     firstget = get_shared_buffer();
       sleep(1); //simulate longer processing
33     secondget = get_shared_buffer2();
       sum += secondget;
35     fprintf(stdout, "Consumer process\t\t\t%2d,%2d,%2d\n",
         firstget, secondget, sum);
37     //end of critical section
       pthread_mutex_unlock(&mutex);
39   }
     fprintf(stdout, "\n%s %d\n%s\n", "Consumer read values total", sum, "Terminating
         Consumer");
41 }

43 void * producer(void *notused) {
     int sum = 0;
45   int r, count;
     for (count = 1; count <= 10; count++) {
47     r = rand() % 4;
       sleep(r);
49
```

```
      pthread_mutex_lock(&mutex);
51    //start of critical section
      set_shared_buffer(count);
53    sum += count;
      fprintf(stdout, "\t%2d\n", sum);
55    //end of critical section
      pthread_mutex_unlock(&mutex);
57  }
    fprintf(stdout, "Producer done\nTerminating Producer\n");
59 }

61 int main() {
    fprintf(stdout, "Action\t\tValue\tSum of Produced\tfirst, second,Sum\n");
63    fprintf(stdout, "------\t\t-----\t---------------\t---------------\n");

65    pthread_attr_t attr;
    struct sched_param sp;
67
    //initializes the thread attributes to their default values
69    pthread_attr_init(&attr);
    pthread_attr_setinheritsched(&attr, PTHREAD_EXPLICIT_SCHED);
71
    //Set the priority to 30 to the producer thread
73    sp.sched_priority = 30;
    pthread_attr_setschedparam(&attr, &sp);
75    // create the producer thread
    pthread_create(NULL, &attr, producer, NULL);
77
    //Set the priority to 20 to the consumer thread
79    sp.sched_priority = 20;
    pthread_attr_setschedparam(&attr, &sp);
81    // create the consumer thread
    pthread_create(NULL, &attr, consumer, NULL);
83
    // let the threads run for a bit
85    sleep(30);
    return 1;
87 }
```

Listing 18.11
Producer-consumer regulated by a mutex.

Listing 18.11 gives an example code. The critical section of the consumer thread starts with a function call for locking the mutex (line 29) and ends with a function call for unlocking the mutex (line 38). The critical section of the producer thread is protected by the mutex similarly.

Figure 18.21 shows a screenshot of a run of the code in Listing 18.11. Clearly, this is a better implementation because the data integrity of the shared variable has been preserved owing to the use of the mutex: in each round the two reads of the shared variable by the consumer are always the same.

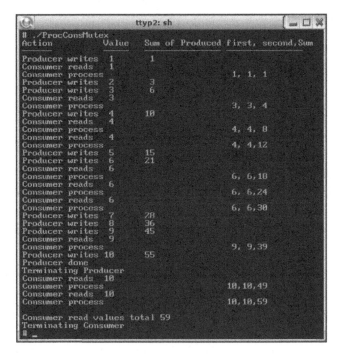

Figure 18.21
Producer-consumer output (using a mutex).

However, the behavior of the producer and that of the consumer are still not synchronized appropriately. For example, the shared variable was duplicately read (e.g., 4, 6, 10) by the consumer before the producer could produce the next value, and the producer had written new data into the shared buffer before the consumer read the previous data (e.g., 2, 5, 7, and 8 were lost).

We will revisit this problem after we have introduced the condition variable in the next section.

18.5 Condition Variable

A *condition variable* is an OS kernel object that is logically associated with a shared resource, allowing a task to wait for a desired condition to occur before using the shared resource. Here, a condition is simply a Boolean expression, or formally a *predicate*. The truth value of a condition may change dynamically because of task operations on the shared resource.

A condition variable is built upon a mutex object called its guarding or binding mutex. Prior to evaluating a condition about the shared resource, a task must have exclusive access to that resource. This is enforced by the guarding mutex. A task blocks and enters into the *waiting list of the guarding mutex* if the mutex is currently locked by another task. At the time of

Figure 18.22
The notation for the condition variable: it has a thread waiting list and is bounded with a mutex.

evaluation, if the condition is not true, the task blocks and enters into the *waiting list of the condition variable*. In Figure 18.22, we give our notation for the condition variable, where the guarding mutex and the two task waiting lists are highlighted. The diamond with a dotted boundary indicates the condition to be evaluated by a task. Depending on the problem at hand, multiple tasks that use the same condition variable may have different conditions (predicates) to evaluate.

Table 18.4 gives a list of POSIX functions for condition variables. The function pthread_cond_wait() allows the calling thread to unlock the guarding mutex[1] and wait for a condition to occur, and pthread_cond_signal() unblocks at least one thread that is waiting for the specified condition variable. If more than one thread is unblocked, especially when pthread_cond_broadcast() is used, the threads that are unblocked need to contend for the guarding mutex. Consequently, whichever thread can successfully lock the mutex will return from its call to pthread_cond_wait(), while all the other thread(s) that were just unblocked from the condition variable will block again on the guarding mutex.

Table 18.4 POSIX.1 function calls for condition variables

POSIX Function	Description
pthread_cond_init pthread_cond_destroy	Initialize a condition variable Destroy a condition variable
pthread_cond_wait	Given that the calling thread currently owns the associated mutex, this function causes the calling thread to unlock the mutex and block on a condition variable until being signaled by the condition variable. Upon successful return, the calling thread once again becomes the owner of the mutex
pthread_cond_signal	Signal to unblock at least one of the threads that are blocked on the specified condition variable
pthread_cond_timedwait	The same as pthread_cond_wait except that the calling thread blocks on a condition variable until being signaled or the specified time-out expires
pthread_cond_broadcast	Signal to unblock all threads currently blocked on the specified condition variable

---

[1] This allows another thread to lock the mutex and enter its critical section.

It is worth noting that the condition variable itself knows nothing about the predicate(s) being monitored by those threads on its waiting list. It is the programmers' responsibility to ensure that *the right condition (predicate) is checked by the right thread at the right time*. First, it is imperative that a task always grabs the guarding mutex (via a call to pthread_mutex_lock) before checking its desired condition. If the desired condition is not true, a call to pthread_cond_wait() is issued to wait for the condition to occur. In addition, a task may have to re-evaluate its desired condition once its call to pthread_cond_wait() returns, because a return from the function pthread_cond_wait() does not imply that its desired condition is either true or false. As good practice, it is recommended that the call to pthread_cond_wait() be enclosed in a *predicate-testing loop*, so that every time the call returns, the thread can re-evaluate the condition to determine whether it can safely proceed or should wait again.

Below, we look at some examples of using condition variables for task synchronization.

18.5.1 Barrier Synchronization

The POSIX.1-2008 standard defines a *barrier* as an OS kernel object used for blocking multiple tasks until the desired number of tasks have each passed a certain point in their execution. Similarly to the mutex and the condition variable, a barrier is normally used for synchronizing multiple threads within the same process, but if the process-shared attribute is set, it can be operated by any thread (within the same or different processes) that has access to the memory where the barrier is allocated.

A barrier object is initialized with an integer N which specifies the number of threads to be synchronized. A thread can define a *synchronization point* via a call to the POSIX function pthread_barrier_wait() on an initialized barrier object. The collection of synchronization points defined in all the tasks that use the same barrier object logically form a "corral" for the tasks. The calling thread of pthread_barrier_wait() blocks until the number of threads that have called pthread _barrier_wait() equals N, at which time all the threads successfully return from the call to pthread_barrier _wait() and proceed to their next instructions. This is illustrated in Figure 18.23.

Instead of working with the POSIX functions for barriers, we look at how to use a condition variable to implement barrier synchronization. A design is given in Figure 18.24, which is referred to as the *barrier synchronization pattern*:

(1) Each task that calls the function barrier_sync_point() first attempts to lock the guarding mutex. If the mutex is already locked, this thread enters into the mutex's waiting list.
(2) A task that has successfully locked the mutex increases the variable barrier_count by 1, indicating that one more task has arrived at the barrier.

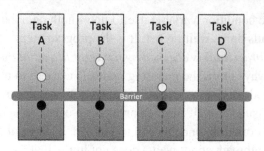

Figure 18.23
Barrier synchronization.

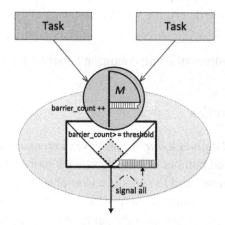

Figure 18.24
Barrier synchronization pattern.

(3) The desired condition for this design is "barrier_count>=threshold,"—that is, the number of tasks that have called barrier_sync_point() is no less than the specified threshold. If a task evaluates the desired condition to be false, it is enqueued into the condition variable's waiting list, and the mutex is unlocked simultaneously.

(4) If the desired condition is true (i.e., the number of tasks being blocked on the condition variable, together with the current task, is no longer less than the threshold), the task will signal the condition variable to unblock all the waiting tasks.

Listing 18.12 gives one such implementation.

```
  #include <stdio.h>
2 #include <pthread.h>
  #include <time.h>
4
  #define THREAD_PLAYER 4
6
```

```
   typedef struct {
 8     pthread_mutex_t barrier_lock;
       pthread_cond_t barrier_condvar;
10     int barrier_count; //how many are ready
       int barrier_threshold; //how many to sync
12 } barrier_t;

14 barrier_t *barrier;
   char msg[] ="   0  \t\t_____\t_____\t_____\t_____\n";
16 int index[] ={ 0, 3, 11, 21, 30, 39 };
   char zero = '0';
18 int step = 0;

20 void print_msg() {
       int myid = pthread_self();
22     step++;
       msg[3] = zero + (step % 10);
24     msg[index[myid]] = '!';
       fprintf(stdout, "\n");
26     fprintf(stdout, &msg);
   }
28
   barrier_t * initialize_barrier(int threshold) {
30     barrier_t *mybarrier = (barrier_t *) malloc(sizeof(barrier_t));
       memset(mybarrier, 0, sizeof(barrier_t));
32
       pthread_mutex_init(&mybarrier->barrier_lock, NULL);
34     pthread_cond_init(&mybarrier->barrier_condvar, NULL);
       mybarrier->barrier_threshold = threshold;
36     mybarrier->barrier_count = 0;
       return mybarrier;
38 }

40 void barrier_sync_point(barrier_t *barrier) {
       pthread_mutex_lock(&barrier->barrier_lock);
42     print_msg();
       barrier->barrier_count++;
44     if (barrier->barrier_count < barrier->barrier_threshold) {
         pthread_cond_wait(&barrier->barrier_condvar, &barrier->barrier_lock);
46     } else { //enough threads are ready
         fprintf(stdout, "\nEnough threads passed the barrier!\n");
48       barrier->barrier_count = 0;
         int i;
50       for (i = barrier->barrier_threshold; i > 0; i--)
           pthread_cond_signal(&barrier->barrier_condvar);
52     }
       pthread_mutex_unlock(&barrier->barrier_lock);
54 }

56 void * usertask() {
```

```
      int r = rand() % 5;
58    sleep(r+1); // simulate task processing
      barrier_sync_point(barrier);
60    sleep(r); // simulate task processing
      fprintf(stdout, "Thread %d is done!\n", pthread_self());
62  }

64  int main() {
      pthread_t thread[THREAD_PLAYER];
66    barrier = initialize_barrier(THREAD_PLAYER);
      pthread_attr_t attr;
68    struct sched_param sp;
      //initializes the thread attributes to their default values
70    pthread_attr_init(&attr);
      pthread_attr_setinheritsched(&attr, PTHREAD_EXPLICIT_SCHED);
72
      //Set the priority to 30
74    sp.sched_priority = 30;
      pthread_attr_setschedparam(&attr, &sp);
76
      int i;
78    for (i = 0; i < 4; i++) {
        pthread_create(&thread[i], &attr, usertask, NULL);
80    }
      fprintf(stdout, "Step\t\tThread %d\tThread %d\tThread %d\tThread %d\n",
82        thread[0], thread[1], thread[2], thread[3]);
      fprintf(stdout, &msg);
84    // let the threads run for a bit
      sleep(20);
86    return 1;
    }
```

Listing 18.12
Example implementation of a barrier.

First, note that a barrier is defined to have four members:

(1) barrier_lock: a guarding mutex object;
(2) barrier_condvar: a condition variable;
(3) barrier_count: how many threads have arrived at the barrier; and
(4) barrier_threshold: how many threads to synchronize.

The function initialize_barrier() is called to initialize a barrier object, where the mutex and condition variable are initialized and the threshold is set. Barrier synchronization is implemented by the function barrier_sync_point().

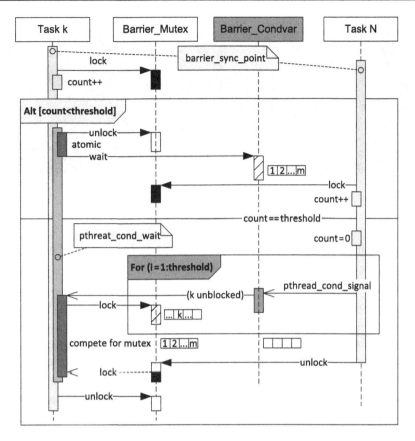

Figure 18.25
Using a condition variable for barrier synchronization.

Figure 18.25 illustrates a potential sequence of operations:

(1) Task T_k, where $1 \le k \le m$ (m is one less than the threshold), calls the function barrier_sync_point().

   (a) T_k first locks the guarding mutex of the condition variable.

   (b) T_k then increases the variable barrier_count by 1.

   (c) The desired condition is false because k is less than the threshold, so task T_k calls the function pthread_cond_wait(). In so doing, it unlocks the mutex, and enters into the waiting list of the condition variable. The call of pthread_cond_wait() will not return until task T_k is dequeued by a signal.

(2) All m tasks will end up with being enqueued in the waiting list of the condition variable.

(3) The last task T_N, assuming N equals the threshold, locks the guarding mutex.

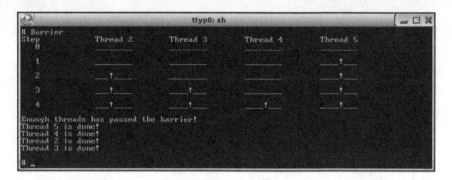

Figure 18.26
A run of the code in Listing 18.12.

(4)   After the variable barrier_count has been increased by 1, the desired condition becomes true. This allows task T_N to take the other branch:

(a)   The value of barrier_count is reset to 0.

(b)   For each of the waiting tasks, T_N calls pthread_cond_signal() to send a signal to the condition variable.

(c)   For each signal, the condition variable will dequeue/unblock a task.

(d)   Each task unblocked from the condition variable attempts to lock the mutex, but ends up being enqueued in the waiting list of the mutex (because the mutex is currently owned by task T_N). Consequently, after the for loop, the waiting list of the condition variable becomes empty, while the waiting list of the mutex is filled with m tasks.

(e)   Task T_N unlocks the mutex.

(f)   Whichever task is able to lock the mutex returns from its call of pthread_cond_wait().

(g)   The last step of barrier_sync_point() is to unlock the mutex. This allows the tasks in the waiting list of the mutex to return one by one from their calls of pthread_cond_wait().

Figure 18.26 shows a screenshot of a run of the code in Listing 18.12, where the threads with the ID numbers 5, 2, and 3 are forced to wait until the thread with ID number 4 has arrived, at which time all four threads proceed to the rest of their execution.

18.5.2 Producer-Consumer Pattern

The barrier implementation we saw in the last section is built upon one condition variable, which is associated with a single predicate to be evaluated by all the tasks. In general, a condition variable can be associated with different predicates by different tasks.

Now, let us reconsider the producer-consumer problem. Figure 18.27 illustrates a design based on a condition variable, which is associated with two predicates that allow the producer and

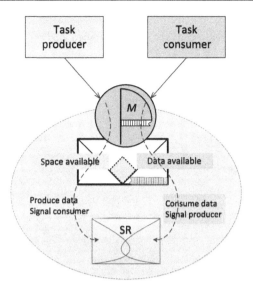

Figure 18.27
Producer-consumer pattern.

the consumer to coordinate their behaviors. Briefly, the producer task, after locking the mutex, checks whether the shared area has sufficient space to hold new data. If it does, it places new data into the shared area and signals the condition variable to unblock a consumer task, if any; otherwise, the producer task blocks on the condition variable. This ensures that the producer will not overwrite the data that have not been read by the consumer task. On the other hand, the consumer task, after locking the mutex, checks whether there are data available in the shared area. If there are, it retrieves the data and then signals the condition variable to unblock a producer task, if any; otherwise, the consumer task blocks on the condition variable.

```
   #include <stdio.h>
 2 #include <pthread.h>
   #include <time.h>

 4
   //Data is shared by child processes producer and consumer
 6 int shared_buffer = -1;

 8 int pred_data_ready = 0;
   pthread_mutex_t mutex = PTHREAD_MUTEX_INITIALIZER;
10 pthread_cond_t condvar = PTHREAD_COND_INITIALIZER;

12 void set_shared_buffer(int value) {
      fprintf(stdout, "Producer writes\t%2d", value);
14    shared_buffer = value;
   }
```

```
16 int get_shared_buffer() {
     fprintf(stdout, "Consumer reads\t%2d\n", shared_buffer);
18   return shared_buffer;
   }
20 int get_shared_buffer2() {
     return shared_buffer;
22 }

24 void * consumer(void *notused) {
     int sum = 0;
26   int firstget, secondget;
     int r, count;
28   for (count = 1; count <= 10; count++) {
       r = rand() % 4;
30     sleep(r);

32     pthread_mutex_lock(&mutex);
       while (!pred_data_ready) {
34       pthread_cond_wait(&condvar, &mutex);
       }
36     //start of critical section
       firstget = get_shared_buffer();
38     sleep(1); //simulate longer processing
       secondget = get_shared_buffer2();
40     sum += secondget;
       fprintf(stdout, "Consumer process\t\t\t%2d,%2d,%2d\n",
42       firstget, secondget, sum);
       pred_data_ready = 0;
44     //end of critical section
       pthread_cond_signal(&condvar);
46     pthread_mutex_unlock(&mutex);
     }
48   fprintf(stdout, "\n%s %d\n%s\n", "Consumer read values total", sum,
                     "Terminating Consumer");
   }
50
   void * producer(void *notused) {
52   int sum = 0;
     int r, count;
54   for (count = 1; count <= 10; count++) {
       r = rand() % 4;
56     sleep(r);

58     pthread_mutex_lock(&mutex);
       while (pred_data_ready) {
60       pthread_cond_wait(&condvar, &mutex);
       }
62     //start of critical section
       set_shared_buffer(count);
64     sum += count;
```

```
         fprintf(stdout, "\t%2d\n", sum);
66       pred_data_ready = 1;
         //end of critical section
68       pthread_cond_signal(&condvar);
         pthread_mutex_unlock(&mutex);
70     }
       fprintf(stdout, "Producer done\nTerminating Producer\n");
72 }


74 int main() {
       fprintf(stdout,"Action\t\tValue\tSum of Produced\tfirst,second,Sum\n");
76     fprintf(stdout, "------\t\t-----\t---------------\t---------------\n");

78     pthread_attr_t attr;
       struct sched_param sp;
80
       //initializes the thread attributes to their default values
82     pthread_attr_init(&attr);
       pthread_attr_setinheritsched(&attr, PTHREAD_EXPLICIT_SCHED);
84
       //Set the priority to 30 to the producer thread
86     sp.sched_priority = 30;
       pthread_attr_setschedparam(&attr, &sp);
88     // create the producer thread
       pthrcad_create(NULL, &attr, producer, NULL);
90
       //Set the priority to 20 to the consumer thread
92     sp.sched_priority = 20;
       pthread_attr_setschedparam(&attr, &sp);
94     // create the consumer thread
       pthread_create(NULL, &attr, consumer, NULL);
96
       // let the threads run for a bit
98     sleep(50);
       return 1;
100 }
```

Listing 18.13

Sample code of producer-consumer pattern.

Listing 18.13 gives an example implementation of the producer-consumer pattern, and a high-level code structure is given by the UML sequence diagram in Figure 18.28.

It is worthwhile noticing the following points:

(1) To be consistent, we still use an integer variable (i.e., shared_buffer) as the shared area between the producer and the consumer. In general, the shared area can be a memory block containing unstructured or structured objects.

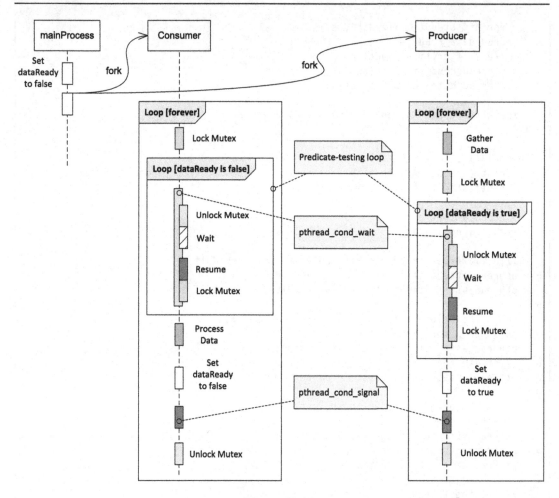

Figure 18.28

UML sequence diagram: producer-consumer pattern (interactions with the mutex and condition variable objects are suppressed).

(2) The variable pred_data_ready is used to form two predicates, one for the consumer and one for the producer. The consumer waits for the condition variable until its predicate (pred_data_ready) is true, while the producer waits for the condition variable until its predicate (!pred_data_ready) is true. When the condition (pred_data_ready) is true, it is safe for the consumer to read from the shared_buffer while the producer is prohibited from writing new data to the shared_buffer. Similarly, when the condition (pred_data_ready) is false, it is safe for the producer to write new data to the shared_buffer while the consumer is prohibited from accessing the shared_buffer.

(3) Both the consumer and the producer have a predicate-testing loop that encloses the function pthread_cond_wait(). This enforces a task to wait until its desired condition becomes true.

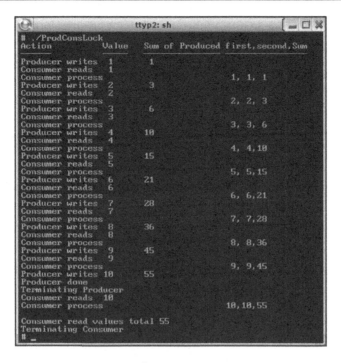

Figure 18.29
A run of the code in Listing 18.13.

(4)   The function pthread_cond_wait() itself contains several steps. The calling thread unlocks the guarding mutex prior to entering into the waiting list of the condition variable. Whenever a blocked thread is woken up by a signal, it resumes its execution and attempts to lock the guarding mutex. Hence, a thread must be the owner of the mutex both before and after the predicate-testing loop.

(5)   Before signaling the condition variable, the consumer (producer) changes the variable prcd_data_ready; this is to make sure that the producer (consumer), once unblocked, is able to jump out of its predicate-testing loop.

Figure 18.29 shows a screenshot of a run of the code in Listing 18.13. Note that the data integrity is preserved and there is no data loss or duplication. The producer-consumer problem introduced in Section 18.1 has been gracefully solved. The condition variable ensures that data are written to the buffer or read from the buffer only when the buffer is in the proper state.

18.5.3 Read-Write Locks

In this section, we examine an implementation of read-write locks which is built upon more than one control variable.

Table 18.5 POSIX.1 function calls for read-write locks

POSIX Function	Description
pthread_rwlock_init	Initialize a read-write lock object
pthread_rwlock_destroy	Destroy a read-write lock object
pthread_rwlock_unlock	Unlock a read-write lock object
pthread_rwlock_rdlock	Lock a read-write lock object for reading
pthread_rwlock_wrlock	Lock a read-write lock object for writing

A resource protected by a read-write lock allows many threads to have simultaneous read-only access to it while allowing only one thread to have *exclusive* write access at any given time. In other words, a reader excludes only writers, whereas a writer excludes readers and any other writers.

Table 18.5 gives a list of POSIX functions for read-write locks. Given a resource protected by an associated read-write lock, one or more readers acquire read access to the resource by performing a pthread_rwlock_rdlock() operation on the read-write lock; a writer acquires exclusive write access by performing a pthread_rwlock_wrlock() operation.

```
1 // '@': a task is started, waiting for mutex
  // '.': a task is waiting for read condition variable
3 // '*': a task is waiting for write condition variable
  // 'r': a task is reading
5 // 'w': a task is writing
  // '!': a task is done
7 #include <stdio.h>
  #include <pthread.h>
9 #include <time.h>

11 #define THREAD_PLAYER 8

13 typedef struct {
   pthread_mutex_t    guard_mutex;
15 pthread_cond_t     read_condvar;
   pthread_cond_t     write_condvar;
17 int          rw_count; //-1 indicates a writer is active
   int          waiting_reader_count;
19 } rwlock_t;

21 rwlock_t *rwlock;
   char msg[] = "   \t   \t   \t   \t   \t   \t   \t   \n";
23 int index[] = { 0, 0, 1, 6, 11, 16, 21, 26, 31, 36 };

25 void acquire_write_privilege(rwlock_t *rwlock) {
   pthread_mutex_lock(&rwlock->guard_mutex);
27 while (rwlock->rw_count != 0) {
```

```
      print_msg('*');
29    pthread_cond_wait (&rwlock->write_condvar, &rwlock->guard_mutex);
   }
31    rwlock->rw_count = -1;
   pthread_mutex_unlock (&rwlock->guard_mutex);
33 }

35 void release_write_privilege(rwlock_t *rwlock) {
   pthread_mutex_lock(&rwlock->guard_mutex);
37    rwlock->rw_count = 0;
   int i;
39    if (rwlock->waiting_reader_count) {
      for (i = rwlock->waiting_reader_count; i>0; i--)
41       pthread_cond_signal (&rwlock->read_condvar);
   } else
43    //writers has lower priority than readers
      pthread_cond_signal (&rwlock->write_condvar);
45    print_msg('!');
   pthread_mutex_unlock (&rwlock->guard_mutex);
47 }

49 void acquire_read_privilege(rwlock_t *rwlock) {
   pthread_mutex_lock(&rwlock->guard_mutex);
51    rwlock->waiting_reader_count ++;
   while (rwlock->rw_count < 0) {
53    print_msg('.');
      pthread_cond_wait (&rwlock->read_condvar, &rwlock->guard_mutex);
55    }
   rwlock->waiting_reader_count --;
57    rwlock->rw_count ++;
   pthread_mutex_unlock (&rwlock->guard_mutex);
59 }

61 void release_read_privilege(rwlock_t *rwlock) {
   pthread_mutex_lock(&rwlock->guard_mutex);
63    rwlock->rw_count --;
   if (rwlock->rw_count == 0)
65    //the last reader sends signal to a waiting writer
      pthread_cond_signal (&rwlock->write_condvar);
67    print_msg('!');
   pthread_mutex_unlock (&rwlock->guard_mutex);
69 }

71 void print_msg(char c) {
   int myid = pthread_self();
73    msg[index[myid]] = c;
   fprintf(stdout, "%s", &msg);
75 }

77 void * reader() {
```

```
     int r = rand() % 5;
 79  sleep(r+1); // simulate task processing
     print_msg('@');
 81  acquire_read_privilege(rwlock);
     print_msg('r');
 83  sleep(r); // simulate task processing
     release_read_privilege(rwlock);
 85  print_msg(' ');
   }
 87
   void * writer() {
 89  int r = rand() % 5;
     sleep(r+1); // simulate task processing
 91  print_msg('@');
     acquire_write_privilege(rwlock);
 93  print_msg('w');
     sleep(r); // simulate task processing
 95  release_write_privilege(rwlock);
     print_msg(' ');
 97 }

 99 rwlock_t * initialize_rwlock() {
     rwlock_t *mylock = (rwlock_t *) malloc(sizeof(rwlock_t));
101  memset(mylock, 0, sizeof(rwlock_t));

103  pthread_mutex_init(&mylock->guard_mutex, NULL);
     pthread_cond_init(&mylock->read_condvar, NULL);
105  pthread_cond_init(&mylock->write_condvar, NULL);
     mylock->rw_count = 0;
107  mylock->waiting_reader_count = 0;
     return mylock;
109 }

111 int main() {
     pthread_t thread[THREAD_PLAYER];
113  rwlock = initialize_rwlock();

115  pthread_create(&thread[0], NULL, reader, NULL);
     pthread_create(&thread[1], NULL, reader, NULL);
117  pthread_create(&thread[2], NULL, writer, NULL);
     pthread_create(&thread[3], NULL, writer, NULL);
119  pthread_create(&thread[4], NULL, reader, NULL);
     pthread_create(&thread[5], NULL, writer, NULL);
121  pthread_create(&thread[6], NULL, reader, NULL);
     pthread_create(&thread[7], NULL, reader, NULL);
123
     fprintf(stdout, " T%d\t T%d\t T%d\t T%d\t T%d\t T%d\t T%d\t T%d\n",
125      thread[0], thread[1], thread[2], thread[3],
         thread[4], thread[5], thread[6], thread[7]);
127  fprintf(stdout, "____\t____\t____\t____\t____\t____\t____\t____\n");
```

```
     // let the threads run for a bit
129  sleep(30);
     return 1;
131 }
```

Listing 18.14
Sample code of a read-write lock.

Read-write locks can be implemented by using a mutex and condition variables. One example is given in Listing 18.14, where a read-write lock object (lines 13-19) contains

- guard_mutex—a guarding mutex;
- read_condvar—a condition variable for readers;
- write_condvar—a condition variable for writers;
- rw_count—an integer (a positive number indicating the number of active readers, −1 indicating an active writer);
- waiting_reader_count—an integer indicating the number of blocked readers.

The function acquire_write_privilege() corresponds to the POSIX function pthread_rwlock_wrlock(). A writer task that calls acquire_write_privilege() locks the guarding mutex first. It then waits for its desired condition to occur: a writer cannot obtain the read-write lock unless there is no active reader or writer—that is, *rw_count* should be 0. Once it has obtained the read-write lock, the task sets *rw_count* to −1 to indicate that a writer is active.

Upon completion of its use of the resource, a writer calls release_write_privilege() to unlock the read-write lock. In so doing, it resets *rw_count* to 0, and signals all the readers blocking on read_condvar and one writer blocking on write_condvar, if any. This is illustrated in Figure 18.30.

Similarly, the function acquire_read_privilege() corresponds to the POSIX function pthread_rwlock_rdlock(). A reader task that calls acquire_read_privilege() locks the guarding mutex first. It then waits for its desired condition to occur: a reader cannot obtain the read-write lock unless there is no active writer—that is, *rw_count* should not be negative. Once it has obtained the read-write lock, the task increases *rw_count* by 1 to indicate that a new reader has become active.

Upon completion of its use of the resource, a reader calls release_read_privilege() to unlock the read-write lock. In so doing, it decreases *rw_count* by 1, and if there is no active reader, it signals one writer blocking on the write_condvar, if any. This is also illustrated in Figure 18.30.

Figure 18.31 shows a screenshot of a run of the code in Listing 18.14.

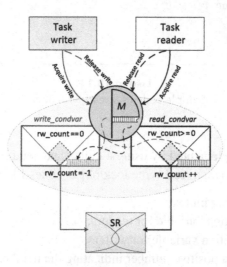

Figure 18.30
Read-write lock pattern.

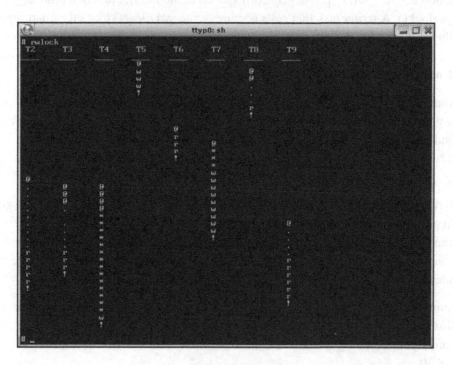

Figure 18.31
A run of the code in Listing 18.14.

Problems

18.1   When multiple tasks use a shared resource, what problems may occur if the resource is not well protected?

18.2   Explain why memory mapping is an efficient way to exchange information between two processes.

18.3   What is a critical section? What are contending critical sections?

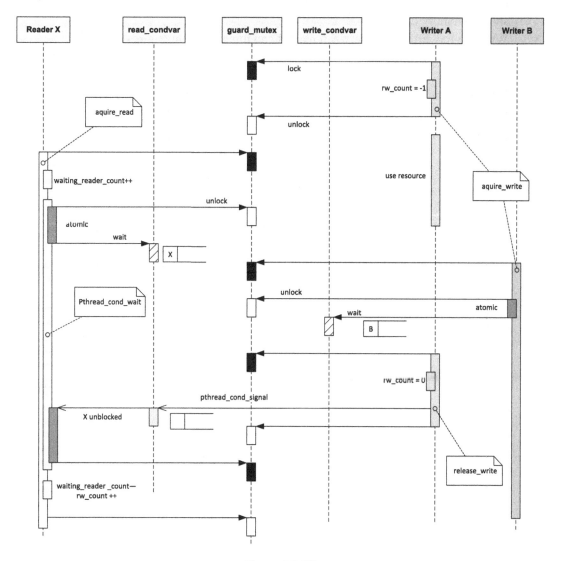

Figure 18.32
The use of a read-write lock.

18.4   As far as a semaphore is concerned, what is a procure operation and what is a vacate operation?

18.5   Explain how to use a semaphore to protect a multi-instance resource.

18.6   Compare the output in Figure 18.4 and the output in Figure 18.14, explain the difference. Study the code in Listings 18.2 and 18.3, what problem(s) may happen to the content of the shared file object?

18.7   Compare the output of the code given in Section 18.2.2 and the output of the code given in Section 18.3.5.2, explain why a semaphore can be employed to protect the shared memory object.

18.8   What are the differences between named and unnamed semaphores?

18.9   Compared with semaphores, what features make mutexes a better way to protect resources?

18.10   What is a condition variable? Why is a guarding mutex needed when using a condition variable? Why is it recommended that the call to pthread_cond_wait() be enclosed in a predicate-testing loop?

18.11   Explain how the condition variable is used in barrier synchronization.

18.12   Explain how the condition variable is used to solve the producer-consumer problem.

18.13   Explain how condition variables are used in the implementation of a read-write lock.

18.14   Figure 18.32 illustrates the use of a read-write lock. Refer to the code given in Listing 18.14, and explain the sequence of operations step-by-step.

18.15   Modify the code in Listing 18.13 such that an integer array is used to contain the data to be shared between the producer and the consumer.

CHAPTER 19

Intertask Communication: Message Queue

Contents

The events in our lives happen in a sequence in time, but in their significance to ourselves they find their own order: the continuous thread of revelation.

Eudora Welty

19.1 Introduction to Message Queues

In a multitasking system there may exist tasks that need to cooperate with each other by exchanging messages. This is called *intertask communication*, which can be classified into two types: synchronous intertask communication and asynchronous intertask communication.

In synchronous intertask communication, the sender and receiver need to wait for each other until the message transmission is complete. In other words, the sender cannot continue until the receiver has received the message. In asynchronous intertask communication, the sender may continue to execute its next instruction while the message is being delivered to the receiver. Since the sender and the receiver do not wait for each other, the two threads of execution are not synchronized.

As compared with synchronous communication, asynchronous communication offers better run-time performance. The downside is that the tasks involved in asynchronous

Real-Time Embedded Systems. http://dx.doi.org/10.1016/B978-0-12-801507-0.00019-5

communication are more difficult to implement, and it is probably much harder to prove the correctness of the system.

The POSIX.1-2008 standard specifies several asynchronous intertask communication mechanisms, including message queues, pipes, sockets, and signaling. The message queue is our focus in this chapter.

A POSIX *message queue* is a buffer-like object that acts as a liaison between a message sender and a message receiver. Messages sent to a message queue are stored in the queue and can be retrieved at any time by a receiver. A POSIX message queue is commonly implemented in real-time operating systems (OSs), including VxWorks and QNX.

As a side note, a message queue is such an important concept that it has been further developed from a single OS node to the Internet level. For example, the Java Message Service offers a distributed message queuing service that allows Web-based applications to exchange messages over the Internet.

19.2 Message Queue Statics and Dynamics

A message queue allows a variable number of messages, each of varying length, to be queued. The sender and the receiver of a message do not need to interact with the message queue at the same time.

Figure 19.1 gives the structure of message queue:

- Each message has an associated integer priority. The body of a message may have a flat (i.e., a byte stream) or a user-defined structure (say, with the sender's ID to be used in a reply message).
- A message queue is a kernel object with several attributes:
    - mq_flags: access permission to the message queue, such as read access, write access, and nonblocking access;
    - mq_maxmsg: the limit on the number of messages that can be placed in the queue;
    - mq_msgsize: the upper limit on the size of each message that may be placed in the queue; and
    - mq_curmsgs: the number of messages that are currently in the queue.
- Multiple tasks can send to and receive from the same message queue.
- Messages are queued in their order of priority via mq_send(). A new message with a higher priority is inserted before messages with lower priorities. A new message is inserted after the queued messages with the same priority, if any.
- Each message queue is associated with a waiting list for senders. If a message queue disallows nonblocking access and it is full (i.e., the number of messages in the queue is

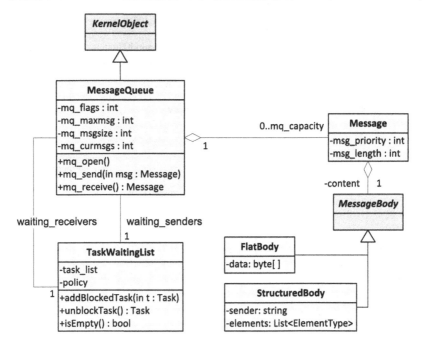

Figure 19.1
UML static structure of a message queue.

equal to its maximum capacity), when a task is trying to send a message to the queue, the task will be blocked and placed in the waiting list for senders.
- Messages are received in the order of priority via mq_receive().
- Each message queue is associated with a waiting list for receivers. If a message queue disallows nonblocking access and it is empty, when a task is trying to receive a message from the queue, the task will be blocked and placed in the waiting list for receivers.

Figure 19.2 gives the dynamic state transitions of a message queue:

- Upon creation, a message queue is in the "Empty" state, with the state variable *mq_curmsgs* set to 0, and *mq_maxmsg* set to a value appropriate for the application on hand.
- Whenever a task sends a message to an empty message queue, it transitions to a new state named "NonEmpty," with *mq_curmsgs* increased by 1.
- Whenever a task attempts to receive a message from a message queue containing exactly one message, it transitions back to the "Empty" state, with *mq_curmsgs* reset to 0.
- When a task sends a message to a nonempty message queue, it stays in the "NonEmpty" state if the number of messages in the queue, including the new message, is less than the queue capacity. Otherwise, it transitions to a new state named "Full."

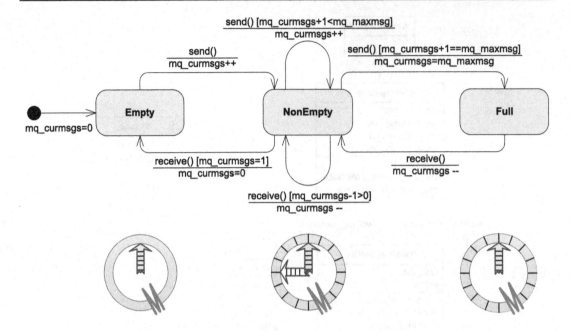

Figure 19.2
Message queue states and message queue notation in different states.

- When a task attempts to receive a message from a nonempty message queue, it stays in the "NonEmpty" state if the number of messages in the queue, excluding the one being received, is more than zero. Otherwise, it transitions to the "Empty" state.
- When a task attempts to receive a message from a full message queue, it transitions to the "NonEmpty" state, with *mq_curmsgs* decreased by 1.

What if a task attempts to receive a message from an empty message queue? We have three cases, depending on whether the "nonblocking access" flag is set or clear in mq_flags:

(1) Nonblocking access: the "nonblocking access" flag is set. In such a case, the receiving task is not blocked and no message is returned.

(2) Blocking access: the "nonblocking access' flag is clear. In such a case, the receiving task is blocked and is placed in the waiting list for receivers. As shown in Figure 19.3, depending on the policy being implemented by the OS, blocked receivers can be placed in the waiting list in FIFO order (say, QNX) or in priority order. If a priority-based policy is implemented, when a message becomes available in the message queue, the task with the highest priority that has been waiting the longest is unblocked first to receive its message.

(3) Timed blocking access: the "nonblocking access" flag is clear and a time-out value is specified. This is similar to the blocking access case, except that the receiving task is unblocked and no message is returned when the specified time-out expires.

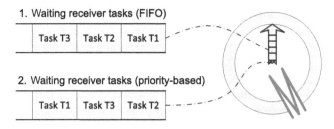

Figure 19.3
Receivers blocked when a message queue is empty.

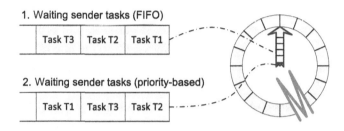

Figure 19.4
Senders blocked when a message queue is full.

What if a task attempts to send a message to a full message queue? We have three cases, depending on whether the "nonblocking access" flag is set or clear in mq_flags:

(1) Nonblocking access: the "nonblocking access" flag is set. In such a case, the sending task is not blocked and the message is lost.

(2) Blocking access: the "nonblocking access" flag is clear. In such a case, the sending task is blocked and is added to the queue's waiting list for senders. As shown in Figure 19.4, depending on the policy being implemented by the OS, blocked senders can be placed in the waiting list in FIFO order or in priority order. If a priority-based policy is implemented (say, QNX), when space becomes available in the message queue, the task with the highest priority that has been waiting the longest is unblocked first to send its message.

(3) Timed blocking access: the "nonblocking access" flag is clear and a time-out value is specified. This is similar to the blocking access case, except that the sending task is unblocked and the message is lost when the specified time-out expires.

19.3 Message Queue Usage Patterns

In the discussion below, if not indicated otherwise, blocking access is assumed (i.e., the "nonblocking access" flag is not set).

19.3.1 Unidirectional Communication

Figure 19.5 illustrates a one-way intertask communication using a message queue, where task T_1 sends messages to task T_2. We refer to this design as a *unidirectional queuing pattern*.

Since T_1 and T_2 may or may not be periodic tasks, we simply use γ_1 to denote the *initial average rate* of sending messages by task T_1, and γ_2 to denote the *initial average rate* of receiving messages by task T_2. Let us consider three scenarios:

(1) $\gamma_1 = \gamma_2$: the initial average rate of sending messages to the queue is the same as the rate of receiving messages from the queue. Ideally, exactly one message is ready in the queue whenever task T_2 is trying to receive a message, and the queue is always empty whenever task T_1 is ready to send a message. Neither of the tasks would be blocked on the queue.

(2) $\gamma_1 > \gamma_2$: the initial average rate of sending messages to the queue is greater than the rate of receiving from the queue. In this case, the message queue becomes full eventually and task T_1 will be blocked. After this point,

    (a)   the number of messages in the queue is either *mq_maxmsg* or *mq_maxmsg* − 1: every time task T_2 receives a message from the queue, the empty space will shortly be occupied by a new message from T_1.

    (b)   it is likely that T_2 is never blocked, but T_1 is blocked every time it attempts to send a message to the queue while it is full. Consequently, the average rate of sending messages is synchronized (reduced) to the rate of receiving messages.

(3) $\gamma_1 < \gamma_2$: the initial average rate of sending messages to the queue is less than the rate of receiving messages from the queue. In this case, the message queue becomes empty eventually and task T_2 will be blocked. After this point,

    (a)   the number of messages in the queue is either 0 or 1: every time task T_1 sends a message to the queue, the message will shortly be retrieved by T_2.

    (b)   it is likely that T_1 is never blocked, but T_2 is blocked every time it attempts to receive a message from the queue while it is empty. Consequently, the average rate of receiving messages is synchronized (reduced) to the rate of sending messages.

In sum, the message queue virtually offers a flow control, and the two tasks' communications behavior has been effectively synchronized.

Figure 19.5
Unidirectional queuing pattern.

19.3.2 Acked-Unidirectional Communication

Figure 19.6 illustrates a one-way intertask communication design that is exactly the same as the one given in Figure 19.5, except that a semaphore is used to close the loop of communication.

Suppose the tasks T_1 and T_2 use the code sequences below to access the message queue:

```
T1()
{ ...
  mq_send(mq, msg_s1);
  sem_procure();
  ...
}

T2()
{ ...
  mq_receive(mq, msg_r2);
  sem_vacate();
  ...
}
```

The constraint placed on the semaphore says that the semaphore count has an initial value of 0 and cannot be greater than 1. We may have the following scenarios:

- Suppose T_2 has a higher priority and runs first. Since the message queue is empty initially, it blocks on receiving a message from the queue. When T_1 is scheduled to run and sends a message to the queue, T_2 will be unblocked and preempts T_1. After receiving a message, T_2 increases the semaphore to 1 by sem_vacate(). When T_1 is scheduled to run, it will successfully procure the semaphore. At this time point, T_1 knows for sure that its message has been received by task T_2.
- Suppose T_1 has a higher priority and runs first. After sending a message to the queue, it blocks on procuring the semaphore because the count is 0. When T_2 is scheduled to run, it can successfully receive a message from the queue. Afterward, T_2 increases the semaphore to 1 by sem_vacate(). This will immediately unblock T_1, which preempts T_2

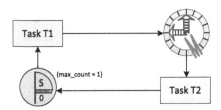

Figure 19.6
Acked-unidirectional queuing pattern.

and resumes its execution. At the time of being unblocked, T_1 knows for sure that its message has been received by task T_2.

In any case, the semaphore serves as an acknowledgment of the message just received by task T_2. In addition, at any point in time, the queue contains at most one outstanding message to be acknowledged. We refer to this design as an *acked-unidirectional queuing pattern*.

Note that acked-unidirectional communication can also be achieved without using a semaphore, as long as the message queue is set to disable nonblocking access and a constraint {mq_maxmsg = 1} is imposed on it. In such a design, T_1 knows for sure that the last message has been successfully received if it is not blocked when sending a new message. In other words, the emptiness of the queue serves as an acknowledgment.

19.3.3 Bidirectional Communication

As shown in Figure 19.7, full-duplex communication between two tasks requires two message queues, one for each direction. We refer to this design as a *bidirectional queuing pattern*.

Let us first consider the situation where each task has the same *fixed* sending and receiving frequency. In particular, we assume that the tasks T_1 and T_2 use the code sequences below to access the message queues (here, the fixed frequency is one per round of execution)[1]:

```
T1()
{ ...
  mq_send(mq2, msg_s1);
  mq_receive(mq1, msg_r1);
  ...
}

T2()
{ ...
```

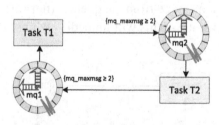

Figure 19.7
Bidirectional queuing pattern.

---

[1] The mq_send and mq_receive statements are at the top level of the task code. In other words, they are not enclosed inside any control structures such as loops.

```
    mq_send(mq1, msg_s2);
    mq_receive(mq2, msg_r2);
    ...
}
```

We may have the following scenarios:

(1) Suppose T_1 has a higher priority and runs first. It sends a message to queue $mq2$, and then blocks on receiving a message from $mq1$. After T_2 has been scheduled to run, as soon as it sends a message to $mq1$, T_1 is unblocked and T_2 is preempted. T_1 is blocked again in its next round of execution on receiving a message from $mq1$. When T_2 is scheduled to run, it receives a message from $mq2$. T_1 is blocked on $mq1$ until after T_2 has sent another message to $mq1$ in its next round of execution.

(2) Suppose T_2 has a higher priority and runs first. It sends a message to queue $mq1$, and then blocks on receiving a message from $mq2$. After T_1 has been scheduled to run, as soon as it sends a message to $mq2$, T_2 is unblocked and T_1 is preempted. T_2 is blocked again in its next round of execution on receiving a message from $mq2$. When T_1 is scheduled to run, it receives a message from $mq1$. T_2 is blocked on $mq2$ until after T_1 has sent another message to $mq2$ in its next round of execution.

In sum, neither task can be blocked on sending as long as $max\_maxmsg \geq 2$, and the higher-priority task can be blocked on receiving for each round of its execution.

What if we modify the constraint on the two message queues used in Figure 19.7 such that $mq\_maxmsg = 1$? In such a case, the intertask communication is interlocked more tightly. In particular, we have the following scenarios:

(1) Suppose T_1 has a higher priority and runs first:
    (a) T_1 sends a message to queue $mq2$.
    (b) T_1 blocks on receiving a message from $mq1$ because it is empty.
    (c) T_2 is scheduled to run and sends a message to $mq1$.
    (d) T_1 is unblocked. It preempts T_2 and receives a message from $mq1$.
    (e) T_1 blocks in its next round of execution on sending a message to $mq2$ because it is full.
    (f) T_2 is scheduled to run and receives a message from $mq2$.
    (g) T_1 is unblocked and T_2 is preempted.
    (h) Repeat from step (a).

(2) Suppose T_2 has a higher priority and runs first:
    (a) T_2 sends a message to queue $mq1$.
    (b) T_2 blocks on receiving a message from $mq2$ because it is empty.
    (c) T_1 is scheduled to run and sends a message to $mq2$.
    (d) T_2 is unblocked. It preempts T_1 and receives a message from $mq2$.
    (e) T_2 blocks in its next round of execution on sending a message to $mq1$ because it is full.

(f) T_1 is scheduled to run and receives a message from $mq1$.

(g) T_2 is unblocked and T_1 is preempted.

(h) Repeat from step (a).

The communications behavior can be rather complicated when the two tasks in each round of execution may send or receive an *arbitrary* number of messages—say, using a loop to receive all the messages available from a queue. As an exercise, Problem 19.6 asks you to analyze the communications behavior when each task uses fixed but different frequencies for sending and receiving.

19.3.4 Client-Server Communication

In client-server architecture, a client task sends service requests to a server task, which usually replies to the client upon service completion. Message queues can be used to transmit service requests and replies between a server and multiple client tasks running on a single OS node.

Figure 19.8 illustrates our first design, where two message queues are reserved for each client, one for sending requests and one for receiving replies.

This design has several drawbacks. The first issue is scalability. It requires that two message queues be reserved for each active client. If the server needs to simultaneously support k number of active clients, then the system would have to maintain a pool of $2k$ message queues, which might consume a huge amount of system memory.

In addition, since the server needs to receive messages from multiple message queues, it has to examine the queues by following a certain execution order in its code— say, checking the

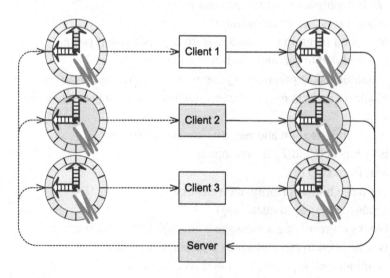

Figure 19.8
Client-server: two message queues for each client.

queue from client 1 first, then client 2, and so on. The server can receive and process messages in two ways:

(1) By applying the round-robin principle at the message queue level: it comes to the next message queue after the current message queue is emptied.
(2) By applying the round-robin principle at the message level: in each round it receives at most one message from each message queue.

In either case, if the message queues disallow nonblocking access, the server could be blocked on a message queue (say, from client 1) if it is or becomes empty. In such a case, the server is no longer responsive, even though the other message queues currently have messages waiting for the server to process them. Of course, this issue can be resolved by adding extra logics to the sever code. One solution is to examine, if permitted, the emptiness of a message queue before receiving from it. Another solution is to allow the server to send one special "guard" message with the lowest possible priority to a message queue immediately before the queue is processed. Whenever the server receives a guard message from a queue, it knows that the queue is already empty and switches to the next queue.

Another issue is that, regardless of whether the message queues allow nonblocking access or not, the server code has coerced an extra priority order: if message queue Q_1 is handled prior to message queue Q_2, then all messages from Q_1 would be received and processed before messages from Q_2. This may cause message priority inversion in the sense that a lower-priority message might be processed before a higher-priority message.

Figure 19.9 illustrates another design, where two message queues are reserved for the server, one for receiving requests and one for sending replies.

This design avoids the scalability issue exhibited in the first design. However, the fact that all the reply messages from the server are contained in a single message queue shared by all the

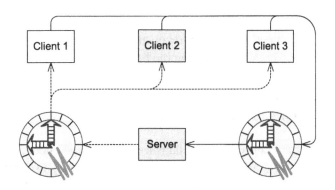

Figure 19.9
Client-server: two message queues for the server.

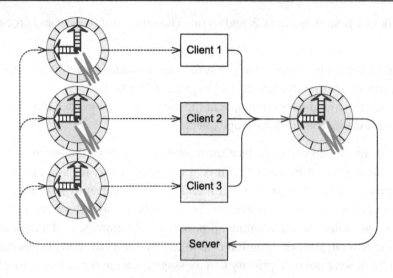

Figure 19.10
Client-server queuing pattern: one message queue for each receiver.

clients raises another issue: when a client receives a message from the queue, it might have taken a message intended for another client. Such a mistake will not be realized until after the message has been processed (and hopefully the message body has the information about the intended receiver's identity). Of course, when this happens, a client can put the message back in the queue. This, however, renders several side effects to the system: the message integrity can be compromised, the message security is hampered, and the timing constraints of client tasks can be broken owing to message loss or receiving delay.

A better design is given in Figure 19.10, where one message queue is reserved for each message receiver. We refer to this design as a *client-server queuing pattern*. In particular, the server task has a message queue to receive request messages from clients, and each client task has a message queue to receive reply messages from the server. Each request message from a client contains information about where to send the reply message—that is, the message queue associated with the client. Since each receiver reads messages only from one dedicated message queue, all the drawbacks exhibited in the first and second designs have been resolved.

19.4 POSIX Functions for Message Queues

The POSIX.1-2008 standard defines several flags for the message queue: O_RDONLY (receive only), O_WRONLY (send only), O_RDWR (send-receive), O_NONBLOCK (nonblocking access), O_EXCL (exclusive queue name), and O_CREAT (create a new message queue if it does not exist).

Table 19.1 POSIX.1 message queue function calls

POSIX Function	Description
mq_close	Close a message queue (descriptor)
mq_getattr	Get the current attributes of the queue
mq_notify	Register the calling process to be notified of message arrival at an empty message queue when it transitions from empty to nonempty
mq_open	Create a new message queue or open an existing message queue.
mq_receive	Receive the oldest of the highest-priority messages from the queue. When O_NONBLOCK is not set, this call blocks if there is no message in the queue
mq_send	Send a message into the queue. When O_NONBLOCK is not set, this call blocks if the number of elements in the queue is equal to its maximum capacity
mq_setattr	Set the attributes of the queue
mq_timedreceive	Similar to mq_receive(), except that if this call blocks, the blocking is terminated when the specified time-out expires
mq_timedsend	Similar to mq_send(), except that if this call blocks, the blocking is terminated when the specified time-out expires
mq_unlink	Remove a message queue by name from the system

Table 19.1 lists the POSIX functions for message queues. The function mq_open is called to create a new message queue or open an existing message queue. A process can create or open more than one message queue. The function mq_open() returns a message queue descriptor, which is a *per-process* handle that refers to a message queue object managed by the OS kernel. When mq_open() is called to create a new queue, queue attributes, including mq_maxmsg, mq_msgsize, and its flags, can be initialized.

Sending messages is done via mq_send or mq_timedsend. If a message queue is not full, a call to mq_send or mq_timedsend on the queue will cause the argument message to be queued according to the specified message priority. In the case that a message queue is full,

- if the flag O_NONBLOCK is set, a call to mq_send or mq_timedsend returns an error and the message is not queued;
- if the flag O_NONBLOCK is not set, a call to mq_send blocks until space becomes available in the message queue, or until the call is interrupted by a signal;
- if the flag O_NONBLOCK is not set, a call to mq_timedsend blocks until space becomes available in the message queue, or until the call is interrupted by a signal, or until the specified time-out expires;
- if more than one thread is blocked when space becomes available,
  - for priority-based implementation, the highest-priority thread that has been waiting the longest is unblocked to send its message;
  - for FIFO-based implementation, the thread that has been waiting the longest is unblocked to send its message.

Receiving messages is done via mq_receive or mq_timedreceive. If a message queue is not empty, a call to mq_receive or mq_timedreceive on the queue will cause the oldest of the highest-priority messages to be dequeued. In the case that a message queue is empty,

- if the flag O_NONBLOCK is set, a call to mq_receive or mq_timedreceive returns an error and no message is dequeued;
- if the flag O_NONBLOCK is not set, a call to mq_receive blocks until a message becomes available in the message queue, or until the call is interrupted by a signal;
- if the flag O_NONBLOCK is not set, a call to mq_timedreceive blocks until a message becomes available in the message queue, or until the call is interrupted by a signal, or until the specified time-out expires;
- if more than one thread is blocked when a message becomes available,
  - for priority-based implementation, the highest-priority thread that has been waiting the longest is unblocked to receive its message;
  - for FIFO-based implementation, the thread that has been waiting the longest is unblocked to receive its message.

POSIX.1 also specifies a notification mechanism that allows a process to be asynchronously notified via a signal event when a message arrives in a previously empty queue. This enables a process to continue its execution while waiting for messages to arrive. Via a call to mq_notify(), a process can register itself to a message queue, specifying a notification (signal event) that will be sent back to the process when an empty queue receives a message and there are no waiting receivers. The notification mechanism can be summarized as follows:

- At any time, a message queue allows only one process to be registered to receive a notification. Attempts to register to a queue will fail if the queue already has a process registered. A registered process can deregister itself by calling mq_notify() with NULL as the notification argument.
- The process registered to a queue is notified only when a message arrival causes the queue to transition from empty to nonempty. If a queue is not empty at the time of the registration, a notification will occur only after the queue has been emptied and a new message has arrived.
- The registered process is notified only if some other process is not currently blocked in mq_receive() or mq_timedreceive() waiting to receive a message. If another process is blocked, it will has the privilege to receive the arriving message, and the registered process will remain registered.
- The registration of a process to a queue is one-shot: it is removed after one notification is sent to the process. In case that a process needs to be notified every time the queue transitions from empty to non-empty, it has to reregister itself after each notification.

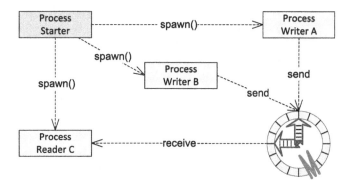

Figure 19.11
An example of using a message queue.

19.5 An Example of Using Message Queues

As an example, we consider the problem illustrated in Figure 19.11, where two writer processes and a reader process exchange messages via a message queue.

Listing 19.1 gives the header file. The user message structure has three fields: a sequence number, a sender name, and a message content. Three utility functions are defined in the header: print_msg for printing formatted outputs, MQ_size for checking the number of messages in a queue, and mq_status for generating a formatted string to represent the status of a message queue.

```
1  /*
   *  myheader.h
3  */
   #ifndef MYHEADER_H_
5  #define MYHEADER_H_

7  #define MQ_CAPACITY 5
   #define PRT_STR_LEN 60
9  #define ASCII_SPACE 32
   #define Hyphen 45

11
   typedef struct {
13   int sequence;
     char content[10];
```

```
15    char sender[10];
    } myMSG_t;
17
    char *senders[2] = {"Writer A", "Writer B"};
19  char *prefix[2] = {"(A,", "(B,"};
    char MQ_NAME[] = "mymq";
21  struct mq_attr *mqstat;
    char pmsg[PRT_STR_LEN];
23  int prt_index[] = { 1, 15, 30, 45 };
    char empty_mq[] = "[.....]";
25  char full_mq[] = "[*****]";

27  void print_msg(int myindex, char *msg1, char *msg2) {
    memset(pmsg, ASCII_SPACE, PRT_STR_LEN-1);
29    memcpy(&pmsg[prt_index[myindex]], msg1, strlen(msg1));
    memcpy(&pmsg[prt_index[3]], msg2, strlen(msg2));
31    fprintf(stdout, "%s", pmsg);
    memset(pmsg, Hyphen, PRT_STR_LEN-1);
33    fprintf(stdout, "%s", pmsg);
    }
35  long MQ_size(mqd_t _MQ){
    if (mq_getattr(_MQ, mqstat) == -1) return -1;
37    long nmsg = mqstat-> mq_curmsgs;
    return nmsg;
39  }
    char *mq_status(long len){
41    char *status = (char *) malloc(MQ_CAPACITY+3);
    memset(status, 0, MQ_CAPACITY+3);
43    strcpy(status, "[");
    long counter;
45    for (counter=1;counter<=len;counter++) strcat(status, "*");
    for (;counter<MQ_CAPACITY+1;counter++) strcat(status, ".");
47    strcat(status, "]");
    return status;
49  }
    #endif /* MYHEADER_H_ */
```

Listing 19.1
Header of using a message queue.

Listing 19.2 gives the starter program. The main() function first calls the function create_msg_queue() to create a message queue with its capacity set to 5. The flags used in the call mq_open() indicates that the message queue is readable, writable, and blockable. Lines 36-44 are used to output header information to the terminal.

Three child processes are created in lines 47-50. Afterward, the starter enters into a loop, which terminates whenever one of the three child processes is terminated.

```c
  #include <stdio.h>
2 #include <stdlib.h>
  #include <spawn.h>
4 #include <unistd.h>
  #include <fcntl.h>
6 #include <sys/wait.h>
  #include <errno.h>
8 #include <mqueue.h>
  #include "myheader.h"
10
  static void create_msg_queue(char *name, int msg_len) {
12   struct mq_attr attr;
     mqd_t queue;
14
     mq_unlink(name);
16   memset((void*) &attr, 0, sizeof(struct mq_attr));
     attr.mq_maxmsg = MQ_CAPACITY;
18   attr.mq_msgsize = msg_len;
     queue = mq_open(name, O_RDWR | O_CREAT, S_IRUSR | S_IWUSR, &attr);
20   if (queue == -1)
       printf("Couldn't open queue: %s(%d)\n", strerror(errno), errno);
22   mq_close(queue);
  }
24
  int main(int argc, char **argv) {
26   char *args0[] = { "mqwriter", "0", NULL };
     char *args1[] = { "mqwriter", "1", NULL };
28   char *args2[] = { "mqreader", "2", NULL };
     struct inheritance inherit;
30   pid_t pid, pidwa, pidwb, pidr;
     int status;
32
     //Creating Message Queue
34   create_msg_queue(MQ_NAME, sizeof(myMSG_t));
36   pmsg[59] = '\n';
     memset(pmsg, ASCII_SPACE, PRT_STR_LEN-1);
38   memcpy(&pmsg[prt_index[0]], "Writer A", 8);
     memcpy(&pmsg[prt_index[1]], "Writer B", 8);
40   memcpy(&pmsg[prt_index[2]], "Reader", 6);
     memcpy(&pmsg[prt_index[3]], "MQ Status", 9);
42   fprintf(stdout, "%s", pmsg);
     memset(pmsg, Hyphen, PRT_STR_LEN-1);
44   fprintf(stdout, "%s", pmsg);
46   // create 3 child processes
     inherit.flags = 0;
48   pidwa = spawn(args0[0], 0, NULL, &inherit, args0, environ);
```

```
     pidwb = spawn(args1[0], 0, NULL, &inherit, args1, environ);
50   pidr =  spawn(args2[0], 0, NULL, &inherit, args2, environ);
     // spawn() is QNX function. Use posix_spawn() for portability
52   // posix_spawn(&pidwa, args0[0], NULL, NULL, args0, environ);
     // posix_spawn(&pidwb, args1[0], NULL, NULL, args1, environ);
54   // posix_spawn(&pidr,  args2[0], NULL, NULL, args2, environ);

56   while (1) {
       if ((pid = wait(&status)) == -1) {
58       perror("Starter: ");
         mq_unlink(MQ_NAME);
60       exit(EXIT_FAILURE);
       }
62     printf("User process %d terminated\n", pid);
     }
64 }
```

Listing 19.2
The starter program.

Listing 19.3 gives the code of a message queue reader, which, for each iteration of the loop, tries to receive a message from the queue and prints the queue status afterward.

```
   #include <stdlib.h>
2  #include <stdio.h>
   #include <mqueue.h>
4  #include "myheader.h"

6  int main(int argc, char *argv[]) {
     mqd_t myMQ;
8    myMSG_t *MQ_msg_ptr;

10     if (argc!=2){
         fprintf(stderr, "Use: mqreader index");
12       exit(EXIT_FAILURE);
       }
14     int myindex = atoi(argv[1]);
     pmsg[PRT_STR_LEN-1] = '\n';
16
     myMQ = mq_open(MQ_NAME, O_RDONLY);
18   mqstat = (struct mq_attr*) malloc(sizeof(struct mq_attr));
     memset(mqstat, 0, sizeof(struct mq_attr));
20   MQ_msg_ptr = (myMSG_t *) malloc(sizeof(myMSG_t));

22   while(1){
       sleep(2); // simulate task processing
```

```
24    if (MQ_size(myMQ)==0) //MQ is empty
        print_msg(myindex, ".", empty_mq);
26
      sleep(1); // simulate task processing
28    memset(MQ_msg_ptr, 0, sizeof(myMSG_t));
      mq_receive(myMQ, MQ_msg_ptr, sizeof(myMSG_t), NULL);
30
      long size = MQ_size(myMQ);
32    print_msg(myindex, MQ_msg_ptr->content, mq_status(size));
    }
34  mq_close(myMQ);
    return EXIT_SUCCESS;
36 }
```

Listing 19.3
Message queue reader.

Listing 19.4 gives the code of a message queue writer, which, for each iteration of the loop, sleeps for a random number of seconds, then tries to send a new message to the queue and prints the queue status afterward.

```
1 #include <stdlib.h>
  #include <stdio.h>
3 #include <mqueue.h>
  #include "myheader.h"
5
  void send_msg(char *sender, int sequence, char *msg, mqd_t _MQ){
7   myMSG_t *_msg = (myMSG_t *) malloc(sizeof(myMSG_t));
    memset(_msg, 0, sizeof(myMSG_t));
9   _msg->sequence = sequence;
    memcpy(_msg->content, msg, strlen(msg));
11  memcpy(_msg->sender, sender, strlen(sender));
    mq_send(_MQ, _msg, sizeof(myMSG_t), 0);
13  free(_msg);
  }
15
  int main(int argc, char *argv[]) {
17  mqd_t myMQ;
    static int seq=0;
19
    if (argc!=2){
21    fprintf(stderr, "Use: mqwriter index");
      exit(EXIT_FAILURE);
23    }
    int myindex = atoi(argv[1]);
25    char *sender = senders[myindex];
    pmsg[PRT_STR_LEN-1] = '\n';
```

```
27    myMQ = mq_open(MQ_NAME, O_WRONLY);
29    mqstat = (struct mq_attr*) malloc(sizeof(struct mq_attr));
      memset(mqstat, 0, sizeof(struct mq_attr));
31
      while(1){
33      int r = rand() % 3 + 1;
        sleep(r); // simulate task processing
35
        char *msg = (char *) malloc(10);
37      memset(msg, 0, 10);
        strcpy(msg, prefix[myindex]);
39      itoa(++seq, msg+strlen(msg), 10);
        strcat(msg, ")");
41
        if (MQ_size(myMQ)==MQ_CAPACITY) //MQ is full
43        print_msg(myindex, ".", full_mq);
        send_msg(sender, seq, msg, myMQ);
45      long size = MQ_size(myMQ);
        print_msg(myindex, msg, mq_status(size));
47    }
      mq_close(myMQ);
49    return EXIT_SUCCESS;
    }
```

Listing 19.4
Message queue writer.

Figure 19.12 shows the output of one run of the program:

(1) At the beginning, the message queue is empty, and Reader is blocked.
(2) Writer A sends a message to the queue. The message is denoted by (A, 1), where the letter indicates the sender's ID, and the number indicates the sequence number of the message. The message content is not printed.
(3) Writer B sends a message to the queue. The queue contains two messages now.
(4) Reader receives a message from the queue. The received message is from Writer A. The queue now has one message left.
(5) Writer A sends its second message to the queue.
(6) Writer B sends its second message to the queue.
(7) Writer A sends its third message to the queue.
(8) Writer B sends its third message to the queue. The queue becomes full.
(9) Reader receives a message from the queue. The received message is from Writer B. The queue now has one space available.
(10) Writer A sends its fourth message to the queue. The queue becomes full.
(11) Writer B sends its fourth message to the queue and is blocked.

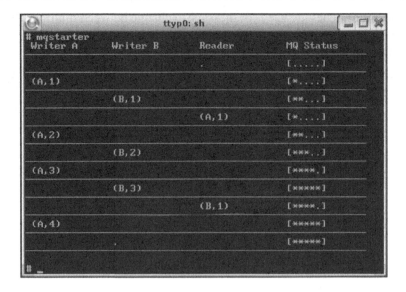

Figure 19.12
Output of mq reader/writer.

Problems

19.1 What is synchronous intertask communication? What is asynchronous intertask communication? What are the pros and cons of asynchronous intertask communication?

19.2 What is a POSIX message queue? Why does a message queue need two waiting lists?

19.3 Describe the state transitions of a message queue.

19.4 What happens when a task attempts to receive a message from an empty message queue?

19.5 What happens when a task attempts to send a message to a full message queue?

19.6 Refer to the design in Figure 19.7, and suppose the tasks T_1 and T_2 use the code sequences below to access the message queues:

```
T1()
{ ...
  mq_send(mq2, msg_s11);
  mq_send(mq2, msg_s12);
  mq_receive(mq1, msg_r1);
  ...
}
T2()
{ ...
  mq_send(mq1, msg_s2);
  mq_receive(mq2, msg_r21);
  mq_receive(mq2, msg_r22);
  ...
}
```

Analyze the communication behavior of the two tasks.

19.7 When message queues are used for message exchange between multiple clients and a server, what is the problem if a message queue is reserved for each sender?

19.8 Explain the notification mechanism featured by a POSIX message queue.

19.9 Implement the client-server design illustrated in Figure 19.10. You should use a starter program to start one server task and four client tasks.

Intertask Communication: Pipe

Nobody can go back and start a new beginning, but anyone can start today and make a new ending.

Maria Robinson

20.1 Introduction to Pipes

Similarly to message queues, pipes offer another mechanism for intertask communication. However, message queues and pipes differ in several aspects:

- A message queue is a container of multiple messages, each of which can be manipulated by a task separately. A pipe is a container of unstructured bytes.
- Each message in a message queue is associated with a priority; consequently, messages can be processed by tasks in priority order. On the other hand, a pipe is a strictly FIFO stream of bytes, and the notion of priority is not applicable.
- A message queue is a *named kernel object*, with a name uniquely defined in the kernel's namespace. On the other hand, a pipe is a kernel object that can be a named pipe or an unnamed pipe.

A named pipe has a unique name and appears in the file system just like a file. It is typically used between processes that are not necessarily related and do not need to coexist at the same time. A named pipe is a "persistent" file object; it can be referenced by name as long as the file is not deleted.

Real-Time Embedded Systems. http://dx.doi.org/10.1016/B978-0-12-801507-0.00020-1

An unnamed or anonymous pipe does not have a name and does not appear in the file system. It is typically used between processes that are related either by a parent-child relationship (say, by fork or posix_spawn) or by a sibling relationship (i.e., child processes of the same parent that defines the pipe). An unnamed pipe exists as long as it is not closed by the processes that use it. Most operating systems implement a shell utility "|," which creates an unnamed pipe between shell commands. For example, by "ls | more," the output of the "ls" command is piped to the "more" command, which shows one page at a time.

20.2 Pipe Statics and Dynamics

Figure 20.1 gives the pipe control structure:

- A pipe is a kernel object with limited capacity indicated by *pipe_length*. Different operating system implementations may have different values for *pipe_length*.
- A pipe has two ends, with data flowing from one end to the other. One end is for reading data from the pipe, referenced by a unique file descriptor that can be opened for reading only; the other end is for writing data to the pipe, referenced by a unique file descriptor that can be opened for writing only.
- Like normal file descriptors, both descriptors of a pipe can be configured to enable or disable *nonblocking access*.
- When the read end of a pipe disables nonblocking access, a task is blocked if it attempts to read from the pipe when it is empty. A blocked reader is placed on the *waiting_readers* list, waiting for some data to be written by other tasks.

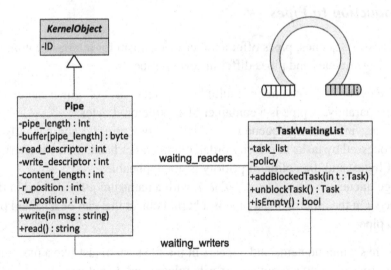

Figure 20.1
UML static structure of a pipe and its notation.

- When the write end of a pipe disables nonblocking access, a task is blocked if it attempts to write to a pipe which has less space than requested. A blocked writer is placed on the *waiting_writers* list, waiting for more space to become available.
- A pipe has two indexes: *r_position* indicates the beginning of the pipe and *w_position* indicates the end of the pipe. Each read request starts from *r_position*, whereas each write request appends data after *w_position*. The amount of readable data, indicated by *content_length*, equals the difference between *w_position* and *r_position*.

Figure 20.2 gives the dynamic state transitions of a pipe with nonblocking access enabled:

- Upon creation, a pipe is in the "Empty" state, with the state variable *content_length* set to 0, and *freespace* set to the pipe capacity.
- When a task attempts to read data from an empty pipe, the pipe remains unchanged.
- When a task attempts to write data of length *len* to an empty pipe, it transitions to the "NonEmpty" state, and
  - if the available space is no less than *len*, the data are completely written to the pipe, and the state variables are updated accordingly;
  - if the available space is less than *len*, the pipe is filled with the first portion of the data, and the state variables are updated accordingly.
- When a task attempts to write data of length *len* to a nonempty pipe,
  - if the available space is no less than *len*, the data are completely written to the pipe, and the state variables are updated accordingly;
  - if the available space is less than *len* and *len* is no more than the pipe capacity, no data are written to the pipe (the writer should try again, if desired);
  - if *len* is more than the pipe capacity, the space remaining in the pipe is filled with the first portion of the data, and the state variables are updated accordingly.
- When a task attempts to read data of length *len* from a nonempty pipe,

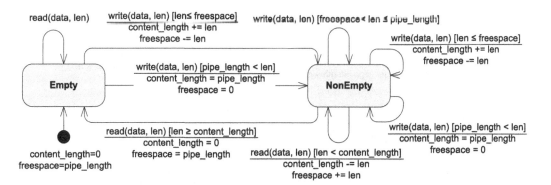

Figure 20.2
UML state model of pipes: nonblocking access.

- if the readable data are more than *len*, *len* bytes of data are removed from the pipe, and the state variables are updated accordingly;
- if the readable data are less than or equal to *len*, all the pipe data are removed, and the pipe transitions to the "Empty" state.

Figure 20.3 gives the dynamic state transitions of a pipe which disables nonblocking access. As far as the pipe transition behavior is concerned, Figure 20.3 differs from Figure 20.2 in two transitions:

(1)   When a task attempts to write data of length *len* to an empty pipe, it remains in the "Empty" state if *len* is greater than the pipe capacity.
(2)   When a task attempts to write data of length *len* to a nonempty pipe, it remains in the "NonEmpty" state if *len* is greater than the pipe capacity.

In other words, regardless of whether the pipe is empty or not, a blocking write request that is above the pipe capacity will not affect the pipe and should be avoided.

20.3 Pipe Usage Patterns

Figure 20.4a illustrates a design where data can flow from one task to another via a pipe. We refer to this design as a *unidirectional piping pattern*. There are three points to remember:

(1)   Given that the "nonblocking access" flag is not set, the intertask communication between task A and task B is asynchronous only if the pipe is neither empty nor full. Task A is synchronized with task B when the pipe becomes full, and task B is synchronized with task A when the pipe becomes empty.
(2)   Since task A is the only writer and task B is the only reader, and they operate respectively at different ends of the pipe, it is not necessary for the two tasks to employ a resource

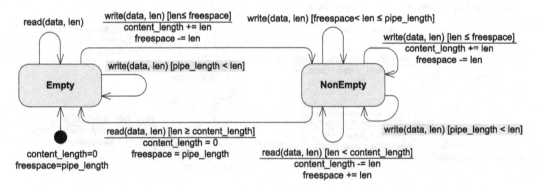

Figure 20.3
UML state model of pipes: blocking access.

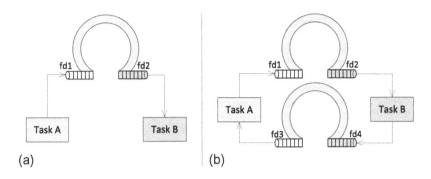

Figure 20.4
Information exchange using pipes. (a) Unidirectional piping pattern. (b) Bidirectional piping pattern.

lock to protect the data integrity. Task B, if desired, can always reassemble the consecutively received streams of bytes together in the right order.

(3) Since the pipe contains unstructured data, how task B makes sense of the received data is application specific.

Figure 20.4b illustrates a design where data can flow back and forth between two tasks via two pipes. We refer to this design as a *bidirectional piping pattern*. When the two pipes enable nonblocking access, this design allows two tasks to exchange application data asynchronously.

20.4 POSIX Functions for Pipes

Table 20.1 lists a few POSIX functions for pipes.

An unnamed pipe can be created by pipe() or popen(), and a named pipe is created by open(). Each end of a pipe has a flag O_NONBLOCK, which can be set or cleared by the function fcntl() to enable or disable nonblocking access.

The function read() is the input operation on pipes (and ordinary files). Via a call to *read(fd, buf, n)*, a task attempts to read *n* bytes from the file associated with the file descriptor *fd* into the buffer pointed to by *buf*.

A read operation is *atomic* if the sequence of bytes *expected to return* when the read operation started is exactly the same as the sequence of bytes *actually returned* when the read operation terminates. The POSIX function read() is not always atomic, because the number of bytes returned might be less than *n* if the read operation was interrupted by a signal, or if there were fewer than *n* bytes in the pipe.

Table 20.1 POSIX.1 pipe function calls

POSIX Function	Description
pipe	Create an unnamed pipe. It returns a file descriptor for the read end and a file descriptor for the write end
popen	Execute a specified command and creates a pipe between this calling process and the executed command
pclose	Close the pipe created by popen
open	Create a named pipe similar to creating a file
fcntl	Get/set the flags and access modes of an open file. In particular, the nonblocking access mode, O_NONBLOCK, can be set or cleared
read	Read a certain number of bytes from a file descriptor
write	Write a certain number of bytes to a file descriptor
select	Examine the given file descriptor sets to see whether some are ready for reading, are ready for writing, or have an exceptional condition pending

When a pipe is open with both reader and writer tasks, if the pipe is empty and the flag O_NONBLOCK of the pipe's read end is clear, a reader task will be blocked until some data are written or the pipe is closed by all the writer tasks.

The function write() is the output operation on pipes (and ordinary files). Via a call to $write(fd, buf, n)$, a task attempts to write n bytes from the buffer pointed to by buf to the file associated with the file descriptor fd.

A write operation is *atomic* if, when the write operation terminates, the pipe ends up with an exact copy of the sequence of bytes in buf with the original byte order preserved. In other words, all the bytes from a single write operation that started out together should end up together in the pipe, without interleaving from other operations.

The POSIX.1-2008 standard defines a maximum value PIPE_BUF (i.e. the pipe_length in Figure 20.1), and requires that write operations of PIPE_BUF or fewer bytes should be atomic in a POSIX-compliant operating system implementation.

Table 20.2 summarizes the behavior of the pipe write operation:

- Blocking is possible only when the flag O_NONBLOCK is clear.
- A write is atomic if there is enough space in the pipe for all the data requested to be written immediately.
- When $len \leq PIPE\_BUF$, partial write is not allowed for the sake of atomicity.
- When $len > PIPE\_BUF$ and the NON_BLOCK flag is set, partial write is allowed so that each write can make progress if there is any room in the pipe. In such a situation, the write operation is *not* atomic. The remaining data, if desired, can be pushed to the pipe by successive write operations.

Table 20.2 A pipe write operation: $r = write(wd, data, len)$

Options		α: available space in $pipe(rd, wd)$		
		$\alpha = 0$	$\alpha < len$	$\alpha \geq len$
O_NONBLOCK clear	$len \leq PIPE\_BUF$	Atomic, blocking	Atomic, blocking	Atomic, $r = len$
	$len > PIPE\_BUF$	Blocking	Blocking	n/a
O_NONBLOCK set	$len \leq PIPE\_BUF$	$r = -1$	$r = -1$	Atomic, $r = len$
	$len > PIPE\_BUF$	$r = -1$	$r = \alpha$, nonatomic	n/a

n/a, not applicable.

In user applications, it is good practice to set the upper bound of pipe I/O operations to be lower than PIPE_BUF.

20.4.1 Multiple Writers and Readers

We first consider Figure 20.5, where there are two writers sending data to a single reader via a pipe.

Concurrent access to a pipe is safe when all the writer tasks issue only blocking write operations: such a write is atomic in the sense that the whole amount written in one operation is not interleaved with data from any other operation.

Concurrent access to a pipe is *not* safe when at least one writer issues nonblocking write requests that are above the pipe capacity (PIPE_BUF). As shown in Table 20.2, such a write operation is not atomic and may end up with "partial write": only the first portion of the data is written into the pipe. Although the writer task could issue another write request to send the leftover data, owing to concurrency, the data from this writer could be separated by data from other writers. Figure 20.6 illustrates such a situation where the data from one writer task is interleaved with data from another writer task. Consequently, the integrity of the data from the former writer task has been violated.

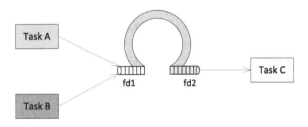

Figure 20.5
A pipe with multiple writers.

One solution is to surround each write operation that is above the pipe capacity with a resource lock (e.g., a semaphore or mutex), and ensure that a writer releases a lock only after the whole data object has been completely written to the pipe.

Now, let us consider the pipe read operation. Read operations on a pipe could segment the pipe data at arbitrary positions. This is easy to handle when there is only one reader, because the data integrity can be recovered by merging the pieces together.

However, when there are several concurrent readers as shown in Figure 20.7, multiple pieces of a whole data object might be scattered among two or more readers.[1] One solution is to fix the length of the data objects to be transmitted over a pipe to be an integer fraction of the pipe capacity, and make sure that all the readers and writer(s) of the pipe strictly use the fixed length to read or write data.

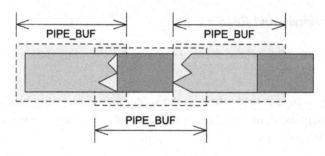

Figure 20.6
Pipe data integrity is violated.

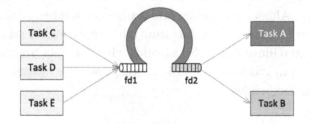

Figure 20.7
A pipe with multiple readers.

---

[1] If desired, this feature could be used to produce random behaviors.

20.4.2 POSIX Select Operation on Pipes

Since each end of a pipe is simply a file descriptor, the POSIX function select()[2] allows a task to block and wait for specific conditions to occur on one or more pipes. The select function and some macros are defined as follows:

```
int select   (int nfds,
              fd_set *restrict readfds,
              fd_set *restrict writefds,
              fd_set *restrict errorfds,
              struct timeval *restrict timeout);
void FD_CLR  (int fd, fd_set *fdset);
int  FD_ISSET(int fd, fd_set *fdset);
void FD_SET  (int fd, fd_set *fdset);
void FD_ZERO (fd_set *fdset);
```

The select operation examines the first *nfds* file descriptors (i.e., from zero to *nfds* − 1), and blocks until

- the time-out occurs;
- it is interrupted by a signal;
- some of the descriptors in *readfds* (and within the range *nfds*) become ready for reading;
- some of the descriptors in *writefds* (and within the range *nfds*) become ready for writing; or
- some of the descriptors in *errorfds* (and within the range *nfds*) have error conditions pending.

Upon successful completion, the select() function modifies the descriptor sets to indicate which file descriptors are ready for reading, are ready for writing, or have an error condition pending, respectively, and returns the total number of ready descriptors in all the sets.

The macro FD_CLR(*fd*, *fdset*) removes *fd* from the descriptor set pointed to by *fdset*. FD_ISSET(*fd*, *fdset*) returns a nonzero value if *fd* is a member of the set pointed to by *fdset*. FD_SET(*fd*, *fdset*) adds *fd* to the set pointed to by *fdset*. FD_ZERO(*fdset*) initializes the descriptor set pointed to by *fdset* to a null set.

Figure 20.8 shows a design with two tasks communicating via four pipes: task A writes to pipes 1 and 2, and reads from pipes 3 and 4; task B writes to pipes 3 and 4, and reads from pipes 1 and 2.

A block of pseudocode for task A is given below. The select operation allows task A to simultaneously wait for any of the following conditions to happen: (a) either pipe 1 or pipe 2

---

[2] The select operation can be generally used to wait for I/O from multiple servers.

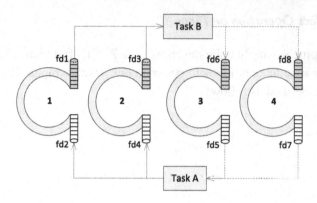

Figure 20.8
Select operation on pipes.

is ready for writing, (b) either pipe 3 or pipe 4 is ready for reading, or (c) an error occurs to one of the four pipes.

```
FD_ZERO (readfds);
FD_SET  (fd5, readfds);
FD_SET  (fd7, readfds);
FD_ZERO (writefds);
FD_SET  (fd2, writefds);
FD_SET  (fd4, writefds);
FD_ZERO (errorfds);
FD_SET  (fd2, errorfds);
FD_SET  (fd4, errorfds);
FD_SET  (fd5, errorfds);
FD_SET  (fd7, errorfds);
nfds = max(fd2, fd4, fd5, fd7)+1;
n = select(nfds,readfds, writefds, errorfds, null);
```

The above select operation blocks, allowing task A to simultaneously wait for three conditions to happen: (1) data become available in either pipe 3 or pipe 4; (2) space becomes available in either pipe 1 or pipe 2; or (3) an error occurs to any of the pipe ends that task A is waiting for.

20.5 An Example of Using Pipes

As an example, the code in Listing 20.1 implements the design given in Figure 20.8.

The function reset_FD_SETS (from line 23 to line 46) resets the read and write file descriptor sets for tasks A and B. The functions pipe_right() and pipe_left() are utilities for generating formatted output strings. Task A calls writechar() to write characters to pipe 1 or pipe 2, and task B calls writenum() to write digital numbers to pipe 3 or pipe 4. The functions print_msg() and print_header() are utilities for printing formatted strings to the console.

In the main() function, from line 130 to line 142, four unnamed pipes are created and the file descriptors of each pipe are maintained by an entry in the array mypipe. The POSIX function fork() is used to create another process: the child process is task B and the parent process is task A. The fork function returns a nonzero process ID to the parent process, which is used in the code to distinguish the two blocks of code to be executed respectively by the parent and child processes.

```
1  #include <stdio.h>
   #include <stdlib.h>
3  #include <spawn.h>
   #include <unistd.h>
5  #include <fcntl.h>
   #include <sys/wait.h>
7  #include <sys/select.h>
   #include <errno.h>
9
   #define PRT_STR_LEN 90
11 #define ASCII_SPACE 32
   #define Hyphen 45
13 struct pipe_t {
     int rfd;
15   int wfd;
   };
17 struct pipe_t mypipe[4];
   fd_set a_rfd, a_wfd, b_rfd, b_wfd;
19 char ch = 'A', num = '0';
   char pmsg[PRT_STR_LEN];
21 int prt_index[] = { 1, 15, 30, 45, 60, 75 };

23 void reset_FD_SETS(int cpid) {
     int i;
25   if (cpid == 0) { //child process B
       FD_ZERO( &b_rfd );
27     FD_ZERO( &b_wfd );
       for (i = 0; i < 4; i++) {
29       if (i < 2)
           FD_SET( mypipe[i].rfd, &b_rfd );
31       else
           FD_SET( mypipe[i].wfd, &b_wfd );
33     }
     } else { //parent process A
35     FD_ZERO( &a_rfd );
       FD_ZERO( &a_wfd );
37     for (i = 0; i < 4; i++) {
         if (i < 2)
39         FD_SET( mypipe[i].wfd, &a_wfd );
         else
41         FD_SET( mypipe[i].rfd, &a_rfd );
```

```
            }
43      }
        memset(pmsg, Hyphen, PRT_STR_LEN-1);
45      fprintf(stdout, "%s", pmsg);
      }
47  char *pipe_right(char ch) {
      char *str = (char *) malloc(4);
49    memset(str, 0, 4);
      strcpy(str, ">");
51    memset(str + 1, ch, 1);
      strcat(str, ">");
53    return str;
      }
55  char *pipe_left(char ch) {
      char *str = (char *) malloc(4);
57    memset(str, 0, 4);
      strcpy(str, "<");
59    memset(str + 1, ch, 1);
      strcat(str, "<");
61    return str;
      }
63
    void writechar(int mindex, int pindex) {
65    int result;
      result = write(mypipe[pindex].wfd, &ch, 1);
67    if (result != 1) {
        perror("write");
69      exit(2);
      }
71    if (mindex == 0)
        print_msg(mindex, pindex + 1, "Write", pipe_right(ch));
73    else
        print_msg(mindex, pindex + 1, "Write", pipe_left(ch));
75    if (ch == 'J') ch = 'A' - 1;
      ch++;
77  }

79  void writenum(int mindex, int pindex) {
      int result;
81    result = write(mypipe[pindex].wfd, &num, 1);
      if (result != 1) {
83      perror("write");
        exit(2);
85    }
      if (mindex == 0)
87      print_msg(mindex, pindex + 1, "Write", pipe_right(num));
      else
89      print_msg(mindex, pindex + 1, "Write", pipe_left(num));
      if (num == '9') num = '0' - 1;
91    num++;
```

```
   }
93
   void readchar(int mindex, int pindex) {
95   char ch;
     int result;
97   result = read(mypipe[pindex].rfd, &ch, 1);
     if (result != 1) {
99     perror("read");
       exit(3);
101   }
     if (mindex == 0)
103     print_msg(mindex, pindex + 1, "Read", pipe_left(ch));
     else
105     print_msg(mindex, pindex + 1, "Read", pipe_right(ch));
   }
107 void print_msg(int myindex, int pindex, char *msg1, char *msg2) {
     memset(pmsg, ASCII_SPACE, PRT_STR_LEN - 1);
109   pmsg[PRT_STR_LEN - 1] = '\n';
     memcpy(&pmsg[prt_index[myindex]], msg1, strlen(msg1));
111   memcpy(&pmsg[prt_index[pindex]], msg2, strlen(msg2));
     fprintf(stdout, "%s", pmsg);
113 }
   void print_header(){
115   memset(pmsg, ASCII_SPACE, PRT_STR_LEN-1);
     memcpy(&pmsg[prt_index[0]], "Process A", 9);
117   memcpy(&pmsg[prt_index[1]], "Pipe 1", 6);
     memcpy(&pmsg[prt_index[2]], "Pipe 2", 6);
119   memcpy(&pmsg[prt_index[3]], "Pipe 3", 6);
     memcpy(&pmsg[prt_index[4]], "Pipe 4", 6);
121   memcpy(&pmsg[prt_index[5]], "Process B", 9);
     fprintf(stdout, "%s", pmsg);
123 }
   int main(int argc, char **argv) {
125   int i, n, result, cpid, maxfd=0;
     int fd[2];
127   pmsg[PRT_STR_LEN - 1] = '\n';

129   /* create the pipes */
     for (i = 0; i < 4; i++) {
131     result = pipe(fd);
       if (result < 0) {
133       perror("pipe ");
         exit(1);
135     }
       mypipe[i].rfd = fd[0];
137     mypipe[i].wfd = fd[1];
       if (mypipe[i].rfd > maxfd)
139       maxfd = mypipe[i].rfd;
       if (mypipe[i].wfd > maxfd)
141       maxfd = mypipe[i].wfd;
```

```
      }
143   //fork() returns 0 to the child (new) process,
      //and returns the PID of the child to the parent process.
145   cpid = fork();
      if (cpid == -1) {
147     perror("fork");
        exit(EXIT_FAILURE);
149   }
      if (cpid == 0) { //this is the child process B
151     sleep(1);
        reset_FD_SETS(cpid);
153     while ((n = select(maxfd + 1, &b_rfd, &b_wfd, 0, 0)) > 0) {
          if (FD_ISSET( mypipe[0].rfd, &b_rfd ))
155         readchar(5, 0);
          if (FD_ISSET( mypipe[1].rfd, &b_rfd ))
157         readchar(5, 1);
          if (FD_ISSET( mypipe[2].wfd, &b_wfd ))
159         writenum(5, 2);
          if (FD_ISSET( mypipe[3].wfd, &b_wfd ))
161         writenum(5, 3);
          reset_FD_SETS(cpid);
163       sleep(2);
        }
165     _exit(EXIT_SUCCESS);
      } else { //this is the parent process A
167     reset_FD_SETS(cpid);
        print_header();
169     sleep(2);
        while ((n = select(maxfd + 1, &a_rfd, &a_wfd, 0, 0)) > 0) {
171       if (FD_ISSET( mypipe[0].wfd, &a_wfd ))
            writechar(0, 0);
173       if (FD_ISSET( mypipe[1].wfd, &a_wfd ))
            writechar(0, 1);
175       if (FD_ISSET( mypipe[2].rfd, &a_rfd ))
            readchar(0, 2);
177       if (FD_ISSET( mypipe[3].rfd, &a_rfd ))
            readchar(0, 3);
179       reset_FD_SETS(cpid);
          sleep(2);
181     }
      wait(NULL); /* Wait for child */
183   exit(EXIT_SUCCESS);
      }
185 }
```

Listing 20.1
Sample code of creating a named semaphore.

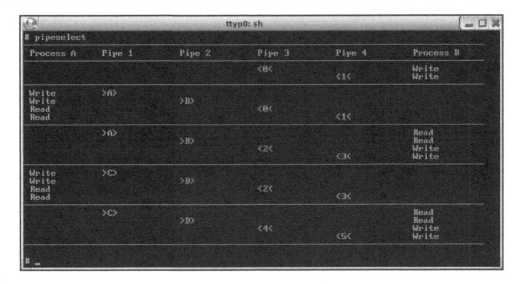

Figure 20.9
Output of the example program.

Figure 20.9 gives a screenshot of a run of the program, showing how the two tasks exchange data via the four pipes.

Problems

20.1 What are the differences between message queues and pipes? When should you use one over the other?

20.2 What are the differences between named pipes and unnamed pipes? When should you use one over the other?

20.3 When nonblocking access is enabled for a pipe, what happens if you try to write data that exceed the pipe capacity?

20.4 When nonblocking access is disabled for a pipe, what happens if you try to write data that exceed the pipe capacity?

20.5 Refer to the bidirectional piping pattern illustrated in Figure 20.3b. If the two pipes disable nonblocking access, would the design allow the two tasks to exchange application data asynchronously?

20.6 Is the POSIX read() operation atomic? Is the POSIX write() operation atomic? Does POSIX allow "partial write" when PIPE_BUF is no less than the data to be written? Why?

20.7 When there are multiple writers to a pipe, is it possible that the integrity of the pipe data can be violated? If so, how can one protect the data integrity?

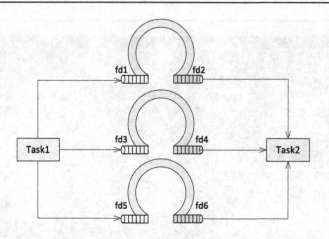

Figure 20.10
Select operation on pipes.

20.8   Refer to Figure 20.8, and write a pseudocode for task B such that it can simultaneously wait until any of the following conditions becomes true: (a) either pipe 1 or pipe 2 is ready for reading, (b) either pipe 3 or pipe 4 is ready for writing, or (c) 100 ms has passed.

20.9   Refer to Figure 20.10, and (a) write a pseudocode for task 1 such that it can simultaneously wait for the three pipes to become ready for writing, and (b) write a pseudocode for task 2 such that it can simultaneously wait for the three pipes to become ready for reading.

20.10   There is a loop in lines 170-181 in Listing 20.1. This loop is used by task A to read from or write to pipes. Why is it necessary to call the function reset_FD_SETS() in each iteration of the loop?

Intertask Communication: Signaling

Contents

No one can whistle a symphony. It takes a whole orchestra to play it.

H.E. Luccock

21.1 Introduction to POSIX Signals

Signaling is an asynchronous interprocess communication mechanism in POSIX-compliant operating systems (OSs). Owing to the asynchronous nature of signals, a task can perform useful work while waiting for a significant event; when the event occurs, the task is notified immediately.

Real-Time Embedded Systems. http://dx.doi.org/10.1016/B978-0-12-801507-0.00021-3

A POSIX process can be viewed as a *virtual machine* with a group of citizens (i.e., threads) living peacefully in a *protected memory space*. The signal concept is then analogous to a *virtual interrupt* to such a virtual machine. Just like hardware interrupts, signals are delivered to a process asynchronously: while a task can specify the particular actions to execute when a signal arrives, the task cannot predict when a signal might arrive.

Table 21.1 gives a list of signals supported in the IEEE POSIX 1003.1-2008 standard [17], where each signal name is defined as a macro (an integer expression) that yields an integer constant referred to as the number of that signal.

Table 21.1 Signals defined in POSIX.1-2008 (IEEE Std 1003.1-2008)

Signal Name	Default Action	Description
SIGABRT	Abort	Process abort signal
SIGALRM	Terminate	Alarm clock
SIGBUS	Abort	Access to an undefined portion of a memory object
SIGCHLD	Ignore	Child process terminated, stopped or continued
SIGCONT	Continue	Continue executing, if stopped
SIGFPE	Abort	Erroneous arithmetic operation
SIGHUP	Terminate	Hang up
SIGILL	Abort	Illegal instruction
SIGINT	Terminate	Terminal interrupt signal
SIGKILL	Terminate	Kill (cannot be caught or ignored)
SIGPIPE	Terminate	Write on a pipe with no one to read it
SIGQUIT	Abort	Terminal quit signal
SIGSEGV	Abort	Invalid memory reference
SIGSTOP	Stop	Stop executing (cannot be caught or ignored)
SIGTERM	Terminate	Termination signal
SIGTSTP	Stop	Terminal stop signal
SIGTTIN	Stop	Background process attempting read
SIGTTOU	Stop	Background process attempting write
SIGUSR1	Terminate	User-defined signal 1
SIGUSR2	Terminate	User-defined signal 2
SIGPOLL	Terminate	Pollable event
SIGPROF	Terminate	Profiling timer expired
SIGSYS	Abort	Bad system call
SIGTRAP	Abort	Trace/breakpoint trap
SIGURG	Ignore	High bandwidth data are available at a socket
SIGVTALRM	Terminate	Virtual timer expired
SIGXCPU	Abort	CPU time limit exceeded
SIGXFSZ	Abort	File size limit exceeded
SIGRTMIN	Terminate	A macro indicating the first real-time signal
SIGRTMAX	Terminate	A macro indicating the last real-time signal

An OS signal can be generated for various reasons. A few causes are listed below:

- System errors. For instance, if a task attempted to divide by zero on an x86 CPU, an exception would be generated. In response, the OS kernel would send the SIGFPE signal to the task. Similarly, if a task attempted to access a memory address outside its virtual memory space, the OS kernel would notify this task via a SIGSEGV signal. Both signals will cause the running program to exit.
- External hardware events. An interrupt service routine (ISR) is responsible for immediately responding to the interrupt events from a certain hardware device. However, it is not uncommon that at the end of an ISR a particular signal is sent out to notify a driver (server) task for further processing. For instance, the key combination Ctrl-C at the controlling terminal of a running process causes the OS to send a SIGINT signal, which typically causes the running process to terminate. The user-defined signals (SIGUSR1, SIGUSR2) can also be used to notify a server about the arrival of input at a serial communication port.
- External software events. For instance, when a task writes to a pipe which has been closed by the reader, the OS kernel typically raises a SIGPIPE signal to notify the writer process. An OS raises a SIGCHLD signal to notify a process about the termination of a child process. A timer manager might use a specific signal to notify its clients upon the expiration of a timer.
- Explicit requests. A task can explicitly invoke some system calls (e.g., kill and sigqueue) to send a signal to itself or other tasks.

Some signals, such as SIGSEGV, indicate fatal errors. These signals will kill the program immediately if no other way has been specified in advance to handle the signal. By default, when an OS timer expires, it sends a signal of type SIGALRM to the process that has created this timer object. If desired, a timer can be configured to raise signals of other types—say, SIGUSR1. Indeed, some signals such as SIGALRM and SIGUSR1 are often used to implement event-oriented services in user applications.

21.2 Signal Handling

Signaling mechanisms may differ in their details on different OSs. In general, signals can be ignored, blocked/unblocked, caught and handled.

First, signals can be ignored, and ignored signals are maintained at the process level. In other words, if a thread ignores or catches a signal, it affects all threads within the process. An ignored signal is discarded immediately, having no effects on the execution of the process and its threads.

A signal can be blocked (masked). Each thread has its own signal mask. A signal is blocked if the appropriate bit in the mask is set to 1. If a thread blocks a signal, it affects only that thread.

For a particular signal in a multithreaded process, it is good practice to have the signal blocked in all threads but one, which is dedicated to handling it.

A process and each of its threads have a separate signal queue. When a signal arrives, it is acted upon immediately unless the signal is currently blocked or another signal is being handled. In such a case, it becomes pending and is put into an appropriate signal queue. When a pending signal is unmasked, it is acted upon immediately.

Signals can be targeted at a process or at a specific thread:

- An unignored signal targeted at a thread will be delivered to that thread alone.
- An unignored signal targeted at a process is delivered to the first thread that does not have the signal blocked. If all threads have the signal blocked, the signal will be queued on the process until any thread unblocks (unmasks) the signal. When this happens, the signal will be moved from the process to the thread that has just unblocked it.

When a task receives a signal, its normal flow of execution is interrupted by the OS,[1] and the control is transferred to the handler preinstalled for that signal.

21.3 Signal Vector Table and Handlers

Signals are the software equivalent of hardware interrupts. Just like interrupt vector numbers, an OS typically uses signal numbers to identify the supported signal types. The ISRs for hardware interrupts, once installed, apply to the whole system. In contrast, most OSs allow each process to have its own signal handlers installed. In such a case, an OS kernel needs to maintain a signal vector table for each process.[2] A *signal vector table* is indexed by signal numbers, each entry of which is simply a pointer to the corresponding signal handler.

The OS kernel typically installs a default handler for each supported signal. As indicated in Table 21.1, the default action of most signals is to terminate the process. If this is not desirable, an application should install user-defined handlers to deal with signals properly.

Many restrictions for coding ISRs are also applicable to signal handlers. For instance, immediately before a signal handler is entered, the signal under consideration has to be masked. This will prevent nested signals of the same type from interrupting the execution of the handler. If the handler returns normally, the OS will automatically restore the signal mask to its original form (before the handler was called). It is also good practice for a task to save the original handler before installing a new one and restore the original handler after the task has finished.

---

[1] Execution can be interrupted during any nonatomic instruction.
[2] Note that signal handlers are attached to processes, not threads.

21.4 POSIX Signal Functions

Table 21.2 gives a list of POSIX functions defined for signal handling.

As an example, the function sigaction is used to examine or specify the action that is associated with a specific signal. It has the following signature:

```
int sigaction( int sig, const struct sigaction * act, struct sigaction * oact),
```

where *sig* is the signal number under consideration, *act* is a pointer to a sigaction structure that specifies how you want to modify the action for the given signal and *oact* is a pointer to a sigaction structure that the function fills with information about the current action for the signal. The sigaction structure has a member named sa_handler (or sa_sigaction) that can take a value of

- Function to install a new signal handler for the signal;
- SIG_DFL to set the signal with the OS default action;

Table 21.2 Some POSIX.1 function calls for signals

POSIX Function	Description
kill	Send a signal to a process
pthread_kill	Send a signal to a specific thread
pthread_sigmask	Examine and/or change the calling thread's signal mask
raise	Send a signal to the executing process. It returns after the signal handler has run to completion
sigaction	Examine and/or change the action to be associated with a specific signal
sigaddset	Add a signal to a signal set
sigdelset	Remove a signal from a signal set
sigemptyset	Create an empty signal set
sigfillset	Create a set with all the POSIX signals
sigismember	Test whether a signal is a member of a signal set
signal	Set a default or user-provided handler for a signal, or set the calling process to ignore a signal
sigpause	Remove a signal from the signal mask and suspend the calling process until a signal is received
sigpending	Examine pending signals
sigprocmask	Examine and/or change the calling process's signal mask
sigqueue	Queue a signal to a process
sigsetjmp	Set a jump point, saving the calling environment as well as the current signal mask
siglongjmp	Restore the environment as well as the signal mask saved by the most recent invocation of sigsetjmp in the same process
sigsuspend	Wait for a signal other than those signals being masked
sigwait	Wait for a specified signal
sigtimedwait	Wait for a specified signal for the specified amount of time
sigwaitinfo	Wait for a specified signal (and related information)

- SIG_IGN to set the process to ignore the signal. All pending and new signals of this type are discarded.

21.5 QNX Implementation of POSIX Signals

QNX supports the standard POSIX signals. The entire range of signals in QNX goes from _SIGMIN (1) to _SIGMAX (64). Signals are delivered in priority order. For a signal with a signal number sn, its priority is derived by ($\_SIGMAX - sn$). In other words, a lower signal number has a higher priority.

Each signal in QNX can also have an associated eight-bit code and a 32-bit value to carry application-specific information.

21.5.1 Example: Handling Signals in Different Processes

Signals can be used by a process to notify other processes that some special events have just occurred. In response, the normal execution of the signaled process is interrupted and preempted by the corresponding signal handler.

Listings 21.1–21.3 give an example, where the same signals are handled differently by the two signaled processes.

Within the loop of the signaling process given in Listing 21.1, a SIGUSR1 signal is sent to two client processes. Note that the value associated with the signal sent to the first process is 1, which is different from the value associated with the signal sent to the second process.

```
1  #include <stdio.h>
   #include <stdlib.h>
3  #include <signal.h>
   #include <sys/siginfo.h>
5  #include <unistd.h>
   #include <time.h>
7
   int main( int argc, char *argv[] )
9  {
     int pid1, pid2, r;
11     struct sigevent event1, event2;

13     if (argc!=3){
       fprintf(stderr, "Use: signaler pid1 pid2");
15       exit(EXIT_FAILURE);
     }
17     SIGEV_SIGNAL_VALUE_INIT( &event1, SIGUSR1, 1 );
     SIGEV_SIGNAL_VALUE_INIT( &event2, SIGUSR1, 0 );
19     pid1 = atoi(argv[1]);
     pid2 = atoi(argv[2]);
```

```
21    srand( time( NULL ) );

23    do {
        r = rand() % 3;
25      if (r == 2){
          fprintf(stderr, "*");
27
          //send a signal
29        //kill( pid1, SIGUSR1 );
          //kill( pid2, SIGUSR1 );
31        sigqueue(pid1, SIGUSR1, event1.sigev_value );
          sigqueue(pid2, SIGUSR1, event2.sigev_value );
33      }
        sleep (1);
35    } while( r != 3 ); /* end do...while */

37    return EXIT_SUCCESS;
    }
```

Listing 21.1

Signaling process: generating SIGUSR1 signals.

Listing 21.2 gives the code for process 1, which, every time upon being signaled prints the integer value "1" associated with the received signal event. Listing 21.3 gives the code for process 2, which prints "_0_" every time upon being signaled.

```
1 #include <stdio.h>
  #include <stdlib.h>
3 #include <signal.h>
  #include <unistd.h>
5
  int main( void )
7 {
      extern void handler(int, siginfo_t*, void*);
9     struct sigaction act;
      sigset_t set;
11

13    printf("%d\n", getpid());
      /*
15     * Define a handler for SIGUSR1.
       */
17    sigemptyset( &set );
      sigaddset( &set, SIGUSR1 );
19    act.sa_mask = set;
      act.sa_sigaction = &handler;
21    act.sa_flags = SA_SIGINFO; // make it a queued signal
      sigaction( SIGUSR1, &act, NULL );
```

```
23
    while (1){
25    fprintf(stderr, ".");
      sleep (1);
27    fprintf(stderr, ".");
      sleep (1);
29    fprintf(stderr, ".");
      sleep (1);
31   }
      return EXIT_SUCCESS;
33   }

35 void handler( int signo, siginfo_t* info, void* other )
     {
37     fprintf(stderr, "%d", info->si_value.sival_int);
     }
```

Listing 21.2
Process 1: handling SIGUSR1 signals.

```
  #include <stdio.h>
2 #include <stdlib.h>
  #include <signal.h>
4 #include <unistd.h>

6 int main( void )
    {
8     extern void handler(int, siginfo_t*, void*);
      struct sigaction act;
10    sigset_t set;

12
      printf("%d\n", getpid());
14    /*
       * Define a handler for SIGUSR1.
16     */
      sigemptyset( &set );
18    sigaddset( &set, SIGUSR1 );
      act.sa_mask = set;
20    act.sa_sigaction = &handler;
      act.sa_flags = SA_SIGINFO; // make it a queued signal
22    sigaction( SIGUSR1, &act, NULL );

24   while (1){
       fprintf(stderr, "^");
26     sleep (1);
       fprintf(stderr, "^");
28     sleep (1);
     }
```

```
30      return EXIT_SUCCESS;
    }

32
  void handler( int signo, siginfo_t* info, void* other )
34  {
      fprintf(stderr, "_");
36      fprintf(stderr, "%d", info->si_value.sival_int);
      fprintf(stderr, "_");
38  }
```

Listing 21.3
Process 2: handling SIGUSR1 signals.

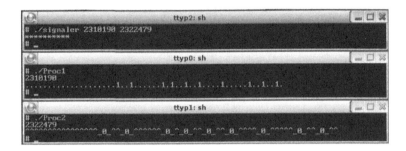

Figure 21.1
Screenshot of a run of the program given in Listings 21.1–21.3.

Figure 21.1 shows a run of the program given in Listings 21.1–21.3.

21.5.2 Example: Controlling a Task Server

A more complicated example is given in Listings 21.4 and 21.5. The code in Listing 21.4 outlines a task server which can be adapted to implement the APERIODIC-SERVER covered in Section 15.1.2.

Briefly, signals of the type SIGUSR1 are used for preemption, and signals of the type SIGUSR2 are used for execution resumption. The task server also uses a semaphore (see Chapter 18 for details) to control its execution. The semaphore *sem* is not available initially (it has a value of 0), so the server is blocked by the call sem_wait() at line 50.

Upon accepting a SIGUSR2 signal, the function sem_wait() is interrupted and the signal handler sig_resumption_handler() is invoked, which calls sem_post() to increase the semaphore value by 1. The execution of the task server comes to line 51 once sig_resumption_handler() returns. The call of sem_wait() at line 51 successfully decreases the semaphore value to 0, and the task server enters into the forever loop.

For each iteration of the loop, the task server simply invokes the function aperiodic_jobs() to process aperiodic jobs, if available. Upon accepting a SIGUSR1 signal, the task server will be interrupted and the signal handler sig_preemption_handler() is invoked. Since the semaphore value at this point in time is 0, the call of sem_wait() at line 72 will block the task server until it is interrupted by a SIGUSR2 signal. Consequently, the execution of the task server has been preempted. Upon accepting a SIGUSR2 signal, the sig_resumption_handler will increase the semaphore value to 1. When the control comes back again to the sig_preemption_handler, the call of sem_wait() at line 75 will successfully decrease the semaphore value to 0, and the task server resumes its execution.

```c
  #include <stdio.h>
2 #include <stdlib.h>
  #include <math.h>
4 #include <termios.h>
  #include <unistd.h>
6 #include <string.h>
  #include <semaphore.h>
8 #include <mqueue.h>
  #include <errno.h>
10 #include <signal.h>

12 sem_t *sem;
  int curvalue = 0;
14
  int main( void )
16  {
      extern void sig_preemption_handler(int, siginfo_t*, void*);
18      extern void sig_resumption_handler(int, siginfo_t*, void*);
      struct sigaction act1, act2;
20      sigset_t set1, set2;
      char SEM_NAME[] = "mysem";
22
      printf("%d\n", getpid());
24      //Create a semaphore
      sem_unlink(SEM_NAME);
26      sem = sem_open(SEM_NAME, O_CREAT, 0644, 0);
      if (sem == SEM_FAILED) {
28          perror("reader:unable to open semaphore");
          sem_unlink(SEM_NAME);
30          exit(1);
      }
32
      sigemptyset( &set1 );
34      sigaddset( &set1, SIGUSR1 ); //Used as Preemption Signal
      act1.sa_mask = set1;
36      act1.sa_flags = SA_SIGINFO; // make it a queued signal
```

```
        act1.sa_sigaction = &sig_preemption_handler;
38      sigaction( SIGUSR1, &act1, NULL );

40      sigemptyset( &set2 );
        sigaddset( &set2, SIGUSR2 ); //Used as Resumption Signal
42      act2.sa_mask = set2;
        act2.sa_flags = SA_SIGINFO; // make it a queued signal
44      act2.sa_sigaction = &sig_resumption_handler;
        sigaction( SIGUSR2, &act2, NULL );

46
        //the calling process of sem_wait blocks until it can
48      //decrement the counter, or the call is interrupted by signal
        //Two consecutive calls are needed for such a reason
50      sem_wait(sem); // to be interrupted by SIGUSR2
        sem_wait(sem); //this effectively decrements semaphore to 0
52      while (1){
            aperiodic_jobs();
54      }
        return EXIT_SUCCESS;
56  }
  int aperiodic_jobs(void){
58      //simulate execution of aperiodic_jobs
        double i=0;
60      fprintf(stderr, "^");
        for (i=0; i<10000000; i++) {
62          double a = pow(i+10, i);
        }
64      fprintf(stderr, "%d", curvalue);
  }
66 void sig_preemption_handler(int signo, siginfo_t* info, void* other)
    {
68      curvalue = info->si_value.sival_int;
        fprintf(stderr, "-O");
70      //make sure SIGUSR2 is not masked in initialization,
        //otherwise SIGUSR2 cannot be nestedly handled
72      sem_wait(sem); // to be interrupted by SIGUSR2
        //now the control is returned from sig_resumption_handler
74      fprintf(stderr, "K-");
        sem_wait(sem); //this effectively decrements semaphore to 0
76  }
  void sig_resumption_handler(int signo, siginfo_t* info, void* other)
78  {
        curvalue = info->si_value.sival_int;
80      fprintf(stderr, ".");
        sem_post(sem);
82  }
```

Listing 21.4
Task server: controlled by signals.

```
   #include <stdio.h>
 2 #include <stdlib.h>
   #include <signal.h>
 4 #include <sys/siginfo.h>
   #include <unistd.h>
 6 #include <time.h>

 8 int main( int argc, char *argv[] )
    {
10     int pid, r, i=0;
       struct sigevent event1, event2;

12     if (argc!=2){
14       fprintf(stderr, "Use: controller pid");
         exit(EXIT_FAILURE);
16     }
       SIGEV_SIGNAL_VALUE_INIT( &event1, SIGUSR1, 1 );
18     SIGEV_SIGNAL_VALUE_INIT( &event2, SIGUSR2, 2 );
       pid = atoi(argv[1]);
20     srand( time( NULL ) );

22     while (1){
         r = 1+ rand() % 8;
24       sleep (r);
         if ((i%2)==1){
26         sigqueue(pid, SIGUSR1, event1.sigev_value );
         }
28       else {
           sigqueue(pid, SIGUSR2, event2.sigev_value );
30       }
         i++;
32     }
       return EXIT_SUCCESS;
34  }
```

Listing 21.5
Controller: sending preemption/resumption signals.

The code in Listing 21.5 implements a simple driver for the task server, interleavingly sending SIGUSR1 and SIGUSR2 signals to the task server to pause and resume its execution. Figure 21.2 shows a run of the program.

Figure 21.2
Task Server accepting preemption and resumption signals.

21.6 Spinlocks and Interrupt Events from ISRs

21.6.1 POSIX Spinlocks

Spinlocks, similarly to semaphores and mutexes, are mutual exclusion devices for protecting critical sections. The calling task blocks if it requests an unavailable semaphore or mutex. In contrast, the calling task of a spinlock will not block if the spinlock is not available; instead, it goes into a *tight loop* where it repeatedly checks the spinlock until it becomes available. This loop gives the "spin" part of a spinlock.

Table 21.3 lists the POSIX functions for spinlocks.

Note that the pthread_spin_lock operation checks and grabs a spinlock in an atomic manner. This ensures that only one spinning thread, even if there are several, can obtain the spinlock. Also, spinlocks are intended for use on preemptive systems. If a task in a nonpreemptive single-processor system ever went spinning on a lock, it would spin forever; no other thread would ever be able to obtain the CPU to release the lock.

Depending on the OS implementation, the above POSIX functions for spinlocks may or may not be used in ISRs.

21.6.2 QNX Event Structure

In addition to POSIX signals, QNX also supports other asynchronous notification mechanisms such as interrupts and pulses (see Section 21.7 for QNX pulses). These are treated in QNX as different types of events, and QNX has implemented a dedicated subsystem to uniformly handle different types of events.

An event object is described by a data structure named `sigevent`. Its members and the values each member can take are given in Table 21.4.

Table 21.3 POSIX.1 function calls for spinlocks

POSIX Function	Description
pthread_spin_destroy	Destroy a thread spinlock object
pthread_spin_init	Initialize a thread spinlock object
pthread_spin_lock	The calling thread acquires the lock if it is not held by another thread; otherwise, the thread spins
pthread_spin_trylock	The calling thread acquires the lock if it is not held by another thread; otherwise, the call returns with failure
pthread_spin_unlock	Release a thread spinlock. A thread spinning on the lock will acquire the lock

Table 21.4 QNX sigevent structure

sigenv_notify	sigev_signo	sigev_value	sigev_coid	sigev_priority	sigev_code
SIGEV_NONE SIGEV_SIGNAL SIGEV_THREAD SIGEV_INTR	Signal	Value			
SIGEV_PULSE SIGEV_SIGNAL_CODE SIGEV_SIGNAL_THREAD SIGEV_UNBLOCK	Signal Signal	Value Value Value	Connection	Priority	Code Code Code

The value of sigenv_notify indicates the event type:

- SIGEV_NONE—a POSIX event indicating a null notification;
- SIGEV_SIGNAL—a POSIX signal event without code or a value;
- SIGEV_THREAD—a POSIX event to create a new thread (say, an OS timer can be set up to raise an event of this type to create a new thread every time it expires);
- SIGEV_INTR—a QNX event originated from a hardware interrupt (to be further handled at the user application level);
- SIGEV_PULSE—a QNX pulse event;
- SIGEV_SIGNAL_CODE—a QNX signal event with code and a value;
- SIGEV_SIGNAL_THREAD—a QNX signal event to a specific thread;
- SIGEV_UNBLOCK—a QNX event to force a thread to become unblocked.

Among the above event types, only the constants SIGEV_NONE, SIGEV_SIGNAL, and SIGEV_THREAD are defined in the POSIX.1-2008 standard. The others are QNX specific. Also, the POSIX `sigevent` structure defines only sigenv_notify, sigev_signo, and sigev_value; the other members are QNX specific.

QNX also provides several macros for initializing various types of events. For example, the macro SIGEV_SIGNAL_INIT(&evt, signo) is used to initialize a signal event *evt* with *signo* as the signal type to be raised. Given an event object *evt*, its type can be returned by calling the macro SIGEV_GET_TYPE(&evt).

21.6.3 Interrupt Handling in QNX Applications

Table 21.5 lists the QNX functions for interrupt handling at the user application level.

By calling the function InterruptAttach(), a user task (process or thread) can attach an interrupt handler (ISR) to a hardware interrupt source. The user task is also called the attaching process (or the attaching thread) of the interrupt handler installed via InterruptAttach(). Upon an interrupt from the hardware source, an interrupt handler can

Table 21.5 QNX function calls for interrupt handling

POSIX Function	Description
InterruptAttach	Attach an interrupt handler to a hardware interrupt source
InterruptDetach	Detach an interrupt handler by ID
InterruptDisable	Disable all hardware interrupts
InterruptEnable	Enable all hardware interrupts
InterruptLock	Guard a critical section by a spinlock object shared between an interrupt handler and a thread. This call disables interrupts, spinning in a tight loop until the spinlock is acquired
InterruptMask	Disable a specific hardware interrupt
InterruptUnlock	Unlock the specified spinlock and re-enable interrupts
InterruptUnmask	Enable a specific hardware interrupt
InterruptWait	The calling thread blocks, waiting for the interrupt handler it attached before to return a hardware interrupt event of type SIGEV_INTR

(selectively) deliver a SIGEV_INTR event to the attaching process (or thread). In such a sense, the attaching process (or thread) of an interrupt handler actually acts as the server task of the handler. The interrupt handler is removed when the attaching process or thread exits.

When a handler is attached, by default it is placed in front of any existing handlers for that interrupt and is called first (see Section 4.5.3 for ISR chaining). By the setting of a flag, an interrupt handler can also be placed at the end of any existing handlers.

The QNX implementation of pthread_spin_lock and pthread_spin_unlock cannot be used in ISRs. Instead, the functions InterruptLock() and InterruptUnLock() allow both ISRs and threads to use a spinlock. The calling task of InterruptLock() tries to acquire a spinlock. If the lock is not immediately available, the task spins in a tight loop until the lock is acquired. It can be used to protect access to shared data structures between an interrupt handler and the attaching thread of the handler.

21.6.4 Example: Interrupt Events from ISRs

Figure 21.3 gives a design showing the use of spinlock objects. A QNX-based implementation of this design is given in Listing 21.6. This example helps us to learn the following points:

(1) The program first performs initialization (lines 233-264):
   (a) It calls setupTerminal() to disable echo of the keyboard inputs.
   (b) It attaches a handler for terminal interrupt signals (SIGINT). Once the program has been interrupted by Ctrl-C, the signal handler sig_handler() is responsible for restoring the system settings changed by this program.
   (c) It initializes the two character arrays *time_comm_ptr* and *key_comm_ptr*.
   (d) It initializes two event objects of type SIGEV_INTR: *keyevent* and *timeevent*.
   (e) It creates a semaphore object referenced by a pointer: *sem*.

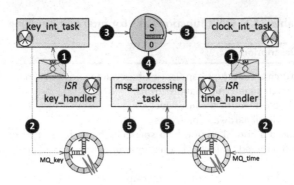

Figure 21.3
Interrupt locks and interrupt events.

(f)   It opens and initializes two message queues: *MQ_key* and *MQ_time*. Both have a capacity of 20 and enable nonblocking access.

(g)   Lastly, it initializes two spinlock objects referenced by the pointers *klock* and *tlock*.

(2)   In addition to the main thread, the process creates three task threads:

(a)   msg_processing_task—for message processing;

(b)   key_int_task—for servicing events resulted from the x86 keyboard interrupts; and

(c)   clock_int_task—for servicing events resulted from the x86 real-time clock interrupts (say, raised every 1 ms).

The two interrupt service tasks have a higher priority than the message processing task.

(3)   The keyboard ISR, defined by the function key_handler(), is attached by the corresponding service task key_int_task. Likewise, the clock ISR, defined by the function time_handler(), is attached by the clock_int_task.

(4)   The communication area *key_comm_ptr*, denoted by the "shared memory object" in Figure 21.3, is shared by both key_handler() and its service task key_int_task(). Hence, their critical sections are protected by the spinlock object *klock* via the function calls InterruptLock() and InterruptUnlock(), respectively. Note that it is not necessary to use a spinlock object when no data are shared between an ISR and its service task.

(5)   Similarly, the communication area *time_comm_ptr* is shared by both time_handler() and its service task clock_int_task(). Their critical sections are protected by the spinlock object *tlock*.

(6)   The keyboard ISR (key_handler) returns a keyevent to its service task every other time that the keyboard has raised interrupts. Since a hardware interrupt is raised for both a key-down event and a key-up event, the overall effect of this keyboard ISR is to ensure that the service task is notified only once for each key press (a key-down event followed by a key-up event). Similarly, the clock ISR (time_handler) returns a timeevent to its service task every 1000 times that the clock has raised interrupts. If the real-time clock raises an interrupt every 1 ms, then the overall effect of this clock ISR is to ensure that the service task is notified each second.

(7) Every time the key_int_task accepts a *keyevent* (via the InterruptWait() call), if the key is a printable ASCII character (with code between 0x20 and 0x7F), the character, together with the information in the shared area *key_comm_ptr* (which is the number of keys being pressed so far), is used to generate a new message, which is sent to the queue *MQ_key*. If mq_send() fails because the queue is full, *lost_kmsg_count*—the number of lost messages—is updated. Note that the key_int_task sends out a new message every time a key is pressed, but only increases the semaphore value by 1 every KEY_BURST_NUM times. This allows msg_processing_task to process messages in a batch mode. Further, clock_int_task has almost the same structure as key_int_task, except that it waits for a *timeevent* and sends messages to the queue *MQ_time*.

(8) When there is no message, msg_processing_task (lines 189-204) should block, which is achieved by waiting for the semaphore. Whenever the semaphore becomes available, the task works on the two queues *MQ_key* and *MQ_time* sequentially, receiving and processing the messages one after another.

```
1  #include <stdio.h>
   #include <termios.h>
3  #include <unistd.h>
   #include <string.h>
5  #include <pthread.h>
   #include <sys/neutrino.h>
7  #include <sys/syspage.h>
   #include <semaphore.h>
9  #include <mqueue.h>
   #include <errno.h>
11 #include <signal.h>

13 #define MQ_MESSAGE_SIZE 12
   #define MQ_SIZE 20
15
   #define COMM_AREA_LEN 11
17 #define X86_KEYBOARD_INTERRUPT 1
   #define KEY_BURST_NUM 3
19 #define TIME_BURST_NUM 5

21 int key_int_id, timer_int_id;
   struct sigevent keyevent, timeevent;
23 volatile int kcounter, tcounter;
   char key_comm_ptr[COMM_AREA_LEN];
25 char time_comm_ptr[COMM_AREA_LEN];
   const char *kmsg = " ";
27 const char *tmsg = " ";

29 mqd_t MQ_key, MQ_time;
   struct mq_attr *mqstat;
```

```
31  char *MQ_msg_ptr;
    char MQK_NAME[] = "kmq";
33  char MQT_NAME[] = "tmq";
    int lost_kmsg_count, lost_tmsg_count;
35
    sem_t *sem;
37  char SEM_NAME[] = "mysem";

39  struct termios org_opts, new_opts;
    intrspin_t *klock;
41  intrspin_t *tlock;
    extern void sig_handler();
43
    const struct sigevent *key_handler(void *area, int id) {
45    ++kcounter;
      if ((kcounter % 2) == 0) {
47      InterruptLock( klock );
        char buffer[MQ_MESSAGE_SIZE];
49      itoa(kcounter/2, buffer, 10);
        strcat(buffer, kmsg);
51      memcpy(area, buffer, strlen(buffer) + 1);
        InterruptUnlock(klock);
53      return (&keyevent);
      } else
55      return NULL;
    }
57
    const struct sigevent *time_handler(void *area, int id) {
59    ++tcounter;
      if ((tcounter % 1000) == 0) {
61      InterruptLock( tlock );
        char buffer[MQ_MESSAGE_SIZE];
63      itoa(tcounter / 1000, buffer, 10);
        strcat(buffer, tmsg);
65      memcpy(area, buffer, strlen(buffer) + 1);
        InterruptUnlock(tlock);
67      return (&timeevent);
      } else
69      return NULL;
    }
71
    void attach_signal_handler() {
73    struct sigaction act;
      sigset_t set;
75    sigemptyset(&set);
      sigaddset(&set, SIGINT);
77
      act.sa_flags = 0;
79    act.sa_mask = set;
      act.sa_handler = &sig_handler;
```

```
81    sigaction(SIGINT, &act, NULL);
   }
83
   void sig_handler() {
85    //------ restore old settings ---------
      tcsetattr(STDIN_FILENO, TCSANOW, &org_opts);
87    InterruptUnlock( klock );
      InterruptUnlock( tlock );
89    InterruptDetach(key_int_id);
 ·    InterruptDetach(timer_int_id);
91    fprintf(stderr,"ISRs detached in signal handler\n");
      fprintf(stderr,"Num of keyboard msgs lost: %d\n", lost_kmsg_count);
93    fprintf(stderr,"Num of timer msgs lost: %d\n",lost_tmsg_count);
      mq_close(MQ_key);
95    mq_close(MQ_time);
      mq_unlink(MQK_NAME);
97    mq_unlink(MQT_NAME);
      kill(getpid(), SIGKILL);
99  }

101 void setupTerminal() {
      if (tcgetattr(STDIN_FILENO, &org_opts) != 0) return;
103
      //---- set new terminal parms --------
105   memcpy(&new_opts, &org_opts, sizeof(new_opts));
      new_opts.c_lflag &= ~(ICANON | ECHO);
107   //new_opts.c_cc[VMIN] = 0;
      //new_opts.c_cc[VTIME] = 0;
109   tcsetattr(STDIN_FILENO, TCSANOW, &new_opts);
   }
111
   void * key_int_task(void *arg) {
113  int mq_return, counter=0;
      // this thread is dedicated to handling interrupts
115  // Request I/O privileges
      ThreadCtl( NTO TCTL IO. 0);
117
      // Attach ISR vector
119  key_int_id = InterruptAttach(X86_KEYBOARD_INTERRUPT,
          &key_handler, &key_comm_ptr, COMM_AREA_LEN, 0);
121
      while (1) {
123    // the thread that attached the interrupt is the one that
        //must wait for the SIGEV_INTR.
125    InterruptWait(0, NULL);
        char key_pressed[2] = "\0\0";
127    key_pressed[0] = getchar();
        if ((key_pressed[0] > 0x20) && (key_pressed[0] < 0x7F)) {
129      counter++;
          char *buffer = (char *) malloc(COMM_AREA_LEN + 4);
```

```
131      memset(buffer, 0, COMM_AREA_LEN + 4);
         InterruptLock( klock );
133      strcpy(buffer, "(");
         strcat(buffer, key_comm_ptr);
135      strcat(buffer, ",");
         strcat(buffer, key_pressed);
137      strcat(buffer, ")");
         mq_return = mq_send(MQ_key, buffer, strlen(buffer), 0);
139      if (mq_return<0) lost_kmsg_count++;
         InterruptUnlock( klock );
141      free(buffer);
         if (counter%KEY_BURST_NUM == 0) sem_post(sem);
143    }
     }
145    return 0;
   }
147 void * clock_int_task(void *arg) {
     int mq_return, counter=0;
149    // this thread is dedicated to handling interrupts
     ThreadCtl(_NTO_TCTL_IO, 0);
151    key_int_id = InterruptAttach(SYSPAGE_ENTRY(qtime)->intr,
         &time_handler, &time_comm_ptr, COMM_AREA_LEN, 0);
153    while (1) {
       InterruptWait(0, NULL);
155      counter++;
       char *buffer = (char *) malloc(COMM_AREA_LEN + 1);
157      memset(buffer, 0, COMM_AREA_LEN + 4);
       InterruptLock( tlock );
159      strcpy(buffer, "(");
       strcat(buffer, time_comm_ptr);
161      strcat(buffer, ",*)");
       mq_return = mq_send(MQ_time, buffer, strlen(buffer), 0);
163      if (mq_return<0) lost_tmsg_count++;
       InterruptUnlock( tlock );
165      free(buffer);
       if (counter%TIME_BURST_NUM == 0)  sem_post(sem);
167    }
     return 0;
169 }

171 void processMsg(char *msg, char *qmsg) {
     strcat(msg, qmsg);
173 }

175 void process_MQ(mqd_t mqt, char *msg) {
     if (mq_getattr(mqt, mqstat) == -1) return;
177    long nmsg = mqstat-> mq_curmsgs;
     if (nmsg == 0) return;
179
     //caller won't block if queue is full or empty.
```

```
181    mqstat->mq_flags |= O_NONBLOCK;
       mq_setattr(mqt, mqstat, NULL);
183    while (mq_receive(mqt,MQ_msg_ptr,MQ_MESSAGE_SIZE,NULL)> 0) {
         //a message received
185      processMsg(msg, MQ_msg_ptr);
         memset(MQ_msg_ptr, 0, MQ_MESSAGE_SIZE + 1);
187    }
     }
189 void * msg_processing_task(void *arg) {
       // this thread is dedicated to processing msgs
191    while (1) {
         sem_wait(sem);
193      // got a signal from ISR; data ready
         char *msg1 = malloc(MQ_MESSAGE_SIZE*KEY_BURST_NUM);
195      memset(msg1, 0, MQ_MESSAGE_SIZE*KEY_BURST_NUM);
         char *msg2 = malloc(MQ_MESSAGE_SIZE*TIME_BURST_NUM);
197      memset(msg2, 0, MQ_MESSAGE_SIZE*TIME_BURST_NUM);
         process_MQ(MQ_key, msg1);
199      process_MQ(MQ_time, msg2);
         printf("%s\n", msg1); free(msg1);
201      printf("\t\t\t\t%s\n", msg2);free(msg2);
       }
203    return 0;
     }
205
   static mqd_t open_MQ(char *name) {
207    mqd_t queue;
       struct mq_attr attr;
209
       memset((void*) &attr, 0, sizeof(struct mq_attr));
211    attr.mq_maxmsg = MQ_SIZE;
       attr.mq_msgsize = MQ_MESSAGE_SIZE;
213
       queue = mq_open(name, O_RDWR|O_CREAT|O_NONBLOCK, S_IRUSR|S_IWUSR, &attr);
215    if (queue == -1)
         printf("Open queue fails %s(%d)\n", strerror(errno), errno);
217
       return queue;
219 }
   void new_thread(void* (*start_routine)(void*), int priority) {
221    pthread_attr_t _t_attr;
       struct sched_param schd_parameter;
223    //initializes the thread attributes to their default values
       pthread_attr_init(&_t_attr);
225    pthread_attr_setinheritsched(&_t_attr, PTHREAD_EXPLICIT_SCHED);

227    //Set the priority
       schd_parameter.sched_priority = priority;
229    pthread_attr_setschedparam(&_t_attr, &schd_parameter);
       // start up a thread
```

```
231    pthread_create(NULL, &t_attr, start_routine, NULL);
     }
233  void initialization() {
       setupTerminal();
235    attach_signal_handler();
       memset(&key_comm_ptr, 0, COMM_AREA_LEN);
237    memset(&time_comm_ptr, 0, COMM_AREA_LEN);
       // Initialize event structure
239    keyevent.sigev_notify = SIGEV_INTR;
       timeevent.sigev_notify = SIGEV_INTR;
241
       //Create a semaphore
243    sem = sem_open(SEM_NAME, O_CREAT, 0644, 0);
       if (sem == SEM_FAILED) {
245      perror("reader:unable to open semaphore");
         sem_unlink(SEM_NAME);
247      exit(1);
       }
249
       MQ_key = open_MQ(MQK_NAME);
251    MQ_time = open_MQ(MQT_NAME);
       mqstat = (struct mq_attr*) malloc(sizeof(struct mq_attr));
253    memset(mqstat, 0, sizeof(struct mq_attr));
       MQ_msg_ptr = (char *) malloc(MQ_MESSAGE_SIZE + 1);
255
       klock = (intrspin_t*) malloc(sizeof(intrspin_t));
257    memset(klock, 0, sizeof(*klock));
       tlock = (intrspin_t*) malloc(sizeof(intrspin_t));
259    memset(tlock, 0, sizeof(*tlock));
261    printf("-------------------\t\t--------------------------\n");
       printf("     Key Burst \t\t\t Timer Burst\n");
263    printf("-------------------\t\t--------------------------\n");
     }
265  int main() {
       ThreadCtl(_NTO_TCTL_IO, 0);
267    initialization();
       new_thread(key_int_task, 60);
269    new_thread(clock_int_task, 60);
       new_thread(msg_processing_task, 30);
271    while (1) sleep(10);
     }
```

Listing 21.6
Sample code of using spinlocks.

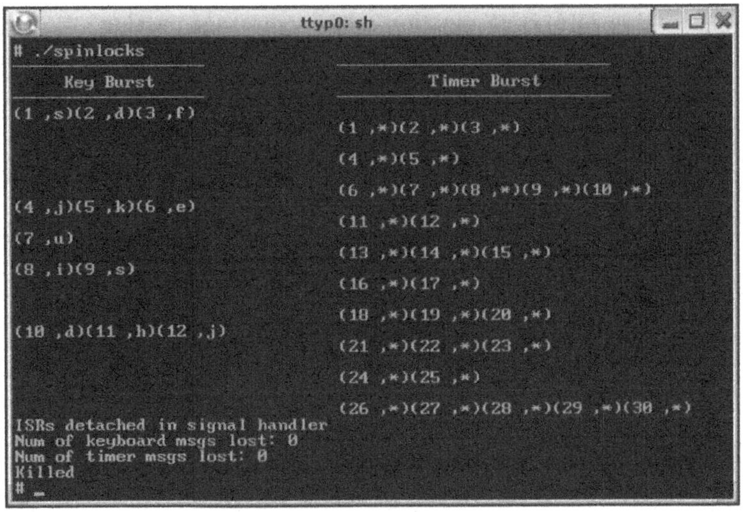

Figure 21.4
A run of the program in Listing 21.6.

Figure 21.4 shows a screenshot of a run of the program in Listing 21.6.

21.7 QNX Pulses

QNX pulsing is a special notification mechanism where a sender sends a lightweight message called a *pulse* to a receiver without waiting for an acknowledgment [8]. The pulsing mechanism further relies on the synchronous message passing subsystem, by which process A can register a pulse event together with a firing condition to process B. Whenever the firing condition occurs, process B will notify process A by delivering the pulse event back.

Below, we first introduce the QNX synchronous message passing, followed by the asynchronous pulsing mechanism.

21.7.1 QNX Synchronous Message Passing

The sender and receiver in asynchronous intertask communication do not need to wait for each other, but asynchronous communication does rely on some kernel objects such as message queues and pipes to serve as the intermediate buffering mechanism. In contrast, each synchronous communication action virtually sets up a synchronization point between the sender and the receiver, where neither the sender nor the receiver can continue until the other party becomes ready.

Table 21.6 gives a few QNX functions [7] for interprocess message passing. Communication channels and connections are two critical concepts in message passing.

Table 21.6 QNX function calls for message passing

QNX Function	Description
ChannelCreate name_attach ConnectAttach name_open	Create a channel that can be used to receive messages and pulses Create a channel and associate a name with it Establish a connection between a process and a channel Locate the channel with the specified name and create a connection to it
MsgSend MsgReceive MsgReply	Send a message to a channel. The calling thread enters the *SenderBlocked* state until a reply is received. The message is enqueued on the channel in the order of the sending thread's priority Receive a message or pulse on a channel. If there is no message waiting in the channel, the calling thread enters the *ReceiverBlocked* state until a message arrives. This call is always followed by MsgReply() Reply to a message and unblock the calling thread of MsgSend()
MsgDeliverEvent MsgSendPulse MsgReceivePulse	A nonblocking function to deliver a pulse event or signal event to a channel A nonblocking function to send a pulse to a channel. Each pulse has a priority, and the pulses are enqueued on a channel in the order of their priorities Receive a pulse on a channel. If there is no pulse waiting in the channel, the calling thread enters the *ReceiverBlocked* state until a pulse arrives

21.7.1.1 Channels and connections

In QNX, message passing is targeted not directly from thread to thread, but through channels and connections. A thread that wishes to receive messages or pulses must create a *channel*, and another thread that wishes to send a message to that thread must do so via a *connection* to that channel.

The channel concept as illustrated in Figure 21.5 has the following features:

- A channel can be created by a thread via name_attach() or ChannelCreate(). The function name_attach() associates a name with the newly created channel, which allows a thread from a different process to reference the channel by name.
- A channel is not bound to the creating thread, but is bound to the owning process. One or more threads of the owning process can receive messages from a channel.
- A channel can be used for both interprocess and intraprocess communication.
- A channel has a *receive buffer*. If there is no receiving thread when messages or pulses arrive at a channel, they are queued in the receive buffer in priority order.
- A receiving thread receives a message or pulse from a channel via MsgReceive() or MsgReceivePulse().
- One or more connections can be attached to a channel. Messages or pulses are sent not directly to a channel, but via an attached connection (refer to the functions MsgDeliverEvent(), MsgSend(), and MsgSendPulse() in Table 21.6). A connection can be

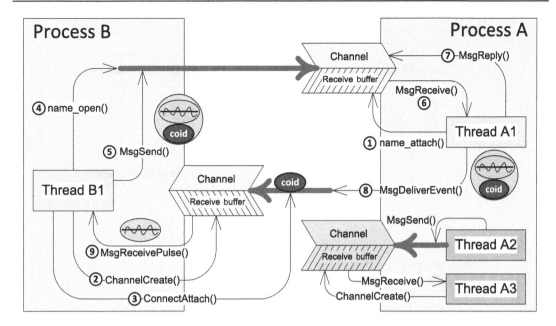

Figure 21.5
QNX messaging function calls and concepts.

created and attached to a channel via name_open() or ConnectAttach(). A thread may create a connection attached to its own channel, and share such a "self-connection" object with other threads (within the same process or in different processes) so that they can send messages or pulses via that connection back to the owner thread.

21.7.1.2 Synchronous communication

QNX synchronous communication is captured by the two UML state diagrams given in Figure 21.6.

Here is one scenario where the receiver remains in the *ReceiverReady* state:

(1) Initially, both the sender and the receiver are in their respective ready state. In particular, the receiver has its channel ready for accepting messages and the sender has a connection attached to the receiver's channel.

(2) The sender calls MsgSend(m) to send a message *m* to the channel. After this, it transitions to the *SenderBlocked* state.

(3) The sender blocks until the receiver calls MsgReceive() to receive the message on the channel. This call causes the sender to transition to the *ReplyBlocked* state.

(4) The sender blocks until the receiver calls MsgReply() to reply to the message *m*. The reply message itself may or may not contain data. This call causes the sender to transition back to the *SenderReady* state.

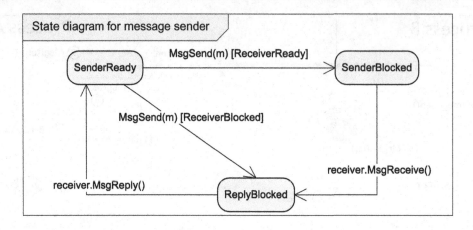

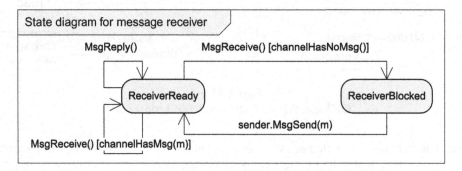

Figure 21.6
QNX synchronous communication.

Here is another scenario:

(1) Initially, both the sender and the receiver are in their respective ready state.
(2) The receiver calls MsgReceive() to receive a message but the channel is empty. The receiver transitions to the *ReceiverBlocked* state.
(3) The sender calls MsgSend(m) to send a message *m* to the channel. This call brings the sender directly to the *ReplyBlocked* state because the receiver is currently in the *ReceiverBlocked* state. This call also causes the receiver to transition back to the *ReceiverReady* state.
(4) The receiver calls MsgReply() to reply to the message *m*. This call unblocks the sender, which transitions back to the *SenderReady* state.

In the above scenarios we have only focused on a single sender-receiver pair. A channel may have multiple receiving threads and multiple senders, each of which may be blocked. QNX uses several lists to manage the senders and receivers blocked on a channel. Specifically, a channel has three thread lists: a receiverBlocked-list, which is a queue of threads waiting for messages; a senderBlocked-list, which is a priority-FIFO queue of threads that are in the *SenderBlocked* state; and a replyBlocked-list, which is a list of threads that are waiting for a reply.

21.7.2 QNX Asynchronous Pulsing Mechanism

A *pulse* is a fixed-size, nonblocking message with the following properties:

- A pulse can have an eight-bit code. The pulse code should be in the range _PULSE _CODE_MINAVAIL to _PULSE_CODE_MAXAVAIL. Since pulses might be used for a variety of purposes, the pulse code can serve as a type or cookie to distinguish one purpose from the other.
- A pulse can carry a small payload (four bytes of data) that may vary from time to time.
- A pulse has a priority, which specifies where the pulse will be inserted in the receive buffer of a channel. For example, suppose a channel contains a message being sent by a thread with a priority of 20. When a pulse with a priority of 30 arrives at the channel, the pulse will be inserted into the receive buffer in front of the message. The next call of MsgReceive() on the channel will not return the message, but will return the pulse instead. A pulse priority can be, but not necessarily, the same as the priority of the creating process.

A pulse event is a type of event with an embedded pulse. Pulses are often used as a notification mechanism within interrupt handlers or timers where blocking is not affordable. They also allow a server or driver to notify a client when new data become available.

Figure 21.5 illustrates a typical use of pulses. The sequence of operations is simplified and structured in Figure 21.7, which has two phases as explained below (Steps 1–9 are also labeled in Figure 21.5):

(1) *Pulse registration.* QNX synchronous communication is used to register a pulse event from the task *PulseReceiver* to the task *PulseSender*.
  - (a) Step 1. *PulseSender* first needs to create a channel to receive messages for pulse registration. The function name_attach() is used so that the name of the channel can be referenced by other tasks.
  - (b) Step 2. *PulseReceiver* also needs to create a channel to receive pulses. The function CreateChannel() is used in this case.
  - (c) Step 3. In addition, *PulseReceiver* needs to attach to its channel a connection over which pulses will be sent. The call to ConnectAttach() returns a connection ID.
  - (d) Step 4. *PulseReceiver* calls name_open() to locate the channel created by *PulseSender* and attach a connection to that channel.
  - (e) Step 5. *PulseReceiver* calls MsgSend() to send a message to the channel owned by *PulseSender*. This message contains a *pulse event*, which is initialized with the ID of

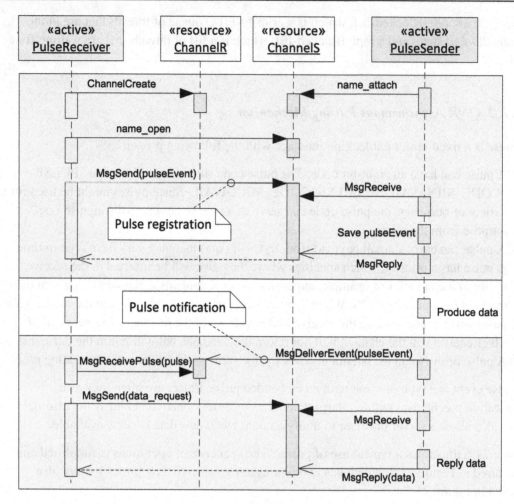

Figure 21.7
QNX asynchronous pulsing mechanism.

the connection attached to *PulseReceiver*'s channel. A pulse event can be initialized by the macro SIGEV_PULSE_INIT(&pulseEvent, coid, priority, code, value).

(f)   Step 6. *PulseSender* calls MsgReceive() to receive the message. The pulse event contained inside the message is saved for later use.

(g)   Step 7. *PulseSender* calls MsgReply() to reply the *PulseReceiver*. After this time point, both tasks are free to perform their own instructions.

(2)   *Pulse notification.*

(a)   Step 8. Whenever *PulseSender* realizes that it is time to notify *PulseReceiver* (say, new data become available for *PulseReceiver*), it calls MsgDeliverEvent() to send the pulse event registered by *PulseReceiver*. In this way, the pulse embedded within the pulse event is enqueued in *PulseReceiver*'s channel.

(b) Step 9. *PulseReceiver* calls MsgReceivePulse() to receive a pulse from its channel. The pulse code or value may indicate that *PulseSender* has new data available for *PulseReceiver*.

(c) Step 10. *PulseReceiver* calls MsgSend() to send a message to *PulseSender*, requesting new data.

(d) Step 11. *PulseSender* calls MsgReceive() to receive the request message.

(e) Step 12. *PulseSender* calls MsgReply() to reply to *PulseReceiver*. This reply message contains the requested data.

Note that in the above scenario both the *PulseSender* and *PulseReceiver* can enter into the *ReceiverBlocked* state when they try to receive a message or a pulse from an empty channel. This issue in general can be resolved by multi-threading: each active task (client or server) can maintain an incoming message queue and an outgoing message queue (which may or may not be the OS message queue objects), having one thread dedicated to the main application logic, one thread dedicated to sending messages or events (moving each message from the outgoing queue to an intended channel), one thread dedicated to receiving and replying messages or events (moving messages from one or more channels into the incoming queue), and one thread dedicated to event or message processing (which may consume messages from the incoming queue and add new messages to the outgoing queue). In so doing, the task could continue its execution in other threads while some threads are temporarily blocked.

21.7.3 Hierarchical Messaging Pattern

A task can be a message sender and receiver at the same time. The blocking nature of MsgSend() may introduce deadlocks if a task T_i sends to another task T_j and T_j simultaneously sends to T_i.

Such a deadlock can be avoided by using the *hierarchical messaging pattern* illustrated in Figure 21.8. The idea is to design a system with a hierarchy of processes such that the

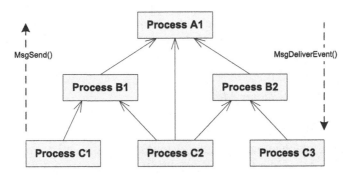

Figure 21.8
QNX hierarchical messaging pattern.

MsgSend operations flow only unidirectionally. Specific to the design in Figure 21.8 are that the processes at the C level can send to the processes at the B or A level, and the processes at the B level can send to the processes at the A level. The MsgSend operations cannot be used in the opposite direction. Instead, the nonblocking function MsgDeliverEvent() should be used, if desired, to deliver notifications along the downward direction.

This pattern is actually used in the design in Figure 21.5, where process A never uses MsgSend() to send a message to process B.

21.7.4 Priority Inheritance by Message Receivers

A receiver task receives messages and pulses in priority order.

Suppose a system has three tasks:

(1) T_H is a server task with a high priority 70;
(2) T_M is a task with a medium priority 40;
(3) T_L is a client task with a low priority 10.

When task T_L sends a message m to T_H, if the message m were processed at the server's priority level, priority inversion would happen, because T_L could get its work done before task T_M.

This can be avoided by setting a flag on the server's channel to enable "message-driven priority inheritance" such that the server's *effective priority* changes to that of the message sender or the pulse. As a result, the server will execute the work entailed by the message at the message sender's priority level.

When a pulse or message is enqueued in its channel, the server immediately boosts its priority if its current priority is lower than that of the pulse or the sender. Otherwise, its priority is changed when the pulse or message is received from its channel.

For example, when server task T_H receives a message from T_L, T_H's effective priority changes to 10. Next, suppose that T_M sends a pulse with a priority of 50 to T_H while it is still at a priority of 10. Since the pulse has a higher priority than T_H, its priority is boosted to 50 as soon as T_M sends the pulse to the channel.

21.7.5 Example: A Simple Timer Manager

As an example of using QNX pulses, Figure 21.9 gives a design where the UTimer server is a simple timer manager, which allows client tasks to install timers and notifies a client whenever its timer expires. In so doing, the UTimer server itself needs to monitor the passing of time, which is achieved by an OS timer. We use UTimer to refer to the timer objects managed by the UTimer server, to distinguish them from the OS timer.

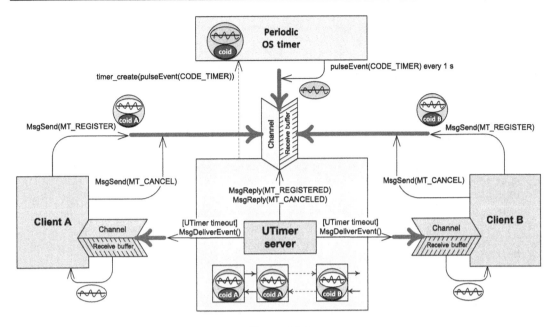

Figure 21.9
A simple user timer manager.

The UTimer server maintains a list of UTimer nodes, each of which contains a pulse event and a UTimer object. A Utimer object keeps a record of the number of seconds after which the client wishes to be notified. This also sets up the firing condition for the corresponding pulse event: as the left time becomes zero, the client which has installed this UTimer will be notified. To install a UTimer, a client uses a message of type MT_REGISTER to register a pulse event to the UTimer server. To monitor the passing of time, the UTimer server registers a pulse event to a periodic OS timer, which is able to send a pulse to the UTimer server every second. Every time it receives a pulse from the OS timer, the UTimer server knows that 1 s has just passed, and updates all the UTimer objects accordingly. Whenever the firing condition of a UTimer becomes true, the UTimer server will send a pulse to the corresponding client.

```
#define MY_SERV "Utimer_Server"
2
  // pulse code from OS timer to Utimer Server
4 #define CODE_TIMER   _PULSE_CODE_MINAVAIL+1
  // pulse code from Utimer Server to Clients
6 #define CODE_SERVER   _PULSE_CODE_MINAVAIL+2

8 struct UTimer{
    int UtimerID; // a unique Utimer ID
10   int priority;
```

```
     int secondsLeft;
12 };

14 typedef struct UTimer_Node{
     struct UTimer_Node* pre;
16   struct UTimer_Node* next;
     struct UTimer utimer;
18   int rcvid; //notify me when timeout
     struct sigevent pulseEvent;
20 } UTimer_Node;

22 struct UTimer_List{
     UTimer_Node* head;
24   UTimer_Node* tail;
   };
26
   //Message types
28 #define MT_REGISTER   8      // message to server
   #define MT_REGISTERED 9      // message to client
30 #define MT_CANCEL     10     // message to server
   #define MT_CANCELED   11     // message to client
32
   //Message structure
34 typedef struct {
     int messageType;  // MT_REGISTERED, MT_CANCELED
36   int UtimerID;
   } ServerMessageT;
38
   typedef struct {
40   int messageType;  // MT_REGISTER, MT_CANCEL
     int UtimerID;
42   int priority;
     int seconds;
44   struct sigevent pulseEvent;
   } ClientMsgT;
46
   typedef union {
48   ClientMsgT      cmsg;   // a message from a client,
     struct _pulse   pulse;  // a pulse from QNX timer
50 } MessageT;
```

Listing 21.7

UTimer.h: UTimer and user message structures.

The structures of UTimer and user messages are given in Listing 21.7:

- Two pulse codes are defined: one for pulses from the OS timer to the UTimer server, and one for pulses from the UTimer server to clients.
- A UTimer object has a unique ID, a priority, and a field for monitoring the number of seconds left.

- UTimer_List is a double-linked list of UTimer_Node. In addition to the two pointers for links, a UTimer_Node has a UTimer object, a pulse event, and a receiver ID. The UTimer has the firing condition for the pulse event, while the receiver ID, together with the connection information contained in the pulse event, tells the UTimer server where to deliver the pulse event.
- Four message types are defined: MT_REGISTER and MT_REGISTERED are used for UTimer installation, and MT_CANCEL and MT_CANCELED are used for UTimer cancellation.

```
   #include <stdio.h>
 2 #include <stdlib.h>
   #include <unistd.h>
 4 #include <errno.h>
   #include <sys/neutrino.h>
 6 #include <sys/iomsg.h>
   #include <time.h>
 8 #include <signal.h>
   #include <sys/siginfo.h>
10 #include <sys/iofunc.h>
   #include <sys/dispatch.h>
12 #include <sys/syspage.h>
   #include "UTimer.h"
14
   int srv_coid, rcvid, chid;
16 struct sigevent pulseEventFromServer;
   struct _pulse    pulseObject;
18 ClientMsgT       cmsg;
   ServerMessageT   smsg;
20
   void jobBeforeCheckingPulse(int timerID){
22   // here is a chance to cancel the installed timer.
     printf("Client processing some job before pulse\n\n");
24 }
   void jobUponPulse(void){
26   printf("Client processing some job upon pulse\n\n");
   }
28
   void registerATimer(int timeoutSec){
30   int myTimerID;
32   cmsg.messageType = MT_REGISTER;
     cmsg.priority = 10;
34   //set the number of seconds to wait for
     cmsg.seconds = timeoutSec;
36
     // The pulse event to be registered to the Utimer Server
38   cmsg.pulseEvent = pulseEventFromServer;
```

```
40    MsgSend( srv_coid, &cmsg, sizeof(cmsg), &smsg, sizeof(smsg) );
      if (smsg.messageType == MT_REGISTERED){
42      myTimerID = smsg.UtimerID; //used for timer canceling
        printf("Client %d has registered a UTimer %d!\n", getpid(), myTimerID);
44    }
      jobBeforeCheckingPulse(myTimerID);
46
      rcvid = MsgReceivePulse(chid, &pulseObject,
48                sizeof(pulseObject), NULL);
      printf("Client %d has received a pulse with code %d\n",
50           getpid(), pulseObject.code);
      printf("My UTimer with the ID %d has expired\n", myTimerID);
52    if (pulseObject.code == CODE_SERVER){
        //timeout, now do some jobs
54      jobUponPulse();
      }
56 }

58 void setupPulseEvent(void){
      int coid;
60
      /* A communication channel*/
62    chid = ChannelCreate( 0 );
      /* A connection to the channel on which to receive pulses*/
64    coid = ConnectAttach( 0, 0, chid, _NTO_SIDE_CHANNEL, 0 );

66    // set up a pulse event to be registered to the server
      SIGEV_PULSE_INIT( &pulseEventFromServer, coid, SIGEV_PULSE_PRIO_INHERIT,
          CODE_SERVER, 0 );
68 }

70 int main( int argc, char **argv)
   {
72    /* find the Utimer Server */
      if ( (srv_coid = name_open( MY_SERV, 0 )) == -1) {
74      printf("failed to find server, errno %d\n", errno );
        exit(1);
76    }
      setupPulseEvent();
78    registerATimer(10);
      return 0;
80 }
```

Listing 21.8
TClient.c: using the UTimer server.

Listing 21.8 is an example client which registers a UTimer to the server and simply performs some activities both before and after being pulsed.

```
   /*
 2 *  UTimer Server: receives pulses from a periodic QNX timer;
   *  and manages Utimer objects installed by clients.
 4 *  It sends a pulse to a client when its Utimer expires.
   */
 6 #include <stdio.h>
   #include <stdlib.h>
 8 #include <time.h>
   #include <signal.h>
10 #include <errno.h>
   #include <unistd.h>
12 #include <sys/siginfo.h>
   #include <sys/neutrino.h>
14 #include <sys/iomsg.h>
   #include <sys/iofunc.h>
16 #include <sys/dispatch.h>
   #include "UTimer.h"
18
   void setupPulseAndTimer (void);
20 void handleAPulse (void);
   void handleAMessage (int rcvid, ClientMsgT *cmsg);
22 void installUTimer(UTimer_Node* ptr);
   void cancelUTimer(int UtimerID);
24
   int chid;
26 int unique_timerID = 50;
   name_attach_t       *attach;
28 struct sigevent     pulseEventFromTimer;
   struct UTimer_List  *tlist;
30
   int main (void)  {
32   int rcvid;  // process ID of the message sender
     MessageT externalMsg;
34
     // set up the pulse and timer
36   setupPulse ();
     setupOSTimer ();
38
     tlist=(struct UTimer_List*)malloc(sizeof(struct UTimer_List));
40   for (;;) {
       rcvid = MsgReceive (attach->chid, &externalMsg,
42             sizeof (externalMsg), NULL);
       if (rcvid == 0) {
```

```
44      handleAPulse ();
     } else {
46      handleAMessage (rcvid, &externalMsg.cmsg);
     }
48  }
   return (EXIT_SUCCESS);
50 }

52 void setupPulse (void) {
   int  coid;        // connection ID
54
   if ( (attach = name_attach( NULL, MY_SERV, 0 )) == NULL){
56    fprintf (stderr, "Couldn't create channel!\n");
     perror (NULL);
58    exit (EXIT_FAILURE);
   }
60  // create a connection back to ourselves for receiving pulses
   coid = ConnectAttach (0, 0, attach->chid, 0, 0);
62  if (coid == -1) {
     fprintf (stderr, "Couldn't ConnectAttach to self!\n");
64    perror (NULL);
     exit (EXIT_FAILURE);
66  }
   // set up a pulse event with code MT_TIMER
68  SIGEV_PULSE_INIT (&pulseEventFromTimer, coid, SIGEV_PULSE_PRIO_INHERIT,
       CODE_TIMER, 0);
 }
70
 void setupOSTimer (void) {
72  timer_t            timerid;    // timer ID for the OS timer
   struct itimerspec  timer;      // the timer data structure
74
   // create an OS timer, binding it to the pulse event
76  if (timer_create (CLOCK_REALTIME, &pulseEventFromTimer,
             &timerid) == -1) {
78    fprintf (stderr, "Timer fails, errno %d\n", errno);
     perror (NULL);
80    exit (EXIT_FAILURE);
   }
82
   // setup the timer to deliver a pulse once per second
84  timer.it_value.tv_sec = 1;
   timer.it_value.tv_nsec = 0;
86  timer.it_interval.tv_sec = 1;
   timer.it_interval.tv_nsec = 0;
88
   // start the OS timer
90  timer_settime (timerid, 0, &timer, NULL);
 }
92
```

```
    void removeUTimer(UTimer_Node *ptr){
94   UTimer_Node *p;

96   p = tlist->head;
     while(p != ptr)
98     p = p->next;

100  if (p == tlist->head)
     {   //remove the old head
102    tlist->head = p->next;
       if (p->next != NULL)
104      p->next->pre = NULL;
     } else {
106    p->pre->next = p->next;
       if (p->next != NULL)
108      p->next->pre = p->pre;
     }
110  p->next = NULL;
     p->pre = NULL;
112 }

114 int isTimeOut(UTimer_Node *ptr){
     if (ptr->utimer.secondsLeft == 0)
116    return 1;
     else return 0;
118 }

120 void handleAPulse (void) {
     UTimer_Node *ptr = NULL;
122  UTimer_Node *psaved = NULL;

124  ptr = tlist->head;
     if (ptr != NULL) psaved = ptr->next;
126  while(ptr != NULL) {
       ptr->utimer.secondsLeft--;
128    if(isTimeOut(ptr)) {
         removeUTimer(ptr);
130      //deliver pulse if necessary
         ptr->pulseEvent.sigev_value.sival_int
132          = ptr->utimer.UtimerID;
         MsgDeliverEvent(ptr->rcvid, &ptr->pulseEvent);
134      free(ptr);
         printf ("The UTimer object %d has expired, pulse the client!\n", ptr->
             utimer.UtimerID);
136    }
       ptr = psaved;
138    if (ptr !=NULL) psaved = ptr->next;
     }
140 }
```

```
142 void handleAMessage (int rcvid, ClientMsgT *cmsg) {
    UTimer_Node *np;
144 ServerMessageT *replyMsg;

146    switch (cmsg -> messageType) {

148 case MT_REGISTER:
    //create a UTimer Node
150 unique_timerID++;
    np = (UTimer_Node*)malloc(sizeof(UTimer_Node));
152 np->utimer.UtimerID = unique_timerID;
    np->utimer.priority = cmsg->priority;
154 np->utimer.secondsLeft = cmsg->seconds;
    np->pulseEvent = cmsg->pulseEvent;
156 np->rcvid = rcvid;
    np->pre = NULL;
158 np->next = NULL;

160    //insert the timer
    installUTimer(np);
162 printf ("The UTimer object %d has been installed for %d seconds!\n", np->utimer
        .UtimerID, np->utimer.secondsLeft);

164 replyMsg = (ServerMessageT *)malloc(sizeof(ServerMessageT));
    replyMsg->messageType = MT_REGISTERED;
166 replyMsg->UtimerID = np->utimer.UtimerID;
    MsgReply(rcvid, EOK, replyMsg, sizeof(*replyMsg));
168 free(replyMsg);
    break;
170
  case MT_CANCEL:
172    //search for the timer and remove it
    cancelUTimer(cmsg->UtimerID);
174
    replyMsg = (ServerMessageT *)malloc(sizeof(ServerMessageT));
176 replyMsg->messageType = MT_CANCELED;
    MsgReply(rcvid, EOK, replyMsg, sizeof(*replyMsg));
178 free(replyMsg);
    break;
180    }
    }
182
  void installUTimer(UTimer_Node* ptr) {
184 UTimer_Node *np;

186 np = tlist->head;
    if(np == NULL)
188 {    //no UTimer Node exists
      tlist->head = ptr;
190    }
```

```
        else if(ptr->utimer.priority > np->utimer.priority)
192     {   //the head node has a lower priority
          ptr->pre = NULL;
194       ptr->next = np;
          np->pre = ptr;
196       tlist->head = ptr;
        }
198     else
        {   //find a spot for it in the list
200       UTimer_Node *prev;
          while(np != NULL) {
202         if(ptr->utimer.priority > np->utimer.priority) {
              np->pre->next = ptr;
204           ptr->next = np;
              ptr->pre = np->pre;
206           np->pre = ptr;
              break;
208         }
            else {
210           prev = np;
              np = np->next;
212         }
          }
214
          if(np == NULL) {
216         prev->next = ptr;
            ptr->pre = prev;
218         ptr->next = NULL;
          }
220     }
      }
222
    void cancelUTimer(int UtimerID){
224     UTimer_Node *p;

226     p = tlist->head;
        while((p!= NULL) && (p->utimer.UtimerID != UtimerID))
228       p = p->next;

230     if (p != NULL) {
          if (p == tlist->head) {
232         tlist->head = p->next;

234         if (p->next != NULL)
              p->next->pre = NULL;
236       } else {
            p->pre->next = p->next;
238         if (p->next != NULL)
              p->next->pre = p->pre;
240       }
```

```
242
        p->next = NULL;
        p->pre = NULL;
        free(p);
244   }
    }
```

Listing 21.9
UTimerServer.c: managing registered UTimer objects.

Listing 21.9 gives the UTimer server code. In setupPulse(), a channel with the name Utimer_Server is created. Note that many connections can be attached to this channel: one is created by the UTimer server itself and will be used by the OS timer to send pulses; one is created by each client via a call to name_open(). When the server calls MsgReceive() in the main() function, depending on the value of *rcvid*, the received object can be a message of type ClientMsgT from a client or a pulse from the OS timer. If a pulse is received, the server updates all the UTimer objects and sends a pulse to a client if one of those UTimer objects registered by that client expires. If a message is received, depending on the message type, the server will add a new UTimer object to its UTimer list, or remove a UTimer object from the list.

Figures 21.10 and 21.11 give screenshots of a run of the client and the UTimer server.

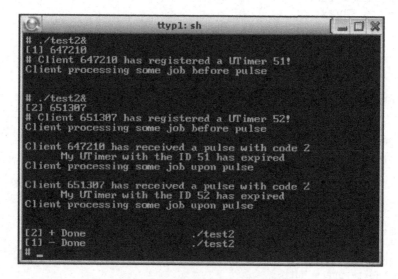

Figure 21.10
A run of the program in Listing 21.8.

Figure 21.11
A run of the program in Listing 21.9.

Problems

21.1　What can cause an OS signal to be generated? By default, what type of signal is raised when an OS timer expires?

21.2　How does a task respond when it receives a signal? Compare the handling process of hardware interrupts and that of OS signals.

21.3　Refer to the example in Section 21.5.2, and explain how signals are used together with a semaphore to achieve execution preemption and resumption.

21.4　Compared with semaphores, what is the main advantage of using a spinlock?

21.5　In QNX, how do you attach an ISR to a hardware interrupt? Can multiple ISRs be attached to the same hardware interrupt source? If so, in what order will the ISRs be invoked when an interrupt event occurs?

21.6　Refer to the example in Section 21.6.4. Why is it necessary to use two spinlocks?

21.7　Explain the QNX channel concept. Explain how the QNX synchronous communication protocol works.

21.8　What is a pulse? What is a pulse event?

21.9　Modify TClient.c given in Listing 21.8 so that a client can cancel the UTimer objects installed by itself.

21.10　Modify TClient.c given in Listing 21.8 so that a client can have multiple dedicated threads as suggested in the end of Section 21.7.2. Draw a UML sequence diagram to illustrate the parallel tasks of the client design.

21.11　Modify UTimerServer.c given in Listing in Listing 21.9 so that the server can have multiple dedicated threads as suggested in the end of Section 21.7.2. Draw a UML sequence diagram to illustrate the parallel tasks of the server design.

Software Timer Management

Contents

As the Wheel of Time turns, places wear many names.... Yet no one knows the Great Pattern the Wheel weaves.

Robert Jordan

22.1 Hardware Timer and Software Timer

To simplify our discussion, we use a personal computer (PC) as an example to explain the timer concept. A modern PC typically has a main system clock that is generated by a base oscillator—a circuit regulated by a quartz crystal that vibrates at a certain frequency when electricity is passed through it.

On the motherboard there are circuits called "frequency dividers" or "frequency multipliers," which are used to split or multiply the system clock to produce various clock signals needed by certain components. In this way, each device is synchronized with the system clock to fulfill its functionality, such as instruction execution and data transmission. Table 22.1 gives some example clocks and how they relate to each other. For instance, a frequency divider is used to reduce the system clock to accommodate the 33 MHz PCI bus; the processor operates at 600 MHz, which is six times the frequency of the system clock.

Real-Time Embedded Systems. http://dx.doi.org/10.1016/B978-0-12-801507-0.00022-5

Table 22.1 Hardware clocks in a reference PC

Clock	Speed (MHz)	Generated as
System clock	100	Base oscillator
Memory bus	100	System clock
Processor	600	System clock × 6
Level 2 cache	300	System clock × 3
PCI bus	33	System clock/3
ISA bus	8.3	System clock/12
PIT source clock	14.31818	PIT oscillator
PIT chip input	1.193181$\overline{6}$	PIT source clock/12

A PC typically has a special chip called a programmable interval timer (PIT). The input clock frequency of the PIT chip runs (roughly) at 1.193182 MHz, which is 1/12 the frequency of a 14.31818 MHz oscillator. A PIT may feature several channels. The output from channel 0 is referred to as the *hardware timer*, which is wired to the interrupt request 0 line of the programmable interrupt controller (PIC) to generate timer interrupt signals.

Each PIT channel is associated with a hardware counter (divider). Let us use `Counter0` to refer to the counter associated with channel 0. Every pulse from the PIT input clock causes `Counter0` to be decreased. When `Counter0` has reached zero, it is reset and a pulse is generated on the output. The hardware timer frequency can be changed by setting `Counter0` to a desired value. Table 22.2 gives some examples. For instance, if `Counter0` is set to 11,931, then the PIT channel 0 output frequency is 100 Hz, and timer interrupt occurs every 10 ms.

As shown in Figure 22.1, when a hardware timer interrupt occurs, the microprocessor honors the interrupt request by executing the timer interrupt service routine (ISR) launched upon system startup. This timer ISR may have the following responsibilities:

- The OS kernel keeps track of the current time by a time-of-day variable. Suppose the hardware timer interrupts occur at exactly 10 ms intervals. Every time the timer ISR runs, it means that 10 ms has just elapsed, and the time-of-day variable is incremented

Table 22.2 Hardware timer examples

PIT Chip Output Speed (Hz)	Generated as	PIC Timer Interrupt
18.2065	PIT chip input/65,536	Every 54.9254 ms
100	PIT chip input/11,931	Every 10 ms
1000	PIT chip input/1193	Every 1 ms
10,000	PIT chip input/119	Every 100 μs
100,000	PIT chip input/12	Every 10 μs

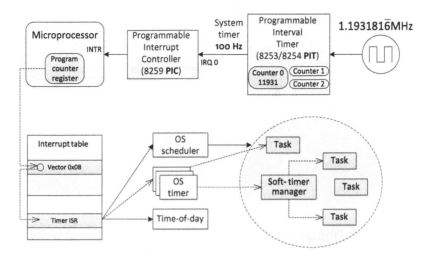

Figure 22.1
Hardware timer and timer interrupt.

by 10 ms. For such a reason, the timer interrupt frequency is also called the system's *timing resolution*.

*   The timer ISR may inform the operating system (OS) scheduler if it needs to schedule tasks at each timer tick.
*   The timer ISR may check each of the OS timers installed by applications and issue time-out signals if required. When at time t an application installs an OS timer with a value of w, the OS calculates the absolute time $t + w$ and inserts it into a *timer queue* sorted in time order. Every time the timer ISR runs, it compares the current time (i.e. time of day) against the head of the timer queue. When the current time is greater than or equal to the head of the queue, a time-out signal is sent to whichever thread installed the timer.

The interrupt latency and overhead can be substantial as the number of OS timers increases. Most real-time applications require only time-out signals with a course granularity— say, on the order of milliseconds or even hundreds of milliseconds. In order to reduce interrupt overhead, a software timer manager is typically implemented to allow client programs to install timers with granularities that are much higher than the hardware timing resolution.

Both the OS timers and the timers monitored by a software timer manager are *software timers*.

Now, suppose a system has a timing resolution of 10 ms, and an application installs a 20 ms OS timer. Can this application get a time-out signal exactly when 20 ms has just elapsed? The answer is no. As shown in Figure 22.2, being unaware of the exact instant when the hardware timer ticks, an application may issue the 20 ms time-out request at any point between two ticks. The immediate next time tick is not counted because there is no way to tell how much

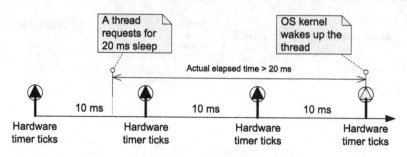

Figure 22.2
Timing resolution and imprecision.

time has just passed since the last timer tick. This uncounted time interval is called the *jitter time*. It is at the third timer tick that a time-out signal is sent to the application, at which time more than 20 ms has just passed. Obviously, such *timing imprecision* is due to the jitter time. In general, the timing of everything in a system is no more accurate than the system's timing resolution.

A system's timing resolution is usually set to tens of microseconds up to tens of milliseconds. As indicated in Table 22.2, when Counter0 is set to 12, the timer interrupt occurs every 10 µs, which also gives us a timing resolution of 10 µs. Then, could we simply set Counter0 to a smaller value or even use a multiplier to obtain a better system timing resolution? The answer is again no, because each time tick imposes an interrupt load on the system. A very small tick size could cause all the processor cycles to be consumed by the timer ISR, leaving no time to process user tasks.

Although there is a lower bound for system timing resolution, some OSs may offer timing services with a granularity that is much finer than the system timing resolution. For example, the QNX OS maintains a 64-bit high-resolution counter that increases by 1 on each processor cycle. For a 1 GHz processor, this counter increases every nanosecond [7]. This counter can be read by ClockCycles(); it is also used to implement the function nanosleep().

22.2 Software Timer Manager

Let us now take a closer look at the implementation of a software timer manager.

Assume that our job is to design a software timer manager with a timing resolution of 100 ms. In other words, it is able to handle timers that are multiples of 100 ms. Obviously, it needs to take appropriate actions every time 100 ms has just passed. The question is, how does this timer manager know that 100 ms has just passed?

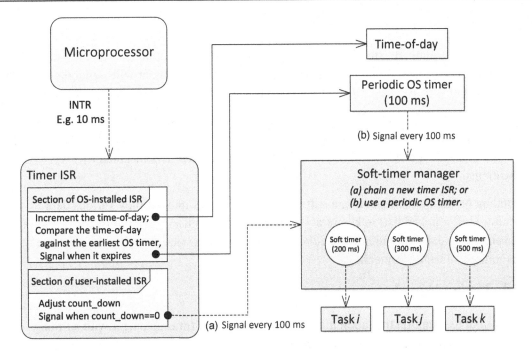

Figure 22.3
Soft-timer architecture.

Two solutions are illustrated in Figure 22.3: one is to chain a new timer ISR dedicated to the software timer manager, and the other is to take advantage of an OS timer.

22.2.1 Chain a Dedicated Timer ISR

We learned about *interrupt chaining* in Section 4.5.3. This approach requires the designer of the software timer manager to attach a new timer ISR that is dedicated to the timer manager.

As illustrated in Figure 22.3, the timer ISR installed by the OS is responsible for updating the time of day and monitoring OS timers. The user-installed timer ISR is responsible for signaling the timer manager when a certain condition becomes true. Specific to our example, the timer manager needs to know the elapsing of 100 ms. Hence, every 100 ms the dedicated timer ISR should send a time-out signal to the timer manager. Suppose the system's timing resolution is 10 ms. Owing to interrupt chaining, the dedicated timer ISR runs every 10 ms. In order to monitor the elapsing of 100 ms, it uses a countdown variable, which is set to 10 initially, decreases by 1 every time the ISR runs, and is reset to 10 whenever it becomes 0. Moreover, the ISR sends a time-out signal whenever the countdown variable becomes 0. Consequently, the timer manager is pulsed in a period of exactly 100 ms.

22.2.2 Use an OS Timer

This approach is relatively simpler. As illustrated in Figure 22.3, upon startup, the timer manager launches an OS timer with a period of 100 ms. This OS timer will signal the software timer manager every 100 ms (note that the OS timer is ultimately monitored by the timer ISR).

Regardless of whether a dedicated timer ISR or a periodic OS timer is used, every time the software timer manager receives a time-out signal, it knows that 100 ms has just passed, and this is the time that it needs to take appropriate actions on the timers installed by user applications.

Depending on the implementation, a software timer manager may employ various data structures (say, a linked list) to keep track of user-installed timers. Inspired by the clocks we use every day, a data structure called a *timing wheel* [48] has been proved to be effective for managing user-installed timers.

22.3 Timing Wheels

Now it is our assumption that upon initialization, a software timer manager with a timing resolution of τ can always get a timer tick every τ units of time has passed (by launching a periodic OS timer or a dedicated timer ISR).

A *timing wheel*, denoted by $\circlearrowright_{(n,\tau)}^{i}$, is a circular structure with n slots, denoted by ϖ_0 to ϖ_{n-1}. The index i is called the wheel's dial (aka clock hand), which always indicates the current slot ϖ_i. The value of the index i changes exactly every τ units of time: $i = 0$ upon initialization, it keeps its current value (i.e., the wheel stays at the current slot) for exactly τ units of time, and then becomes $(i + 1) \pmod{n}$.

A wheel-based timer manager with a resolution of τ is a software timer manager equipped with a timing wheel $\circlearrowright_{(n,\tau)}^{i}$, where n is a fixed number and each slot of the wheel is associated with a list of software timers installed upon requests from client tasks. On each timer tick, the manager moves the dial of the wheel to the next slot, and checks the list of timers associated with that slot. If any timer expires, it removes the timer and sends a time-out signal to the requesting task.

Figure 22.4 shows two situations of a timing wheel $\circlearrowright_{(10,100\,\text{ms})}^{i}$, where $i = 0$ in Figure 22.4(a) and $i = 3$ in Figure 22.4(b). The wheel's timing range per round is $100\,\text{ms} \times 10 = 1\,\text{s}$.

A timing wheel $\circlearrowright_{(n,\tau)}^{i}$ can be used to accommodate software timers that are multiples of τ. Let us use $\phi(k, l)$ to denote a timer, where k is the timing requirement and l indicates the time left before this timer expires. We have $k = l$ when the timer is requested. When a new timer is installed, the current dial is always used as the *reference point* to determine the slot in which the timer should be placed.

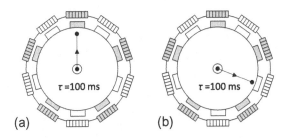

Figure 22.4
A timing wheel with a resolution of 100 ms. (a) Initial setting. (b) 300 ms passed.

Now, let us assume that the timer manager receives a request to install a timer
$\phi(300\,\text{ms}, 300\,\text{ms})$ when its timing wheel $\circlearrowleft^0_{(10,100\,\text{ms})}$ is as shown in Figure 22.4(a). Since the
reference point $i = 0$, the manager will install $\phi(300\,\text{ms}, 0)$ in slot ϖ_3. Note that the time left
for the installed timer is 0; this is because 300 ms would have passed (actually slightly less
than 300 ms, the reason will soon be clear) as the dial moves to slot ϖ_3.

It is a bit complicated to install a timer that is over the wheel's timing range. One solution is
explained below for the installation of a timer $\phi(1200\,\text{ms}, 1200\,\text{ms})$. Assume that
Figure 22.4(b) shows the setting of the timing wheel when the timer manager receives a
request to install the 1200 ms timer.

1. The timer $\phi(1200\,\text{ms}, 200\,\text{ms})$ is temporarily installed in the current slot ϖ_3. Note
   that the time left for the installed timer is 200 ms; this is because when this timer is
   examined the next time, the wheel would have undergone a whole round (i.e., 1000 ms has
   passed).
2. As the dial wraps around and reaches slot ϖ_3 again, the timer manager checks
   the timer and realizes that there is still 200 ms left. It then reinstalls the timer
   $\phi(1200\,\text{ms}, 0)$ in slot ϖ_5. The reader should now be clear why the time left for this timer
   becomes 0.
3. As the dial moves to slot ϖ_5, the timer manager checks the timer and realizes that it is
   now time to notify the user that installed the original 1200 ms timer.

The timing wheel $\circlearrowleft^i_{(10,100\,\text{ms})}$ as shown in Figure 22.4 can be implemented by using the
following structures.

```
struct MilliTimer{
    int timerID; //unique ID
    int priority;
    int left_time;
};
```

```
typedef struct Timer_Node{
    struct Timer_Node* pre;
    struct Timer_Node* next;
    struct MilliTimer  mtimer;
    int     rcvid;       //notify me when timeout
    struct sigevent    pulseEvent;
} Timer_Node;

struct Timer_Slot{
    struct Timer_Node* head;
    struct Timer_Node* tail;
};

struct TimingWheel{
    struct Timer_Slot* milli[10];
    int current_slot;
    int resolution = 100; // 100 ms
};
```

Specifically, the timing wheel $\circlearrowright^i_{(10,100\,ms)}$ is a TimingWheel object with 10 Timer_Slot objects, each of which is a list of Timer_Node objects. A Timer_Node contains a MilliTimer object, pointers to its neighbors, and information for notifying the installer as the timer expires.

A wheel-based timer manager can be implemented in the same way as the UTimer manager described in Section 21.7.5. For instance, the pseudocode in HALDLETIMERTICK is different from, but corresponds to, the function handleAPulse() in the UTimer manager. As far as the software architecture is concerned, it is obvious that the UTimer manager is similar to WHEEL-MANAGER given below.

INSTALL-TIMER (*TimingWheel* : *tw*, *TimerNode* : *tn*)

1   $index = tw.current\_slot$
2   $N = tw.Num\_Of\_Slots$
3   $\tau = tw.resolution$
4   $tvalue = tn.left\_time$
5   **if** $tvalue < N \times \tau$
6       $tn.left\_time = 0$
7       $future\_slot = (index + \lceil tvalue/\tau \rceil) \pmod{N}$
8   INSTALL-TIMER-AT (*tn*, *future_slot*)
9   **else**   // the timer needs to be reinstalled later
10      $tn.left\_time = tvalue - N \times \tau$
11   INSTALL-TIMER-AT (*tn*, *index*)

HALDLETIMERTICK (*TimingWheel* : *tw*)

1   FORWARDWHEELS (*tw*)
2   *index* = *tw.current_slot*
3   *tlist* = GET-TIMER-LIST (*index*)
4   **while** *tlist* ≠ ∅
5   *tNode* = GET-NEXT-TIMER-NODE (*tlist*)
6   **if** *tNode.left_time* == 0
7   SIGNAL-TASK (*tNode.taskID*)
8   **else**   // the timer is not expired yet
9   INSTALL-TIMER (*tw*, *tNode*)
10 **end**

WHEEL-MANAGER (N, τ)

1   *tw* = INITIALIZE-TIMING-WHEEL (N, τ)
2   REGISTER-TO-TIMER-INTERRUPT (τ)   // or use periodic OS timer
3   **for** (; ;)
4   *msg* = RECEIVE-MESSAGE ()
5   **if** *msg* is a timeout signal
6   HANDLETIMERTICK (*tw*)
7   **clsc**   // *msg* is a user request for timer installation
8   [*taskID*, *tvalue*] = GETTIMERREQUEST (*msg*)
9   *tNode* = CREATE-NEW-TIMER (*taskID*)
10   *tNode.left_time* = *tvalue*
11   INSTALL-TIMER (*tw*, *tNode*)
12   **end**
13 **end**

22.3.1 Precision Error

As we have mentioned before, software timers may suffer from *timing imprecision* for the following reasons:

(1) Timer expiration is checked only at timer ticks.
(2) Timers can be installed at any arbitrary time instant. In particular, when a timer is installed at a time instant *t*, it is highly likely that *t* is not exactly at a timer tick, but in between two ticks.

Consequently, when a software timer manager is implemented, time-out signals are sent out to the corresponding users either slightly earlier or slightly later than expected. One way or the other, such a bias has to be taken.

In line 7 of INSTALL-TIMER,

$$(current\_slot + \lceil tvalue/\tau \rceil) \pmod{N}$$

is assigned to the variable *future_slot*.

By this, the implementation has actually taken a bias toward signaling users earlier. If in line 7

$$(current\_slot + \lceil tvalue/\tau \rceil + 1) \pmod{N}$$

were assigned instead to the variable *future_slot*, the algorithm would have a bias toward warning users later.

Let us use an example to explain this further. Suppose a user asks to install a 200 ms timer in the timing wheel $\circlearrowright_{(10,100\,ms)}^{0}$ as shown in Figure 22.4(a). Consider two cases as illustrated in Figure 22.5.

In the case shown in Figure 22.5(a), the timer is installed slightly after a timer tick. More formally, δ ms after the manager has moved the dial to slot ϖ_0, the timer is installed in slot

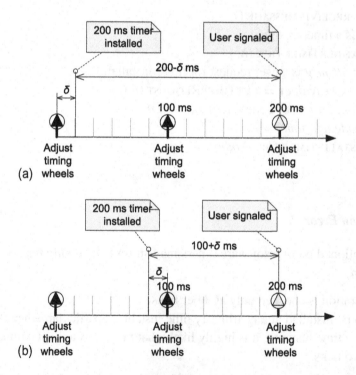

Figure 22.5

Precision errors when a timing wheel is used. (a) User timer installed right after entering into a new time slot. (b) User timer installed right before entering into a new time slot.

$(current\_slot + \lceil tvalue/\tau \rceil)$ (mod N) $= (0 + \lceil 200/100 \rceil)$ (mod 10) $= 2$. This means this timer will expire in $(200 - \delta)$ ms, which is slightly earlier than 200 ms.

In the case shown in Figure 22.5(b), the timer is installed slightly before a timer tick. More formally, δ ms before the manager moves the dial to slot ϖ_1 (or equivalently, $(100 - \delta)$ ms after the manager has moved the dial to slot ϖ_0), the timer is installed in slot ϖ_2. This means this timer will expire in $(100 + \delta)$ ms, which is still earlier than 200 ms.

In both cases, the user that has installed the 200 ms timer would get a time-out signal earlier than expected. The actual timing duration for a 200 ms timer is in the range $(100 \text{ ms}, 200 \text{ ms}]$. In general, the actual timing duration for a timer $\phi(k, l)$ installed in a timing wheel with a resolution τ is in the range $(k - \tau, k]$. In other words, the maximum precision error is the same as the wheel's timing resolution τ.

According to the above analysis, the timing wheel as shown in Figure 22.4 is not usable if a user needs to install timers with a 50 ms tolerance of precision error. In order to satisfy the user's timing requirements, the timing wheel has to be redesigned— say, by increasing the wheel's timing resolution.

22.3.2 Timers with Wide Ranges

Let us now consider a different issue. Given the timing wheel $\circlearrowleft^3_{(10,100 \text{ ms})}$ in Figure 22.4(b), how do we install a timer of 4 min, 5 s, and 200 ms?

One solution is to reinstall the timer with lesser and lesser *left_time* values. Let us denote the original timer by $\phi(245{,}200 \text{ ms}, 245{,}200 \text{ ms})$.

1. The timer $\phi(245{,}200 \text{ ms}, l)$ is first installed in slot ϖ_3, where
   $l = 245{,}200 - 1000 = 244{,}200$ ms.
2. As the dial wraps around and reaches slot ϖ_3 again (i.e., 1 s has just passed), the timer is removed and reinstalled in slot ϖ_3 with $l = l - 1000$ ms. This step repeats 244 times.
3. As the dial wraps around and reaches slot ϖ_3, the timer is removed and reinstalled in slot ϖ_5 with $l = 0$ ms.
4. As the dial reaches slot ϖ_5, the timer expires and a time-out signal is sent to the user.

Another solution is to trade space for time: redesign the timing wheel such that it has at least 2452 slots with each still representing 100 ms. This huge timing wheel makes it unnecessary to reinstall timers that are smaller than 245,200 ms. However, this is definitely not a good solution because of memory consumption. A more scalable solution is to use a hierarchical structure—hierarchical timing wheel.

22.4 Hierarchical Timing Wheels

A clock used in our real world typically has more than one hand, and for different hands, one tick indicates different time units.

Similarly, a k-layer *hierarchical timing wheel* involves k timing wheels, denoted by $\circlearrowleft_{(n_k,\tau_k)}^{i_k} \cdots \circlearrowleft_{(n_2,\tau_2)}^{i_2} \circlearrowleft_{(n_1,\tau_1)}^{i_1}$, where

- $n_j \times \tau_j = \tau_{j+1}$ ($1 \leq j < k$). In other words, the time duration represented by one round of a wheel at the jth layer equals the duration represented by one tick of the wheel at the $(j+1)$th layer.
- the dial i_j ($1 \leq j \leq k$) takes integer values from $[0, n_j - 1]$.
- its timing resolution (timer tick period) is τ_1.
- its timing range is $n_k \times \tau_k$.

As an example, Figure 22.6 shows a four-layer hierarchical timing wheel

$$\circlearrowleft_{(60,1\,\text{min})}^{i_4} \circlearrowleft_{(60,1\,\text{s})}^{i_3} \circlearrowleft_{(10,100\,\text{ms})}^{i_2} \circlearrowleft_{(10,10\,\text{ms})}^{i_1},$$

where its timing resolution is 10 ms and its timing range is 60 min.

After the k-layer hierarchical timing wheel starts, on each timer tick, the wheel at the first layer moves its dial to its next slot. Whenever the dial of the first-layer wheel enters its slot ϖ_0, the wheel at the second layer moves its dial to its next slot. Similarly, whenever the dial of

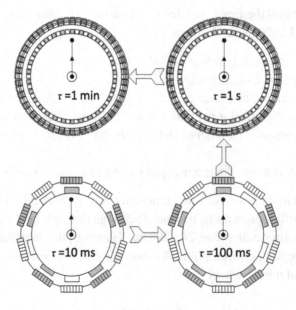

Figure 22.6
A four-layer hierarchical timing wheel.

the second-layer wheel enters its slot ϖ_0, the wheel at the third layer moves its dial to its next slot, and so on. The whole wheel system starts over again as the dial of the kth-layer wheel re-enters its slot ϖ_0.

22.4.1 Reference Context and Timer Management

Given a k-layer hierarchical timing wheel $\circlearrowleft_{(n_k,\tau_k)}^{i_k} \cdots \circlearrowleft_{(n_2,\tau_2)}^{i_2} \circlearrowleft_{(n_1,\tau_1)}^{i_1}$, its current context is denoted by $[i_k, \ldots, i_2, i_1]$. For example, the current context of the four-layer hierarchical timing wheel in Figure 22.6 is $[0, 0, 0, 0]$, while the current context of the four-layer hierarchical timing wheel in Figure 22.7 is $[41, 21, 2, 8]$.

Whenever a new timer is installed, the current context is always used as a *reference context*. Let us use an example to show how to install a timer in a hierarchical timing wheel. Suppose a timer of 4 min 5 s 200 ms is to be installed in the wheel as shown in Figure 22.7. Let us denote the original timer by $\phi(4 \min 5 \text{ s } 200 \text{ ms}, 4 \min 5 \text{ s } 200 \text{ ms})$. The timer installation procedure is as follows:

(1) The reference context $[41, 21, 2, 8]$ is saved for this timer.
(2) Start with the minute wheel $\circlearrowleft_{(60,1 \min)}^{i_4}$. The timer $\phi(4 \min 5 \text{ s } 200 \text{ ms}, 5 \text{ s } 200 \text{ ms})$ is installed in slot ϖ_{45} of the wheel, four ticks ahead of the current slot ϖ_{41}.
(3) At the time instant when the dial of the minute wheel moves to slot ϖ_{45}, the current context becomes $[45, 0, 0, 0]$. Is this the time to process the timer $\phi(4 \min 5 \text{ s } 200 \text{ ms}, 5 \text{ s } 200 \text{ ms})$? The answer is no, because the time elapsed since the reference context $[41, 21, 2, 8]$ is still less than 4 min.

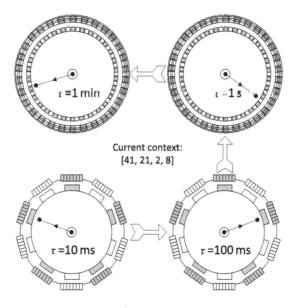

Current context:
[41, 21, 2, 8]

Figure 22.7
An example of using a hierarchical timing wheel.

(4)   As time goes on, at the timer tick when the current context becomes $[45, 21, 2, 8]$, the manager would know that 4 min has just passed since the timer installation. At this point, it is clear that the reference context is used to suppress the timing precision error under τ_1. Now, it is the time to consider the time left for the timer. The timer $\phi(4\,\text{min}\,5\,\text{s}\,200\,\text{ms}, 5\,\text{s}\,200\,\text{ms})$ is removed from the minute wheel and the timer $\phi(4\,\text{min}\,5\,\text{s}\,200\,\text{ms}, 200\,\text{ms})$ is added in slot ϖ_{26} of the second wheel $\circlearrowleft_{(60,1\text{s})}^{i3}$, five ticks ahead of the current slot ϖ_{21}.

(5)   At the time instant when the dial of the second wheel moves to slot ϖ_{26}, the current context becomes $[45, 26, 0, 0]$. Is this the time to process the timer $\phi(4\,\text{min}\,5\,\text{s}\,200\,\text{ms}, 200\,\text{ms})$? The answer is no, because the time elapsed since the reference context $[41, 21, 2, 8]$ is still less than 4 min 5 s.

(6)   As time goes on, at the timer tick when the current context becomes $[45, 26, 2, 8]$, the manager would know that 4 min 5 s has just passed since the timer installation. Now, it is the time to consider the time left for the timer. The timer $\phi(4\,\text{min}\,5\,\text{s}\,200\,\text{ms}, 200\,\text{ms})$ is removed from the minute wheel and the timer $\phi(4\,\text{min}\,5\,\text{s}\,200\,\text{ms}, 0)$ is added in slot ϖ_4 of the 100 ms wheel $\circlearrowleft_{(10,100\,\text{ms})}^{i2}$, two ticks ahead of the current slot ϖ_2.

(7)   At the time instant when the dial of the 100 ms wheel moves to slot ϖ_4, the current context is $[45, 26, 4, 0]$. Is this the time to process the timer $\phi(4\,\text{min}\,5\,\text{s}\,200\,\text{ms}, 0)$? The answer is no, because the time elapsed since the reference context $[41, 21, 2, 8]$ is still less than 4 min 5 s 200 ms.

(8)   As time goes on, at the timer tick when the current context becomes $[45, 26, 4, 8]$, the manager would know that 4 min 5 s 200 ms has just passed since the timer installation. Now, it is the time to send a time-out signal to the whichever user installed this timer.

Installing a periodic timer is also straightforward. The only difference is that when a periodic timer expires, it is reinstalled as soon as a time-out signal is sent out.

22.4.2 Implementation

As an example, the following data structures can be used to implement a three-layer hierarchical timing wheel that offers a resolution of 50 ms and a timing range of 5 min.

```
#define minute_slots    5
#define second_slots    60
#define millis_slots    20

struct WheelContext{
    int minute_index;
    int second_index;
    int millis_index;
};
```

```
struct MyTimer{
    int timerID; //unique ID
    int priority;
    int minutesLeft;
    int secondsLeft;
    int millisLeft;
};

typedef struct Timer_Node{
    struct Timer_Node*   pre;
    struct Timer_Node*   next;
    struct WheelContext  reference_context;
    struct MyTimer       mytimer;
    int     rcvid;        //notify me when timeout
    struct sigevent      pulseEvent;
} Timer_Node;

struct Timer_Slot{
    struct Timer_Node* head;
    struct Timer_Node* tail;
};

struct HierarchicalTimingWheel{
    struct Timer_Slot*   millis_wheel[millis_slots];
    struct Timer Slot*   second_wheel[second_slots];
    struct Timer_Slot*   minute_wheel[minute_slots];
    struct WheelContext  current_context;
    int     resolution = 50; // 50 ms
};
```

Notice the following points:

- This hierarchical timing wheel is composed of three wheels: millis_wheel, second_wheel, and minute_wheel, which have 20, 60, and five timer slots, respectively.
- Each timer slot contains a timer list, which is represented by two pointers that, respectively, point to the list's head and tail—two Timer_Node objects.
- A Timer_Node object contains a timer, a reference context (the dial location of each wheel when the timer is installed), and information for notifying the owner of the timer.
- A timer object has a unique ID and information about how much time is left before this timer expires.

Obviously, this design is an extension of the wheel design given in Section 22.3. Likewise, a timer manager based on a hierarchical timing wheel can be implemented the same way as the UTimer manager described in Section 21.7.5. The main difference is the mechanism for forwarding the dials upon timer ticks, and the use of reference contexts for monitoring timer time-outs. A pseudocode is given in HALDLETIMERTICK() below.

HALDLETIMERTICK (*HierarchicalTimingWheel* : *hw*)

```
1   FORWARD-WHEELS (hw)
2   if CURRENTSLOTHASTIMERS(hw->minute_wheel)
3       ptimer = hw->minute_wheel[hw->current_context.minute_index]->head;
4       while (ptimer != NULL)
5           pnext = ptimer->next;
6           if ((ptimer->reference_context.second_index ==
                       hw->current_context.second_index) &
                   (ptimer->reference_context.millis_index ==
                       hw->current_context.millis_index))
7               REMOVETIMERFROMWHEEL(ptimer, hw->minute_wheel);
8               if (ISTIMEOUT(ptimer))
9                   SENDTIMEOUTEVENT(ptimer->taskID);
10              else INSTALLTIMER(ptimer, hw->second_wheel);
11          ptimer = pnext;
12      end
13  end
14  if CURRENTSLOTHASTIMERS(hw->second_wheel)
15      ptimer = hw->second_wheel[hw->current_context.second_index]->head;
16      while (ptimer != NULL)
17          pnext = ptimer->next;
18          if (ptimer->reference_context.millis_index ==
                       hw->current_context.millis_index)
19              REMOVETIMERFROMWHEEL(ptimer, hw->second_wheel);
20              if (ISTIMEOUT(ptimer))
21                  SENDTIMEOUTEVENT(ptimer->taskID);
22              else INSTALLTIMER(ptimer, hw->millis_wheel);
23          ptimer = pnext;
24      end
25  end
26  if CURRENTSLOTHASTIMERS(hw->millis_wheel)
27      ptimer = hw->millis_wheel[hw->current_context.millis_index]->head;
28      while (ptimer != NULL)
29          pnext = ptimer->next;
30          REMOVETIMERFROMWHEEL(ptimer, hw->millis_wheel);
31          SENDTIMEOUTEVENT(ptimer->taskID);
32          ptimer = pnext;
33      end
34  end
```

FORWARD-WHEELS (*HierarchicalTimingWheel* : *hw*)

1 hw->current_context.millis_index++;
2 **if** (hw->current_context.millis_index % millis_slots == 0)
3   hw->current_context.millis_index = 0;
4   hw->current_context.second_index++;
5   **if** (hw->current_context.second_index % second_slots == 0)
6   hw->current_context.second_index = 0;
7   hw->current_context.minute_index++;
8   **if** (hw->current_context.minute_index % minute_slots == 0)
9   hw->current_context.minute_index = 0;

Figurc 22.8 gives a UML class diagram showing the design of a general hierarchical timing wheel.

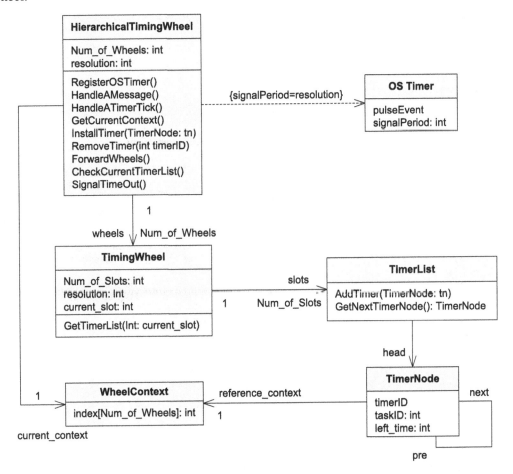

Figure 22.8
UML class diagram of a hierarchical timing wheel.

Problems

22.1 What is the major difference between hardware timers and software timers?

22.2 Suppose a system has applications that need to install timers of 200, 350, and 400 ms. If a software timer manager is designed to support those timer requirements, what would be the timing resolution of the timer manager? (Hint: find the greatest common divisor.)

22.3 Explain how to install a 600 ms timer in the timing wheel as given in Figure 22.4(b).

22.4 What is a hierarchical timing wheel? How does it work?

22.5 Explain how to install a 3 min 45 s 360 ms timer in the timing wheel as given in Figure 22.7.

22.6 Figure 22.9 shows an architecture design of a timer manager based on a hierarchical timing wheel (You may want to refer to Figure 21.9 for the use of communication channels). Implement the timer manager and test it by installing the following timers:

(a) a timer of 200 ms;

(b) a timer of 4 s 400 ms; and

(c) a timer of 6 min 28 s 700 ms.

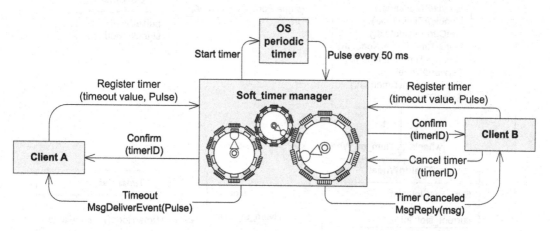

Figure 22.9
A soft-timer manager using a timing wheel.

QNX Resource Management

Contents

Management is doing things right; leadership is doing the right things.

Peter F. Drucker

23.1 Introduction to QNX Resource Management

In Chapter 13 we briefly introduced QNX—a POSIX-compliant operating system (OS) that is built upon a *microkernel architecture*. The microkernel is optimized to provide the minimal services used by a team of optional cooperating processes, which in turn provide the higher-level OS functionality [8].

A key design philosophy of QNX is the use of resource managers for managing a wide variety of devices that may be found in embedded systems. A *resource manager* is a user-level server program that presents an abstract view of some service(s). There are two types of resource managers: device resource managers and filesystem resource managers.

Device resource managers may deal with actual hardware devices, such as serial ports, parallel ports, network cards, and disk drives. They are traditionally called *device drivers*, handling

Real-Time Embedded Systems. http://dx.doi.org/10.1016/B978-0-12-801507-0.00023-7

hardware interrupts and managing the flow of data between devices and user applications. Device resource managers may also deal with virtual devices, such as pseudoterminals, message queues, and pipes.

A filesystem resource manager deals with a collection of resources of the same kind, presenting them *visually* as a directory in the filesystem. For example, a filesystem resource manager can be implemented to render the contents of a tar file as a filesystem with files and subfolders.

A resource manager is known to the QNX microkernel by a mechanism called pathname space mapping. In particular, the QNX kernel is paired with a process manager, which is responsible for managing processes, memory, and pathnames.

- A device resource manager, at startup, registers a specific pathname to the process manager. It creates only a single-file entry in the filesystem to represent the device under consideration.
- A filesystem resource manager, at startup, registers a specific *mountpoint*—a pathname prefix—to the process manager. It creates a directory at the mountpoint and is responsible for all the resources represented by the entries under the mountpoint.

By pathname space mapping, devices and resources are rendered as filesystem entries. A direct benefit is that all the functionality of a resource manager can be accessed using file-descriptor-based POSIX function calls, such as the open(), read(), and write() functions. This allows a client application to interact with multiple servers through a common interface. Moreover, for simple testing, a user can easily interact with a server by command-line POSIX utilities, such as cat and echo.

All QNX device drivers and filesystems are implemented as resource managers.

23.2 Resource Manager Architecture

As shown in Figure 23.1, the process manager maintains a *pathname tree* to track the pathname space. The process manager creates a record for each active resource manager; the record associates the specified pathname/mountpoint with the information for locating the manager (i.e., a node descriptor, a process ID, a channel ID, and a handle[1]).

Figure 23.2 illustrates a scenario of using a serial communication port managed by the devc-ser8250 process:

(1) The client calls the POSIX function open(), attempting to open the file "/dev/ser1" with the O_RDWR permission.

---

[1] Handles are used to distinguish multiple names when a resource manager registers more than one name.

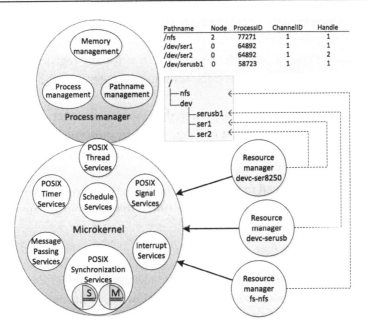

Pathname	Node	ProcessID	ChannelID	Handle
/nfs	2	77271	1	1
/dev/ser1	0	64892	1	1
/dev/ser2	0	64892	1	2
/dev/serusb1	0	58723	1	1

Figure 23.1

Pathname space mapping.

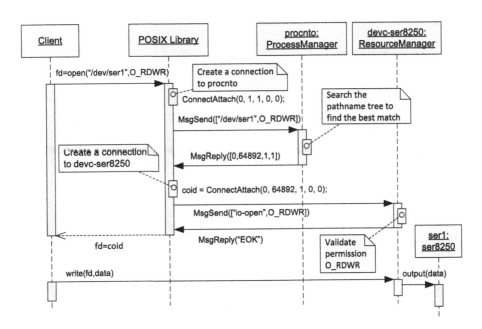

Figure 23.2

A scenario of using a resource manager.

(2) The POSIX function creates a connection to the process manager and sends a request message to the process manager.

(3) The process manager looks through its pathname tree for any registered pathname prefixes that match the query name. Typically, the resource manager with the longest prefix match is returned.

(4) The POSIX function creates a connection to the resource manager.

(5) The POSIX function sends a request message to the resource manager.

(6) The resource manager validates the access mode specified in the open( ) call, and returns EOK if permitted.

(7) The connection ID is returned to the client, which can then write to or read from the hardware device just like working with a file (through the resource manager).

23.2.1 Control Structure

Figure 23.3 illustrates the control flow of a resource manager. It has a loop structure, continuously waiting for well-defined messages from clients. In general, a resource manager follows the procedure below to handle client messages:

(1) It receives a client message from its channel's receive buffer.

(2) Depending on the message type, it dispatches the message to a default or user-defined message handler.

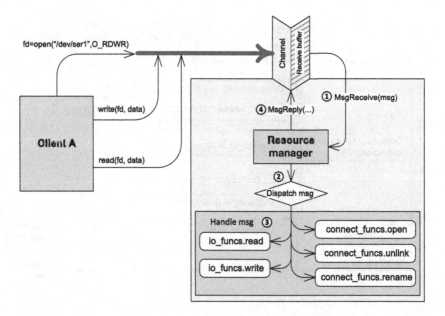

Figure 23.3
A scenario of using a resource manager.

(3) The message is processed by the message handler.

(4) It replies to the client.

23.2.2 Key Data Structures

Figure 23.4 gives the attributes and relationships of three key structures defined for manipulating a resource manager [7].

- The `iofunc_ocb_t` (open control block) structure defines a single usage context. Such an object is created by a resource manager in response to a client's open() call, used throughout the session between this client and the resource manager, and is destroyed upon a close() from the client. The structure contains per-open information such as the open modes as passed to the function open() by a client, and the current position in a file. In particular, it has a pointer to the `iofunc_attr_t` structure.
- The `iofunc_attr_t` (attribute) structure defines the characteristics of a single resource, containing per-name information such as the flags of resource manipulation, the number of bytes in the resource (nbytes), content modification time (mtime), access time(atime), and change of status time (ctime). Such an object is created when the resource manager is launched and exists until the resource manager is terminated.
- The `iofunc_mount_t` (mount) structure contains per-mountpoint information. It is only optionally used in a filesystem resource manager.

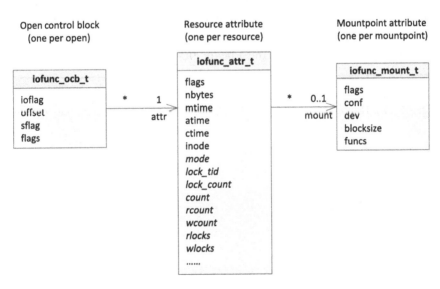

Figure 23.4

Key data structures of resource managers.

In general, a device resource manager has a single iofunc_attr_t object referenced by one or more iofunc_ocb_t objects. A filesystem resource manager typically has several iofunc_attr_t objects, each of which is referenced by one or more iofunc_ocb_t objects. A filesystem resource manager may optionally have one iofunc_mount_t object.

23.3 Example 1: Calculator as a Resource Manager

As an example, we take a look at a simple calculator implemented as a filesystem resource manager.

23.3.1 Superstructure

The resource manager design is straightforward: when the calculator starts to run, it manages four resources under the mountpoint /dev/mathoz—o1, operator, o2, and result:

- /dev/o1: the first operand from a client.
- /dev/operator: any of "+," "-," "X," of "%" from a client.
- /dev/o2: the second operand from a client.
- /dev/result: the result to the client.

Since the four resources under the mountpoint /dev/mathoz are not real files, our first question is where to maintain the resource contents. The best practice is to define a superstructure as recommended by QNX. For this example, we define my_attr—a superstructure for iofunc_attr_t. As shown in Listing 23.1, my_attr has two members: base and content, where the former is of type iofunc_attr_t and the latter points to the resource content. In general, a superstructure for iofunc_attr_t can have any number of user-defined members but a member of type iofunc_attr_t must be the first.

```
   /*
 2 *   math_of_OZ.c
   *   mountpoint: /dev/mathoz
 4 */
   #include <stdio.h>
 6 #include <stddef.h>
   #include <stdlib.h>
 8 #include <errno.h>
   #include <dirent.h>
10 #include <unistd.h>
   #include <limits.h>
12 #include <string.h>
   #define ALIGN(x)  (((x) + 3) & ~3)
14 #define NUM_ENTS    4
   #define CAPACITY  10
16
```

```
     struct my_attr;
18  #define IOFUNC_ATTR_T struct my_attr
     #include <sys/iofunc.h>
20  #include <sys/dispatch.h>
     struct my_attr {
22    iofunc_attr_t base; /* must always be first */
      char *content; /* the file content */
24  };

26  static IOFUNC_ATTR_T math_attrs [NUM_ENTS];
     const char *fname[4] = { "o1", "operator", "o2", "result" };
28  const char defent[4] = { '1', '+', '1', '2' };
```

Listing 23.1
Source code of Mathoz: declarations.

In line 26, math_attrs is defined as an array of four elements of the superstructure, one for each of the four resources. Also, in Listing 23.2, another variable of the superstructure, dir_attr, is declared for the mountpoint directory itself.

```
     int main(int argc, char **argv) {
222    dispatch_t *dpp;
       resmgr_attr_t resmgr_attr;
224    resmgr_context_t *ctp;
       resmgr_connect_funcs_t connect_func;
226    resmgr_io_funcs_t io_func;
       IOFUNC_ATTR_T dir_attr;
228
       // Setup the attributes for the directory
230    iofunc_attr_init(&dir_attr.base, S_IFDIR | 0555, 0, 0);
       dir_attr.base.inode = NUM_ENTS + 1; //
232    dir_attr.base.nbytes = NUM_ENTS; //
       // Setup the attributes for each resource under /dev/mathoz
234    int i;
       for (i = 0; i < 4; i++) {
236      iofunc_attr_init(&math_attrs[i].base, S_IFREG | 0444, 0, 0);
         math_attrs[i].base.inode = i + 1;
238      math_attrs[i].base.nbytes = 1;
         math_attrs[i].content = (char*) malloc(CAPACITY);
240      math_attrs[i].content[0] = defent[i];
       }
```

Listing 23.2
Source code of Mathoz: attribute initialization.

The code in Listing 23.2 also initializes dir_attr and math_attrs. In particular, for each resource, the length (in bytes) of its content is held by nbytes, and the value of inode is used to determine its offset under the mountpoint. The right portion of Figure 23.5 shows the attributes of the superstructure objects after initialization.

23.3.2 Handle Messages from Clients

As shown on the left in Figure 23.5, a client interacts with the resource manager by sending messages to the communication channel associated with the manager. A resource manager needs to handle two types of messages: connect messages and I/O messages.

Some POSIX calls from a client can raise connect messages to the resource manager. As indicated in Table 23.1, each POSIX call is handled by a corresponding message handler defined in the structure `resmgr_connect_funcs_t`. This structure can be initialized by the function iofunc_func_init(), which sets the function pointers to the default handlers defined in the QNX library, if applicable. A designer can also set the function pointers to his/her own connect message handlers.

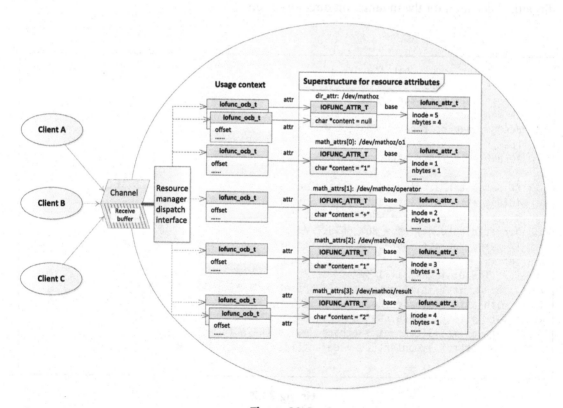

Figure 23.5
Resource manager design.

Table 23.1 Mapping client calls to connect message handlers

POSIX Call from Client			Connect Message Handler (Function Pointer Defined in the Structure resmgr_connect_funcs_t)	Default Handler by iofunc_func_init()
open(),	fopen()		open	_iofunc_open_default()
unlink()			unlink	No
rename()			rename	No
mknod(),	mkdir(),	mkfifo()	mknod	No
readlink()			readlink	No
link()			link	No
mount()			mount	No

Table 23.2 Mapping client calls to I/O message handlers

POSIX Call from Client			I/O Message Handler (Function Pointer Defined in the Structure resmgr_io_funcs_t)	Default Handler by iofunc_func_init()
read(),	readdir()		read	_iofunc_read_default()
write(),	fwrite()		write	_iofunc_write_default()
close()			close_ocb	_iofunc_close_ocb_default()
stat(),	lstat(),	fstat()	stat	_iofunc_stat_default()
select()			notify	No
pathconf(),	fpathconf()		pathconf	_iofunc_pathconf_default()
lseek(),	fseek(),	rewinddir()	lseek	_iofunc_lseek_default()
chmod(),	fchmod()		chmod	_iofunc_chmod_default()
chown(),	fchown()		chown	_iofunc_chown_default()
utime()			utime	_iofunc_utime_default()
lockf(),	fcntl()		lock	_iofunc_lock_default()
chsize(),	fcntl(),	ftruncate()	space	No
mmap(),	munmap()		mmap	_iofunc_mmap_default()
dup(),	dup2(),	fork()	dup	Handled automatically

Some POSIX calls from a client can raise I/O messages to the resource manager. As indicated in Table 23.2, each POSIX call is handled by a corresponding message handler defined in the structure `resmgr_io_funcs_t`. Likewise, this structure can be initialized by the function iofunc_func_init(), which sets the function pointers to the default handlers defined in the QNX library, if applicable. A designer can also set the function pointers to his/her own I/O message handlers.

```
244    // bind default message handler functions
       iofunc_func_init(_RESMGR_CONNECT_NFUNCS, &connect_func, _RESMGR_IO_NFUNCS, &
           io_func);
       // override with our own message handlers;
246    connect_func.open = my_open;
       io_func.read = my_read;
248    io_func.write = my_write;
```

Listing 23.3
Source code of Mathoz: set up message handlers.

The code in Listing 23.3 shows the use of iofunc_func_init() to initialize connect_func and io_func with the default handlers. In lines 246-248, we override the default handlers with our own handlers for open, write, and read messages.

The handler for open messages is given in Listing 23.4. When the mountpoint /dev/mathoz is opened, iofunc_open_default() is invoked; when a resource under the mountpoint is opened, its offset is determined first, followed by a call to iofunc_open_default(). In either case, the function iofunc_open_default() will create an object of type `iofunc_ocb_t` to be used throughout this session until it is closed by the client. Note that in line 46, math_attrs[dir_offset].base is used, which will be referenced by the `iofunc_ocb_t` object to be created by iofunc_open_default().

```
30  static int my_open(resmgr_context_t *ctp, io_open_t *msg, iofunc_attr_t *attr,void
        *extra) {
        static char *path_buf;
32
        if (msg -> connect.path[0] == 0) { // open the directory
34          return (iofunc_open_default(ctp, msg, attr, extra));
        } else {
36          printf("my_open: resource name is %s.\n", msg -> connect.path);
            path_buf = (char *) malloc(msg -> connect.path_len + 1);
38          if (path_buf == NULL) return (ENOMEM);
            strcpy(path_buf, msg -> connect.path);
40          path_buf[msg -> connect.path_len] = '\0';
            int dir_offset = 0;
42          for (dir_offset = 0; dir_offset < NUM_ENTS; dir_offset++) {
              if (strcmp(path_buf, fname[dir_offset]) == 0)
44              break;
            }
46          return (iofunc_open_default(ctp, msg, &math_attrs[dir_offset].base, extra));
        }
48  }
```

Listing 23.4
Source code of Mathoz: handle open messages.

The handler for write messages is given in Listing 23.5. In line 69, the content of the write message is read to buf. In line 78, the resource offset, dir_offset, is derived from the inode attribute of the `iofunc_attr_t` object referenced by the current `iofunc_ocb_t` object. If the client is writing to o1, o2, or operator, the previous resource content is cleared (lines 81-82), the new content is written (line 83), and the value of nbytes is updated (line 84) to reflect the new content length. The code in lines 87-90 is used to update the resource attributes with regard to the modification/change time.

```
50  int my_write(resmgr_context_t *ctp, io_write_t *msg, iofunc_ocb_t *ocb) {
      int status;
52    static char *buf;
      int dir_offset;
54
      // verify that the resource is writable
56    if ((status = iofunc_write_verify(ctp, msg, ocb, NULL)) != EOK)
        return (status);
58    // not expecting messages of extended types
      if ((msg->i.xtype & _IO_XTYPE_MASK) != _IO_XTYPE_NONE)
60      return (ENOSYS);

62    /* set up the number of bytes (from client's write()) */
      _IO_SET_WRITE_NBYTES (ctp, msg->i.nbytes);
64    buf = (char *) malloc(msg->i.nbytes + 1);
      if (buf == NULL) return (ENOMEM);
66
      // Read a message from a client. Use sizeof(msg->i) as offset
68    // because data immediately follows the _io_write structure.
      if (resmgr_msgread(ctp, buf, msg->i.nbytes, sizeof(msg->i)) == -1) {
70      free(buf);
        return (errno);
72    }

74    // Process the message
      buf[msg->i.nbytes] = '\0';
76    printf("my_write: input is %s\n", buf);

78    dir_offset = ocb -> attr -> base.inode - 1;
      if (dir_offset < 3) { //cannot write to "result"
80
        if (math_attrs[dir_offset].content != NULL)
82        memset(math_attrs[dir_offset].content, 0, CAPACITY);
        strcpy(math_attrs[dir_offset].content, buf);
84      math_attrs[dir_offset].base.nbytes = strlen(buf);

86      // if any data written, update POSIX structures
        if (msg->i.nbytes > 0) {
88        ocb->attr->base.flags |= IOFUNC_ATTR_MTIME | IOFUNC_ATTR_CTIME;
          iofunc_time_update(ocb->attr);
```

```
90      }
     }
92   // free the buffer
     free(buf);
94   return (_RESMGR_NPARTS (0));
   }
```

Listing 23.5
Source code of Mathoz: handle write messages.

The handler for read messages is given in Listing 23.6. Depending on whether the client is reading the mountpoint directory or a resource under the mountpoint, the function my_read() (lines 204-219) invokes my_read_dir() and my_read_file(), respectively, to handle the read message.

In the function my_read_dir(), the reply message is constructed from the resource names under the directory and is returned to the client through a call to MsgReply().

In the function my_read_file(), if this is the first time that the resource /dev/mathoz/result is read after being opened (lines 179-182), the function calculate_result() is invoked to populate the content of the resource with the operation result. In line 185, nleft is set to be the number of bytes after the current file offset. In line 188, nbytes is set to be the minimum of nleft and the number of bytes requested in the current read message from the client. If there is any byte to return (lines 189-195), a message is sent to the client, the file offset is moved forward by the number of bytes returned, and the relevant time attributes are updated.

```
   int dirent_size(char *fname) {
98     return (ALIGN (sizeof (struct dirent) - 4 + strlen (fname)));
   }
100
   struct dirent *
102   dirent_fill(struct dirent *dp, int inode, int offset, char *fname) {
       dp -> d_ino = inode;
104     dp -> d_offset = offset;
       strcpy(dp -> d_name, fname);
106     dp -> d_namelen = strlen(dp -> d_name);
       dp -> d_reclen = ALIGN (sizeof (struct dirent) - 4 + dp -> d_namelen);
108     return ((struct dirent *) ((char *) dp + dp -> d_reclen));
   }
110
   static int my_read_dir(resmgr_context_t *ctp, io_read_t *msg, iofunc_ocb_t *ocb) {
112     int nbytes;
       int nleft;
114     struct dirent *dp;
       char *reply_msg;
```

```
116
    // allocate a buffer for the reply
118 reply_msg = calloc(1, msg -> i.nbytes);
    if (reply_msg == NULL)
120   return (ENOMEM);

122 // assign output buffer
    dp = (struct dirent *) reply_msg;
124
    nleft = msg -> i.nbytes;
126 while (ocb -> offset < NUM_ENTS) {
      // the size of the current entry
128   nbytes = dirent_size(fname[ocb -> offset]);

130   if (nleft - nbytes >= 0) {
        // fill the dirent, and advance the dirent pointer
132     dp = dirent_fill(dp, ocb -> offset + 1, ocb -> offset,
          fname[ocb -> offset]);
134     // move the OCB offset
        ocb -> offset++;
136     // account for the bytes we just used up
        nleft -= nbytes;
138   } else // don't have any more room, stop
        break;
140 }

142 // return info back to the client
    MsgReply(ctp -> rcvid,  (char *) dp - reply_msg,   reply_msg,   (char *) dp -
        reply_msg);
144
    // release our buffer
146 free(reply_msg);
    // tell resource manager library we already did the reply
148 return (_RESMGR_NOREPLY);
  }
150 static void calculate result() {
    int op1 = atoi(math_attrs[0].content);
152 int op2 = atoi(math_attrs[2].content);
    int result = 0;
154 if (strncmp(math_attrs[1].content, "+", 1) == 0)
      result = op1 + op2;
156 else if (strncmp(math_attrs[1].content, "-", 1) == 0)
      result = op1 - op2;
158 else if (strncmp(math_attrs[1].content, "X", 1) == 0)
      result = op1 * op2;
160 else if (strncmp(math_attrs[1].content, "%", 1) == 0) {
      if (op2 == 0)
162     op2 = 1;
      result = op1 % op2;
164 }
```

```
      printf("Result: %d %c %d = %d.\n\n", op1, math_attrs[1].content[0],op2, result);
166   if (math_attrs[3].content != NULL)
         memset(math_attrs[3].content, 0, CAPACITY);
168   itoa(result, math_attrs[3].content, CAPACITY);
      math_attrs[3].base.nbytes = strlen(math_attrs[3].content);
170 }
    static int my_read_file(resmgr_context_t *ctp, io_read_t *msg, iofunc_ocb_t *ocb)
         {
172   // not expecting messages of extended types
      if ((msg -> i.xtype & _IO_XTYPE_MASK) != _IO_XTYPE_NONE) {
174     return (ENOSYS);
      }
176
      int dir_offset = ocb -> attr -> base.inode - 1;
178
      if ((dir_offset == 3) & (ocb -> offset == 0))
180   { //this is the first time /dev/mathoz/result is read
         calculate_result();
182   }

184   // check the resource to see how many bytes are left
      int nleft = math_attrs[dir_offset].base.nbytes - ocb -> offset;
186
      // return the minimum of nleft and the number requested
188   int nbytes = min (nleft, msg -> i.nbytes);
      if (nbytes > 0) {
190     MsgReply(ctp -> rcvid, nbytes, math_attrs[dir_offset].content+ ocb-> offset,
               nbytes);
       ocb -> offset += nbytes;
192     // update flags & time attributes
       ocb -> attr -> base.flags |= IOFUNC_ATTR_ATIME | IOFUNC_ATTR_DIRTY_TIME;
194     iofunc_time_update(ocb->attr);
      }
196   else {
         // nothing to return, indicate End Of File
198     MsgReply (ctp -> rcvid, EOK, NULL, 0);
      }
200   // already done the reply ourselves
      return (_RESMGR_NOREPLY);
202 }

204 static int my_read(resmgr_context_t *ctp, io_read_t *msg, iofunc_ocb_t *ocb) {
      // verify the resource is readable
206   int sts;
      if ((sts = iofunc_read_verify(ctp, msg, ocb, NULL)) != EOK) {
208     return (sts);
      }
210
      // decide if we should perform the "file" or "dir" read
212   if (S_ISDIR (ocb -> attr ->base.mode)) {
```

```
      return (my_read_dir(ctp, msg, ocb));
214 } else if (S_ISREG (ocb -> attr ->base.mode)) {
      return (my_read_file(ctp, msg, ocb));
216 } else {
      return (EBADF);
218 }
   }
```

Listing 23.6
Source code of Mathoz: handle read messages.

23.3.3 Register to Process Manager

The code in Listing 23.7 shows the initialization and registration of the resource manager.

```
250 // initialize attributes for the resource manager
    memset(&resmgr_attr, 0, sizeof(resmgr_attr));
252 resmgr_attr.nparts_max = 1;
    resmgr_attr.msg_max_size = 2048;
254
    // create the dispatch structure
256 if ((dpp = dispatch_create()) == NULL) {
      perror("Unable to dispatch_create\n");
258   exit(EXIT_FAILURE);
    }
260 // register the filesystem resource manager at /dev/mathoz in the pathname
        space
    if (resmgr_attach(dpp, &resmgr_attr, "/dev/mathoz", _FTYPE_ANY,_RESMGR_FLAG_DIR
        ,&connect_func,&io_func,&dir_attr)==-1){
262   perror("Unable to resmgr_attach\n");
      exit(EXIT_FAILURE);
264 }
    // allocate a context
266 ctp = resmgr_context_alloc(dpp);
    // wait here forever, handling messages
268 while (1) {
      if ((ctp = resmgr_block(ctp)) == NULL) {
270     perror("Unable to resmgr_block\n");
        exit(EXIT_FAILURE);
272   }
      resmgr_handler(ctp);
274 }
   }
```

Listing 23.7
Source code of Mathoz: pathname registration.

In line 261, the function resmgr_attach() is called to register the mountpoint /dev/mathoz to the process manager, where the resource manager attribute resmgr_attr and the dispatch structure dpp are initialized in lines 251-253 and lines 256-259, respectively. After registration, the resource manager enters into a loop (lines 268-274) to handle the incoming messages from clients.

23.3.4 Use Resource Manager

Listing 23.8 shows a client of the Mathoz resource manager. Note that from the client perspective, each resource is treated as a normal file that can be opened, read from, and written to. Figure 23.6 shows a screenshot of a run of the tester.

```c
1 #include <stdlib.h>
  #include <stdio.h>
3 #include <time.h>
  #include <sys/types.h>
5 #include <sys/stat.h>
  #include <fcntl.h>
7
  int Test_operator(int num1, char *op, int num2) {
9   int fd1 = open("/dev/mathoz/o1", O_WRONLY);
    char * input1 = (char*)malloc(10);
11  itoa(num1, input1, 10);
    write(fd1, input1, 1);
13  int fd2 = open("/dev/mathoz/operator", O_WRONLY);
    write(fd2, op, 1);
15  int fd3 = open("/dev/mathoz/o2", O_WRONLY);
    char * input2 = (char*)malloc(10);
17  itoa(num2, input2, 10);
    write(fd3, input2, 1);
19  int fd4 = open("/dev/mathoz/result", O_RDONLY);
    char *result = malloc(10);
21  read(fd4, result, 10);
    close(fd1); close(fd2); close(fd3); close(fd4);
23  printf ("Result: %s.\n", result);
    return 0;
25 }

27 int main(int argc, char *argv[]) {

29  Test_operator(7, "+", 4);
    Test_operator(7, "-", 4);
31  Test_operator(7, "X", 4);
    Test_operator(7, "%", 4);
33  return EXIT_SUCCESS;
  }
```

Listing 23.8
Source code of a Mathoz tester.

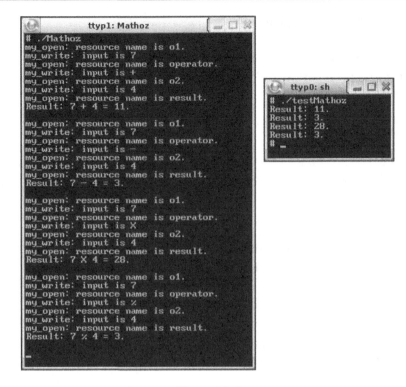

Figure 23.6
Use POSIX functions to interact with the resource manager.

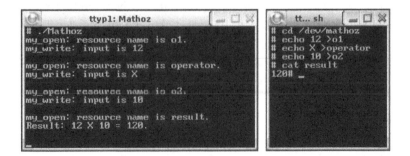

Figure 23.7
Use POSIX utilities to interact with the resource manager.

As mentioned before, a user can also interact with a resource manager by POSIX utilities. Figure 23.7 shows such a scenario, where a person uses the echo and cat utilities to interact with the Mathoz resource manager.

23.4 Example 2: Device Drivers

Device drivers are system-specific. We here take the AT91SAM9G45 evaluation board as an example context to explain how to apply the resource manager concept to develop device drivers.

23.4.1 Joysticks

The AT91SAM9G45 evaluation board has two pushbuttons and a joystick that can be pushed down or turned left, right, up, or down. We can implement a resource manager to keep track of the status of the pushbuttons and joysticks. For example, we can launch the manager at the mountpoint /dev/Joysticks that contains the following resources:

- Button_L: the status of the left pushbutton.
- Button_R: the status of the right pushbutton.
- Joy_Down: whether the joystick is turned downward.
- Joy_Left: whether the joystick is turned leftward.
- Joy_Push: whether the joystick is pushed down.
- Joy_Right: whether the joystick is turned rightward.
- Joy_Up: whether the joystick is turned upward.

The implementation of this resource manager is similar to that of the one examined in the last section. The main difference is that the resources here are readable only because the button or joystick status is supposed not to be changed by code but by direct physical manipulations.

The code in Listing 23.9 gives the function my_read_file(), where the register PIO_PER is set to enable the input pins for the pushbuttons and joysticks, and the register PIO_PDSR is read and scanned to detect the hardware event that has just happened to the buttons or joysticks.

```
 1  static int my_read_file(resmgr_context_t *ctp, io_read_t *msg,
 2      iofunc_ocb_t *ocb) {

 4    static char flag = 0;

 6    // not expecting messages of extended types
    if ((msg -> i.xtype & _IO_XTYPE_MASK) != _IO_XTYPE_NONE)
 8      return (ENOSYS);
    if (!flag) {
10      uint32_t input;
      uintptr_t buttonEnablePtr, buttonInputPtr;

12
        // Gain access to Pushbutton and Joystick registers
14      buttonEnablePtr = mmap_device_io(4, PIOB_BASE + PIO_PER);
      buttonInputPtr = mmap_device_io(4, PIOB_BASE + PIO_PDSR);

16
        //enable PIOB input pins for joysticks
18      out32(buttonEnablePtr, 0x0007C0C0);
```

```
20      //read inputs
        input = in32(buttonInputPtr);
22      input ^= 0xFFFFFFFF;

24      char* string = "0";
        int fileoffset = ocb -> attr -> base.inode - 1;
26      switch (fileoffset) {
        case 0: //Push Button Left
28          if (input & 0x40) string = "1";
            break;
30      case 1: //Push Button Right
            if (input & 0x80) string = "1";
32          break;
        case 2: //Joystick Down
34          if (input & 0x20000) string = "1";
            break;
36      case 3: //Joystick Left
            if (input & 0x4000) string = "1";
38          break;
        case 4: //Joystick Push
40          if (input & 0x40000) string = "1";
            break;
42      case 5: //Joystick Right
            if (input & 0x8000) string = "1";
44          break;
        case 6: //Joystick Up
46          if (input & 0x10000) string = "1";
            break;
48      }
        MsgReply(ctp->rcvid, strlen(string), string, strlen(string));
50      flag = 1;
        } else {
52      MsgReply(ctp -> rcvid, EOK, NULL, 0);
        flag = 0;
54      }
        // already done the reply ourselves
56      return (_RESMGR_NOREPLY);
        }
```

Listing 23.9
Source code of Joydriver.c.

23.4.2 LEDs

The AT91SAM9G45 evaluation board also has two LEDs. Likewise, we can launch a resource manager at the mountpoint /dev/bLED that contains the following resources:

- /dev/bLED/D1: the status of the first LED.
- /dev/bLED/D2: the status of the second LED.

The implementation of this resource manager is similar to that of the one examined in the last section. It is implemented to be writable only because the LED status is directly visible to human users. The code in Listing 23.10 gives the function my_write(), where the register PIO_OER is set to enable the LED output, and the registers PIO_CODR and PIO_SODR are used to turn on or turn off the LED requested by a client.

```
   int my_write (resmgr_context_t *ctp, io_write_t *msg, iofunc_ocb_t *ocb)
2  {
     int status;
4    static char *buf;

6    // verify that the device is opened for write
     if ((status = iofunc_write_verify(ctp, msg, ocb, NULL)) != EOK)
8      return (status);
     // check for and handle an XTYPE override
10   if ((msg->i.xtype & _IO_XTYPE_MASK) != _IO_XTYPE_NONE)
       return(ENOSYS);
12
     /* set up the number of bytes (from the client's write()) */
14   _IO_SET_WRITE_NBYTES (ctp, msg->i.nbytes);
     buf = (char *) malloc(msg->i.nbytes + 1);
16   if (buf == NULL)
       return(ENOMEM);
18
     // Read data from the message buffer.
20   if (resmgr_msgread(ctp,buf,msg->i.nbytes,sizeof(msg->i))== -1){
       free (buf);
22     return (errno);
     }
24   buf [msg->i.nbytes] = '\0';

26   // Gain access to LED registers
     uintptr_t OERPtr, CODRPtr, SODRPtr;
28   OERPtr = mmap_device_io(4, PIOD_BASE+PIO_OER);
     CODRPtr = mmap_device_io(4, PIOD_BASE+PIO_CODR);
30   SODRPtr = mmap_device_io(4, PIOD_BASE+PIO_SODR);

32   // enable LED output
     out32(OERPtr, 0x80000001);
34   // clear registers PIO_CODR and PIO_SODR
     out32(CODRPtr, 0x0);
36   out32(SODRPtr, 0x0);

38   int offset = ocb -> attr -> base.inode - 1;
     switch(offset){
40   case 0: //LED D1
       if(buf[0] == '0') //turn off LED
42       out32(SODRPtr, 0x1);
       else if(buf[0] == '1')
```

```
44      out32(CODRPtr, 0x1);
     break;
46  case 1:  //LED D2
     if(buf[0] == '0') //turn off LED
48      out32(SODRPtr, 0x80000000);
     else if(buf[0] == '1')
50      out32(CODRPtr, 0x80000000);
     break;
52  }
   // update time attributes
54  if (msg->i.nbytes > 0) {
     ocb->attr->base.flags |= IOFUNC_ATTR_MTIME | IOFUNC_ATTR_CTIME;
56    iofunc_time_update(ocb->attr);
   }
58  return (_RESMGR_NPARTS (0));
 }
```

Listing 23.10
Source code of LEDdriver.c.

23.4.3 Polling-Based Input Event Detection

Listing 23.11 gives a simple client of the two device managers we have just implemented. Every 500 ms, the code tries to read the resources representing the pushbuttons and joysticks. Whenever a hardware event is detected, it is acknowledged by turning the two LEDs on or off.

```
  #include <stdlib.h>
2 #include <stdio.h>
  #include <time.h>
4 #include <sys/types.h>
  #include <sys/stat.h>
6 #include <fcntl.h>

8 void msleep(unsigned long msec) {
   struct timespec interval;
10   struct timespec remainder;

12   interval.tv_sec = msec / 1000;
   interval.tv_nsec = (msec % 1000) * (1000 * 1000);

14
   if (nanosleep(&interval, &remainder) == -1) {
16     perror("nanosleep");
   }
18 }

20 int Button_Left_Pressed() {
    int fd = open("/dev/Joysticks/Button_L", O_RDONLY);
```

```
22   char *text = malloc(sizeof(char));
     read(fd, text, 1);
24   close(fd);
     if (text[0] == '1') return 1;
26   return 0;
   }
28 int Button_Right_Pressed() {
     int fd = open("/dev/Joysticks/Button_R", O_RDONLY);
30   char *text = malloc(sizeof(char));
     read(fd, text, 1);
32   close(fd);
     if (text[0] == '1') return 1;
34   return 0;
   }
36 int Joy_Right_Pressed() {
     int fd = open("/dev/Joysticks/Joy_Right", O_RDONLY);
38   char *text = malloc(sizeof(char));
     read(fd, text, 1);
40   close(fd);
     if (text[0] == '1') return 1;
42   return 0;
   }
44 int Joy_Left_Pressed() {
     int fd = open("/dev/Joysticks/Joy_Left", O_RDONLY);
46   char *text = malloc(sizeof(char));
     read(fd, text, 1);
48   close(fd);
     if (text[0] == '1') return 1;
50   return 0;
   }
52 int Joy_Up_Pressed() {
     int fd = open("/dev/Joysticks/Joy_Up", O_RDONLY);
54   char *text = malloc(sizeof(char));
     read(fd, text, 1);
56   close(fd);
     if (text[0] == '1') return 1;
58   return 0;
   }
60 int Joy_Down_Pressed() {
     int fd = open("/dev/Joysticks/Joy_Down", O_RDONLY);
62   char *text = malloc(sizeof(char));
     read(fd, text, 1);
64   close(fd);
     if (text[0] == '1') return 1;
66   return 0;
   }
68 int Joy_Push_Pressed() {
     int fd = open("/dev/Joysticks/Joy_Push", O_RDONLY);
70   char *text = malloc(sizeof(char));
     read(fd, text, 1);
```

```
72    close(fd);
      if (text[0] == '1') return 1;
74    return 0;
   }
76 void turnoff_LED_D1() {
      int fd = open("/dev/bLED/D1", O_WRONLY);
78    char *text = malloc(sizeof(char));
      text[0] = '0';
80    write(fd, text, 1);
      close(fd);
82 }
   void turnon_LED_D1() {
84    int fd = open("/dev/bLED/D1", O_WRONLY);
      char *text = malloc(sizeof(char));
86    text[0] = '1';
      write(fd, text, 1);
88    close(fd);
   }
90 void turnoff_LED_D2() {
      int fd = open("/dev/bLED/D2", O_WRONLY);
92    char *text = malloc(sizeof(char));
      text[0] = '0';
94    write(fd, text, 1);
      close(fd);
96 }
   void turnon_LED_D2() {
98    int fd = open("/dev/bLED/D2", O_WRONLY);
      char *text = malloc(sizeof(char));
100   text[0] = '1';
      write(fd, text, 1);
102   close(fd);
   }
104
   void setoff_LED() {
106   turnoff_LED_D1();
      turnoff_LED_D2();
108 }
   void seton_LED() {
110   turnon_LED_D1();
      turnon_LED_D2();
112 }

114 int main(int argc, char *argv[]) {
      while (1) {
116     msleep(500);
        if (Button_Left_Pressed()>0) {
118       seton_LED();
          continue;
120     }
        if (Button_Right_Pressed()>0) {
```

```
122        setoff_LED();
           continue;
124    }
       if (Joy_Left_Pressed()>0) {
126        turnon_LED_D1();
           continue;
128    }
       if (Joy_Up_Pressed()>0) {
130        turnon_LED_D2();
           continue;
132    }
       if (Joy_Right_Pressed()>0) {
134        turnoff_LED_D2();
           continue;
136    }
       if (Joy_Down_Pressed()>0) {
138        turnoff_LED_D1();
           continue;
140    }
       if (Joy_Push_Pressed()>0) {
142        setoff_LED();
       }
144    }
    return EXIT_SUCCESS;
146 }
```

Listing 23.11
Source code of a joystick tester.

It is worth noting that the hardware input events are detected by a polling-based approach (see Section 14.2.1.1). The polling frequency used here is 500 ms. In a real case, the polling frequency has to be adjusted appropriately: on the one hand, an event might be missed if the polling frequency is too low, but on the other hand, the system may suffer from the polling overhead if the frequency is too high.

Problems

23.1 What is a QNX resource manager?

23.2 What will happen when a client calls the POSIX function open() to open a resource /dev/any/r1 managed by a resource manager mounted at /dev/any?

23.3 What POSIX calls from a client may raise connect messages to a resource manager?

23.4 What POSIX calls from a client may raise I/O messages to a resource manager?

23.5 Extend the Mathoz resource manager to support the power operation—say, $8 \wedge 2 = 64$.

References

[1] Intel 8086 16-bit HMOS Microprocessor, <http://www.intel.com/>, Santa Clara, CA, 1990.

[2] Executable and Linking Format (ELF) Specification Version 1.2, Tool Interface Standard (TIS), 1995.

[3] Intel Pentium Processor, <http://www.intel.com/design/pentium/datashts/24199710.pdf>, Santa Clara, CA, 1997.

[4] Microchip PIC18F8720 Data Sheet, <http://www.Microchip.com/>, Microchip Technology, Chandler, AZ, 2004.

[5] UML Profile for Schedulability, Performance, and Time Specification Version 1.1, <http://www.omg.org/spec/SPTP/1.1/>, Object Management Group, Needham, MA, 2005.

[6] QNX Neutrino Realtime Operating System 6.4: Building Embedded Systems, <http://www.qnx.com/>, QNX Software Systems, 2009.

[7] QNX Neutrino Realtime Operating System 6.4: Library Reference, <http://www.qnx.com/>, QNX Software Systems, 2009.

[8] QNX Neutrino Realtime Operating System 6.4: System Architecture, <http://www.qnx.com/>, QNX Software Systems, 2009.

[9] Pc16550d Universal synchronous Receiver/Transmitter with FIFOs, <http://www.ti.com/lit/ds/symlink/pc16550d.pdf>, Texas Instruments, Dallas, TX, 2011.

[10] UML 2.4.1, <http://www.omg.org/spec/UML/2.4.1/>, Object Management Group, Needham, MA, 2011.

[11] UML Infrastructure 2.4.1, <http://www.omg.org/spec/UML/2.4.1/>, Object Management Group, Needham, MA, 2011.

[12] ARM926EJ-S Technical Reference Manual, <http://infocenter.arm.com/>, ARM, 2012.

[13] AT91SAM9G45-EKES User Guide, <http://www.atmel.com/Images/doc6481.pdf>, Atmel Corporation, San Jose, CA, 2012.

[14] ELF for the ARM Architecture, <http://infocenter.arm.com/>, ARM IHI 0044E, 2012.

[15] OCL 2.3.1, <http://www.omg.org/spec/OCL/2.3.1/>, Object Management Group, Needham, MA, 2012.

[16] Thumb-based Microcontrollers AT91SAM9G45 Preliminary, <http://www.atmel.com/>, Atmel Corporation, San Jose, CA, 2012.

[17] Posix.1-2008 (IEEE Std 1003.1-2008), 2013 ed., The IEEE and The Open Group, 2013.

[18] S.W. Ambler, The Object Primer: Agile Model Driven Development with UML 2, Cambridge University Press, New York, USA, 2004.

[19] F. Armour, G. Miller, Advanced Use Case Modeling: Software Systems, Addison Wesley, Boston, 2000.

[20] B. Baruah, R. Howell, L. Rosier, Feasibility problems for recurring tasks on one processor, Theoret. Comput. Sci. 118 (1) (1993) 3-20.

[21] M. Blaha, J. Rumbaugh, Object-Oriented Modeling and Design with UML, Pearson-Prentice Hall, Upper Saddle River, 2005.

[22] G.C. Buttazzo, Hard Real-Time Computing Systems: Predictable Scheduling Algorithms and Applications, Springer, New York, 2011.

[23] S.M. Carroll, From Eternity to Here: The Quest for the Ultimate Theory of Time, Dutton, New York, 2009.

[24] W. Cedeno, P.A. Laplante, An overview of real-time operating systems, J. Lab. Automat. 12 (1) (2007) 40-45.

[25] M. Cline, G. Lomow, M. Girou, C++ FAQs, Addison Wesley, Boston, 1998.

[26] A. Cockburn, Writing Effective Use Cases, Addison Wesley, Boston, 2000.

[27] A. Cockburn, Use cases, ten years later, Software Test. Qual. Eng. Mag. 4 (March/April(2)) (2002) 37-40.

[28] E. Coffman, M. Elphick, A. Shoshani, System deadlocks, Comput. Surv. 3 (2) (1971) 67-78.

[29] E. Coffman, A. Shoshani, Prevention, detection, and recovery from system deadlocks, in: Proceedings of 4th Annual Princeton Conference on Information Sciences and Systems, 1970.

[30] L. Copeland, A Practitioner's Guide to Software Test Design, Artech House, Norwood, MA, 2004.

[31] L. Crispin, J. Gregory, Agile Testing: A Practical Guide for Testers and Agile Teams, Addison-Wesley Professional, Boston, 2009.

[32] S.P. Dandamudi, Fundamentals of Computer Organization and Design, Springer, New York, 2003.

[33] B.P. Douglass, Doing Hard Time: Developing Real-Time Systems with UML, Objects, Frameworks, and Patterns, Addison Wesley, Boston, 1999.

[34] B.P. Douglass, Real Time UML, Addison Wesley, Boston, 2006.

[35] J. Fitzpatrick, An interview with Steve Furber, Commun. ACM 54 (5) (2011) 34-39.

[36] M. Fowler, Refactoring: Improving the Design of Existing Code, Addison Wesley, Boston, 1999.

[37] M. Fowler, UML Distilled, second ed., Addison-Wesley, Boston, 2000.

[38] M. Harbour, M. Klein, J. Lehoczky, Timing analysis for fixed-priority scheduling of hard real-time systems, IEEE Trans. Software Eng. 20 (1) (1994) 13-28.

[39] D. Harel, Statecharts: a visual formalism for complex systems, Sci. Comput. Program. 8 (1987) 231-274.

[40] R. Hyde, The Art of Assembly Language, No Starch Press, San Francisco, CA, 2010.

[41] I. Jacobson, Object oriented development in an industrial environment, in: OOPSLA'87: Object-Oriented Programming Systems, Languages and Applications, ACM SIGPLAN, 1987, pp. 183-191.

[42] M. Jaffe, N. Leveson, M. Heimdahl, B. Melhart, Software requirements analysis for real-time process-control systems, IEEE Trans. Software Eng. 17 (3) (1991) 241-258.

[43] M.B. Jones, D. Roşu, M.C. Roşu, CPU reservations and time constraints: efficient, predictable scheduling of independent activities, SIGOPS Oper. Syst. Rev. 31 (5) (1997) 198-211.

[44] G. Kiczales, J. Lamping, A. Mendhekar, C. Maeda, C. Lopes, J. Marc Loingtier, J. Irwin, Aspect-oriented programming, in: Proceedings of the European Conference on Object-Oriented Programming (ECOOP), Springer Verlag, New York, 1997, pp. 220-242.

[45] H. Kopetz, Real-Time Systems: Design Principles for Distributed Embedded Applications, Springer, New York, 2011.

[46] D. Kulak, E. Guiney, Use Cases: Requirements in Context, Addison Wesley, Boston, 2000.

[47] P.A. Laplante, S.J. Ovaska, Real-Time Systems Design and Analysis: Tools for the Practitioner, Wiley-IEEE Press, Hoboken, NJ, 2011.

[48] Q. Li, Real-Time Concepts for Embedded Systems, CMP Books, San Francisco, CA, 2003.

[49] C.L. Liu, J. Layland, Scheduling algorithms for multiprogramming in a hard real-time environment, J. ACM 20 (1) (1973) 46-61.

[50] J.W.S. Liu, Real-Time Systems, Prentice Hall, Upper Saddle River, 2000.

[51] H.D. Meikle, Modern Radar Systems, second ed., Artech House Publishers, Norwood, MA, 2008.

[52] B. Meyer, Object-Oriented Software Construction, Prentice Hall, Upper Saddle River, 1988.

[53] T.P. Morgan, Arm holdings eager for pc and server expansion record 2010, Tech. Rep., The Register, 2011.

[54] G.J. Myers, The Art of Software Testing, Wiley, Hoboken, NJ, 2004.

[55] R. Patton, Software Testing, Sams Publishing, Indianapolis, IN, 2005.

[56] G. Reeves, Re: What really happened on mars? Risks-Forum Digest 19 (54) (1998).

[57] J. Regehr, J. Stankovic, Augmented CPU reservations: towards predictable execution on general-purpose operating systems, in: Real-Time Technology and Applications Symposium 2001, Proceedings, Seventh IEEE, 2001, pp. 141-148.

[58] M.A. Richards, Fundamentals of Radar Signal Processing, McGraw-Hill, New York, 2005.

[59] J. Rumbaugh, I. Jacobson, G. Booch, The Unified Modeling Language Reference Manual, second ed., Addison-Wesley, Boston, 2005.

[60] D. Saks, Scope regions in C and C++, Embed. Syst. Design 15 (2007) 15.

[61] D. Saks, Storage class specifiers and storage duration, Embed. Syst. Design 29 (2007) 9.

[62] D. Saks, Linkage in C and C++, Embed. Syst. Design 3 (2008) 9-14.

[63] L. Sha, J.P. Lehoczky, R. Rajkumar, Solutions for some practical problems in prioritized preemptive scheduling, in: IEEE Real-Time Systems Symposium, IEEE Computer Society, 1986, pp. 181-191.

[64] L. Sha, R. Rajkumar, J. Lehoczky, Priority inheritance protocols: an approach to real-time synchronization, IEEE Trans. Comput. 39 (9) (1990) 1175-1185.

[65] L. Sha, R. Rajkumar, S. Sathaye, Generalized rate-monotonic scheduling theory: a framework for developing real-time systems, Proc. IEEE 82 (1) (1994) 68-82.

[66] D.E. Simon, An Embedded Software Primer, Addison-Wesley, Boston, MA, 2000.

[67] B. Sprunt, L. Sha, J.P. Lehoczky, Aperiodic task scheduling for hard real-time systems, Real-Time Syst. J. 1 (1) (1989) 27-60.

[68] J. Strosnider, J. Lehoczky, L. Sha, The deferrable server algorithm for enhanced aperiodic responsiveness in hard real-time environments, IEEE Trans. Comput. 44 (1) (1995) 73-91.

[69] A.S. Tanenbaum, A.S. Woodhull, Operating Systems: Design and Implementation, Pearson Prentice Hall, Upper Saddle River, NJ, 2011.

[70] L. Varshney, Radar system components and system design, Tech. Rep. November 22, Syracuse Research Corporation, NY, 2002.

[71] R. Williams, Real-Time Systems Development, Elsevier, Oxford, 2006.

[72] D. Marshall, Programming in C: UNIX System Calls and Subroutines Using C, http://www.cs.cf.ac.uk/Dave/C/CE.html, 2014.

Index

Note: Page numbers followed by *f* indicate figures and *t* indicate tables.